D0320959

ESTUARY RESTORATION and MAINTENANCE

The National Estuary Program

The CRC Marine Science Series is dedicated to providing state-of-the-art coverage of important topics in marine biology, marine chemistry, marine geology, and physical oceanography. The Series includes volumes that focus on the synthesis of recent advances in marine science.

CRC MARINE SCIENCE SERIES

SERIES EDITOR

Michael J. Kennish, Ph.D.

PUBLISHED TITLES

The Biology of Sea Turtles, Peter L. Lutz and John A. Musick
Chemical Oceanography, Second Edition, Frank J. Millero
Coastal Ecosystem Processes, Daniel M. Alongi
Ecology of Estuaries: Anthropogenic Effects, Michael J. Kennish
Ecology of Marine Bivalves: An Ecosystem Approach, Richard F. Dame
Ecology of Marine Invertebrate Larvae, Larry McEdward
Environmental Oceanography, Second Edition, Tom Beer
Intertidal Deposits: River Mouths, Tidal Flats, and Coastal Lagoons, Doeke Eisma
Morphodynamics of Inner Continental Shelves, L. Donelson Wright
Ocean Pollution: Effects on Living Resources and Humans, Carl J. Sindermann
Physical Oceanographic Processes of the Great Barrier Reef, Eric Wolanski
The Physiology of Fishes, Second Edition, David H. Evans
Pollution Impacts on Marine Biotic Communities, Michael J. Kennish
Practical Handbook of Estuarine and Marine Pollution, Michael J. Kennish
Practical Handbook of Marine Science, Second Edition, Michael J. Kennish

ESTUARY RESTORATION and MAINTENANCE

The National Estuary Program

edited by

Michael J. Kennish, Ph.D.

Institute of Marine and Coastal Sciences
Rutgers University
New Brunswick, New Jersey

CRC Press

Boca Raton London New York Washington, D.C.

Library of Congress Cataloging-in-Publication Data

Kennish, Michael J.
Estuary restoration and maintenance : the National Estuary Program / Michael J. Kennish.
p. cm. -- (Marine science series)
Includes bibliographical references (p.).
ISBN 0-8493-0720-1 (alk. paper)
1. National Estuary Program (U.S.) 2. Estuarine area conservation--United States. I. Title. II. Series.
QH76.K46 1999
333.91′64--dc21

99-31011
CIP

International Standard Book Number 0-8493-0720-1
Library of Congress Card Number 99-31011
Printed in the United States of America 1 2 3 4 5 6 7 8 9 0
Printed on acid-free paper

Preface

Effective management and protection of estuarine environments require a holistic approach: (1) assessing conditions across a complex array of interconnected and interdependent systems from upland habitats of watersheds to the open waters of embayments and (2) formulating strategies and implementing plans to restore and/or maintain system components. Many of the problems encountered in estuaries originate from escalating population growth, urbanization, and industrial expansion in nearby watersheds. The cumulative impacts of human activities and development are manifested in a myriad of ways in estuaries, such as the destruction and alteration of habitats, modification of river flow, increase in nonpoint source pollutant inputs (e.g., toxic substances), enrichment of nutrients, introduction of pathogens and exotic species, and overexploitation of fisheries. Management programs, therefore, must address both watershed and estuary functions when determining the status and long-term viability of estuarine resources. Highly integrated, watershed-based strategies are essential for mitigating human impacts on estuarine systems.

Section 320 of the amended Clean Water Act of 1987 provides a vehicle for revitalizing impacted U.S. estuaries. This legislation established the National Estuary Program (NEP) with the goal of restoring and maintaining the chemical, physical, and biological integrity of nationally significant estuaries. The success of the NEP depends on the unselfish cooperation of concerned citizens, public interest groups, academic institutions, business and industry, and government agencies all working together to protect and enhance the water quality, natural resources, and economic vitality of these critically important coastal systems.

Since the 1960s, there have been numerous attempts to restore damaged estuarine habitats to their natural state. Many of these efforts have been unsuccessful. In the NEP, restoration implies improving the health of estuaries rather than attempting to completely restore them to their historical state. In this context, the NEP aims to achieve the highest restoration or target goals possible while concomitantly supporting recreational and other beneficial uses of the system. Although the improvement of system health is the primary motivation of the NEP, resource sustainability and economically viable development also remain priority elements.

The purpose of this book is to examine the multitude of anthropogenic problems affecting estuaries nationwide and to investigate the management actions that have been devised to remedy them. Emphasis is placed on the NEP; four case studies from this valuable program are presented. Considering the serious challenges faced by estuarine scientists and resource managers as recounted in the volume, the compelling need for the NEP will become clearly evident to all readers. It is hoped that, by increasing the awareness of the ongoing destructive human activities and practices threatening U.S. coastal waters, additional public support will be generated for environmental initiatives such as the NEP.

Estuary Restoration and Maintenance: The National Estuary Program is principally designed as a reference for estuarine scientists, resource managers, decision-makers, and other professionals dealing with anthropogenic impacts and remediation programs in estuaries and coastal marine waters. Individuals engaged in applied research on habitat restoration will find the volume useful. Finally, the book can also be used for the development of advanced undergraduate and graduate courses in marine pollution.

I have been involved in the Barnegat Bay National Estuary Program (BBNEP) since 1996, serving as a member of the Science and Technical Advisory Committee (STAC) and as the data synthesis coordinator for the program. This volume is an outgrowth of my involvement in the BBNEP. I thank the following individuals who are overseeing various aspects of the BBNEP: M.P. DeLuca (STAC Co-Chair); R. Dieterich (USEPA Program Coordinator); E. Evenson (STAC); P. Griber (STAC Co-Chair); R. Lathrop (STAC); R. Nicholson (STAC); R. Scro (Program Director); and S. Seitzinger (STAC). All of these individuals have been very helpful to me on national estuary program matters during the past three years.

Several people are acknowledged for their assistance in the preparation of the book. At CRC Press, I am indebted to John B. Sulzycki for overseeing all editorial and production activities. The editorial staff of CRC Press provided technical guidance. In particular, I acknowledge the editorial expertise of Christine Andreasen. I am also grateful to my colleagues at Rutgers University who have worked with me on various investigations of estuarine and coastal marine environments that are relevant to this project. In the Institute of Marine and Coastal Sciences, I express my appreciation to K. W. Able, M. P. DeLuca, J. F. Grassle, R. A. Lutz, and N. P. Psuty. Special thanks are extended to R. A. Lutz for his collaboration on many marine science projects. Finally, I express my love and devotion to my wife, Jo-Ann, and sons, Shawn and Michael, for their support and understanding during the preparation of the volume.

This is Publication Number 99-12 of the Institute of Marine and Coastal Sciences, Rutgers University.

Michael J. Kennish

Editor

Michael J. Kennish, Ph.D., is a research marine scientist, laboratory manager, and graduate faculty member in the Institute of Marine and Coastal Sciences at Rutgers University, New Brunswick, New Jersey.

He graduated in 1972 from Rutgers University, Camden, New Jersey, with a B.A. degree in Geology and obtained his M.S. and Ph.D. degrees in the same discipline from Rutgers University, New Brunswick, in 1974 and 1977, respectively.

Dr. Kennish's professional affiliations include the American Fisheries Society (Mid-Atlantic Chapter), American Geophysical Union, American Institute of Physics, Atlantic Estuarine Research Society, New Jersey Academy of Science, and Sigma Xi. He is the President of the New Jersey Academy of Science.

Dr. Kennish has conducted biological and geological research on coastal and deep-sea environments for more than 25 years. While maintaining a wide range of research interests in marine ecology and marine geology, Dr. Kennish has been most actively involved with studies of marine pollution in estuarine and coastal marine ecosystems as well as with biological and geological investigations of deep-sea hydrothermal vents and seafloor spreading centers. He is the author or editor of nine books dealing with various aspects of estuarine and marine science. In addition to these books, Dr. Kennish has published more than 90 research articles and book chapters and has presented papers at numerous conferences. His biographical profile appears in *Who's Who in Frontiers of Science and Technology, Who's Who Among Rising Young Americans, Who's Who in Science and Engineering,* and *American Men and Women of Science.*

Contents

Introduction

I. Value of estuaries

Estuaries rank among the most important ecosystems on earth in terms of their ecologic and economic value. The biotic productivity of these coastal ecotones is extremely high, rivaling the most intensely cultivated farmlands. Several factors account for this:

- Abundant nutrients
- Conservation, retention, and efficient recycling of nutrients among benthic, wetland, and pelagic habitats (i.e., coupling of subsystems)
- Consortia of phytoplankton, benthic microalgae and macroalgae, seagrasses, mangroves, and fringing saltmarsh vegetation that maximize available light and space
- Tidal energy and circulation[1]

Estuaries, together with the nearshore oceanic realm, are responsible for ~50% of the world fisheries harvest, although they comprise only ~8% of the total area covered by marine waters. Their yield is comparable to the most productive upwelling regions of the ocean.[2] Numerous finfish and shellfish populations of commercial and recreational importance depend on estuaries for survival during at least a portion of their life cycle. Estuarine nursery grounds play a major role in the maintenance of commercial offshore finfish stocks, particularly along the Atlantic and Gulf Coasts. More than 90% of the total fisheries landings in the Gulf of Mexico consists of estuarine-dependent species.

Having large food supplies, estuaries typically support high densities and biomasses of organisms.[3] However, they are physically controlled systems subject to wide fluctuations in environmental conditions as well as multiple anthropogenic stresses. As a result, biotic communities living there are generally characterized by low species richness. Nevertheless, diverse habitats in open waters, tidal flats, and fringing wetlands act as a refuge for various freshwater and wildlife species, as a haven for a number of specially adapted forms, and as critical sites for development of many marine populations. These organisms primarily utilize the estuarine habitats for nesting, feeding, reproduction, or shelter.

Estuaries also serve important chemical and physical functions. For example, they trap nutrients, filter toxic pollutants, and transform wastes that enter from watersheds, the nearshore ocean, and atmosphere. Bottom sediments are a repository for numerous contaminants derived from these sources. Important physical functions include the amelioration of storm impacts, the attenuation of flooding, and the mitigation of erosion on bordering landmasses. To properly assess the value of estuaries, it is necessary to consider all of the applicable functions — biological, chemical, and physical.

From an economic perspective, estuaries support multibillion dollar commercial and recreational activities, including shipping, marine transportation, oil and gas recovery, electric power generation, marine biotechnology, aquaculture and mariculture, fisheries production, tourism, and other pursuits. U.S. commercial fisheries produced nearly $4 billion in revenues to fishers at U.S. ports in 1992. Furthermore, nearly 20 million

Americans participated in marine recreational fishing — many in estuaries — contributing more than $7 billion dollars to the U.S. economy.[4] Commercial and recreational fisheries together yield ~22.4 billion kilograms of fish each year. Coastal and estuarine fisheries alone inject more than $23 billion annually into the economy. They also supply employment for more than a million people.[5] Marine biotechnology generates annual revenues of $100 million. More than 98% of U.S. commerce (by volume) is transported by water, with estuaries being principal routes for the movement of goods.

Human activities have created serious environmental problems that threaten estuarine resources and the commercial and recreational uses dependent on them. Most of these problems are directly coupled to poorly planned coastal development, pollution, and accompanying modification and destruction of habitats. Today, ~60% of the worldwide human population inhabits the coastal zone.[6] Greater urbanization and industrialization associated with this burgeoning population are responsible for an array of estuarine impacts. For example, point and nonpoint source pollution promotes nutrient overloading, oxygen depletion, chemical contamination (i.e., inputs of halogenated hydrocarbons, polycyclic aromatic hydrocarbons, trace metals, and other substances), pathogen impairment of estuarine waters, and damage to biotic communities.[7-9] The construction of levees, embankments, and bulkheads as well as the excavation of mosquito ditches, implementation of dredging projects, and the reclamation of wetlands destroy or irretrievably alter habitats, including tidal marshes, submerged aquatic vegetation, and portions of the estuarine floor. Consequently, many estuarine habitats are in various stages of degradation.[10,11] Freshwater diversions in watersheds modify natural flow regimes in estuarine basins, which can have dramatic effects on water quality, salinity patterns, and the abundance and distribution of living resources. The introduction of exotic species commonly disrupts the ecological balance in these systems, culminating at times in marked reductions of biodiversity. Overharvesting may seriously deplete fisheries stocks.[12]

Pollution can be devastating to the coastal economy because it tends to restrict commercial and recreational uses of resources. For example, coastal pollution in New Jersey accounted for the loss of nearly $1 billion in marine recreational business revenues in 1989. Estuarine water quality must be improved at many locations nationwide — particularly in urbanized and heavily industrialized systems — if valuable living resources are to be protected and sustained.

The many, varied, and increasingly complex problems arising from anthropogenic use of estuaries have led to a concerted effort to protect estuarine resources and to characterize each system's physical, chemical, and biological conditions. Due to heightened public concern related to estuarine and coastal marine pollution, a series of federal and state environmental legislative acts was enacted during the 1970s and 1980s to mollify pollution impacts as well as to protect and maintain system resources. In response to this legislation, numerous quantitative studies were initiated that yielded crucial information on the processes operating in estuaries and the effects of human activities on system components.

Since the mid-1980s, two national pollution monitoring and assessment programs have been actively investigating the environmental quality of estuarine and coastal marine waters nationwide. These include the National Status and Trends (NS&T) Program of the National Oceanic and Atmospheric Administration (NOAA) and the Environmental Monitoring and Assessment (EMAP) Program of the U.S. Environmental Protection Agency (USEPA). The principal goal of the NS&T Program is to determine the status and long-term trends of toxic contamination in these coastal marine waters.[13] It consists of four components: (1) the National Benthic Surveillance Project, which analyzes contaminants in bottom fish and sediments from nearly 150 sampling sites; (2) the Mussel Watch Project, which analyzes contaminants in tissues of mussels and oysters as well as surface sediments, from more than 250 sampling sites; (3) Biological Effects Surveys and Research, which treats in greater detail those regions where laboratory analysis of samples indicates a potential for substantial environmental degradation and biological impacts of contaminants; and (4) Historical Trends Assessment, which examines more closely the environmental conditions in different coastal regions of the country. The NS&T Program assesses the concentrations of more than 70 chemical contaminants and certain associated effects in biota and sediment samples.

The objectives of EMAP are: (1) to estimate the current status and trends in the condition of the nation's ecological resources on a regional basis with known confidence; (2) to seek associations between human-induced stress and ecological condition; and (3) to generate periodic statistical summaries and interpretive reports on ecological status and trends to resource managers and the public. EMAP-Estuaries, the near-coastal component of EMAP, measures the status and change in selected indicators of ecological condition and provides a quantitative assessment of the regional extent of estuarine environmental problems, most notably eutrophication, hypoxia, sediment contamination, and habitat loss. Both the NS&T and EMAP-Estuaries Programs are identifying emerging environmental problems in the nation's estuaries that must be the focus of comprehensive and cooperative initiatives, such as the National Estuary Program (NEP).

II. The National Estuary Program: origin and purpose

Section 320 of the amended Clean Water Act of 1987 established the NEP, a federally sponsored pollution abatement initiative designed to identify nationally significant estuaries threatened by pollution, development, or overuse and to recommend management actions to restore, maintain, or enhance the environmental quality and living resources of these critically important systems. The USEPA manages the NEP, which currently includes 28 estuaries: Albemarle-Pamlico Sounds, NC; Barataria-Terrebonne Estuarine Complex, LA; Barnegat Bay, NJ; Buzzards Bay, MA; Casco Bay, ME; Charlotte Harbor, FL; Columbia River, OR and WA; Corpus Christi Bay, TX; Delaware Estuary, DE, NJ, and PA; Delaware Inland Bays, DE; Galveston Bay, TX; Indian River Lagoon, FL; Long Island Sound, CT and NY; Maryland Coastal Bays, MD; Massachusetts Bays, MA; Mobile Bay, AL; Morro Bay, CA; Narragansett Bay, RI; New Hampshire Estuaries, NH; New York-New Jersey Harbor, NY and NJ; Peconic Bay, NY; Puget Sound, WA; San Francisco Bay/Delta Estuary, CA; San Juan Bay, PR; Santa Monica Bay, CA; Sarasota Bay, FL; Tampa Bay, FL; and Tillamook Bay, OR. These estuaries are not only extremely valuable ecologically but also yield innumerable recreational, economic, and aesthetic benefits for millions of people. Hence, the NEP supports economic and recreational initiatives while concurrently attempting to resolve priority environmental problems in the systems.

The central component of each NEP is a Management Conference convened by the Administrator of the USEPA.[14] Consisting of representatives from local, state, and federal government agencies, business and industry, citizens groups, and academic institutions working cooperatively on multiple environmental issues, the Management Conference describes conditions in the estuary and develops a Comprehensive Conservation and Management Plan (CCMP) that outlines a strategy for conserving and managing the estuary's resources, including corrective actions to protect and improve the system. To develop a CCMP, the Management Conference follows four phases:

Phase 1: Establishes the structure of committees and procedures for conducting the group's work.

Phase 2: Characterizes the estuary to determine its health, reasons for its decline, and trends for future conditions; assesses the effectiveness of existing efforts to protect the estuary; and defines the highest priority problems to be addressed in the CCMP.

Phase 3: Specifies action plans in the CCMP to address priority problems identified through characterization and public input, with the CCMP building as much as possible on existing local, state, and federal programs.
Phase 4: Monitors and implements the CCMP, reviews progress, and redirects efforts where appropriate.

In addition to identifying the probable causes and sources of the priority problems, the action plans must state the program goals related to these problems, set objectives to attain the goals, delineate the universe of possible management activities for consideration, and state the activities that should be implemented to deal with the problems.[15] Because each NEP requires a systematic, detailed, basinwide assessment of pollution and other anthropogenic impacts, the action plans typically cover a wide range of priority problems. For example, the actions "… may address water and sediment quality, living resources, land and water resources, population growth, public access, or governance. They may focus on toxicants, pathogens, eutrophication, habitat loss or modification, another specific problem, or even a cause … Implementing action plans is a key to estuary cleanup."[16]

III. Plan of the volume

Chapter 1 provides an overview of the NEP. It discusses the principal environmental legislation responsible for its creation (i.e., the amended Clean Water Act of 1987) and also examines other legislative controls implemented during the past 50 years that have played a vital role in the protection of water quality and the enhancement of living resources in estuarine systems nationwide. In addition, Chapter 1 details the types and sources of priority pollutants and habitat modifications encountered in these complex systems and chronicles their effects on biotic communities. Impacts related to humans uses are assessed, such as watershed development, habitat loss and degradation, and altered freshwater flow. Environmental changes associated with human activities have had a dramatic influence on the abundance and distribution of aquatic and wildlife populations inhabiting these nationally significant waterbodies.

Chapters 2 to 5 are case studies of the following NEPs:

- Long Island Sound Estuary Program
- Delaware Estuary Program
- Galveston Bay National Estuary Program
- San Francisco Estuary Program

In the Long Island Sound Estuary Program (Chapter 2), the Management Conference has formulated six priority issues: (1) hypoxia; (2) pathogens; (3) toxic contamination; (4) floatable debris; (5) water quality degradation and habitat loss; and (6) impacts on living resources.[17] In Long Island Sound, some of the most acute environmental impacts have been linked to nutrient enrichment, excessive plant growth, and oxygen depletion (hypoxia). This system has been plagued for years by the development of late summer hypoxia in bottom waters over extensive areas of the western and central basins. Living resource effects attributed to the hypoxia have been significant and include a range of population responses from reduced abundance and growth to physiological stress and death. Efforts are under way to effectively manage the hypoxia by controlling nutrient inputs to the system from both point and nonpoint sources.

Pathogen contamination in Long Island Sound has periodically caused the closure of bathing beaches and shellfishing areas. There have been some incidences of human illness ascribed to pathogen exposure. Several management actions have been advanced to minimize pathogen inputs. These primarily involve improving source control of the pathogens by reducing combined sewer overflows, nonpoint source runoff, sewage treatment plant malfunctions, and vessel discharges.

Long Island Sound also receives an array of chemical contaminants (e.g., halogenated hydrocarbons, polycyclic aromatic hydrocarbons, heavy metals, etc.) that are potentially toxic to estuarine organisms and pose a threat to sensitive habitats. Because both recreational and commercial fisheries are significant in the sound, remedial actions to limit chemical contaminant inputs are critically important. One recommendation for controlling the input of toxic substances is to continue and, where appropriate, to enhance existing regulatory and pollution prevention programs.

Water quality degradation, destruction and alteration of habitat, and overharvesting of fish and wildlife populations have contributed to the overall decline in abundance and diversity of living resources in the sound. Misuse of this natural treasure is also reflected in the large amounts of floatable debris that persist in its waterways. Chapter 2 describes the action plans that have been implemented to address all of these concerns.

The Delaware Estuary Program (Chapter 3) has dealt with various pollution, water quality, habitat, land use, and water use problems in the Delaware Estuary through a cooperative environmental initiative jointly undertaken by the states of New Jersey, Delaware, and Pennsylvania. This heavily industrialized system, which supports the second largest refining-petrochemical center in the U.S., receives wastewaters from more than 150 industrial and municipal dischargers and about 300 combined sewer overflows.[18] The numerous pollutants that enter the estuary from nonpoint sources are a main target of management plans. More than 6 million people residing in areas surrounding the estuary place considerable demands on the system in terms of land and water uses. Habitat losses and alteration over the years have been substantial.

Biological communities in the estuary and bordering wetlands are comprised of dense populations. More than 3×10^5 ha of wetlands and open waters provide habitat for millions of plants and animals. However, elevated levels of some toxic substances — notably trace metals, organochlorine contaminants (e.g., DDT and PCBs), and polycyclic aromatic hydrocarbons — have been detected in bottom sediments, the water column, and tissues of organisms, especially in urban areas. Pathogens (i.e., coliform bacteria) are the major cause of water quality impairment in the upper nontidal river, whereas low dissolved oxygen levels are the principal concern in the tidal river and estuary.

Development and related human activities have impacted most habitats in the estuary and watersheds. Habitat loss and alteration over the past several decades have been greatest in the uplands, although the modifications of wetlands habitat have been historically most profound. Due to tighter regulatory controls and more efficient management strategies, the rate of wetlands destruction has declined appreciably since the 1970s. In addition, some habitats in the estuary have been restored or enhanced through the implementation of effective action plans proposed by the Management Conference.

Chapter 4 is devoted to the Galveston Bay National Estuary Program. From 1989 through 1994, this program conducted an intense investigation of the principal threats to Galveston Bay resources ascribable to human pollution, development, and overuse. Ranking among the nation's most economically valuable estuaries, Galveston Bay supports numerous tourism activities, commercial and recreational fisheries, shipping operations as well as petroleum, gas, and petrochemical industries. As such, it is responsible for billions of dollars of revenue for the State of Texas. Thus, one of the compelling challenges to the Galveston Bay National Estuary Program is to implement effective management strategies to ensure the conservation of resources and the improvement of environmental conditions in the estuary. Clearly, this will require an integrated, ecosystem-level management approach.

Galveston Bay experiences many of the same problems as those afflicting other nationally significant estuaries. For example, there has been substantial habitat loss (particularly wetlands) associated with development in watersheds. Sewage bypasses and overflows, municipal wastewater discharges, urban runoff and agricultural nonpoint source inputs deliver nutrients, pathogens, and toxic substances to estuarine waters and sediments. The most degraded water quality is evident in urbanized tributaries and the upper Houston Ship Channel.[19] Lowest dissolved oxygen levels in the system have been documented in these areas, although conditions have rebounded during the past 2 decades. The combination of pollutant inputs and habitat loss appears to have adversely affected some living resources. Certain species of marine organisms and birds show declining trends in abundance.

Chapter 5 recounts the state of the San Francisco Bay/Delta Estuary, the priority problems affecting the system, and the management plans being implemented to restore and maintain its chemical, physical, and biological integrity. The San Francisco Bay/Delta Estuary is the largest estuarine system on the Pacific Coast, and it supports a rich diversity of flora and fauna. However, many populations have decreased in abundance during the past several decades due to multiple anthropogenic impacts. Currently, more than 130 fish species and nearly 400 wildlife species inhabit the estuary region. Many of these are introduced forms.

There are several priority problems faced by the San Francisco Estuary Project. These include: (1) intensified land use; (2) decline of biological resources; (3) freshwater diversions and altered flow regime; (4) increased pollutants; and (5) accelerated dredging and waterway modification.[20] Management actions have resulted in improvements on several fronts. For example, the rate of wetlands loss has diminished markedly, and the conditions of numerous watershed streams have been upgraded. There is greater control of municipal and industrial point source pollution inputs. In addition, dredging and dredged material disposal activities are now effectively regulated.

Despite the aforementioned improvements, some serious problems persist in the estuary. Freshwater diversions, which induce systemic ecological changes and significant impacts on recreational and commercial fish populations and other biological resources, are on the rise. The introduced Asian clam (*Potamocorbula amurensis*) is cropping immense numbers of phytoplankton that sustain many invertebrate and young fish populations. As foreign species invasions increase, the alteration of benthic communities and fish assemblages escalates.[21,22] Large inputs of pollutants from nonpoint sources continue to degrade water quality and pose a potentially serious problem for biotic communities. Urban expansion threatens to accelerate runoff of contaminants into tributary systems and the open waters of the estuary.

The CCMP has recommended 145 actions that form the core of management strategies to remedy the aforementioned problems. The goal is to restore and maintain adequate water quality and a balanced indigenous population of fish, shellfish, and wildlife while also assuring that the beneficial uses of the bay and delta are protected. Successful implementation of the CCMP requires a long-term concerted effort on the part of government agencies, business and industry, academic institutions, environmental groups, and the general public to effectively manage the estuary and its resources.

References

1. Alongi, D. M., *Coastal Ecosystem Processes*, CRC Press, Boca Raton, FL, 1998.
2. Valiela, I., *Marine Ecological Processes*, 2nd ed., Springer-Verlag, New York, 1995.
3. Pinet, P. R., *Invitation to Oceanography*, Jones and Bartlett Publishers, Boston, MA, 1998.
4. Sissenwine, M. P. and Rosenberg, A. A., Marine fisheries at a critical juncture, in *Oceanography: Contemporary Readings in Ocean Sciences*, 3rd ed., Pirie, R. G., Ed., Oxford University Press, New York, 1996, 293.
5. Pirie, R. G., U.S. ocean resources 2000: a national plan for growth, in *Oceanography: Contemporary Readings in Ocean Sciences*, 3rd ed., Pirie, R. G., Ed., Oxford University Press, New York, 1996, 283.
6. Goldberg, E. D., *Coastal Zone Space — Prelude to Conflict?*, UNESCO Tech. Rept., UNESCO, Paris, 1994, 5.
7. Kennish, M. J., *Ecology of Estuaries: Anthropogenic Effects*, CRC Press, Boca Raton, FL, 1992.
8. Sindermann, C. J., *Ocean Pollution: Effects on Living Resources and Humans*, CRC Press, Boca Raton, FL, 1996.
9. Walker, C. H. and Livingstone, D. R., *Persistent Pollutants in Marine Ecosystems*, Pergamon Press, Oxford, UK, 1992.
10. National Research Council, *Priorities for Coastal Ecosystem Science*, National Academy Press, Washington, D.C., 1994.
11. Viles, H. and Spencer, T., *Coastal Problems, Geomorphology, Ecology, and Society at the Coast*, Edward Arnold, London, 1995.
12. Kennish, M. J., Ed., *Practical Handbook of Estuarine and Marine Pollution*, CRC Press, Boca Raton, FL, 1997.
13. O'Connor, T. P., The National Oceanic and Atmospheric Administration (NOAA) National Status and Trends Mussel Watch Program: national monitoring of chemical contamination in the coastal United States, in *Environmental Statistics, Assessment, and Forecasting*, Cothern, C. R. and Ross, N. P., Eds., CRC Press, Boca Raton, FL, 1994, 331.
14. USEPA, *National Estuary Program Guidance: Technical Characterization in the National Estuary Program*, EPA 842-B-94-006, U.S. Environmental Protection Agency, Washington, D.C., 1994.
15. USEPA, *National Estuary Program Guidance: Comprehensive Conservation and Management Plans Content and Approval Requirements*, EPA 842-B-92-002, U.S. Environmental Protection Agency, Washington, D.C., 1992.
16. USEPA, *Saving Bays and Estuaries: A Primer for Establishing and Managing Estuary Projects*, EPA 503/8-89-001, U.S. Environmental Protection Agency, Washington, D.C., 1989.
17. USEPA, *Long Island Sound Study Comprehensive Conservation and Management Plan*, Tech. Rept, U.S. Environmental Protection Agency, Stony Brook, NY, 1994.
18. Sutton, C. C., O'Herron, J. C., III, and Zappalorti, R. T., *The Scientific Characterization of the Delaware Estuary*, Tech. Rept., Delaware Estuary Program, (DRBC Project No., 321; HA File No. 93.21), U.S. Environmental Protection Agency, New York, 1996.
19. Shipley, F. S. and Kiesling, R. W., Eds., *The State of the Bay: A Characterization of the Galveston Bay Ecosystem*, Galveston Bay National Estuary Program Publication GBNEP-44, Webster, TX, 1994.
20. Monroe, M. W. and Kelly, J., *State of the Estuary*, Tech. Rept., San Francisco Estuary Project, Oakland, CA, 1992.
21. San Francisco Estuary Project, *State of the Estuary: 1992–1997*, Tech. Rept., San Francisco Estuary Project, Oakland, CA, 1998.
22. Cohen, A. N. and Carlton, J. T., Accelerating invasion rate in a highly invaded estuary, *Science*, 279, 555, 1998.

chapter one

Anthropogenic impacts and the National Estuary Program

I. Introduction

The U.S. government has enacted various legislative controls during the past 50 years to protect the marine environment, including the following major environmental legislation: the Federal Water Pollution Control Act of 1948; Federal Water Pollution Control Act Amendments of 1972; Marine Protection, Research, and Sanctuaries Act of 1972; Clean Water Act of 1977; and the Clean Water Act Amendments of 1987 (also known as the Water Quality Act). The Federal Water Pollution Control Act regulates all discharges into navigable waters of the U.S. The Marine Protection, Research, and Sanctuaries Act and the Clean Water Act (Section 404) are the principal tools used by the U.S. Environmental Protection Agency (USEPA) and the U.S. Army Corps of Engineers to regulate the disposal of dredged or fill material into wetlands and marine waters of the U.S.[1] The amendments to the Federal Water Pollution Control Act, reauthorized in 1977 and 1987 as the Clean Water Act and Clean Water Act Amendments, respectively, provide broad protection for the control of both point and nonpoint sources of pollution into natural waters.[2] While the Clean Water Act of 1977 established a technology-based approach to regulate individual point source discharges through National Pollutant Discharge Elimination System (NPDES) permits, the Clean Water Act Amendments of 1987 identified the remaining serious pollution problems, including nonpoint source impacts associated with eutrophication, hydrologic modification, accumulation of toxic pollutants, sedimentation, and increased turbidity.[3]

The Federal Water Pollution Control Act Amendments of 1972 established a national water pollution control program and provided the framework for the current approach to wastewater management in coastal areas. It required the formulation of uniform minimum federal standards for municipal and industrial wastewater treatment, set compliance deadlines, initiated a national discharge permit system, and authorized federal control over the quality of navigable waters. Key goals and policies of the 1972 act included

- Development of technologies necessary to eliminate the discharge of pollutants into navigable waters, waters of the contiguous zone, and the oceans
- Elimination of discharge of pollutants into navigable waters by 1985
- Prohibition of discharge of toxic pollutants in toxic amounts
- Protection and propagation of fish, shellfish, and wildlife, and provision of recreation in and on water whenever attainable
- Development of areawide wastewater treatment plans
- Provision of federal assistance to construct publicly owned treatment works

Of the aforementioned legislation, the Marine Protection, Research, and Sanctuaries Act, the Clean Water Act, and the Clean Water Act Amendments are the principal federal statutes controlling waste disposal in marine environments. Table 1.1 shows the major provisions of these statutes. The open ocean, primarily regulated by the Marine Protection, Research, and Sanctuaries Act, is reasonably well protected. The Marine Protection, Research, and Sanctuaries Act controls the transportation and dumping of wastes in waters seaward of the baseline (inner boundary) of the territorial sea, whereas the Clean Water Act regulates discharges from point sources into all U.S. waters, including the territorial sea and beyond. The Marine Protection, Research, and Sanctuaries Act governs all wastes except oil, sewage from vessels, and pipeline discharges which are regulated under the Clean Water Act. This statute also considers economic feasibility, requiring the balancing of all relevant factors (e.g., alternative land-based disposal methods, socioeconomic conditions, etc.). In 1974, the Marine Protection, Research, and Sanctuaries Act was amended to be consistent with the goals and constraints set forth in the 1972 London Dumping Convention, thereby aligning this legislation with international efforts to protect the marine environment.

Prior to enactment of the Clean Water Act, estuaries and coastal marine waters may have received less protection than open ocean waters. When enacted in 1977, the Clean Water Act was the most comprehensive and expensive environmental legislation, having jurisdiction over all U.S. waters. The primary objectives of this statute were to restore and maintain the chemical, physical, and biological integrity of U.S. water resources, to attain fishable and swimmable waters, and to eliminate the discharge of pollutants into navigable waters by 1983 (zero discharge). Two components were developed: (1) a pollution control program comprised of regulatory requirements that apply to industrial and municipal dischargers and (2) a federal grant program to help municipalities build sewage treatment plants. Hence, a combined federal and state system of pollution control programs was implemented. Major revisions to the Clean Water Act were completed in 1977 and 1981, and the amended Clean Water Act of 1987 further modified the act.

The Clean Water Act delineates two types of pollution sources: point and nonpoint. Point source pollution is that which is discharged from a discernible, confined, and discrete conveyance such as a pipe, conduit, channel, or tunnel. In contrast, nonpoint source pollution enters receiving waters from dispersed, diffuse, and uncontrolled sources such as general surface runoff, stormwater drainage, groundwater seepage, and atmospheric fallout. Nonpoint source pollution is characterized by (1) widely distributed pollution sources and (2) irregular rates of pollutant delivery to estuaries and coastal marine waters. The diffuse nature of nonpoint source pollution typically requires broad-based pollution control strategies (e.g., tighter land use controls, stricter zoning laws, subdivision regulations, erosion and sediment control ordinances, and improved stormwater drainage systems). The seriousness of nonpoint pollution is becoming more evident as the quality of point source discharges continue to improve nationally.

Section 208 of the Clean Water Act is a measure designed to control stormwater, domestic and industrial wastewaters, and other residual wastes through the formulation of effective state and areawide water quality management plans. The 208 plans focus on two strategies for improving water quality and achieving "fishable and swimmable" waters. The first strategy entails greater control of point source pollution. The second strategy consists of improved control of nonpoint source pollution.

One of the principal goals of 208 programs is to eliminate many small, inefficiently operated wastewater treatment plants and transfer their flow to more efficient regional municipal treatment facilities. Section 201 of the Clean Water Act lists the requirements that all wastewater facility plans must address to provide the sewer infrastructure to meet demands of a rapidly developing region. These include: (1) determining 20-year needs for sewer service based on population and economic projections, land use, and other local

Table 1.1 Major Legislative Provisions Affecting Waste Disposal in Marine Waters

Statute and Section	Purpose
	Marine Protection Research and Sanctuaries Act
Sec. 101	Prohibits, unless authorized by permit, the transportation of wastes for dumping and/or the dumping of wastes into the territorial seas of the contiguous zones.
Sec. 102	Authorizes EPA[a] to issue permits for dumping of nondredged materials into the contiguous zone and beyond as long as the materials will not "unreasonably degrade" public health or the marine environment, following criteria specified in statute or established by the Administrator.
Sec. 103	Authorizes Corps of Engineers to issue permits for dumping dredged material, applying EPA's environmental impact criteria to ensure action will not unreasonably degrade human health or the marine environment.
Sec. 104	Specifies permit conditions for waste transported for dumping or to be dumped, issued by EPA or the Coast Guard.
Sec. 107	Authorizes EPA and Corps of Engineers to use the resources of other agencies, and instructs the Coast Guard to conduct surveillance and other appropriate enforcement activities as necessary to prevent unlawful transportation of material for dumping or unlawful dumping.
	Clean Water Act[b]
Sec. 104 (n)	Directs EPA to establish national estuary programs to prevent and control pollution and to conduct and promote studies of health effects of estuarine pollution.
Sec. 104 (q)	Establishes a national clearinghouse for the collection and dissemination of information developed on small sewage flows and alternative treatment technologies.
Sec. 201, 202, 204	Specifies sewage treatment construction grants program eligibility and federal share of cost.
Sec. 208	Authorizes a process for states and regional agencies to establish comprehensive planning for point and nonpoint source pollution.
Sec. 301	Directs states to establish and periodically revise water quality standards[c] for all navigable waters; effluent limitations for point sources requiring BPT should be achieved by July 1, 1977; timetable for achievement of BAT and other standards set. Compliance deadlines for publicly owned treatment works (POTWs) to achieve secondary treatment also set.
Sec. 301 (h)	Authorizes waivers for POTWs in coastal municipalities from secondary treatment for effluent discharged into marine waters if criteria to protect the marine ecosystem can be met.
Sec. 301 (k)	Allows industrial dischargers to receive a compliance extension from BAT requirements until July 1, 1987, for installation of an innovative technology, if it will achieve the same or greater effluent reduction than BAT at a significantly lower cost.
Sec. 302	Allows EPA to establish additional water quality–based limitations once BAT is established, if necessary to attain or maintain fishable/swimmable water quality (for toxics, the *NRDC* v. *EPA* consent decree sets terms).
Sec. 303	Requires states to adopt and periodically revise water quality standards; if they determine that technology-based standards are not sufficient to meet water quality standards, they must establish total maximum daily loads and waste load allocations, and incorporate more stringent effluent limitations into Sec. 402 permits.
Sec. 303 (e)	Requires states to establish water quality management plans for watershed basins, to provide for adequate implementation of water quality standards by basin to control nonpoint pollution; Section 208 areawide plans must be consistent with these plans.

Table 1.1 (continued) Major Legislative Provisions Affecting Waste Disposal in Marine Waters

Statute and Section	Purpose
Sec. 304	Requires EPA to establish and periodically revise water quality criteria to reflect the most recent scientific knowledge about the effects and fate of pollutants, and to maintain the chemical, physical, and biological integrity of navigable waters, groundwater, and ocean waters and establish guidelines for effluent limitations.
Sec. 304 (b)	Outlines factors to be considered when assessing BPT and BAT to set effluent limitation guidelines, including accounting for "non-water quality impact," age of equipment, etc.
Sec. 305 (b)	Sets state water quality reporting requirements.
Sec. 306	Sets new source performance standards for a list of categories of sources.
Sec. 307	Requires EPA to issue categorical pretreatment standards for new and existing indirect sources; POTWs required to adopt and implement local pretreatment programs; toxic effluent limitation standards must be set according to the best available technology economically achievable.
Sec. 308	Requires owners or operators of point sources to maintain records and monitoring equipment, do sampling, and provide such information or any additional information.
Sec. 309	Gives enforcement powers primarily to state authorities. Civil penalties, however, and misdemeanor sanctions can be issued by EPA in U.S. district courts for violation of the act, including permit conditions or limitations; EPA also is authorized to issue criminal penalties for violations of Sections 301, 302, 306, 307, and 308. EPA may take enforcement action for violations of Section 307 (d) which introduce toxic pollutants into POTWs.
Sec. 402	Establishes National Pollutant Discharge Elimination System (NPDES), authorizing EPA Administrator to issue a permit for the discharge of any pollutant(s) to navigable waters that will meet requirements of Sections 301, 302, 306, 307 and other relevant sections; states can assume administrative responsibility of the permit program.
Sec. 403	Directs EPA to establish Ocean Discharge Criteria as guidelines for permit issuance for discharge into territorial seas, the contiguous zone, and open ocean.
Sec. 404	Directs Secretary of the Army to issue permits for dredged or fill material; EPA must establish criteria comparable to Section 403 (c) criteria for dredged and fill material discharges into navigable waters at specified disposal sites.
Sec. 405	Requires EPA to issue sludge use and disposal regulations for POTWs.
Sec. 504	Grants emergency powers to the Administrator to assist in abating pollutant releases; establishes a contingency fund, and requires the Administrator to prepare and publish a contingency plan to respond to such emergencies.
Sec. 505	Citizen suit provision allows citizens to bring civil action in district court against any person in violation of an effluent standard or limitation of an order by the Administrator for failing to perform a nondiscretionary act.

[a] Unless otherwise noted, the Environmental Protection Agency (EPA) is responsible for implementing provision(s).

[b] Relevant provisions of the Clean Water Act Amendments of 1987, which reauthorized and amended the Clean Water Act, are discussed in text.

[c] Water quality standards are ambient standards designed to achieve certain uses of water; these now play a secondary role. Technology-based effluent standards are given the primary role and are designed to reduce pollutants so that ultimately all water is "fishable, swimmable." Effluent standards are performance standards and specify the maximum permissible discharge of a pollutant from a type of source and usually specify the degree of technology to be used ("best available," "best practicable," "reasonably available," etc.), but not the particular method needed to comply. Effluent limitation guidelines, on the other hand, apply to individual sources and specify their particular performance levels. Water quality standards (Sec. 303) are now the benchmarks by which to measure the success of the effluent standards in meeting clean water goals.

Source: Office of Technology Assessment, 1987.

or regional factors; (2) planning area boundaries for sewer service; (3) evaluating the technically feasible treatment options to meet the required effluent limits; (4) conducting a cost effective and environmental analysis of the most viable alternatives; and (5) identifying and implementing a selected alternative.

Section 402 of the Clean Water Act prohibits point source discharges of pollutants to navigable waters in the U.S. from all facilities — industrial and municipal — unless a National Pollutant Discharge Elimination System (NPDES) permit is obtained. An NPDES permit specifies effluent limitations on specific pollutants present in the discharge. The effluent standards are either technology based as set forth in Sections 301 and 304 of the Clean Water Act or water quality based as set forth in Section 302 of the statute.

The Clean Water Act provides broad protection for the control of both point and nonpoint sources of pollution into natural waters.[3,4] During the 1970s, the Clean Water Act shifted the program emphasis toward technology-based standards for discharge effluent and away from discharge standards based solely on receiving water quality. Hence, prior to being released to receiving waters, all point source discharges must now meet standards based on achievable pollutant treatment technologies. Technology-based standards are formulated from estimates of the removal of pollutants that could be achieved through application of best practicable technology, best available technology, or best conventional technology.[5] The Clean Water Act of 1977 focused on a technology-based approach to regulate individual point source discharges through NPDES permits. The amended Clean Water Act of 1987 addressed other serious pollution problems including nonpoint source impacts associated with eutrophication, hydrologic modification, accumulation of toxic pollutants, sedimentation, and increased turbidity.[6]

A primary goal of the amended Clean Water Act of 1987 is the development of nonpoint source pollution control programs. Much of the legislative authority of the USEPA to regulate and protect the quality of surface waters derives from the Clean Water Act. The USEPA began developing stormwater management regulations and permitting requirements as a result of the amended Clean Water Act. Prior to 1987, local and state governments often avoided the subject of stormwater drainage. To abate, eliminate, or prevent stormwater drainage problems, municipal governments began to devise stormwater management programs, usually by adopting local ordinances and/or regulations. Developers and builders have been required by ordinances to design and install systems to minimize or negate stormwater impacts. Additional guidelines and regulations were advanced during the 1990s. Today, local, state, and federal government agencies are all involved in regulating stormwater drainage so that communities can continue quality development and growth while protecting coastal resources.

Aside from stormwater drainage, the Clean Water Act regulates waste dumping from vessels in estuaries and coastal waters as well as pipeline discharges into these waters. Long outfalls from land-based facilities and discharges from stationary drilling platforms are also regulated under the Clean Water Act. However, this statute does not apply to the dumping of waste from vessels in waters beyond the territorial sea, which is regulated under the Marine Protection, Research, and Sanctuaries Act.

The Clean Water Act classifies pollutants into three categories. Conventional pollutants (category 1) consist of total suspended solids, biochemical oxygen demand, pH, oil and grease, and fecal coliform bacteria. Toxic pollutants (category 2) comprise metals and organic chemicals. Nonconventional pollutants (category 3) include any additional substances that may require regulation such as nutrients (total nitrogen, nitrates, and total phosphorus), chlorine, fluoride, and certain metals.[6]

Examining the relative contribution of pollutants to marine waters by major sources, several generalizations can be proposed at the national level. For example, municipal point sources have been significant contributors of certain conventional pollutants in marine waters during the past several decades, notably biochemical oxygen demand, total nitrogen, and oil and grease. Industrial point sources are the principal contributors of many organic chemicals (e.g., chlorinated hydrocarbon compounds) and some heavy metals (e.g., arsenic, cadmium, and mercury). Nonpoint runoff is the dominant source of fecal coliform bacteria, suspended solids, as well as certain nutrients (i.e., total phosphorus), oxygen-demanding pollutants, and heavy metals (i.e., chromium, copper, iron, lead, and zinc).[5] Large amounts of oil and grease, chromium, and lead commonly derive from urban runoff, whereas substantial quantities of herbicides and pesticides typically originate from agricultural runoff.

Administered by the USEPA, the Clean Water Act specifies that individual states apply and coordinate water quality programs. Section 319 of the amended Clean Water Act of 1987 requires the states to develop assessment reports and management plans describing their nonpoint source pollution problems and to formulate a program to remediate these problems. To complete their regional management plans, the states must develop strategies for controlling nonpoint pollution. Some of the major provisions of the amended Clean Water Act of 1987 that are pertinent to waste disposal and pollutants in marine waters include: (1) construction grants and state revolving loans for the construction of sewage treatment facilities; (2) grants provided to states for reducing nonpoint source pollution; (3) requirements of states to identify "hot-spot" waters not expected to meet water quality standards because of toxic pollutants in discharges, and to develop control strategies for these pollutants; and (4) greater penalties for civil and criminal violations of clean water laws.

The Federal Water Pollution Control Act Amendments of 1972, the Clean Water Act of 1977, and the Clean Water Act Amendments of 1987 have led to notable improvements in the water quality of estuaries and coastal marine waters despite continued population growth in the coastal zone. More than 100 million people live in U.S. coastal counties. Although this federal legislation has resulted in significant water quality improvements along much of the nonurban coastal zone, many urbanized estuaries have not experienced such benefits. These heavily used systems will likely encounter continuing pollution problems in the future.

II. Legislation and the National Estuary Program

Section 320 of the amended Clean Water Act of 1987 also established the National Estuary Program (NEP). The purpose of the NEP is to identify nationally significant estuaries threatened by pollution, development, or overuse and to promote the preparation of comprehensive management plans to ensure their ecological integrity. To control the sources of pollution, the NEP uses existing authorities under the Clean Water Act (Section 104), other federal statutes, and state legislative authorities. An integrated management program is developed utilizing existing programs at the federal, state, and local levels to maximize pollution abatement efforts. Each program characterizes the conditions and trends in the system and proposes actions to remediate existing pollution problems.

The long-term goal of the NEP is to protect and restore the health of estuaries and enhance their living resources while supporting economic and recreational initiatives. To accomplish this goal, the USEPA manages the NEP and facilitates development of local NEPs by

- Fostering communication among federal, state, and local governments
- Forging partnerships between government agencies that oversee estuarine resources and the people who depend on estuaries for their livelihood and quality of life
- Transferring scientific and management information, experience, and expertise to program participants

- Increasing public awareness of pollution problems and ensuring public participation in consensus building
- Promoting basinwide planning to control pollution and manage living resources
- Overseeing the development and implementation of pollution abatement and control programs[7]

For each estuary selected to the NEP, the Administrator of the USEPA convenes a Management Conference to examine environmental conditions and trends in the estuary, to identify the most significant problems, and to develop an action-oriented Comprehensive Conservation and Management Plan (CCMP) to address high-priority problems. The conference assures full participation by government agencies (federal, state, and local), educational institutions, industries, user groups, and the general public. In so doing, it provides a forum for consensus building and problem-solving among these entities. The CCMP, in turn, represents a blueprint for revitalizing the estuary by identifying the most pressing problems and recommending priority corrective actions and compliance schedules that address point and nonpoint sources of pollution. The primary goals are to restore and maintain the chemical, physical, and biological integrity of the estuary, including restoration and maintenance of water quality, a balanced indigenous population of shellfish, fish, and wildlife, and recreational activities in the estuary, and to assure that the designated uses of the estuary are protected. Each NEP receives funds over a 3- to 5-year period from the USEPA to produce a CCMP.

A major part of every NEP is a technical assessment of the state of the estuary, termed "characterization," which serves as the basis for defining and selecting the problems to be addressed in the CCMP. The technical characterization process involves an evaluation of the conditions of the resources and uses of the estuary, the priority problems experienced by those resources and uses, and the causes of the priority problems (Figure 1.1). Results of this process are contained in a Characterization Report that provides sound scientific justification for management actions recommended in the CCMP. The Characterization Report typically addresses the following points:

- The status and trends of the water quality, natural resources, and uses of the estuary
- The linkages between pollutant loadings and changes in the water quality, natural resources, and uses of the estuary
- Description of human impacts on the water quality, natural resources, and uses of the estuary
- The identification of the priority problems in the estuary and the selection criteria used to determine them
- Hypotheses of cause-effect relationships for the priority problems and the research necessary to establish relationships
- The likely causes of the priority problems, examining databases on nutrients, chemical contaminants, and natural resources
- The final list, historic description, and background information on priority problems to be addressed in the CCMP
- The environmental quality goals and objectives established for the estuary, which form the basis for the monitoring program developed to evaluate the effectiveness of actions implemented under the CCMP
- Knowledge of uncertainties in the databases can be used to direct further data gathering and research efforts and is important to the development of an effective sampling design in the post-CCMP monitoring program[7]

Since 1987, 28 estuaries (comprising 5 tiers) have been named as NEP sites (Table 1.2).

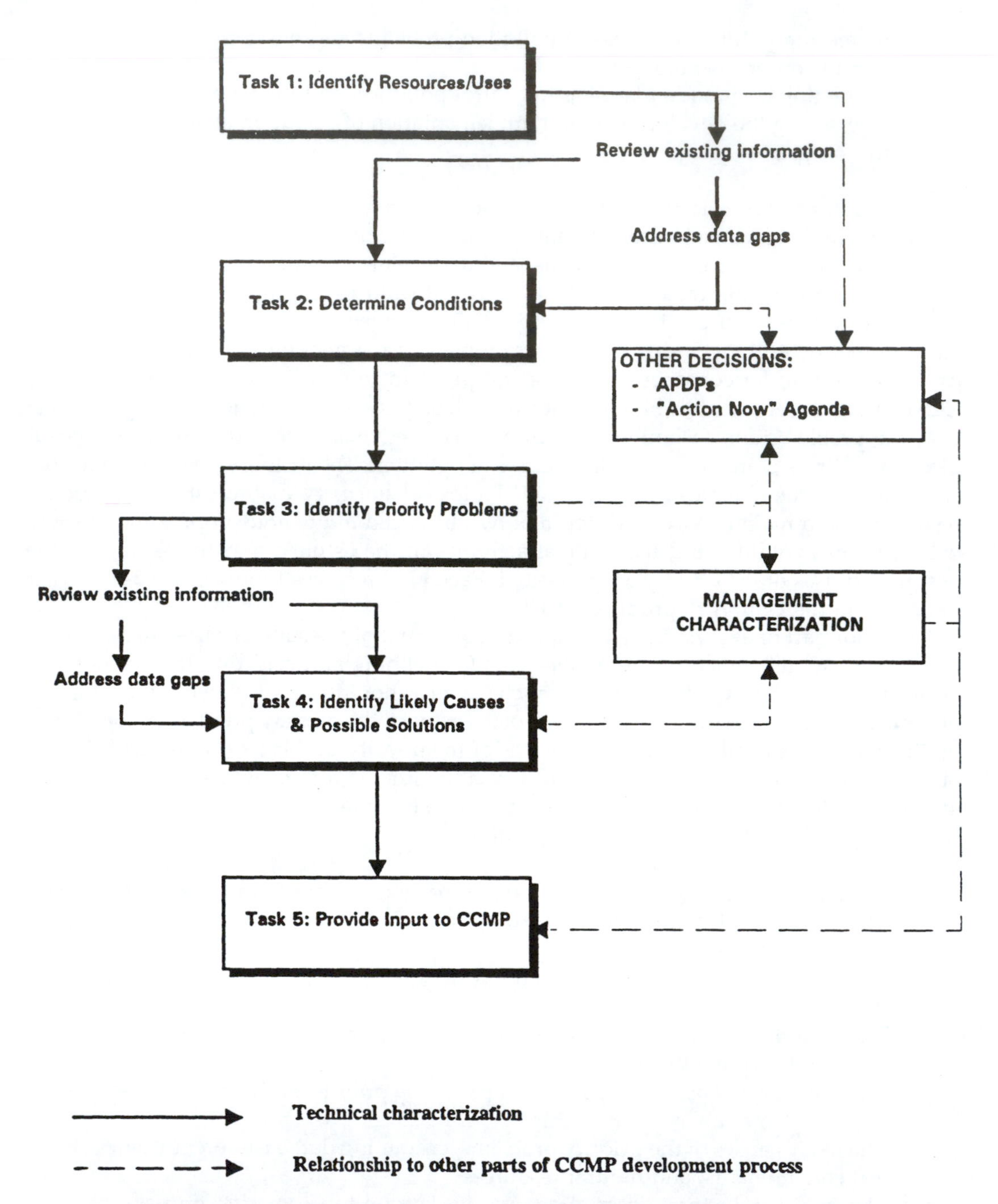

Figure 1.1 Relationships among the technical characterization tasks in the National Estuary Program. (From U.S. Environmental Protection Agency, A National Estuary Program Guidance: Technical Characterization in the National Estuary Program, EPA 842-B-94-006, USEPA Region 2, New York, 1994.)

III. Pollution sources

Table 1.3 provides a description of the principal types of point and nonpoint pollution sources in estuarine and marine environments. Waste disposal operations involving the intentional release of materials to marine waters either via direct dumping or pipeline

Table 1.2 List of National Estuary Programs

Albemarle/Pamlico Sound, NC
Barataria-Terrebonne, LA
Barnegat Bay, NJ
Buzzards Bay, MA
Casco Bay, ME
Charlotte Harbor, NC
Corpus Christi Bay, TX
Delaware Bay, DE/NJ/PA
Delaware Inland Bays, DE
Galveston Bay, TX
Indian River Lagoon, FL
Long Island Sound, CT/MA/NY/RI
Lower Columbia River, WA
Maryland Coastal Bays, MD
Massachusetts Bays, MA
Mobile Bay, AL
Morro Bay, CA
Narragansett Bay, RI/MA
New Hampshire Estuaries, NH
New York/New Jersey Harbor, NY/NJ
Peconic Estuary, NY
Puget Sound, WA
San Francisco Bay/Sacramento-San Joaquin Delta, CA
San Juan Bay, PR
Santa Monica Bay, CA
Sarasota Bay, FL
Tampa Bay, FL
Tillamook Bay, OR

discharges constitute point sources of pollution. The dumping of municipal sewage sludges, dredged spoils, and industrial wastes (e.g., acid-iron wastes, alkali chemicals, and pharmaceuticals) as well as the discharge of municipal and industrial effluents from outfalls are the primary point source categories responsible for the introduction of pollutants to U.S. marine waters. Major contaminant loadings of U.S. marine waters from point sources have decreased substantially during the past 2 decades due to tighter state and federal government regulations and improved industrial controls of point source discharges. Consequently, emphasis has shifted more recently to the assessment of detrimental effects ascribable to secondary, less-easily regulated, but recurrent pollutants originating from nonpoint sources.[8] However, the input of nonpoint source pollutants is more problematical, and acute and insidious biological effects of these pollutants are extremely difficult to assess.[9]

Apart from the input of pollutants to estuaries that may cause considerable ecological harm, other human activities can be equally damaging to biotic communities and habitats in these systems. For example, dredging and filling, dredge material disposal, oil and gas development, and freshwater diversions have all been associated with impacts on estuarine environments. Other anthropogenic factors that have been implicated in the degradation of these environments are overfishing, the introduction of nonindigenous species, shipping, and recreational boating. The magnitude of the effects of these anthropogenic activities is often difficult to determine because of large natural variations in environmental conditions and biotic communities. The occurrences of natural catastrophic events, such as major hurricanes, are frequently overlooked by environmental scientists when assessing

Table 1.3 Point and nonpoint sources of pollution in estuarine and marine waters

Sources	Common Pollutant Categories
	Point Sources
Municipal sewage treatment plants	BOD, bacteria, nutrients, ammonia, toxic chemicals
Industrial facilities	Toxic chemicals, BOD
Combined sewer overflows	BOD, bacteria, nutrients, turbidity, total dissolved solids, ammonia, toxic chemicals
	Nonpoint Sources
Agricultural runoff	Nutrients, turbidity, total dissolved solids, toxic chemicals
Urban runoff	Turbidity, bacteria, nutrients, total dissolved solids, toxic chemicals
Construction runoff	Turbidity, nutrients, toxic chemicals
Mining runoff	Turbidity, acids, toxic chemicals, total dissolved solids
Septic systems	Bacteria, nutrients
Landfills/spills	Toxic chemicals, miscellaneous substances
Silvicultural runoff	Nutrients, turbidity, toxic chemicals

Source: U.S. Environmental Protection Agency, National Water Quality Inventory, Washington, D.C., 1986.

insidious pollution impacts and human activities in estuaries. However, these events may also dramatically alter estuarine environments.

Examples of potentially significant nonpoint sources of pollution from land-based systems include runoff from urban areas, mining and construction sites, and farm lands; leachate from landfills; septic tank leakage; and groundwater transport. Nonpoint runoff is a major source of pollutants to rivers and estuaries (Figure 1.2). Nonpoint source pollutants also originate from human activities at sea associated with accidental releases (e.g., oil spills), marine mining, marine transportation, and the operation of pleasure craft. Although nonpoint pollution represents a rather diffuse source of contaminants in marine waters, it is quantitatively important, particularly on developed coastlines. Nonpoint pollution occurs in virtually all estuarine and coastal marine waters along developed shorelines, but it varies dramatically both spatially and temporally. Because nonpoint runoff is so diffuse, widespread, and variable, it is usually difficult to accurately quantify. Comprehensive data are available primarily for urban and suburban runoff, with large information gaps still existing for other nonpoint sources. Many state assessment programs in the U.S. suffer from inadequate funding, which typically translates into a lack of information gathering, inadequate systematic analyses of gathered data, and ineffective dissemination of results. Therefore, information needed for accurate national assessment and the determination of the relative inputs of pollutants from point and nonpoint sources is usually incomplete.

Marine waste disposal activities continue to be overwhelmingly concentrated in estuarine and coastal marine waters, which receive 80 to 90% of all wastes released to marine environments worldwide. More than half of all industrial and municipal pipelines discharge directly into estuaries, and more than half of all dredged material dumpsites lie in these coastal ecotones. Dredged material, sewage wastes from municipal treatment plants, liquid industrial wastes, and land runoff are the sources of most pollutants released to coastal marine waters in the U.S. (Table 1.4).

Although the relative contribution of pollutants from pipeline discharges, dumping, and nonpoint sources in estuarine and coastal marine waters varies with the type of pollutant and the location, outfall discharges and runoff generally deliver greater concentrations of pollutants to these coastal systems than does dumping. However, in some cases (e.g., Liverpool Bay) dumping is the major source of pollutant entry. Between 1970 and

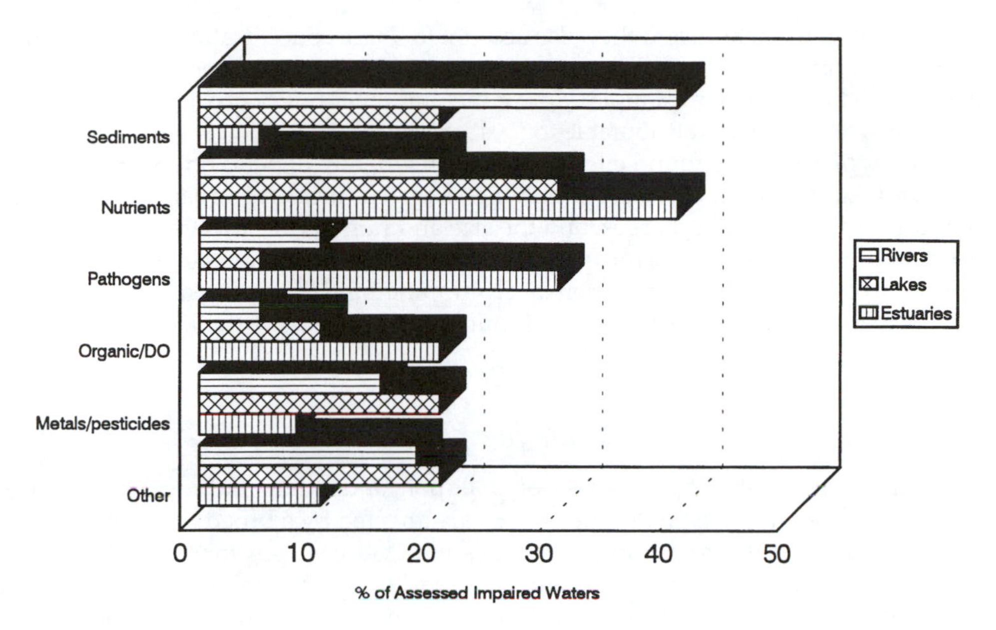

Figure 1.2 Causes of surface water impairment by nonpoint source pollution. (From U.S. Environmental Protection Agency, Managing Nonpoint Source Pollution, EPA 506/9-60, Office of Wetlands, Oceans, and Watersheds, Washington, D.C.)

Table 1.4 Major sources of coastal marine pollution in the U.S.

Type of Pollutant	Source		
	Sewage Treatment	Industrial Facilities	Land Runoff
Nutrients (N and P)	41	7	52
Bacteria	16	<1	84
Oil	41	10	47
Toxic metals	6	46	47

Note: Values in percent. Excludes dredged spoils.

Source: Data from Natural Resources Defense Council, Washington, D.C.

1990, the general trend for U.S. marine waters was a gradual increase in sewage sludge dumping and a dramatic decrease in industrial waste dumping. The greatest volume by far of waste material dumped in shallow marine waters of the U.S. during these 2 decades consisted of dredged material. Most (approximately two thirds) of the dredged material currently dumped in U.S. marine waters is in estuaries, the remainder being divided nearly equally between waters within the 4.8-km territorial boundary and those beyond.

Because of the great variability in composition of marine-dumped wastes and the intermittent and localized nature of dumping operations, it is difficult to compare pollutant inputs derived from sewage wastes and dredged material dumping and those resulting from pipeline discharges and runoff. However, based on a comprehensive national database dealing with relative contributions of pollutants from major sources in the U.S.,[10] some general observations are possible (Table 1.4). Nonpoint runoff is the major contributor of fecal coliform bacteria to U.S. estuaries. It also represents the principal source of suspended solids, total phosphorus, and certain heavy metals (e.g., copper, chromium, iron, lead, and zinc). In contrast, municipal point sources are the dominant contributors of pollutants that raise the biochemical oxygen demand, and they also account for large concentrations of total nitrogen, oil, and grease. Industrial discharges are the principal sources of some heavy metals and many organic chemicals. For example, 90% or more of the input of cadmium, mercury, and chlorinated hydrocarbons is attributable to these discharges.

The types of pollutants found either locally or regionally depend on several important factors. For instance, the relative contributions of agricultural and urban runoff, the type and volume of industrial discharges, and the size and number of sewage treatment plants that release effluent all influence the quantity and quality of pollutants in receiving waters. The amount of harbor or port maintenance, shipping, and recreational and commercial activity in marine waters also affects the input of pollutants and their subsequent biotic impacts on both local and regional scales.

IV. Estuarine pollutants

All NEPs require detailed assessments of pollution and other anthropogenic impacts. Estuaries, together with coastal marine waters, are affected by a broad array of pollutants, many of which originate from human activities in developed portions of the watershed. More than 75% of the total pollutant inputs to estuarine and coastal marine environments derives from runoff, other land sources, and atmospheric deposition. These inputs commonly result in multiple impacts on estuarine and marine organisms (Figure 1.3). Maritime transportation (accidents and discharges) contributes an additional 12% of all pollutants to these environments. A close coupling exists among emerging pollution problems in estuaries, contamination of attendant watersheds, and aerosols in the lower atmosphere.[11] Pollutants commonly reported in estuaries and the coastal ocean include:

1. Excessive nutrients causing progressive enrichment and periodic eutrophication problems
2. Sewage and other oxygen-demanding wastes (principally carbonaceous organic matter) that promote anoxia or hypoxia of coastal waters subsequent to microbial degradation
3. Pathogens (e.g., certain bacteria, viruses, and parasites) and other infectious agents often associated with sewage wastes
4. Petroleum hydrocarbons originating from oil tanker accidents and other major spillages, routine operations during oil transportation, effluent from nonpetroleum industries, municipal wastes, and nonpoint runoff from land
5. Polycyclic aromatic hydrocarbons (PAHs) entering estuarine and marine ecosystems from sewage and industrial effluents, oil spills, creosote oil, combustion of fossil fuels, and forest fires
6. Halogenated hydrocarbon compounds (e.g., organochlorine pesticides) derived principally from agricultural and industrial sources
7. Heavy metals accumulating from smelting, sewage-sludge dumping, ash and dredged material disposal, antifouling paints, seed dressings and slimicides, power station corrosion products, oil refinery effluents, and other industrial processes
8. Radioactive substances generated by uranium mining and milling, nuclear power plants, and industrial, medical, and scientific uses of radioactive materials
9. Calefaction of natural waters, owing primarily to the discharge of condenser cooling waters from electric generating stations
10. Litter and munitions introduced by various land-based and marine activities
11. Fly ash, colliery wastes, flue-gas desulfurization sludges, boiler bottom ash, and mine tailings
12. Drilling muds and cuttings

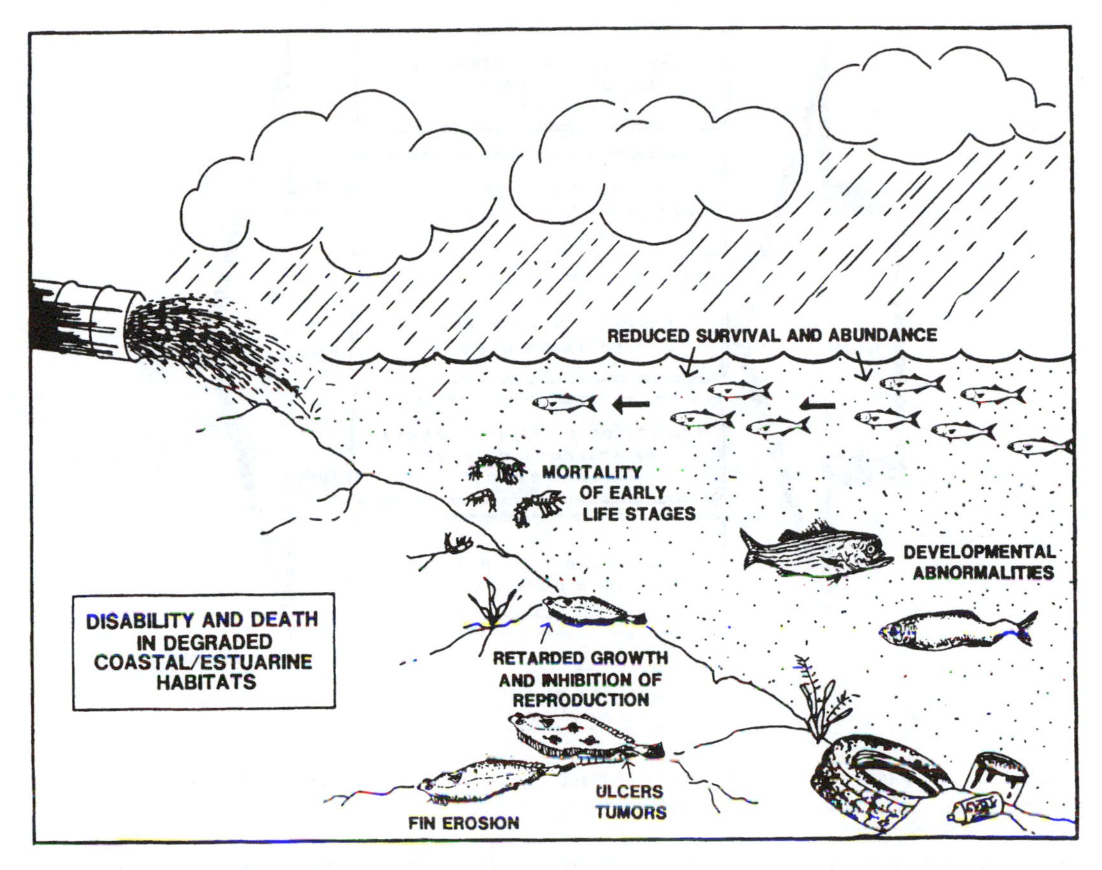

Figure 1.3 Some examples of the effects of contaminants on marine organisms. (From Sindermann, C.J., *Ocean Pollution: Effects on Living Resources and Humans,* CRC Press, Boca Raton, FL, 1996.)

13. Acid-iron and alkali chemicals
14. Pharmaceuticals
15. Suspended solids and turbidity

The NEPs are focusing on pollutants of priority concern. Most notable in this regard are those pollutants that compromise water quality (e.g., nutrient overenrichment, organic loading, pathogens, etc.), directly endanger organisms and communities (e.g., halogenated hydrocarbon compounds, PAHs, and heavy metals), and threaten habitats (e.g., petroleum hydrocarbons). These pollutants are a major concern because they can induce diseases and interfere with the physiologic functions of organisms. Thus, they pose potentially serious problems for estuarine populations (Figures 1.4 to 1.6).

A. Nutrient loading

Although nutrients are essential for growth and reproduction of estuarine organisms, and in moderation promote increased phytoplankton growth that enhances fish and shellfish production, excessive amounts can be detrimental to ecosystem health, causing shifts in the biomass of major plant groups and eutrophication problems (Figure 1.7). Increased loading of nutrients and organic carbon has been correlated with enhanced eutrophication, or hypereutrophication, which leads to imbalances in trophic systems. The addition of

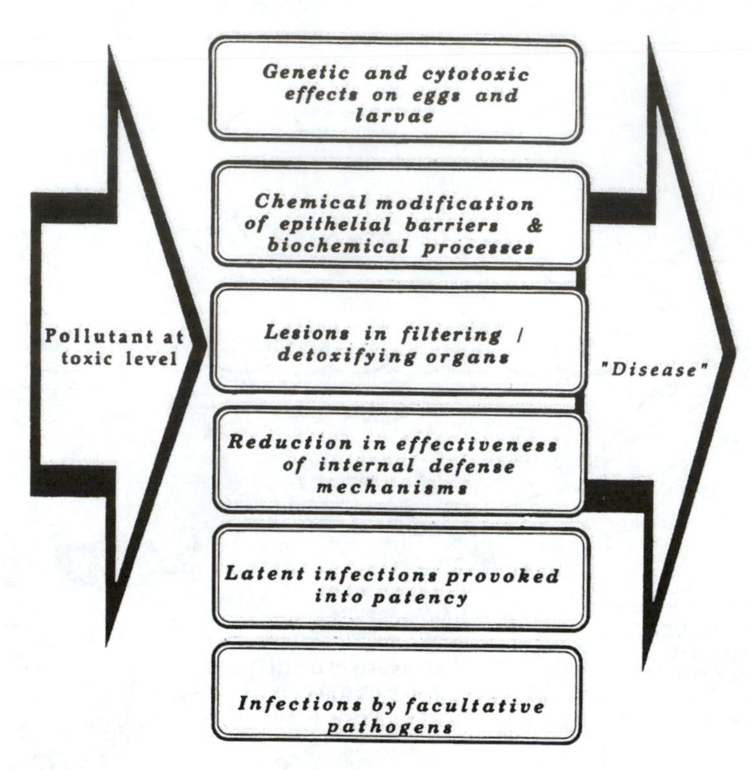

Figure 1.4 Some effects of pollutants on disease processes. (From Sindermann, C.J., *Ocean Pollution: Effects on Living Resources and Humans*, CRC Press, Boca Raton, FL, 1996.)

large nutrient concentrations (e.g., nitrogen and phosphorus) to estuarine waters often fosters progressive organic enrichment, primarily due to accelerated phytoplankton production and biomass accumulation followed by changes in estuarine food webs and periodic hypoxia and anoxia. Red-tide blooms of toxic dinoflagellates (e.g., *Protogonyaulax catenella*, *Pthychodiscus brevis*, and *Pyrodinium bahamense*) as well as nuisance brown-tide blooms (e.g., *Aureococcus anophagefferens*), occasionally develop. Some of these blooms endanger finfish and shellfish populations. Toxic algal blooms also cause severe illnesses (e.g., neurotoxic shellfish poisoning, paralytic shellfish poisoning, and scromboid poisoning) in humans who consume toxin-infested finfish and shellfish.

Nitrogen is usually the chief limiting nutrient to primary production in estuarine and coastal marine waters.[11-15] When nutrient levels remain too high, however, ecological effects can be devastating. Severe eutrophication may result in the following:[5,11]

- Depletion of dissolved oxygen in bottom waters caused by the bacterial decomposition of accumulated phytoplankton and submerged aquatic vegetation
- Persistent hypoxia or anoxia of estuarine waters, commonly inflicting heavy mortality on benthic and finfish populations
- Changes in species abundance, distribution, and diversity, leading to an alteration of estuarine community structure
- Replacement of typically abundant species with less desirable forms, thereby modifying trophic interactions in the system
- Accelerated phytoplankton growth increasing turbidity and light attenuation in the water column, while decreasing production of submerged aquatic vegetation
- A shift from a seagrass-dominated to a phytoplankton-dominated system in shallow estuaries

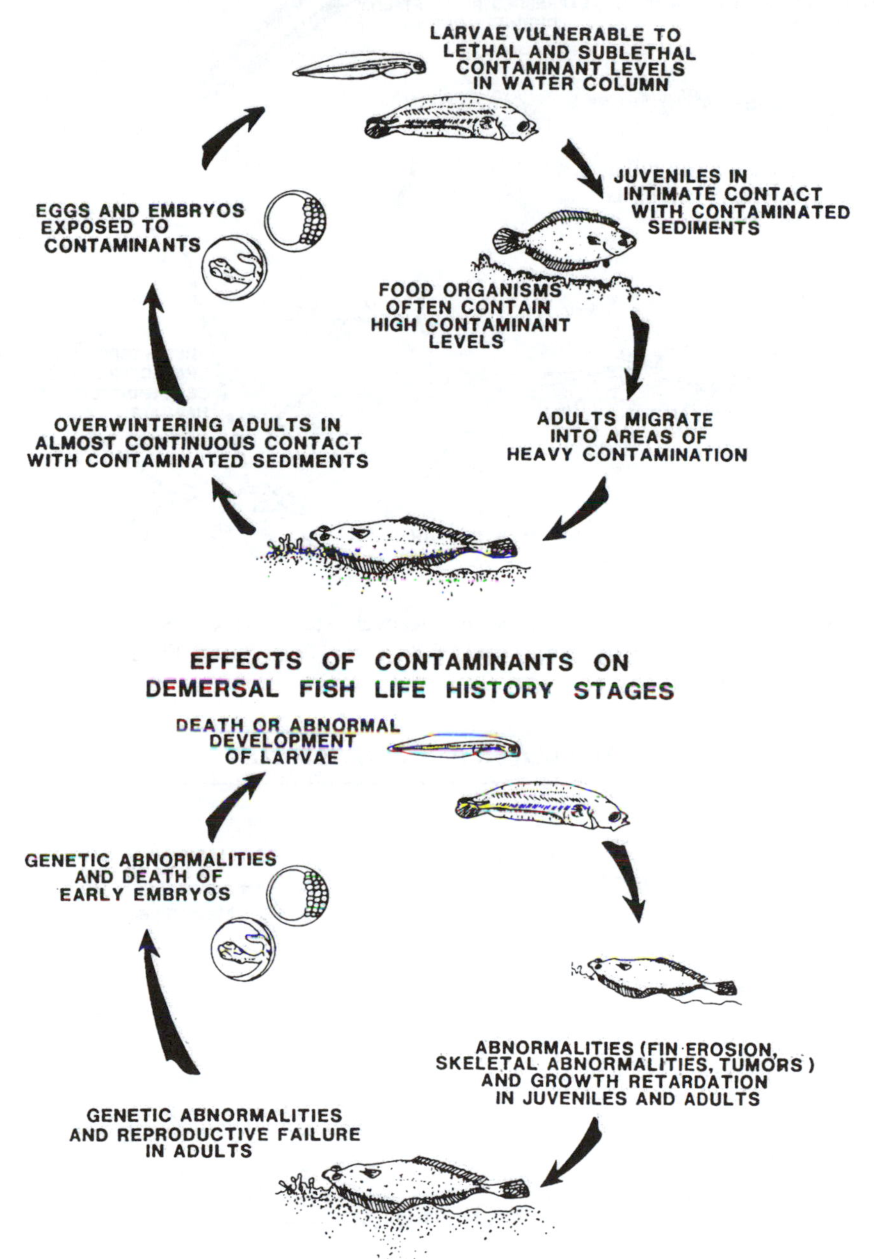

Figure 1.5 Life cycle of demersal fish (e.g., winter flounder, *Pseudopleuronectes americanus*) with potential pollutant impact points (above) and the effects of pollutants (below). (From Sindermann, C.J., *Ocean Pollution: Effects on Living Resources and Humans*, CRC Press, Boca Raton, FL, 1996.)

Nutrients enter estuaries from numerous sources such as surface water (streams, rivers, and direct runoff), groundwater discharge, and atmospheric deposition (both wet and dry deposition). Agricultural and suburban runoff provides substantial concentrations of nutrients in rural areas, as does urban runoff in heavily populated regions. Municipal and industrial wastewaters deliver considerable amounts of nutrients to estuaries, especially in close proximity to metropolitan centers.

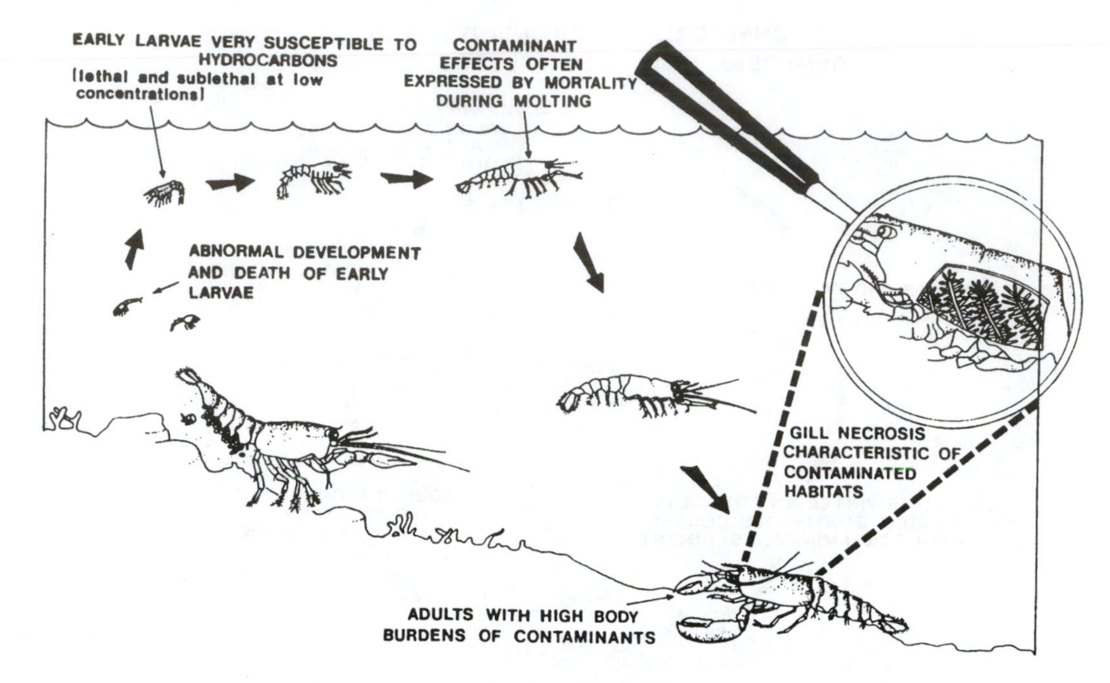

Figure 1.6 Contaminant effects on crustacean life cycle stages. (From Sindermann, C.J., *Ocean Pollution: Effects on Living Resources and Humans*, CRC Press, Boca Raton, FL, 1996.)

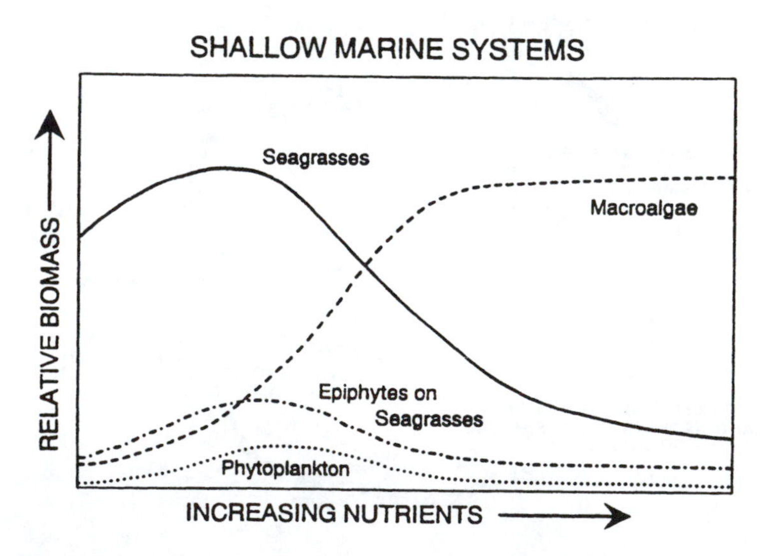

Figure 1.7 Generalized shift in biomass of major plant groups with increasing nutrient input to shallow marine systems. Occasionally the phytoplankton dominate, but usually macroalgae — especially general green taxa — take off, whereas submerged aquatic vegetation decline through competition either for nitrogen or light. (From Harlin, M.M., Changes in major plant groups following nutrient enrichment, in *Eutrophic Shallow Estuaries and Lagoons*, McComb, A.J., Ed., CRC Press, Boca Raton, FL, 1995, 173.)

There are more than 1400 municipal wastewater treatment plants nationwide that discharge more than 37.9 billion liters of treated effluent every day to estuaries and coastal marine waters. More than 85% of this effluent is discharged to estuaries. In addition, about

1300 industrial facilities discharge approximately 42.8 billion liters of treated wastewater and spent cooling water daily to marine waters.[16] Estuarine waters also receive nutrients directly from boats, marinas, and dredge material disposal operations. Away from urbanized areas with heavy development, natural sources of nutrients (e.g., organic mineralization, waterfowl, and wildlife) can be significant.[15,17] Shallow estuaries with restricted circulation are most susceptible to eutrophication problems. However, increasing numbers of larger estuaries show signs of eutrophication — escalating productivity, shifts in community dominants, development of hypoxia or anoxia in bottom waters, and the loss of submerged aquatic vegetation and benthic fauna.[18]

Nutrient problems in estuaries are usually closely coupled to human activities associated with development and land-use practices throughout coastal watersheds. Because of differences in land-use patterns, management practices, levels of wastewater treatment, and local and regional hydrology from one watershed to another, the relative contribution of nutrients from point and nonpoint sources varies appreciably across the coastal zone. However, in most coastal states, nonpoint pollution is the most important source of contaminants.

Pollution control programs must be tailored to the particular needs of a region because it is not possible to prescribe a specific technology or approach at the national level that will satisfactorily address all water quality issues at all locations.[18] In recent years, regional approaches have been devised to control nutrient inputs and improve water quality in estuaries because of the many diffuse nutrient sources in watershed areas.[15] An example is the Chesapeake Bay Program initiated and coordinated by the USEPA. This multifaceted program has implemented several best management practices in watersheds to control nutrient influx. Some have targeted agricultural practices, and others have addressed industrial practices. Detailed nutrient budgets have been developed for the bay (Figures 1.8 and 1.9). In addition, the State of Maryland has adopted more stringent regulations to control pollution at its source. These measures have been necessary to counter the growing problem of anoxic bottom waters in the system. Although these efforts have proven to be beneficial to water quality improvement in the bay, it is becoming evident that the best approach to pollution control is pollution prevention.[18]

B. *Organic loading*

Organic matter accumulating from various anthropogenic and natural sources (e.g., sewage wastes, wildlife excretions, and vascular plant remains) exacerbates hypoxia and anoxia in stressed systems by creating biochemical oxygen demand and chemical oxygen demand. When water column stratification is strong and vertical mixing reduced (e.g., parts of Chesapeake Bay and Long Island Sound), bottom water hypoxia may persist for months. Nuisance organisms and opportunists (e.g., *Capitella* spp.) commonly dominate benthic communities in these impacted waters.

In pristine estuaries, most oxidizable carbon originates from natural sources such as animal remains, biodeposits (feces and pseudofeces), decaying submerged vascular plants (saltmarsh grasses, seagrasses, and mangroves), and benthic macroalgae. More than 90% of the primary production of benthic macrophytes passes to detritus — approximately one third as dissolved organic matter and two thirds as particulate organic matter — because of low exploitation rates of grazing herbivores on live plants. The concentration of detritus in estuaries generally ranges from 0.1 to greater than 125 mg/l. The accumulation of fresh organismal remains and detritus in bottom sediments is generally of most ecological concern in estuaries with high biochemical oxygen demand.

The highest concentrations of organic carbon occur in waters receiving sewage wastes and other anthropogenic carbon inputs, where both dissolved and particulate organic carbon levels exceed 100 mg/l. The organic carbon may enter these systems via river

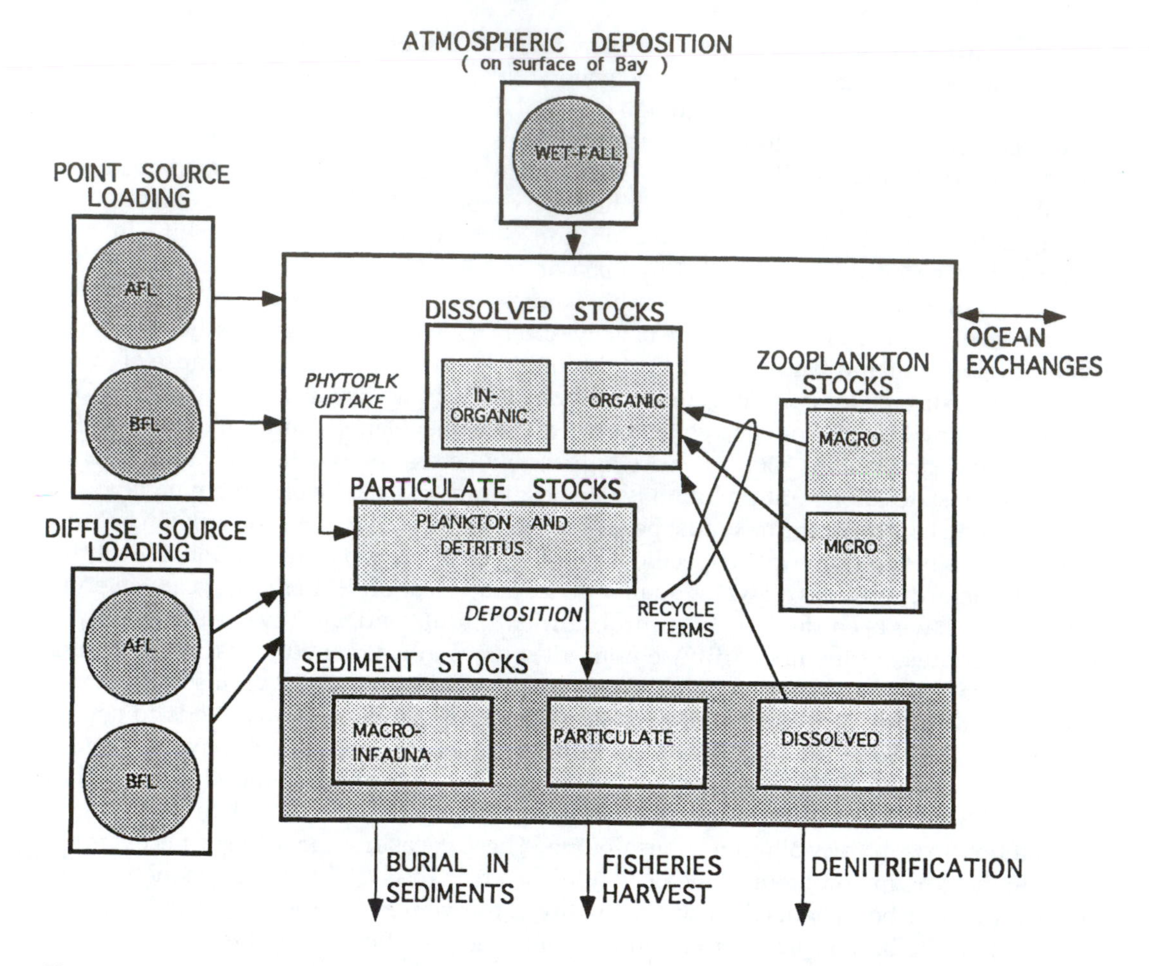

Figure 1.8 A schematic diagram of the Chesapeake Bay nutrient budget. Nutrient sources, storages, recycle pathways, internal losses, and exchanges across the seaward boundary are indicated. (From Boynton, W.R., Garber, J.H., Summers, R., and Kemp, W.M., *Estuaries*, 18, 285, 1995.)

discharges, pipeline outfalls, malfunctioning septic tanks, marinas, boats, dredged materials, and other sources. Oxygen depletion due to biochemical oxygen demand from wastewaters discharged through well-designed outfalls is generally of less ecological concern in well-flushed estuaries and open coastal waters. When oxygen depletion occurs in these cases, it is typically minor and localized in extent compared to that in systems impacted by widespread eutrophication. The depletion of oxygen associated with biochemical oxygen demand can be a serious problem in estuaries with limited exchange. It is important to note that ecological harm in these estuaries may arise when dissolved oxygen concentrations fall to levels of 3 to 4.3 mg/l.[19]

Urbanized estuaries near metropolitan centers (e.g., New York, Boston, Baltimore, Los Angeles, and Seattle) that receive large concentrations of organic wastes often exhibit the most acute organic loading and oxygen depletion problems. Many older cities have combined sewer systems that carry both stormwater and municipal sewage. During periods of precipitation when runoff and sewage flows exceed the capacity of a treatment system, combined sewer overflows discharge untreated sewage, industrial wastewater, and urban runoff at predesignated locations into adjacent waterways. The greatest concentration of organic carbon and chemical contaminants generally is contained in the first flush of

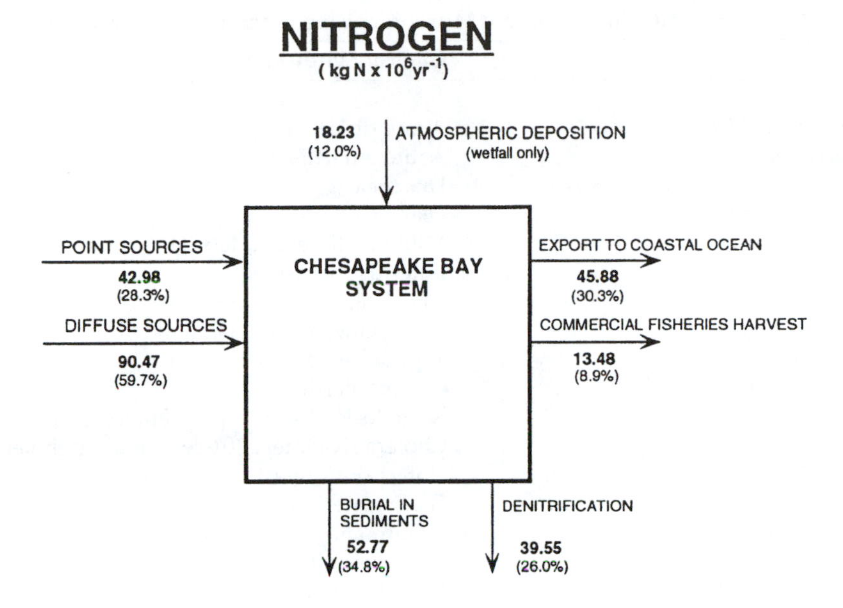

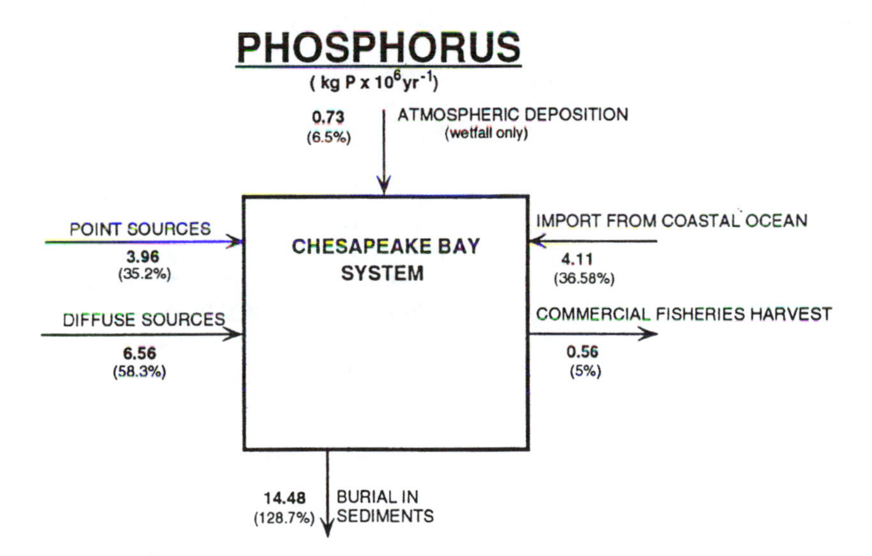

Figure 1.9 Simplified annual total nitrogen (TN) and total phosphorus (TP) budgets for the entire Chesapeake Bay system. (From Boynton, W.R., Garber, J.H., Summers, R., and Kemp, W.M., *Estuaries*, 18, 285, 1995.)

stormwater.[18] Remedial pollution programs must deal with the pulsed nature of these inputs to receiving waters.

C. Pathogens

Pathogenic microorganisms (i.e., bacteria, viruses, protozoa, and helminths) enter sewage sludge during the processing of human and animal waste, and they pose a health risk to humans who either ingest contaminated seafood products or swim in contaminated waters. More than 100 enteric pathogens may be found in treated municipal wastewater and urban stormwater runoff (Table 1.5).[16] Total coliforms and a subset of this group, the

Table 1.5 Human Pathogenic Microorganisms Potentially Waterborne

Pathogen	Clinical Syndrome
Bacteria	
Aeromonas hydrophila	Acute diarrhea
Campylobacter spp.	Acute enteritis
Enterotoxin *Clostridium perfringens*	Diarrhea
Enterotoxin *Escherichia coli*	Diarrhea
Francisella tularensis	Mild or influenzal, febrile, typhoidal illness
Klebsiella pneumoniae	Enteritis (occasional)
Plesiomonas shigelloides	Diarrhea
Pseudomonas aeruginosa	Gastroenteritis (occasional)
Salmonella typhi	Typhoid fever
Other salmonellae	Gastroenteritis
Shigella spp.	Shigellosis ("bacillary dysentery")
Vibrio cholerae	Cholera dysentery (01 serovars) or cholera-like infection (non-01)
V. fluvialis	Gastroenteritis
Lactose-positive *Vibrio*	Pneumonia and septicemia
V. parahaemolyticus	Gastroenteritis
Yersinia enterocolitica	Enteritis, ileitis
Cyanobacteria	
Cylindrospermopsis spp.	Hepatoenteritis
Viruses	
Enteroviruses	Aseptic meningitis, respiratory infection, rash, fever
Poliovirus	Paralysis, encephalitis
Coxsackie virus A	Herpangina, paralysis
Coxsackie virus B	Myocarditis, pericarditis, encephalitis, epidemic pleurodynia, transient paralysis
Echovirus	Meningitis, enteritis
Types 68–71	Encephalitis, acute hemorrhagic conjunctivitis
Hepatitis A	Infectious hepatitis type A
Hepatitis non-A, non-B	Hepatitis type non-A, non-B
Influenza A	Influenza
Norwalk and other parvovirus-like agents	Epidemic, acute nonbacterial gastroenteritis
Rotavirus	Nonbacterial, endemic, infantile gastroenteritis: epidemic vomiting and diarrhea
Protozoa	
Balantidium coli	Balantidiasis (balantidial dysentery)
Cryptosporidium	Cryptosporidiosis
Entamoeba histolytica	Amoebiasis (amoebic dysentery)
Giardia lamblia	Giardiasis — mild, acute, or chronic diarrhea
Helminths	
Ascaris	Ascariasis (roundworm infection)
Ancylostoma	Hookworm infection
Clonorchis	Clonorchiasis (Chinese liver fluke infection)
Diphyllobothrium	Diphyllobothriasis (broadfish tapeworm infection)
Dracunculus mediensis	Dracontiasis (Guinea worm infection)
Fasciola	Fascioliasis (sheep liver fluke infection)
Fasciolopsis	Fasciolopsiasis (giant intestinal fluke infection)
Paragonimus	Paragonimiasis (Oriental lung fluke infection)
Spirometra mansoni	Sparganosis (plerocercoid tapeworm larvae infection)
Taenia	Taeniasis (tapeworm infection)
Trichostrongylus	Trichostrongyliasis
Trichuris	Trichuriasis (whipworm infection)

Table 1.5 (continued) Human Pathogenic Microorganisms Potentially Waterborne

Pathogen	Clinical Syndrome
Bacteria	
Aeromonas hydrophila	Wound and ear infections, septicemia, meningitis, endocarditis, corneal ulcers
A. sobria	Wound and ear infections
Chromobacterium violaceum	Septicemia
Clostridium perfringens	Wound infection — gas gangrene
Klebsiella pneumoniae	Pneumonia, bacteremia
Legionella spp.	Legionellosis (Legionnaires' disease)
Leptospira spp.	Leptospirosis (Weil's disease — jaundice, hemorrhages, aseptic meningitis)
Mycobacterium marinum	Skin infection ("swimming pool granuloma")
M. ulcerans	Skin infection (progressive subcutaneous ulceration)
Pseudomonas aeruginosa	Otitis externa and media: follicular dermatitis (pruritic pustular rash)
P. pseudomallei	Meliodosis (glanders-like infection)
Staphylococcus aureus	Wound and skin infections
Halophilic vibrios (including *Vibrio parahaemolyticus*, *Vibrio alginolyticus*, lactose-positive *Vibrio*)	Wound and ear infections, conjunctivitis, salpingitis, pneumonia, septicemia
Viruses	
Adenovirus	Pharyngoconjunctivitis (swimming pool conjunctivitis), respiratory infection
Adenosatellovirus	Associated with adenovirus type 3 conjunctivitis and respiratory infection in children but etiology not clearly established
Protozoa	
Naegleria fowleri	Primary amoebic meningoencephalitis (PAME)
Helminths	
Schistosoma spp.	Schistosomiasis (bilharzia)
Avian schistosomes (*Trichobilharzia, Austrobilharzia*)	Schistosome dermatitis (swimmer's itch)
Ancylostoma duodenale	Hookworm infection
Necator americanus	Hookworm infection

Source: Compiled from McNeil, A. R., Australia Water Resources Tech. Paper No. 85, Australian Government Publishing Service, Canberra, 1985, 561.

fecal coliforms, are used as an indicator of the potential occurrence of human and/or animal pathogens in estuarine waters and of overall microbial water quality. These bacteria can enter estuaries via sewage treatment plant discharges, faulty septic tanks, and industrial effluent. Wildlife populations can be a significant natural source of indicator bacteria (total coliforms, fecal streptococci, and enterococci) in these coastal systems.[15]

The viability of the pathogens is closely coupled to their survival in the estuarine environment, which is influenced by several key factors (e.g., nutrients, temperature, dissolved oxygen, and sediments). Many of the pathogens die quickly when exposed to the vagaries of environmental conditions in estuaries. Mortality rates are higher in saline water than freshwater. The survival period of coliforms in estuarine waters is shorter than that of other enteric microorganisms. However, some pathogens can be persistent, particularly when sorbed to particulates, which afford some degree of protection.[5] It is common for the pathogens to survive for several days in estuarine water and for even longer periods in fish and shellfish tissue.[16] The concentration of pathogens in contaminated seafood can be high because the microorganisms often survive in estuarine organisms without harming them.

A number of serious human illnesses are linked to pathogenic exposure in estuarine environments. Among the pathogenic bacteria that pose a health hazard to humans are members of the genus *Salmonella,* including the organisms responsible for typhoid; *Shigella* spp., which cause dysentery; and some species of the *Clostridia,* which produce potent exotoxins. Certain viruses in contaminated waters give rise to infectious hepatitis. The consumption of raw, viral-tainted shellfish is known to cause hepatitis A and viral gastroenteritis. Sewage treatment lowers the number of pathogens in wastewater effluent, specifically during the sludge-forming process, although it may not inactivate them completely. Thermophilic digestion and other sewage treatment techniques have the potential to further reduce the abundance of these pathogenic organisms.[12]

Aside from enteric bacterial and viral pathogens, parasites are commonly associated with waterborne diseases. Helminths and protozoa are of particular health concern for wastewater exposure. Hookworms, roundworms, tapeworms, and whipworms are helminths whose transmission is commonly associated with untreated sewage and untreated sludges. Pathogenic enteric protozoa, such as *Entamoeba histolytica,* also impact humans primarily via raw sewage contamination.

Coastal communities are formulating various management strategies to address pathogen contamination of estuarine and coastal marine waters. One approach targets pathogenic sources. For example, efforts are under way in many communities to reduce or eliminate the discharge of raw or inadequately treated sewage due to malfunctions of sewage treatment plants and illegal connections. Other communities are working to replace septic tanks with central sewers and attempting to reduce loadings from stormwater discharges, combined sewer overflows, and more diffuse nonpoint source runoff. Watershed protection efforts are being implemented including land-use controls for new development, improved agricultural practices and livestock management, stormwater runoff mitigation, and open marsh water management. Sewage waste associated with boats and marinas is a priority problem. Programs are being conducted nationwide to establish no-discharge zones and marina-pumpout facilities to mitigate impacts of vessel discharges. Finally, more comprehensive environmental monitoring, assessment, and research programs have been formulated by numerous coastal communities to identify and remediate pathogen problems in estuarine systems.

D. Toxic chemicals

Several classes of toxic chemicals commonly found in estuarine environments are potentially damaging to habitats and hazardous to biotic communities. Among the chemical pollutants of greatest concern are halogenated hydrocarbons, PAHs, and heavy metals. Oil, which contains a wide array of toxic substances, also poses a serious threat to these coastal ecosystems. More than 70,000 synthetic chemicals have been introduced into marine environments during the past 50 years, many of which are toxic to marine life even in minute concentrations.

Because of the persistence of numerous toxic chemicals in estuarine environments and their accumulation by successive levels of food chains, a human health risk may exist from ingestion of carcinogen-contaminated seafood products.[20] Estuarine organisms assimilate toxic chemicals via respiratory, dermal, and oral routes.[21] Uptake of the contaminants is strongly dependent on their bioavailability. Bioaccumulation of the contaminants is a function of many factors, such as temperature, salinity, diet, spawning period, and the ability of an organism to regulate chemical contaminants in its body. The toxicity of the contaminants, in turn, is also influenced by environmental factors.[22]

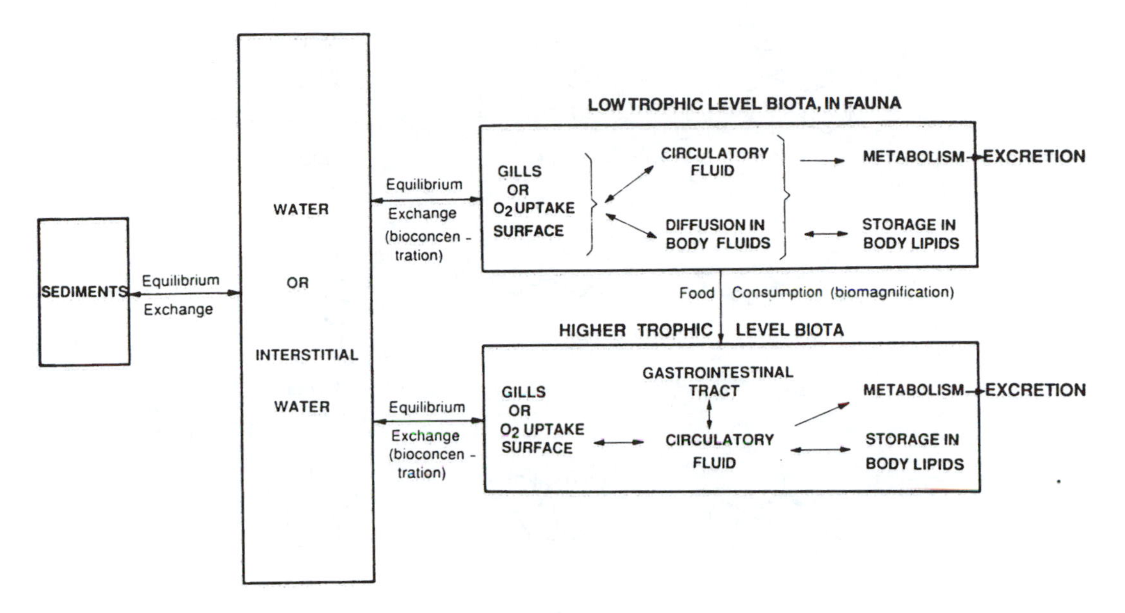

Figure 1.10 Diagrammatic representation of routes of uptake and clearance of lipophilic chemicals by aquatic biota. (From Connell, D.W., Ed., *Bioaccumulation of Xenobiotic Compounds*, CRC Press, Boca Raton, FL, 1989, 100.)

The exposure of estuarine organisms to toxic levels of chemical contaminants elicits a range of pathological responses, notably the lack of repair and regeneration of damaged tissue, inflammation and degeneration of tissue, neoplasm formation, and genetic derangement. Growth inhibition may arise as well as changes in physiology, reproduction, and development. Feeding behavior, respiratory metabolism, and digestive efficiency may also be compromised. At elevated toxin concentration, death often ensues.

1. *Halogenated hydrocarbons*

Some of the most persistent, ubiquitous, and chronically toxic contaminants observed in estuaries worldwide are halogenated hydrocarbons, a large group of low- to high-molecular-weight compounds that have been linked to various environmental and public health concerns. The higher-molecular-weight halocarbons are considered to be a serious threat to estuarine ecosystems because they degrade very slowly in environmental media and tend to accumulate in biota. Due to their fat solubility, these compounds concentrate in lipid-rich tissues and biomagnificate in organisms occupying successive levels of food chains (Figure 1.10). Hence, the highest contaminant levels occur in upper-trophic-level organisms that often serve as a food source for humans. The sublethal effects of long-term exposure of estuarine organisms to low levels of halogenated hydrocarbon compounds are poorly documented.

A long list of organochlorine biocides (insecticides, herbicides, and fungicides) has been detected in water, sediment, and biotic media of many U.S. estuaries. The best known chlorinated hydrocarbon insecticides are DDT and its metabolites (DDD and DDE), aldrin, chlordane, dieldrin, endosulfan, endrin, lindane, heptochlor, chlordecone, methoxychlor, mirex, perthane, and toxaphene (Figure 1.11). DDT remains a persistent problem in systems where elevated concentrations have been recorded in bottom sediments and organisms (Figures 1.12 and 1.13).

Important organochlorine herbicides include chlorophenoxy compounds (e.g., 2,4-D and 2,4-T). Chlorinated benzenes and phenols (e.g., hexachlorobenzene) are notable

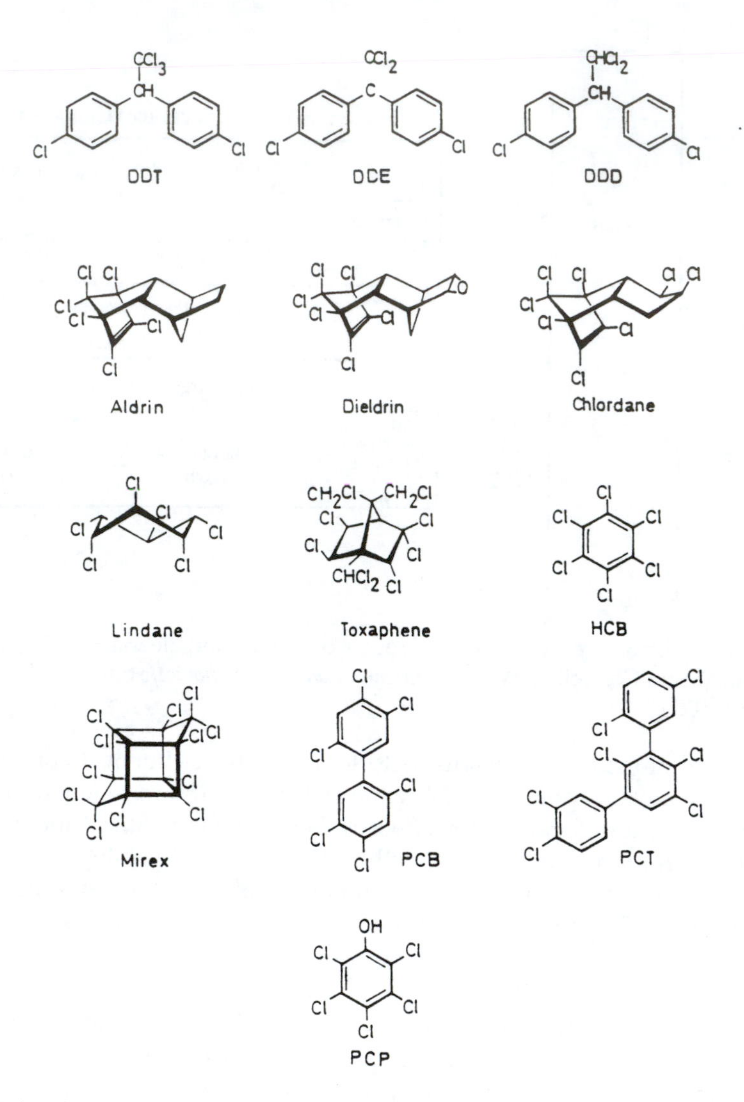

Figure 1.11 Molecular structure of selected organochlorine compounds. (From Reutergardh, L., Chlorinated hydrocarbons in estuaries, in *Chemistry and Biogeochemistry of Estuaries*, Olausson, E. and Cato, I., Eds., John Wiley & Sons, Chichester, UK, 1980, 349.)

organochlorine fungicides. The chlorinated biocides are nerve poisons that have been shown to cause neurological and reproductive failure in nontarget organisms. Most of these compounds have been banned from use in the U.S., although they continue to be utilized by other countries, particularly developing nations in low latitude regions.

Of all industrial contaminants reported in estuarine systems, polychlorinated biphenyls (PCBs) have received the greatest attention. These highly stable, synthetic halogenated aromatic hydrocarbons are deleterious to marine life. For example, they are toxic to fish, which exhibit a greater incidence of altered immune response, blood anemia, epidermal lesions, and fin erosion when exposed to sufficiently high contaminant concentrations. Other organisms (e.g., marine birds and mammals) also experience an array of reproductive abnormalities subsequent to assimilating PCBs. Finally, PCBs are suspected human carcinogens and mutagens in addition to being linked to chronic diseases (e.g., liver malfunctions, reproductive disorders, and skin lesions). Primarily because of human

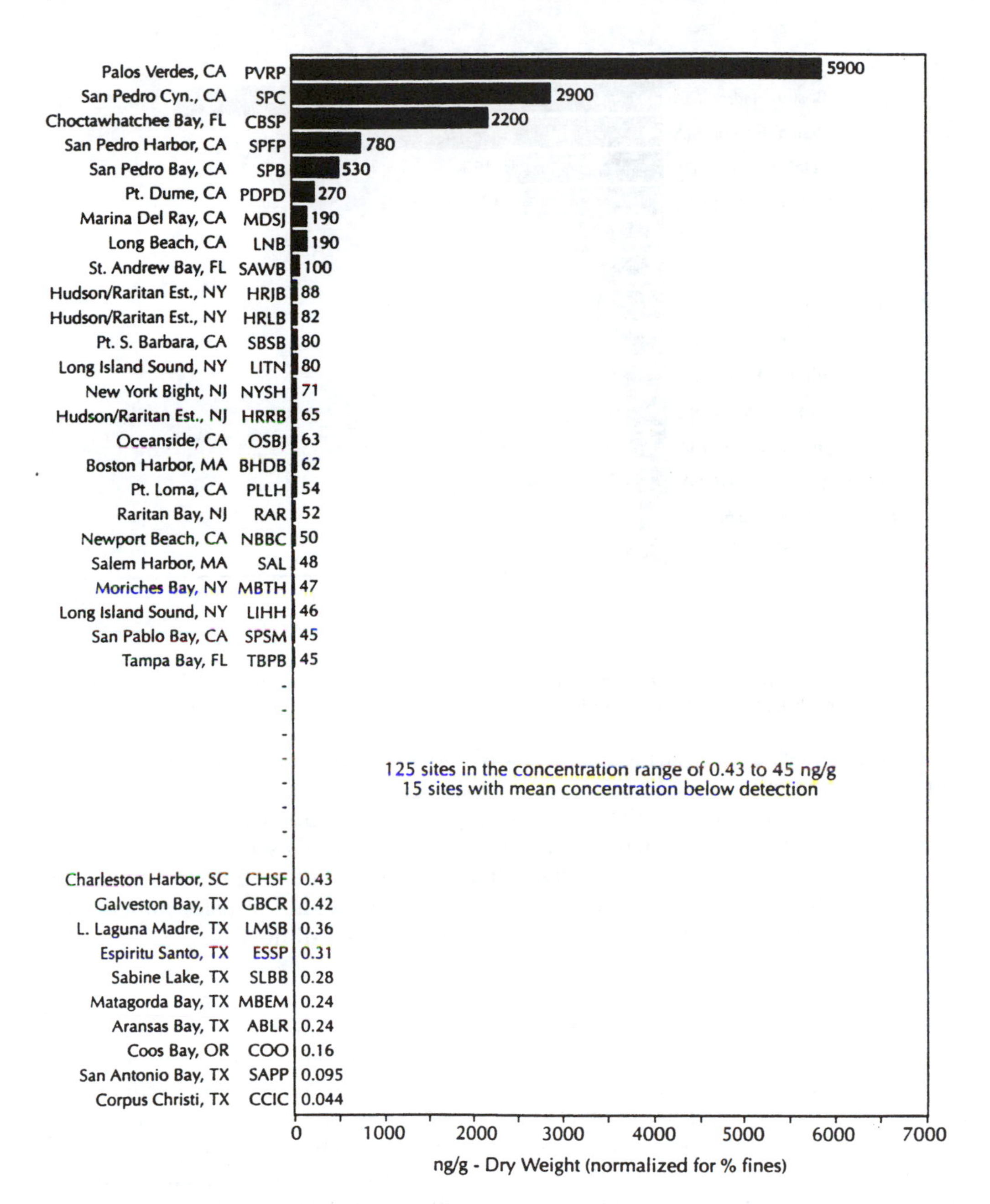

Figure 1.12 Total DDT in sediments of various estuarine and coastal systems of the U.S. (From NOAA, A Summary of Data on Individual Organic Contaminants in Sediments Collected during 1984, 1985, 1986, and 1987, NOAA Tech. Mem. NOS OMA 47, National Oceanic and Atmospheric Association, Rockville, MD, 1989.)

health concerns, PCBs were banned from production in the U.S. in 1977. Nevertheless, PCBs continue to be a serious contamination problem in some estuaries due to their great stability, persistence, and high concentrations in sediments and organisms (Figures 1.14 and 1.15).

Chlorinated dibenzo-*p*-dioxins (CDDs) and chlorinated dibenzofurans (CDFs) are two related classes of aromatic heterocyclic compounds deemed to be responsible for multiple

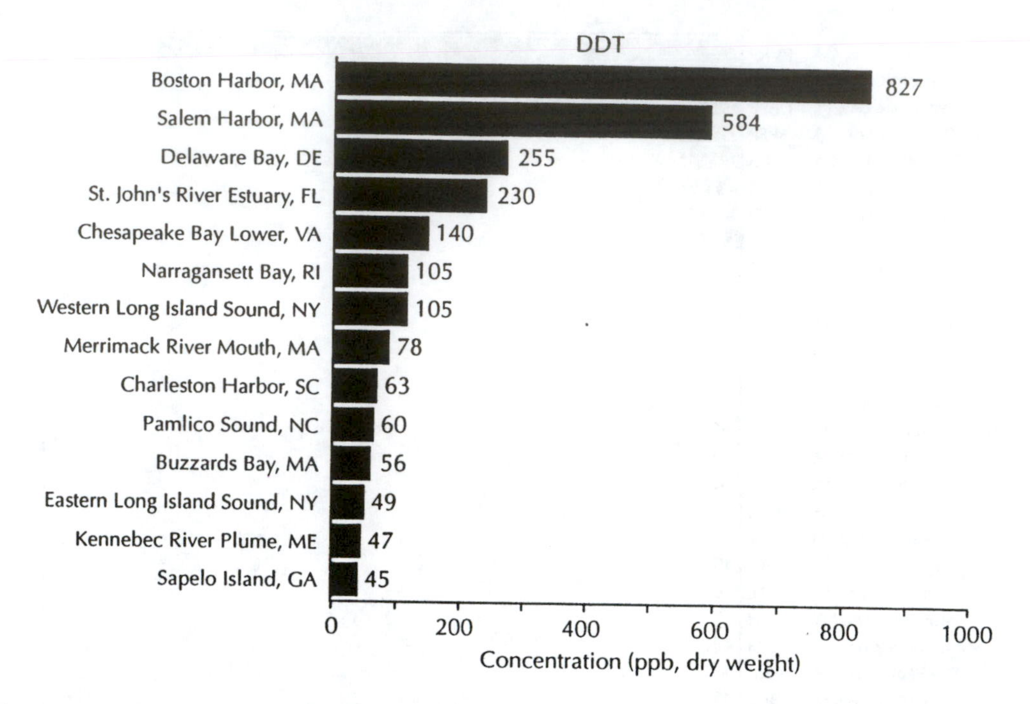

Figure 1.13 Concentrations of DDT in Atlantic Coast fish liver tissue sampled from various U.S. estuaries by the NOAA Status and Trends Program. (From Larsen, P.F., *Rev. Aquat. Sci.*, 6, 67, 1992.)

impacts on estuarine and marine organisms. These environmentally persistent, lipophilic compounds are highly toxic to estuarine organisms and induce serious sublethal effects (e.g., developmental abnormalities, immunosuppression, hormonal and histopathological alterations, liver disorders, impaired reproduction, and cardiovascular changes). Investigations have focused on CDD and CDF impacts on fish and mammals because these contaminants bioaccumulate in marine organisms by food chain transfer rather than by direct uptake from seawater, suspended matter, or bottom sediment.

2. *Oil*

Some of the most devastating pollution impacts in estuaries are associated with the environmental release and accumulation of oil, which not only kills organisms directly but also destroys habitats that support aquatic communities. Polluting oil decimates estuarine populations by physical (smothering, reduced light), habitat (altered pH, decreased dissolved oxygen, decreased food availability), and toxic actions. These effects are often most evident in intertidal and shallow subtidal environments subsequent to major oil spills. However, even small volume releases of oil can be detrimental to organisms and sensitive habitat areas.

Polluting oil in estuaries originates from various point and nonpoint sources. Of greatest importance in this regard are municipal and industrial wastewaters, urban and river runoff, leakage from marine vessels and fixed installations (e.g., production facilities, refineries, and marine terminals) as well as tanker spills and atmospheric deposition.[23] Inputs associated with routine marine transportation activities can be significant. Most of the oil entering estuarine waters, particularly urban systems, does not result from accidents or spills but from routine human activities in watersheds and the estuarine basins themselves. Although locally significant, major oil spills account for a relatively small fraction of the total amount of oil released to estuaries each year.[13,23,24]

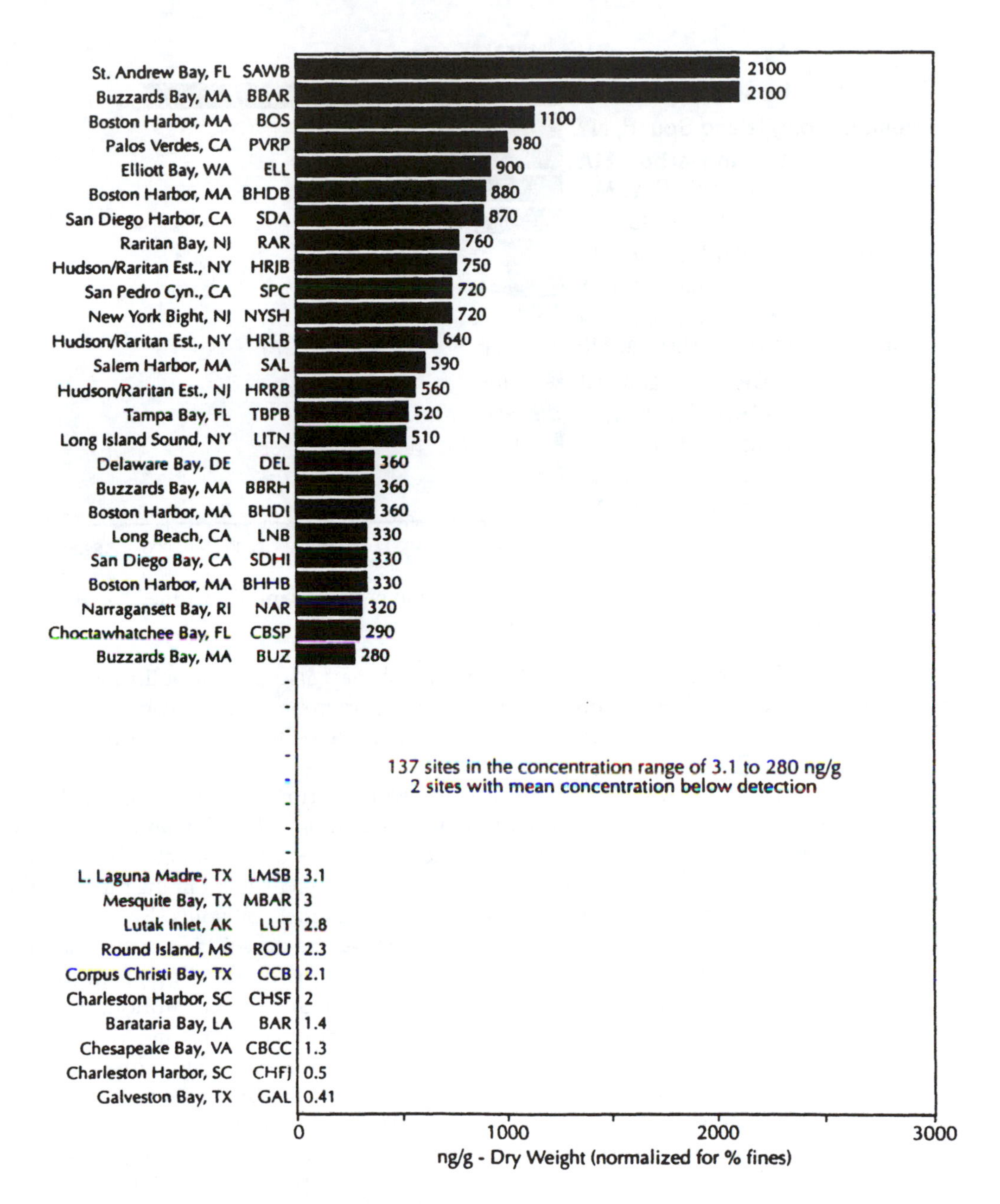

Figure 1.14 Total PCB concentrations in sediments at various National Status and Trends sites during the 1984 to 1987 survey period. (From NOAA, A Summary of Data on Individual Organic Contaminants in Sediments Collected during 1984, 1985, 1986, and 1987, NOAA Tech. Mem. NOS OMA 47, National Oceanic and Atmospheric Association, Rockville, MD, 1989.)

Thousands of hydrocarbon and nonhydrocarbon compounds occur in crude oil. The hydrocarbons, which generally constitute more than 75% of crude and refined oils, are divided into four major classes: (1) straight-chain alkanes (*n*-alkanes or *n*-paraffins); (2) branched alkanes (isoparaffins); (3) cycloalkanes (cycloparaffins); and (4) aromatics (Figure 1.16). Toxicity increases along the series from the alkanes, cycloalkanes, and alkenes to the aromatics. Low-molecular-weight aromatics (e.g., benzene, toluene, and xylene) are considerably toxic, as are acids (e.g., carboxylic acid), phenols, and sulfur compounds

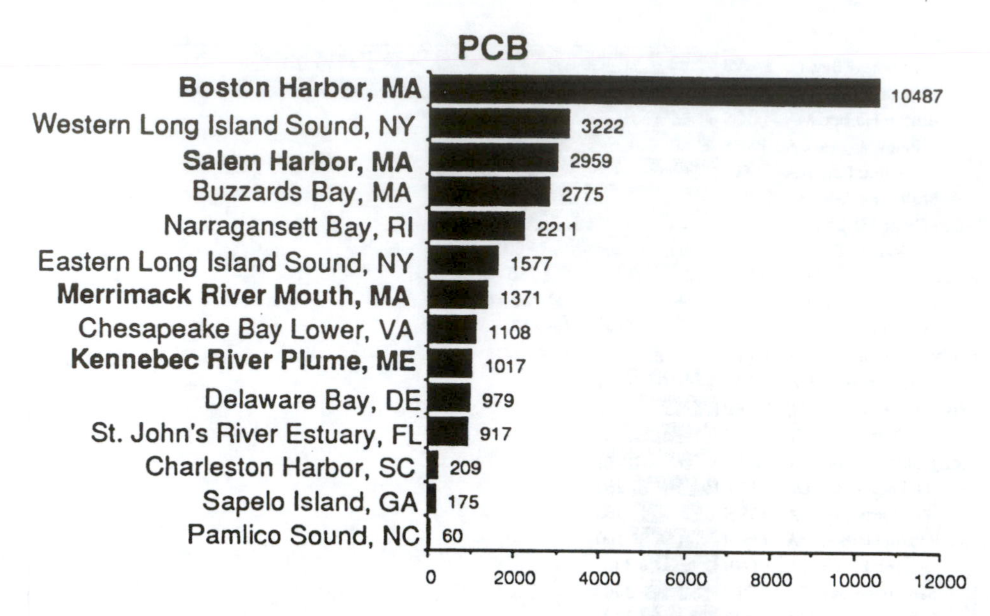

Figure 1.15 Concentrations of PCBs in Atlantic Coast fish liver tissue sampled by the NOAA Status and Trends Program. (From Larsen, P.F., *Rev. Aquat. Sci.*, 6, 67, 1992.)

(e.g., sulfides, thiols, and thiophenes). Metals in the oil may also pose a pollution threat. PAH compounds (e.g., 1,2-benzanthracenes, 3,4-benz(a)pyrene, 1,2-benzphenanthrene, diphenylmethane, fluorene, and phenanthrene), which are toxic to many estuarine organisms, compose 0.2 to 7% of crude oils.[25]

The composition and toxicity of oil spills are altered by various physical–chemical processes, notably evaporation, dissolution, photochemical oxidation, advection and dispersion, emulsification, and sedimentation (Figure 1.17). Bacteria, fungi, and yeast actively degrade oil in marine environments, with bacteria being the critical biological agent in the breakdown process. Collectively, these microbes decompose 40 to 80% of crude oil spills.[13]

The severity of oil pollution impacts on estuarine organisms depends on many factors. Chief among these are: (1) the composition of the oil; (2) volume of the oil; (3) form of the oil (i.e., fresh, weathered, or emulsified); (4) occurrence of the oil (i.e., in solution, suspension, dispersion, or adsorbed onto particulate matter); (5) duration of exposure; (6) involvement of neuston, plankton, nekton, or benthos in the spill or release; (7) juvenile or adult forms involved; (8) previous history of pollutant exposure of the biota; (9) season of the year; (10) natural environmental stresses associated with fluctuations in temperature, salinity, and other variables; (11) type of habitat affected; and (12) cleanup operations (e.g., physical methods of oil recovery and the use of chemical dispersants).[13,26-28] In addition, chemical dispersants, solvents, and other treatments used to clean up oil spills in the environment are also toxic to estuarine life.

Both lethal and sublethal effects of polluting oil on estuarine and marine organisms are well chronicled.[13,24,25] Organisms smothered by a thick layer of oil generally suffer immediate lethal effects. Shorebirds frequently die from drowning or hypothermia after oil covers their plumage. Other organisms, which survive the physical impact of polluting oil, typically have decreased long-term survivorship due to a variety of sublethal effects that alter their reproduction, growth, distribution, and behavior. For example, individuals may lose normal physiological or behavioral function if coated with oil, rendering them more susceptible to disease and predation. Sublethal concentrations of petroleum hydrocarbons can impact the feeding, migration, reproduction, swimming, schooling, and burrowing

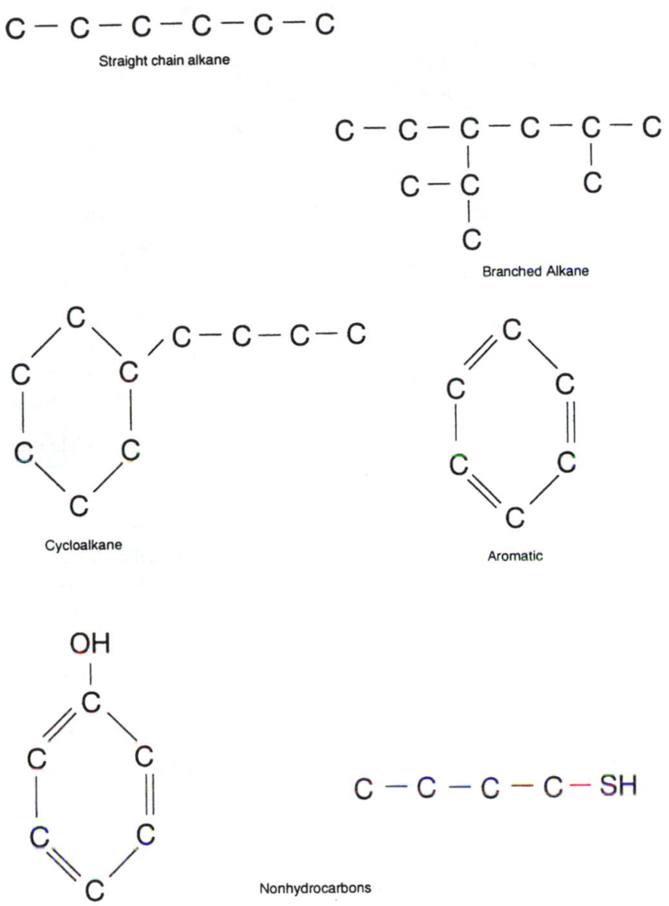

C – C – C – C – C – C

Straight chain alkane

Branched Alkane

Cycloalkane

Aromatic

C – C – C – C – SH

Nonhydrocarbons

Figure 1.16 Types of molecular structures found in petroleum. Hydrogen atoms bonded to carbon atoms are omitted. (From Albers, P.H., Petroleum and individual polycyclic aromatic hydrocarbons, in *Handbook of Ecotoxicology*, Hoffman, D.J., Rattner, B.A., Burton, G.A., Jr., and Cairns, J., Jr., Eds., Lewis Publishers, Boca Raton, FL, 1995, 330.)

behavior of fish. Marine mammals and birds commonly experience lesions, eye irritations, renal deficiencies, respiratory problems, gastrointestinal and blood disorders, and modified enzymatic activity when exposed to sublethal levels of polluting oil.

Benthic communities are most susceptible to oil pollution because of the immobility of rooted vegetation and the limited mobility of most epifauna and infauna. Oil spills can rapidly alter the structure of benthic communities by shifting their species composition, abundance, and diversity. Furthermore, sublethal doses of toxins to eggs, larvae, and juveniles may cause long-term degradation of benthic communities. Acute adverse impacts of oil spills on intertidal and shallow subtidal communities and habitats occasionally persist for as long as a decade or more.

Saltmarshes and other fringing wetlands commonly are the final repositories of polluting oil. Because of thick deposits of fine sediments and organic material in these intertidal habitats, the oil may be sequestered for long periods of time, thereby hindering the recovery of biotic communities.[15,24] Deep oil penetration in the sediments typically causes decimation of biota. When the oil surrounds the roots and rhizomes of wetlands vegetation, the damage is usually pervasive and recovery slow. Similar impacts may be observed in subtidal environments. The spill of Number 2 fuel by the barge *Florida* in

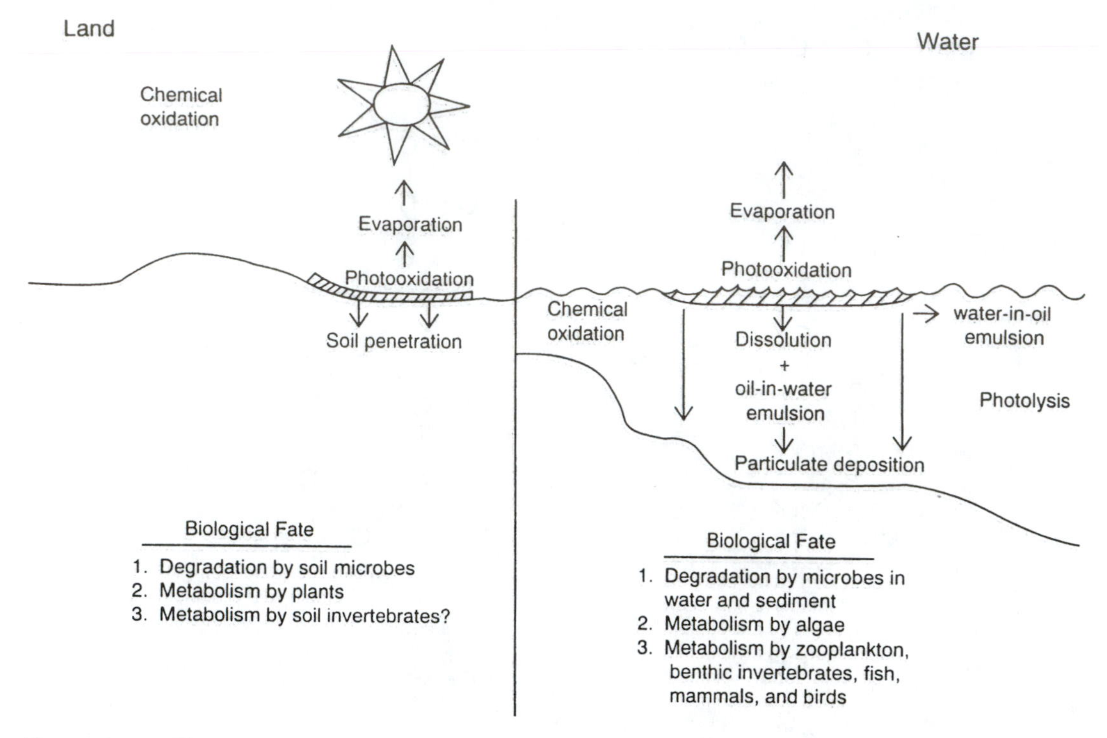

Figure 1.17 Chemical, physical, and biological fate of oil on land vs. in the sea. (From Albers, P.H., Petroleum and individual polycyclic aromatic hydrocarbons, in *Handbook of Ecotoxicology*, Hoffman, D.J., Rattner, B.A., Burton, G.A., Jr., and Cairns, J., Jr., Eds., Lewis Publishers, Boca Raton, FL, 1995, 330.)

September 1969 off West Falmouth, Massachusetts, provides an example. Oil from this spill penetrated deeply into sediments (up to 115 cm), resulting in extensive mortality of the benthic biota. Biotic impacts were protracted as the oil slowly leached out of the sediments back into the overlying water column, creating chronic chemical contamination of the local environment.

3. *Polycyclic aromatic hydrocarbons*

Oil is a major source of PAHs, a class of chemical carcinogens, mutagens, and teratogens suspected of being significantly toxic to estuarine and marine organisms. PAH compounds also originate from a number of other anthropogenic sources, such as municipal and industrial wastewaters, urban and suburban runoff, fossil fuel combustion and waste incineration as well as creosote and asphalt production. Atmospheric deposition of PAH compounds derived from incomplete combustion of organic matter, especially in the high temperature range (500 to 800°C), is also an important pathway of contaminant input to estuaries. Most PAH compounds are produced by a process of thermal decomposition of organic molecules and subsequent recombination of the organic particles (pyrolysis). Important natural sources of PAHs include volcanic eruptions, forest and brush fires, oil seeps, and chlorophyllous and nonchlorophyllous (bacteria, fungi) plants (Figure 1.18).[13,24,25] Of these natural sources, brush and forest fires are quantitatively most significant, delivering 0.19×10^5 mt (metric tons)/yr of PAHs to the atmosphere. Aquatic environments receive an estimated total of 2.3×10^5 mt/yr of PAHs, principally from polluting oil (1.7×10^5 mt/yr) and atmospheric deposition (0.5×10^5 mt/yr).[23]

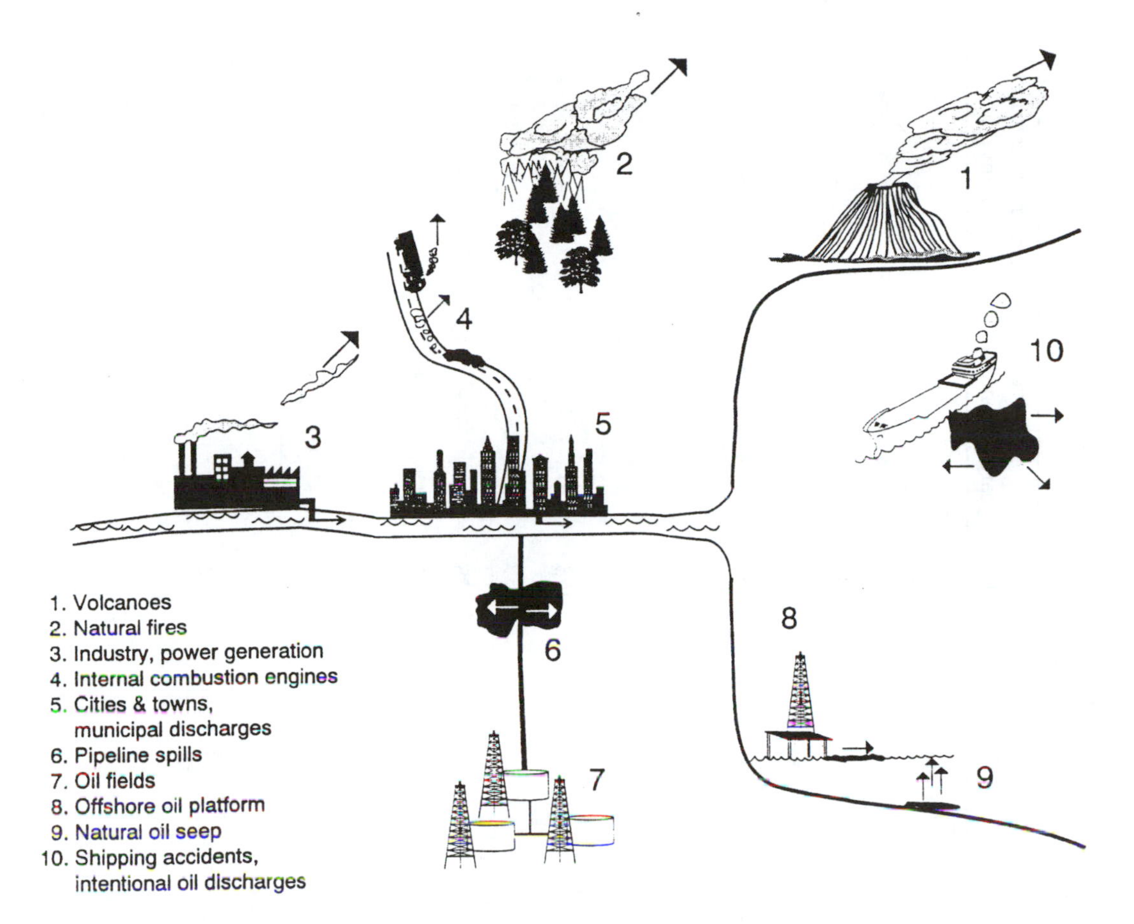

Figure 1.18 Sources of oil and polycyclic aromatic hydrocarbons in the marine environment. (From Albers, P.H., Petroleum and individual polycyclic aromatic hydrocarbons, in *Handbook of Ecotoxicology*, Hoffman, D.J., Rattner, B.A., Burton, G.A., Jr., and Cairns, J., Jr., Eds., Lewis Publishers, Boca Raton, FL, 1995, 330.)

PAHs are aromatic hydrocarbons consisting of hydrogen and carbon arranged in the form of two or more fused benzene rings in linear, angular, or cluster arrangements, with substituted groups possibly attached to the rings.[29] Ranging from naphthalene ($C_{10}H_8$, two rings) to coronene ($C_{24}H_{12}$, seven rings), PAHs encompass a widely diverse group of compounds (Table 1.6). As such, they comprise a homologous series of fused aromatic ring compounds of increasing environmental concern.

The toxicity of PAH compounds varies considerably. The unsubstituted lower molecular weight PAHs containing two or three rings are acutely more toxic than the higher molecular weight PAHs containing four to seven rings. Although the higher molecular weight PAHs are less toxic than the lower molecular weight forms, they are demonstrably carcinogenic, mutagenic, and teratogenic to many organisms. Some of these compounds only become carcinogenic or mutagenic when metabolically activated.

PAHs are relatively insoluble in water, sorb strongly to particulate matter, and accumulate in bottom sediments (Figure 1.19). Hence, benthic organisms are generally exposed to the highest concentrations of PAH compounds, especially in urbanized estuaries that often serve as major repositories of the contaminants. For example, McLeese et al.[30] recorded total PAH concentrations of 120 mg/kg in bottom sediments of Boston Harbor.

Table 1.6 Examples of PAH Compounds Without Attached Methyl Groups or Nitrogen, Sulfur, and Oxygen (Hydrogen Atoms Bonded to Carbon Atoms Are Omitted)[a]

Structure	1957 I.U.P.A.C. name	Other names	Mol. wt.	Relative carcinogenicity[b]	Common abbreviation (if any)
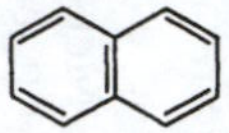	Naphthalene	—	128	—	—
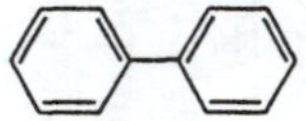	Biphenyl	—	154	—	—
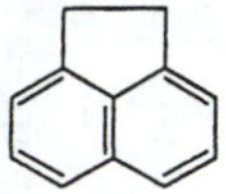	Acenaphthene	—	154	—	—
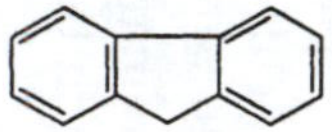	Fluorene	—	166	—	—
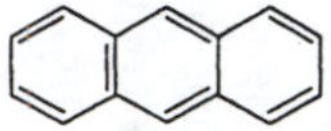	Anthracene	—	178	—	—
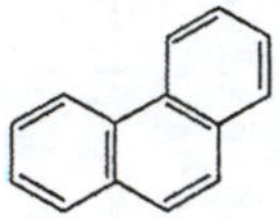	Phenanthrene	—	178	—	—
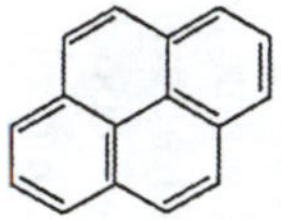	Pyrene	—	202	—	—
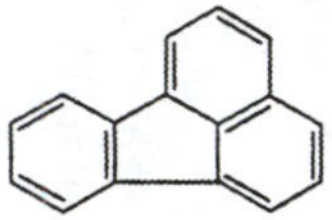	Fluoranthene	—	202	—	—
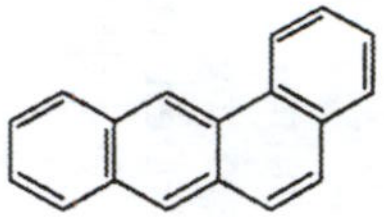	Benzo(a)anthracene	1,2 Benzanthracene	228	<+	B(a)A

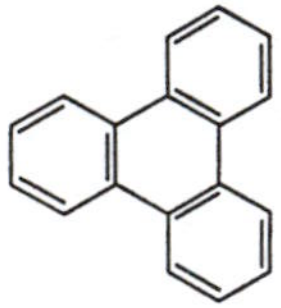 Triphenylene	—	228	—	—
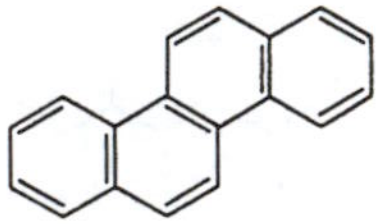 Chrysene	—	228	<+	—
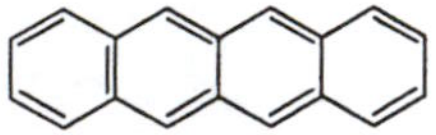 Naphthacene	Tetracene	228	—	—
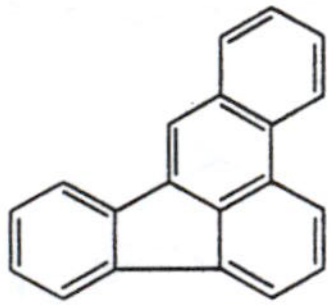 Benzo(b)fluoranthene	3,4 Benzfluoranthene	252	++	B(b)F
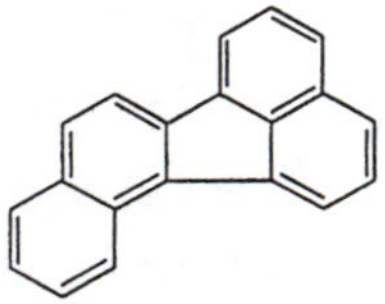 Benzo(j)fluoranthene	10,11 Benzfluoranthene	252	++	B(j)F
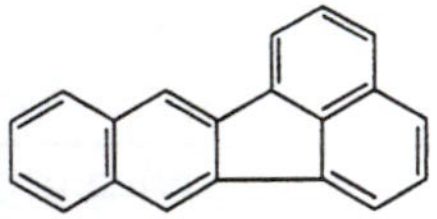 Benzo(k)fluoranthene	11,12 Benzfluoranthene	252	—	B(k)F
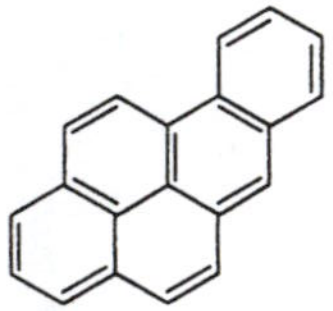 Benzo(a)pyrene	3,4 Benzopyrene	252	++++	B(a)P

Table 1.6 (continued) Examples of PAH Compounds Without Attached Methyl Groups or Nitrogen, Sulfur, and Oxygen (Hydrogen Atoms Bonded to Carbon Atoms Are Omitted)[a]

Structure	1957 I.U.P.A.C. name	Other names	Mol. wt.	Relative carcinogenicity[b]	Common abbreviation (if any)
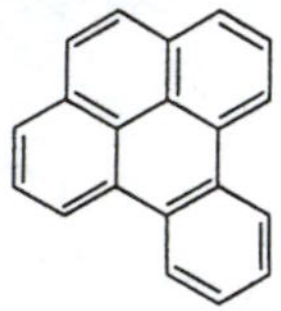	Benzo(e)pyrene	1,2 Benzopyrene	252	<+	B(e)P
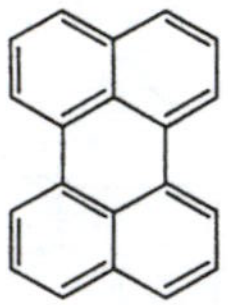	Perylene	—	252	—	—
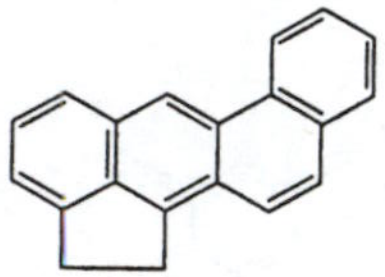	Cholanthrene	—	254	—	—
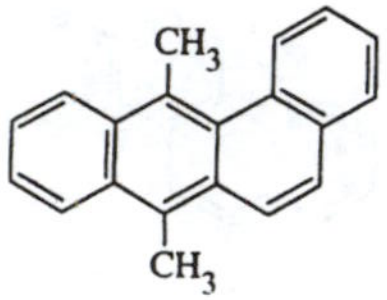 CH_3 CH_3	7,12 Dimethylbenz(a)anthracene	7,12 Dimethyl-1,2-benzanthracene	256	+++++	—
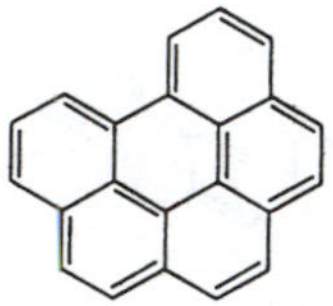	Benzo(ghi)perylene	1,12 Benzperylene	276	—	—

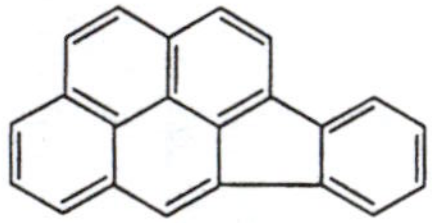 Indeno(1,2,3-cd)pyrene	*o*-Phenylenepyrene	276	+	IP
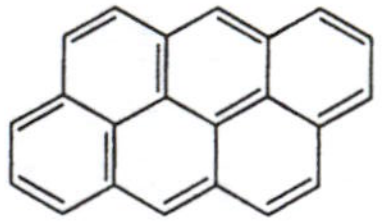 Anthanthrene	—	276	<+	—
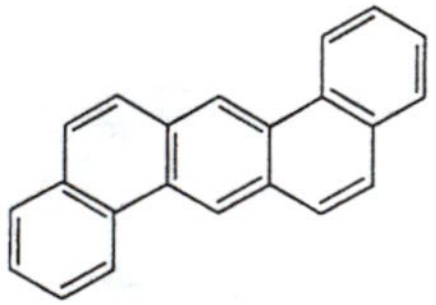 Dibenz(a,h)anthracene	1,2,5,6 Dibenzanthracene	278	+++	—
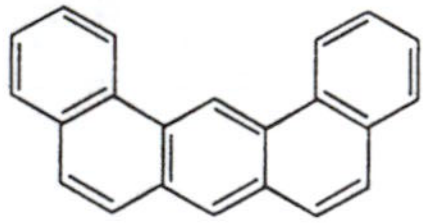 Dibenz(a,j)anthracene	1,2,7,8 Dibenzanthracene	278	—	—
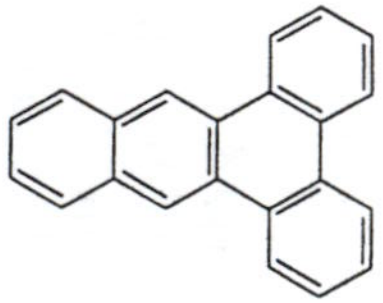 Dibenz(a,c)anthracene	1,2,3,4 Dibenzanthracene	278	—	—
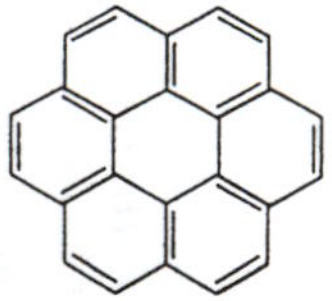 Coronene	—	300	—	—

[a] From Albers, P. H., Petroleum and individual polycyclic aromatic hydrocarbons, in *Handbook of Ecotoxicology*, Hoffman, D. J., Rattner, B. A., Burton, G. A., Jr., and Cairns, J., Jr., Eds., Lewis Publishers, Boca Raton, FL, 1995, 330.

[b] +++++ = extremely active; ++++ = very active; +++ = active; ++ = moderately active; + = weakly active; < = less than; — = inactive or unknown.

Source: Futoma, D. J., Smith, S. R., Smith, T. E., and Tanaka, J., *Polycyclic Aromatic Hydrocarbons in Water Systems*, CRC Press, Boca Raton, FL, 1981, 2.

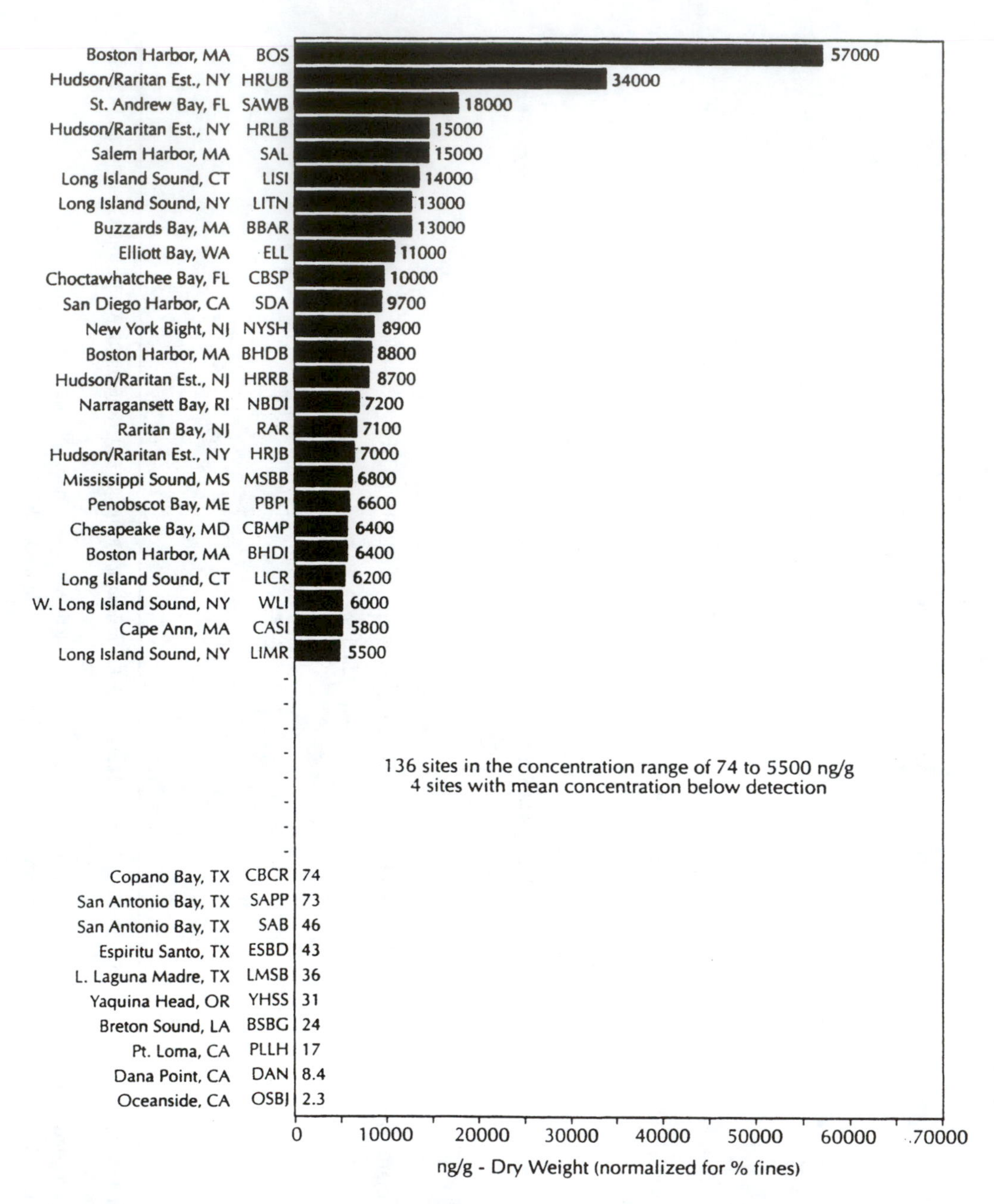

Figure 1.19 Total PAH concentrations in sediments of various estuarine and coastal systems of the United States. (From NOAA, A Summary of Data on Individual Organic Contaminants in Sediments Collected during 1984, 1985, 1986, and 1987, Tech. Mem. NOS OMA 47, National Oceanic and Atmospheric Administration, Rockville, MD, 1989.)

Barrick and Prahl[31] found total combustion-derived PAH levels of 16 to 2400 ng/g in Puget Sound sediments.

Sediment-sorbed PAHs have only limited bioavailability, which mitigates their toxicity potential. In addition, estuarine organisms have different capacities to metabolize PAH compounds. For instance, bivalves and echinoderms have poorly developed mixed function oxygenase (MFO) capability and do not metabolize PAHs efficiently. Consequently,

PAH compounds tend to accumulate in these organisms. In contrast, annelids and fish have well-developed MFO systems, enabling them to rapidly metabolize the PAHs. They accumulate PAHs from the environment only when exposed to very high concentrations, such as in heavily polluted systems.

PAHs are associated with a number of abnormalities in critical metabolic sites of organisms (e.g., livers). Pathological changes are frequently observed in the cells and tissues of these organs. For example, high concentrations of PAHs in the liver of fish commonly lead to hepatomas (liver tumors). Hepatic neoplasms (e.g., hepatocellular carcinoma, hepatocellular adenoma, and cholangiocellular carcinoma) and other diseases in bottom-dwelling fish have been correlated with PAH uptake from sediments.[29] Neoplasm formation in mollusks has also been correlated with PAH exposure in contaminated benthic habitats.[20]

The myriad of sublethal effects manifested in estuarine biota exposed to PAHs complicates assessment programs. This is so because many responses of biota to PAH contamination require months to develop in individual organisms, and population and community impacts are commonly detected years after initial PAH exposure. Sublethal effects on the organisms often involve biochemical, behavioral, physiological, and pathological aberrations. The severity of sublethal and lethal impacts on organisms controls to a large degree many changes in the structure and dynamics of estuarine communities in heavily contaminated systems.

4. Heavy metals

Anthropogenic input of heavy metals to estuarine and marine waters is a potentially serious problem because these contaminants are toxic to organisms above a threshold availability and at elevated concentrations can adversely affect the structure and function of biotic communities. Apart from being highly toxic, heavy metals are extremely persistent in aquatic environments, and they tend to bioaccumulate (Figure 1.20). Accumulation of heavy metals by estuarine organisms involves three phases: (1) metal uptake; (2) metal transport, distribution, and sequestration within the bay; and (3) metal excretion (which may not take place).[32] As noted by Rainbow,[33] major routes of metal uptake by marine invertebrates are from solution and food, although metals may also be taken up pinocytotically, as in the lamellibranch bivalve gill or ascidian pharynx, by esoteric routes including the blood sinuses of the foot of some gastropods, or via the nephridiopores of narcotized polychaetes. Heavy metals also accumulate in fish, waterfowl, and mammals inhabiting estuaries and may pose a health risk to humans who regularly consume these contaminated organisms.[13]

The toxicity of a given metal varies in estuarine and marine organisms for several reasons. The capacity of the organisms to take up, store, remove, or detoxify heavy metals differs considerably. Many intrinsic and extrinsic factors influence heavy metal uptake: (1) intra- and interspecifically variable intrinsic factors such as surface impermeability, nutritional state, stage of molt cycle, and throughput of water by osmotic flux and (2) extrinsic physiochemical factors, such as dissolved metal concentration, temperature, salinity, presence or absence of chelating agents, and presence or absence of other metals. Some of these factors also may influence the accessibility of metals to organisms (i.e., metal bioavailability), which in toxicity assessment is more important than the absolute concentrations of the metals in the environment.[33] Unfortunately, many processes controlling bioavailability of trace metals are poorly understood.

Estuarine and marine organisms exhibit a range of pathological responses when exposed to toxic levels of heavy metals, including tissue inflammation and degeneration, lack of tissue repair and regeneration, enzyme inhibition and cell membrane impacts,

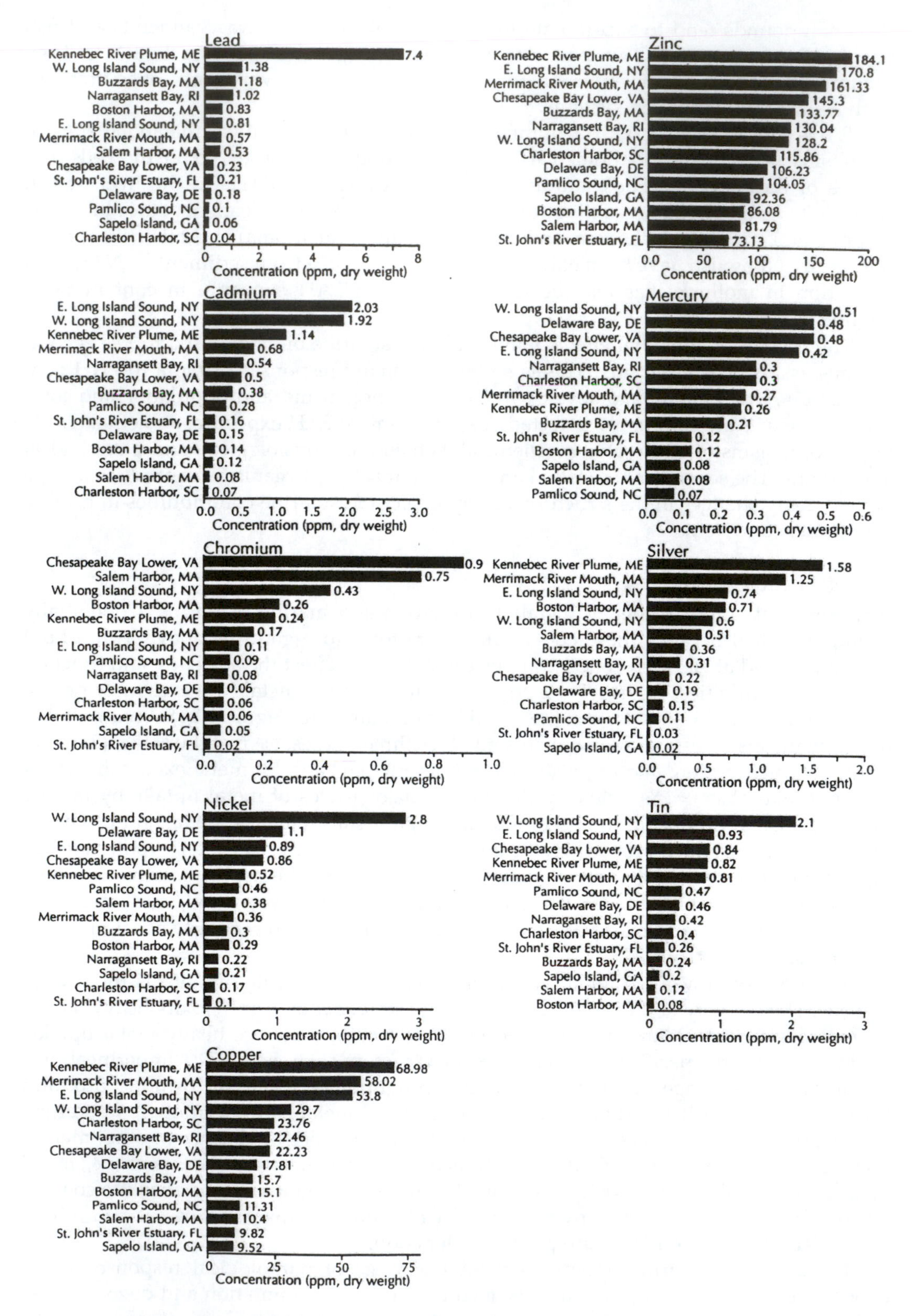

Figure 1.20 Concentrations of heavy metals in Atlantic Coast fish liver tissue sampled from various U.S. estuaries by the NOAA Status and Trends Program. (From Larsen, P.F., *Rev. Aquat. Sci.*, 6, 67, 1992.)

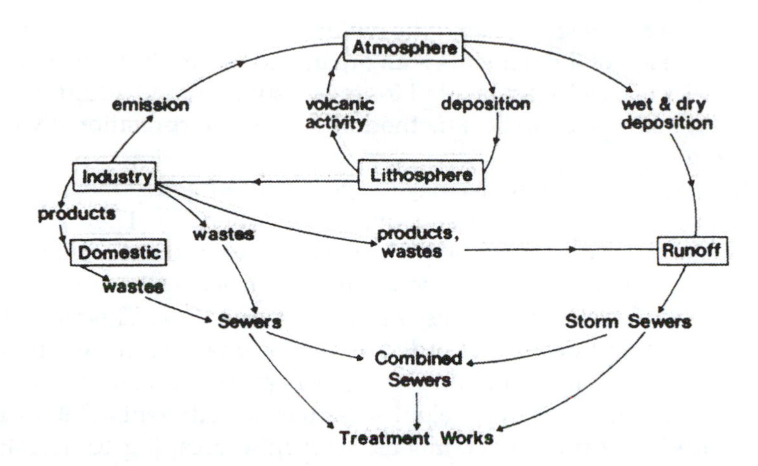

Figure 1.21 Sources and pathways of heavy metals entering wastewater-treatment processes. (From Stephenson, T., in *Heavy Metals in Wastewater and Sludge Treatment Processes*, Vol. 1, *Sources, Analysis, and Legislation*, Lester, J.N., Ed., CRC Press, Boca Raton, FL, 1987, 33.)

neoplasm formation, and genetic derangement.[34] Changes in physiology, reproduction, and development often occur. Feeding behavior, digestive efficiency, and respiratory metabolism may be significantly altered. Growth inhibition and central nervous system disorders can be dramatic.

Heavy metals are subdivided into two groups: (1) transition metals (e.g., cobalt, copper, iron, and manganese), which are essential to metabolism of estuarine organisms at low concentrations but may be toxic at high concentrations, and (2) metalloids (e.g., arsenic, cadmium, lead, mercury, selenium, and tin), which are generally not required for metabolic function but are toxic at low concentrations.[12,13,23,24,33] In addition to transition metals and metalloids, organometals (e.g., tributyl tin, alkylated lead, and methylmercury) represent a group of chemicals highly toxic to marine organisms and potentially deleterious to humans consuming contaminated seafood. An example is methylmercury contamination of Minamata Bay, Japan, during the 1950s, which resulted in the deaths of more than 100 Japanese who had consumed mercury-tainted seafood. Methylmercury contamination is an imposing problem because, in contrast to most metals, methylmercury undergoes biomagnification in food chains, attaining very high levels in upper-trophic-level organisms.

The principal pathways of heavy metal input to estuaries are riverine inflow, wastewater discharges, and atmospheric deposition (Figure 1.21). Where "hot-spot" locations of heavy metals exist in estuaries, they often are coupled to mining, smelting, refining, electroplating, or other types of industrial activity. However, landfill leachates, dredged material dumping, ash disposal, boating and shipping, and marina operations can also contribute large amounts of the contaminants.[13,23,24] Heavy metals that have been linked to specific anthropogenic sources include: (1) cadmium associated with industrial and municipal effluents; (2) copper and zinc related to industrial wastewater discharges; (3) lead originating from smelting activities and additives in gasoline; and (4) mercury derived from coal-fired power plants and pesticide use. Urban and suburban runoff particularly from roadway surfaces may be an important nonpoint source of heavy metals, especially near metropolitan centers.

As heavy metals enter estuaries, they tend to become associated with fine-grained sediments and other particulate matter as a result of ionic sorption and flocculation, hydrophobic interactions with particle surfaces, sorption and co-precipitation with manganese-iron hydroxides, organic complexation, and bioaggregation. The precipitation of hydrated oxides of iron and manganese plays an important role in heavy metal exchanges in estuaries. In addition, the speciation of heavy metals in these coastal systems depends in part on the complexing of the elements with organic compounds. However, heavy metal chemical speciation is not well documented for most estuaries, and additional work needs to be conducted on both the analytical methodology and interpretational view of speciation in these systems.

Estuarine bottom sediments act as major repositories of heavy metals and serve as a source of the contaminants for biota and overlying waters.[13,35] They also exert strong control on the biogeochemical cycling of the elements. Heavy metals, being particle reactive, rapidly sorb to suspended particulate matter or particulates at the sediment–water interface, and they ultimately accumulate on the estuarine floor. Therefore, heavy metal concentrations range from three to five orders of magnitude greater in bottom sediments than in overlying waters.[36,37] It is common to observe heavy metal concentrations of several thousand micrograms per gram dry weight in estuarine sediments (Tables 1.7 and 1.8). Many estuaries effectively trap heavy metals, with little escaping to nearshore oceanic environments.[38,39] The bottom sediments are critical to heavy metal assessment programs because they are long-term integrators of metal inputs.

Because bottom sediments contain from three to five times the concentration of heavy metals found in the overlying waters of estuaries, the bioavailability of even a small fraction of the total sediment burden assumes considerable importance. The concentrations and bioavailabilities of metals in estuarine sediments depend on many different processes, such as: (1) the mobilization of the metals to interstitial waters and their chemical speciation; (2) the transformation of the metals; (3) the control exerted by major sediment components (e.g., iron oxides and organics) to which metals are preferentially bound; (4) the competition between sediment metals for uptake sites in organisms; and (5) the influence of bioturbation, salinity, redox, or pH on these processes.[35]

Bioturbation activity by benthic organisms and physical mixing by currents and waves roil the sediments and facilitate exchange of surface grains with those in deeper layers, thereby increasing the storage capacity of the sediments for particle-reactive metals. This activity also promotes the release of the metals to interstitial and overlying waters. Storms and other turbulent events typically resuspend sediment-sorbed heavy metals from the bottom. While in suspension, the reentrant particulates often release the metals via desorption, and these elements are then remobilized in the estuary. Some metals are removed from solution in the water column by coprecipitation with particulate matter, sorption onto suspended sediment surfaces, and complexing with organic matter. The interaction of heavy metals and sediments increases significantly in the turbidity maximum zone, a highly reactive area of elevated suspended particulate concentrations. In this area, heavy metals undergo varying degrees of recycling.

Several transfer processes greatly influence the cycling of heavy metals in estuaries. At the air–sea surface interface, dry and wet deposition, gaseous exchange, and the role of surface films must be considered. Among important transfer processes in the water column are advection, diffusion, turbulent mixing, and particle settling. In bottom sediments, diagenesis, pore-water diffusion, physical and biological mixing, and resuspension are significant.

Organisms also play an important role in the cycling of heavy metals in estuaries. Phytoplankton utilize these elements in metal-requiring and metal-activated enzyme systems and release them upon death and decomposition. Zooplankton consume phytoplankton and consolidate heavy metals into fecal material that settles to the estuarine floor. Aside from their bioturbating activity, benthic fauna take up the contaminants and release them in feces and pseudofeces. Benthic macroflora (e.g., seagrasses) also assimilate heavy metals, later releasing them to the environment. Fish, birds, and mammals accumulate

Table 1.7 Heavy Metals in Sediments from Selected U.S. Estuaries (μg/g dry wt)

Estuary	Chromium	Copper	Lead	Zinc	Cadmium	Silver	Mercury
Casco Bay, ME	92.10	16.97	29.13	76.27	0.15	0.09	0.12
Merrimack River, MA	41.15	6.47	23.25	35.75	0.07	0.05	0.08
Salem Harbor, MA	2296.67	95.07	186.33	238.00	5.87	0.88	1.19
Boston Harbor, MA	223.67	148.00	123.97	291.67	1.61	2.64	1.05
Buzzards Bay, MA	73.66	25.02	30.72	97.72	0.23	0.37	0.12
Narragansett Bay, RI	93.60	78.95	60.25	144.43	0.35	0.56	0.00
East Long Island Sound. NY	37.63	11.26	22.13	58.83	0.11	0.15	0.09
West Long Island Sound, NY	131.50	111.00	69.75	243.00	0.73	0.68	0.48
Raritan Bay, NJ	181.00	181.00	181.00	433.75	2.74	2.06	2.34
Delaware Bay, DE	27.76	8.34	15.04	49.66	0.24	0.11	0.09
Lower Chesapeake Bay, VA	58.50	11.32	15.70	66.23	0.38	0.08	0.10
Pamlico Sound, NC	79.67	14.13	30.67	102.67	0.33	0.09	0.11
Sapelo Sound, GA	51.80	5.93	16.00	38.33	0.09	0.02	0.03
St. Johns River, FL	37.67	9.77	26.00	67.67	0.18	0.11	0.07
Charlotte Harbor, FL	26.47	1.17	4.33	7.20	0.08	0.01	0.02
Tampa Bay, FL	23.70	4.97	4.67	9.10	0.15	0.08	0.03
Apalachicola Bay, FL	69.17	16.93	30.67	111.67	0.05	0.06	0.06
Mobile Bay, AL	93.00	17.40	29.67	161.00	0.11	0.11	0.12
Mississippi River Delta, LA	72.27	19.40	22.67	90.00	0.47	0.17	0.06
Barataria Bay, LA	52.07	10.50	18.33	59.33	0.19	0.09	0.05
Galveston Bay, TX	41.13	8.03	18.33	33.97	0.05	0.09	0.03
San Antonio Bay, TX	39.43	5.57	11.33	32.00	0.07	0.09	0.02
Corpus Christi Bay, TX	31.43	6.63	13.00	56.00	0.19	0.07	0.04
Lower Laguna Madre, TX	24.53	5.83	11.33	36.00	0.09	0.07	0.03
San Diego Harbor, CA	178.00	218.67	50.97	327.67	0.99	0.76	1.04
San Diego Bay, CA	49.70	7.67	11.61	58.67	0.04	0.76	0.04
Dana Point, CA	39.80	10.03	18.80	53.67	0.22	0.80	0.13
Seal Beach, CA	108.33	26.00	27.37	125.00	0.17	1.27	0.59
San Pedro Canyon, CA	106.50	31.33	17.33	118.33	1.17	1.20	0.32
Santa Monica Bay, CA	53.53	10.53	33.37	46.67	0.18	0.51	0.01
San Francisco Bay, CA	1466.67	160.71	67.39	501.66	0.51	0.37	0.25
Bodega Bay, CA	246.33	0.06	2.17	38.33	0.18	1.74	0.14
Coos Bay, OR	110.30	1.47	4.65	32.00	0.62	0.31	0.11
Columbia River Mouth, OR/WA	29.53	17.00	15.90	107.67	0.86	2.14	0.25
Nisqually Reach, WA	118.07	13.33	24.57	105.33	0.68	2.62	0.32
Commencement Bay, WA	69.50	51.33	34.63	101.00	0.77	5.90	0.01
Elliott Bay, WA	114.37	96.00	20.23	166.00	0.84	1.18	0.11
Lutak Inlet, AK	58.27	26.67	15.90	180.33	0.96	0.09	0.24
Nahku Bay, AK	23.27	9.80	43.30	191.33	1.09	4.37	0.23
Charleston, SC	86.33	16.03	27.33	72.67	—	—	—

Source: Young, D. and Means, J., in *National Status and Trends Program for Marine Environmental Quality. Progress Report on Preliminary Assessment of Findings of the Benthic Surveillance Project*, 1984. National Oceanic and Atmospheric Administration, Rockville, MD, 1987; U.S. Geological Survey, National Water Summary, 1986.

Table 1.8 Heavy Metal Concentrations (μg/g) in Contaminated Estuarine and Marine Sediments[a]

Metal	Baltic Sea[b] (Various Sources)	Bristol Channel/Severn Estuary, U.K.[c] (Industry Sewage)	Mersey Estuary, U.K.[d] (Sewage, Industry including Chlor-Alkali)	Los Angeles Outfall, California[e] (Sewage)	Derwent Estuary, Tasmania[f] (Refinery, Chlor-Alkali)	Restronguet Creek, U.K.[c] (Mining)	Port Pirie, Australia[g] (Smelter)
As		8	71			2520 (13)	151 (1.0)
Cd	8.1 (<0.01)	1.1	3.9	66 (0.3)	862	1.2 (0.3)	267 (0.5)
Cu	283 (1.0)	54	144	940 (8.3)	>400	2540 (19)	122 (3.0)
Hg	9 (0.01)	0.48	6.2	5.4 (0.04)	1130	0.22 (0.12)	8
Ni	920 (1.0)	33	44	130 (9.7)	42	32 (28)	19.4 (12)
Pb	400 (2)	88	205	580 (6.1)	>1000	400 (2)	5270 (2)
Zn	2090 (6)	255	255	2900 (43)	>10,000	2090 (6)	16,667 (11)

[a] Maximum concentrations shown together with local background values (in parentheses), where given.

[b] Data from Brugman, L., *Mar. Pollut. Bull.*, 12, 214, 1981.

[c] Data from Bryan, G. W. et al., *J. Mar Biol. Assoc. U.K.*, Pub. 4, 1985.

[d] Data from Langston, W. J., *Estuarine Coastal Shelf Sci.*, 23, 239, 1986.

[e] Data from Mason, A. Z. and Simkiss, K., *Exp. Cell Res.*, 139, 383, 1982.

[f] Data from Kojoma, Y. and Kagi, J. H. R., *Trends Biochem. Sci.*, 3, 403, 1978.

[g] Data from Ward, T. J. et al., in *Environmental Impacts of Smelters*, Nriagu, J. O., Ed., John Wiley & Sons, New York, 1984, 1.

Source: Langston, W. J., in *Heavy Metals in the Marine Environment*, Furness, R. W. and Rainbow, P. S., Eds., CRC Press, Boca Raton, FL, 1990, 115.

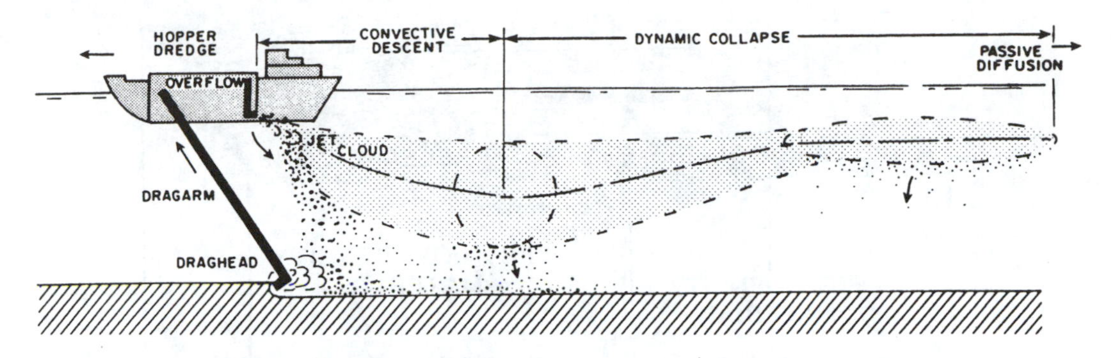

Figure 1.22 Schematic diagram of hopper dredging in showing an upper turbidity plume produced by overflow discharge and a near-bottom turbidity plume produced by draghead agitation and settling of particulates from the upper plume. The conceptual model depicts three transport phases for hopper overflow discharge: convective descent, dynamic collapse, and passive diffusion. Dredging can impact biotic communities and habitats in estuaries. (From Nichols, M., Diaz, R.J., and Schaffner, L.C., *Environ. Geol. Water Sci.*, 15, 31, 1990.)

heavy metals largely by consuming contaminated organisms and, because of their great mobility, can redistribute the metals both within and outside of the estuary.

V. Habitat loss and alteration

Human development and disturbance, coupled to population growth in watershed areas, underlie many of the habitat problems occurring in estuarine environments. It is estimated that 75% of the total population in the U.S. will live within 80 km of the coastline by the turn of the century.[40] Increased land development along the immediate shoreline and in the coastal watershed, together with associated human activities, has led to accelerated degradation of wetlands, rivers, bays, barrier islands, and beaches during the past several decades. Various human activities have physically altered habitat in upland, shoreline, and aquatic habitats. For example, dredging and filling, dredged material disposal, construction of levees, diking, damming, bulkheading, mosquito ditching, and mining have destroyed many hectares of habitat this century (Figures 1.22 and 1.23). Furthermore, the resulting changes in land use have commonly accelerated the influx of sediments, nutrients, and toxic chemical contaminants to aquatic systems.

A consequence of coastal development is resource exploitation (Table 1.9). As people settle along the coast, private and public land development increases, as manifested by escalating domestic, commercial, and industrial construction. Concomitant with community growth is rapid resource exploitation both in watersheds and neighboring estuaries. Mariculture, mining, electric power generation, and other ventures invariably arise; they are often in direct conflict with each other. Commercial and recreational fishing and shellfishing typically increase, as do pleasure boating, shipping, and numerous tourist activities. When poorly planned and mismanaged, these activities can inflict heavy environmental damage, destroying (sometimes irreversibly) the biodiversity, integrity, and stability of impacted systems. In severe cases, the natural limits and carrying capacities of these systems are exceeded, causing a loss of their resiliency and ultimately ecological collapse.

Development adjacent to or in wetlands can partition or isolate habitats, which may interfere with resource utilization by wildlife populations. For example, the fragmentation of habitats is an impediment to wildlife attempting to use wetlands as corridors for movement, such as during seasonal migration or when proceeding from resting areas to feeding areas. Under these conditions, the wildlife populations may be more susceptible

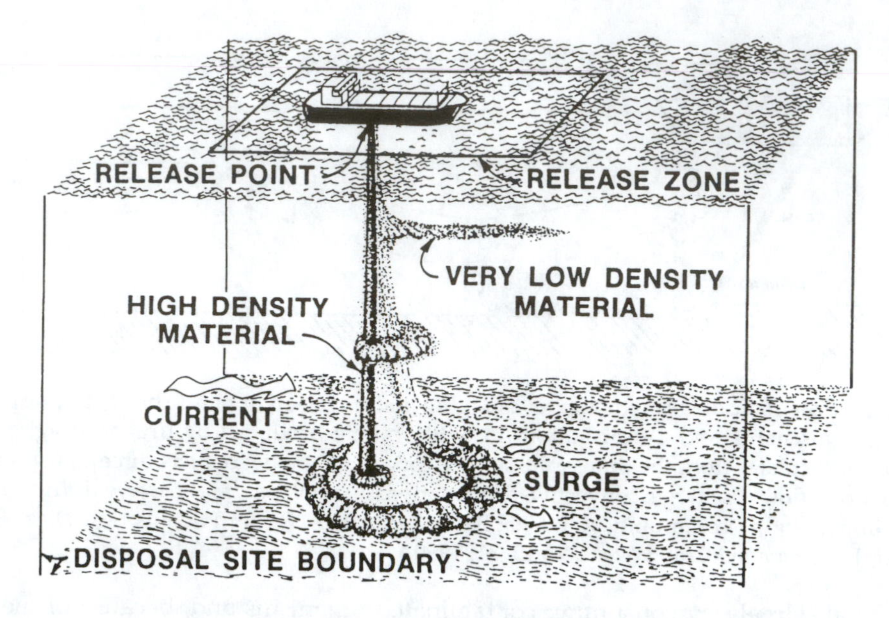

Figure 1.23 Sediment transport processes operating during open-water dredge disposal. (From Pequegnat, W.E., Pequegnat, L.H., James, B.M., Kennedy, E.A., Fay, R.R., and Fredericks, A.D., Procedural Guide for Designation Surveys of Ocean Dredged Material Disposal Sites, Tech. Rept. EL-81-1, U.S. Army Corps of Engineers, Washington, D.C., 1981.)

Table 1.9 Major Uses of Coastal Resources by Humans

Fishing
Sand and gravel dredging
Aquaculture
Oil, gas, and mineral exploration
Recreation
Collecting and hunting (e.g., shells)
Fuel
Timber
Harbors and marinas
Housing and reclamation
Tourism
Waste disposal (including sewage)
Canals and other diversions

Source: Alongi, D. M., *Coastal Ecosystem Processes*, CRC Press, Boca Raton, FL, 1998.

to injury and death during extreme environmental events (e.g., floods, droughts, storm surges, fires, etc.).

The disruption of wetlands hydrology is a common byproduct of coastal development and habitat fragmentation. It originates either through excessive drainage or reduced water supply. There is an ongoing trend in some regions toward continued fragmentation of landscape with increasing human development of watersheds. Sprawled settlement patterns contribute to fragmentation, ecosystem isolation, and functional degradation of upland and wetland complexes. Separated settlement and land uses promote the isolation of wetlands from surrounding uplands, waters, and biological resources of the watershed.

Coastal urbanization (commercial and residential development) may significantly ameliorate ecological functions of watersheds by increasing pollutant inputs, modifying hydrologic regimes, and altering the quality, quantity, and timing of sediment and organic carbon flux in wetlands. Hence, despite regulatory programs that preclude wetlands development, the multiple and far-reaching impacts of coastal settlement continue to degrade wetlands habitat.[41] Many wetlands even remotely impacted by coastal development have diminished ecological value.[42]

As residential and commercial development encroaches on estuarine shorelines, some devastating habitat changes frequently occur. The conversion of forested upland habitat for commercial development and agriculture commonly exacerbates erosion problems and facilitates sediment input to waterways. The introduction of pollutants degrades water quality and, along with the operation of recreational and commercial vessels in shallow estuaries, can severely damage beds of submerged aquatic vegetation (SAV). Excess nutrient inputs foster phytoplankton growth, periodic algal blooms, and elevated water column turbidity, which reduce light penetration and shift primary production from SAV-dominated to phytoplankton-dominated systems. In addition, overenrichment of nutrients may lead to organic loading problems and dissolved oxygen levels that are too low to sustain healthy estuarine communities.

Careless human activities in estuarine basins themselves invariably degrade valuable aquatic habitats either by physical modification or water quality alteration. The human impacts are not only measured in terms of habitat loss and degradation but also in declining biodiversity due to the elimination of various species which have specific habitat requirements. The survival of threatened and endangered species may be reduced. The cost, then, must be placed in the context of ecological damage.

The ramifications of habitat loss and degradation are also economically significant. Such changes generally reduce the recreational and commercial value of wetlands and estuaries by decreasing fish opportunities and eradicating wildlife and waterfowl populations. In addition, the loss of critical habitat, such as wetlands, decreases the land area available to buffer pollution inputs and mitigate effects of major storms. It can also increase the probability of flood inundation and damage associated with hurricanes, storm surges, and other coastal events.

A. Wetlands

Section 404 of the Clean Water Act defines wetlands as those areas that are inundated or saturated by surface or ground water at a frequency and duration sufficient to support, and that under normal circumstances do support, a prevalence of vegetation typically adapted for life in saturated soil conditions. The U.S. Fish and Wildlife Service, in turn, defines wetlands as lands transitional between terrestrial and aquatic systems where the water table is usually at or near the surface or the land is covered by shallow water. Wetlands exhibit one or more of the following attributes: (1) hydrophytes (water-loving vegetation) supported part or all of the year; (2) substrates consisting principally of undrained hydric (water-saturated) soils; or (3) substrates composed predominantly of nonsoil that is saturated with water or covered by shallow water at some time during the growing season each year. Hence, these transitional habitats (e.g., marshes, swamps, etc.) are essentially neither terrestrial nor aquatic but exhibit characteristics of both.[43] They are typically situated in low-lying areas or shallow depressions adjacent to bays, rivers, and streams. Five wetlands systems exist, including marine, estuarine, riverine, lacustrine, and palustrine. Wetlands have a wide range of functional attributes because they vary appreciably in size, shape, hydrology, water chemistry, soils, vegetation, and position in the landscape.

Table 1.10 Important Functions and Values of Natural Wetlands

Flood water storage	Nutrient removal/retention
Flooding reduction	Chemical and nutrient sorption
Erosion control	Food chain support
Sediment stabilization	Fish and wildlife habitat
Sediment/toxicant retention	Migratory waterfowl usage
Groundwater recharge/discharge	Recreation
Natural water purification	Uniqueness/heritage values

Source: Corbitt, R. A. and Bowen, R. T., Constructed wetlands for wastewater treatment, in *Applied Wetlands Science and Technology*, Kent, D. M., Ed., Lewis Publishers, Boca Raton, FL, 1994, 222.

Several factors control the establishment and maintenance of wetlands. Chief among these is the hydrology of an area, which greatly influences specific types of wetlands habitat. The topographical relief often dictates wetlands hydrology sources. The vertical extent of coastal wetlands vegetation and the distribution of different types of wetlands habitat are strongly affected by tidal ranges, time of submergence, and tidal flushing. The salinity of overlying waters influences the productivity and species distribution of wetlands vegetation. Meteorological conditions (e.g., amounts of precipitation and prevailing winds) can also have an impact on the type and distribution of wetlands and various wetland processes. Other factors that should be considered include the types of soils, surficial and bedrock geology, and stratigraphy.

Prior to the passage of the Clean Water Act in 1977, the degradation of wetlands habitat was great, with an estimated 4.68×10^7 ha of wetlands lost in the contiguous U.S. between 1780 and 1980.[43] Approximately 30% of the original wetlands area of the U.S. has been lost. The coterminous U.S. lost about 1.49×10^5 ha of vegetated estuarine wetlands between the mid-1950s and mid-1970s alone, a period when more than half of the original area of saltmarshes and mangroves was destroyed.[44] The most acute impacts have occurred in California, where about 90% of the original wetlands area has been lost.[45]

The NEP has been particularly sensitive to habitat loss and degradation of coastal wetlands, which lie within the realm and effects of salt water (i.e., saltmarshes, seagrasses, and mangroves).[46] Saltmarshes have the broadest distribution, being found on all coasts from the high intertidal zone to the oligohaline habitats upstream on tidal tributaries. Mangroves are restricted to tropical and subtropical regions. Seagrass meadows, although not a traditional type of wetlands habitat, meet the criteria for protection under Section 404 of the Clean Water Act.[47]

Wetlands are generally recognized for a wide range of natural functions and values (Table 1.10). As areas that drain to surface waterbodies (i.e., lakes, streams, rivers, and estuaries), wetlands are now recognized as vital entities that cannot be uncoupled from the estuaries which they influence. Representative biological functions include habitats for reproduction, feeding, and resting of organisms.[48] As such, wetlands provide food and habitat for survival of numerous fish, invertebrate, and wildlife populations. Value assessment must consider the type of vegetation (species, coverage, and survival), fauna (species, density, and habitat quality), sanctuary benefits for fish, wildlife, and waterfowl as well as food chain production and production export to adjacent ecosystems. Representative physical functions include flood and erosion control, sediment entrapment, shoreline stabilization, and groundwater recharge. Representative chemical functions, in turn, include excess nutrient removal, toxic chemical decontamination, retention of disease-causing microbes, and water quality enhancement.

Historically, the value of wetlands has also been based on economic criteria. For example, the monetary value of cranberries, forest products, peat, and other items generated in wetlands has been considered. Therefore, the basics for regulations regarding changes that may be made to wetlands have commingled wetlands functions and values.[48] Because of the long-term loss and alteration of wetlands habitat by development activities, a stringent regulatory framework has evolved for wetlands management.[44] The promulgation of a series of laws and regulations for protecting wetlands and regulating their use offers hope for future management of these habitats.

Prior to the enactment of numerous federal wetlands legislation during the past 25 years (Table 1.11), human activities greatly impacted wetlands systems, causing widespread deterioration of wetland values. Indeed, more than 20 years after passage of the Clean Water Act, the total area and functional quality of coastal wetlands continue to decline, albeit at a slower rate than prior to 1970. Darnell[49] identified three principal types of construction-related human impacts on wetlands: (1) outright destruction of wetlands habitat; (2) elevated loads of suspended solids; and (3) alteration of surface water levels and stream flow patterns. Much wetlands habitat destruction has resulted from draining and filling operations to create land area for residential, industrial, and agricultural development. The impoundment of streams, reduction of water levels, leveeing of flood plains, and siltation of bottoms have also contributed to wetlands habitat destruction. The increase in suspended solids stems from accelerated surface runoff from urban centers, farm lands as well as mining and construction sites. Projects involving channelization, canalization, damming, land drainage, and freshwater diversion generally alter surface water levels and stream flow patterns.

Apart from the radical destruction of saltmarshes associated with land reclamation activities, some of the most visible effects of habitat alteration in these systems are ascribable to grid ditching for mosquito control and marsh diking for construction of causeways for roads and railroads, for flood control, for the creation of pleasure boating and swimming areas, and for the creation of impoundments for wildlife. Ditching increases the amount of tidal water flowing into high marsh areas. It also increases the heterogeneity of the marsh in terms of physical characteristics and the biotic communities. Diking often alters the water quality of marsh creeks and species composition of marsh flora and fauna.[45] Ditching and diking will continue to be factors in the management of saltmarshes and other coastal wetlands into the 21st century.

The loss and degradation of wetlands habitat have profound ecological consequences. Perhaps most importantly, these impacts inevitably cause a reduction of biodiversity due to the loss of species. Clearly, undisturbed coastal wetlands are rich habitats for fish, invertebrates, and wildlife, and they play a significant role in filtering contaminants, thereby enhancing the water quality of coastal waters. Davies[50] describes other adverse effects of wetlands destruction:

- Increased stream flows and the loss of natural water storage capacity in the river basin lead to flooding downstream and the need for flood control projects that require further stream channelization.
- Reduced groundwater recharge that contributes to the loss of stable stream flows and a lowering of the water table.
- Because of reduced water storage capacity, dams are constructed, thus creating a whole new array of problems (e.g., reservoir siltation), which require new impact control measures.

1. *Habitat values*

Saltmarsh communities — halophytic grasses, sedges, and succulents — dominate coastal wetlands vegetation along the intertidal shores of mid- and high-latitude regions. Mangroves are associations of halophytic trees, shrubs, palms, and creepers that form an

Table 1.11 Significant Federal Wetlands-Related Legislation

Legislation	Date	Effect on Wetlands
Section 10, Rivers and Harbors Act	1899	Requires permits from U.S. Army Corps of Engineers for dredge and fill activities in navigable waterways and wetlands
Migratory Bird Hunting and Conservation Stamp Act	1934	Proceeds of duck stamps used to acquire habitat
Federal Aid to Wildlife Restoration Act	1937	Assistance to states and territories for restoring, enhancing, and managing wildlife
Fish and Wildlife Act	1956	Established U.S. Fish and Wildlife Service
U.S. Fish and Wildlife Coordination Act	1958	Requires all federal projects and federally permitted projects to consider wildlife conservation
Land and Water Conservation Fund Act	1965	Purchase of natural areas at federal and state levels
National Wildlife Refuge System Administration Act	1966	Established National Wildlife Refuge System
National Flood Insurance Act	1968	Requires communities to develop floodplain management programs
National Environmental Policy Act	1969	Requires Environmental Impact Statements for federal actions
Water Bank Act	1970	Purchase easements on wetlands
Endangered Species Act	1973	Prohibits federal agencies from undertaking or funding projects which threaten rare or endangered species
Resource Conservation and Recovery Act	1976	Controls disposal of hazardous waste, reducing threat of contamination to wetlands
Section 402, Federal Water Pollution Control Act	1977	Authorized national system for regulating sources of water pollution
Section 404, Federal Water Pollution Control Act	1977	Regulates dredge and fill activities
Coastal Barrier Resources Act	1982	Prohibits federal expenditure or assistance for development on coastal barriers
Food Security Act	1985	
Swampbuster		Discourages farming on wetlands
Conservation Reserve Program		Removes erodible cropland from use
Emergency Wetlands Resources Act	1986	Promotes conservation through intensified cooperation and acquisition efforts
Agricultural Credit Act	1987	Preserves land reverting to Department of Agriculture's Farmers Home Administration
Everglades National Park Protection and Expansion Act	1989	Increased water flow to park and acquired more land
North American Wetlands Conservation Act	1989	Increased protection and restoration of wetlands under the North American Waterfowl Plan
Coastal Wetlands Planning, Protection and Restoration Act	1990	Restoration of coastal wetlands and funds North American Waterfowl Management projects
Coastal Zone Management and Improvement Act	1990	Sets guidelines and provides funding for state coastal zone management programs
Food, Agriculture, Conservation and Trade Act	1990	Established wetlands reserve program which purchases easements on wetlands
Water Resources Development Act	1990	Requires federal agency development of action plan to achieve no-net loss

Source: Kent, D. M., Introduction, in *Applied Wetlands Science and Technology*, Kent, D. M., Ed., Lewis Publishers, Boca Raton, FL, 1994, 7.

interface of dense thickets or forests between marine and terrestrial habitats in subtropical and tropical zones between 25°N and 25°S latitude. Seagrasses comprise a widespread group of vascular plants occupying shallow subtidal waters, where they form a transition between emergent vegetation and unvegetated estuarine bottom. All of these plant communities constitute vital feeding, spawning, and nursery grounds for many estuarine and marine organisms.

Saltmarshes, seagrasses, and mangroves share similar functional roles. For example, their biomass and organic productivity values are all high. Although high standing crops characterize these biotopes, herbivory remains low, resulting in the generation of large amounts of detritus that support detritus-based food chains. The export of a portion of the detritus accumulating in saltmarshes and mangroves stimulates production of contiguous waters. Leaves, stems, and prop roots effectively increase the substrate surface area for epiphytic flora and fauna, thereby enhancing the primary and secondary production of the habitats. The epibiota also provide a rich food supply for fish and invertebrates. A canopy of dense leaves typically found in these coastal wetlands causes lower insolation than in surrounding unvegetated areas. Protected from excessive illumination and insolation, the shaded habitats appear to be particularly beneficial to the benthos. Their lateral zonation also presents a diversity of habitats for the protection and proliferation of fishes, birds, and other organisms.

Aside from generating high production, coastal wetlands are effective in removing excess nutrients, metals, and other pollutants from surface waters. The filtering and sequestering of nutrients and chemical contaminants appear to play a significant role in protecting estuarine water quality. Adsorption onto suspended particulates and their subsequent deposition are the processes primarily responsible for removal of the contaminants. Leaves, stems, and roots reduce wave and current action and facilitate sediment deposition. In addition, the roots and rhizomes bind sediments, stabilize the substrate, and mitigate erosion. Saltmarshes and mangroves may be the most significant habitats for flood control in some watersheds because of their ability to slow and hold water. The broad expanses of saltmarshes and massive structure of mangroves stabilize banks and protect the shoreline.

Reimold[48] examined the ecologic- and economic-based methods for assessing the function and value of wetlands. There are four principal ecologic-based indirect methodologies: (1) community; (2) habitat; (3) Wetlands Evaluation Technique (WET); and (4) risk assessment. Community models (i.e., literature reviews of wildlife community ecology) form the basis of the habitat evaluation concept. The components and approaches of these community models are discussed elsewhere.[51,52]

Several habitat-based assessment methods have been developed. The U.S. Fish and Wildlife Service[53] formulated the Habitat Evaluation Procedure (HEP), which compares the relative value of wildlife habitat in terms of different areas compared at the same point in time or the relative value of the same area at two points in time (present and future). Habitat Suitability Indices (HSIs) are used to make the comparisons; HSI values are derived from wildlife models based on species distribution, life history, and specific habitat requirements of populations. The U.S. Fish and Wildlife Service has developed HSI models for many species, with the models incorporating habitat values, life requisites, and resultant habitat values.

The Habitat Gradient Model (HGM) is another useful habitat-based approach.[54] It is contingent upon the assumption of what plant community would become established in a certain period of time, given satisfactory growing conditions. Grasslands and woodlands are typical categories.

Adamus et al.[55] devised an integrated approach for the assessment of wetlands functions known as the Wetlands Evaluation Technique (WET). A qualitative assessment tool,

WET incorporates eleven functions and values of wetlands systems (i.e., groundwater recharge and discharge, floodflow alteration, sediment stabilization, sediment/toxicant retention, nutrient removal/transformation, production export, aquatic diversity/abundance, wildlife diversity/abundance, recreation, and uniqueness/heritage functions). The WET approach is useful when comprehensive data sets are available, and it represents a valuable screening tool to determine whether or not one of the more quantitative assessment methods (e.g., HEP) should be applied.

The risk assessment approach provides additional measured indicators of wetland functions and values. Johnson[56] used quantitative measures (i.e., multiple linear regression and discriminant analysis) to establish diagnostic criteria as predictors of ecological risk. A drawback of this approach is that extensive data sets are necessary to obtain reliable estimates of risk functions.

The three most commonly used methods to estimate the economic value of coastal wetlands are the cumulative assessment approach, the opportunity cost method, and the contingent valuation method. In the cumulative assessment approach, specific societal goals and objectives associated with or served by wetlands are evaluated against a particular development scenario for a wetland. As noted by Reimold[48] (p. 68): "In such an assessment, persons and agencies are forced to publicly weigh the potential valuation of certain goals and objectives associated with the wetland against the benefits to be derived from implementation of the proposed action impacting the wetland."

The opportunity cost method deals with wetland conservation values. It is used to determine the net monetary benefit derived from the best use of an area had the area not been regulated as a wetland. Without market-oriented transaction data, some problems occur when applying this approach to wetlands. First, data required to properly evaluate wetlands are often not collected. Second, most wetland values are not directly priced. Nevertheless, the economic analyses of wetlands based on opportunity costs have been effectively conducted in the past.[57-59]

Mitchell and Carson[60] developed the contingent valuation method. In this approach, surveys are performed to obtain opinions from respondents concerning wetlands conservation. In particular, respondents are polled for their willingness to pay for wetlands conservation. Thus, the contingent valuation method demonstrates the economic interest of the respondents to protect wetlands area.

Despite the different valuation schemes used in economic assessment of coastal wetlands, it has been difficult to determine their true worth. For example, the value of wetlands based on their natural function as a filter of nutrients and chemical contaminants, as a nursery and reproductive habitat for recreationally and commercially important finfish and shellfish, and as an agent for flood control, groundwater recharge, and sediment entrapment may be subjective and highly variable. In addition, there is no marketplace for these functions. It has been less difficult to determine the economic value of wetlands based on their absolute production of cranberries, forest products, peat, and so on. In the future, more emphasis must be placed on the value of wetlands in terms of ecological function to truly assess their overall worth for decision-making purposes.

2. *Habitat restoration*

Efforts to restore damaged coastal wetlands have been ongoing for decades.[61,62] The purpose of wetlands restoration projects is to return a habitat from an altered or disturbed condition to a previously existing natural condition or altered condition by some human action. Most coastal wetlands in the U.S. have been disturbed to some extent, and numerous experimental restoration projects have been attempted to revitalize these habitats. The main goal of wetlands restoration projects is the replacement of the same type of habitat in an impacted area to compensate for adverse human effects (i.e., mitigation). The success of the restoration projects is evaluated by comparing the flora and fauna of the improved community with those of the natural community as a reference. The probability of success increases as the communities become more similar in composition and function.[46]

As shown by Zedler,[61] the principal objective of saltmarsh restoration (i.e., to replace what has been lost through habitat destruction) is difficult to define for a couple reasons. First, this wetlands type is characterized by dynamic structure and function and thus is far from being stable. The biotic communities must constantly respond to variations in tidal and riverine influences, and nutrient and sediment fluxes, as well as storms and other erosive forces. Hence, it is difficult to set a specific historic time or condition as a target for habitat restoration. Second, it is often impossible to accurately define what has been lost by the impacted habitat because most systems have experienced substantial disturbance. Similar problems of defining restoration goals may be extended to mangrove forests and seagrass beds. Regardless of these practical problems, however, restoration projects must state goals to develop specific plans, which then can be employed to evaluate the success of the restoration projects.

Because restoration of wetlands is essentially the re-creation of a specific vegetation association on a site where that association once naturally occurred, some degree of habitat manipulation is necessary. The restoration, enhancement, and creation of wetlands involve construction actions. Zentner[62] describes a five-step construction process inherent to designing, building, and maintaining wetlands landscapes. They are site analysis, goal setting, construction design, implementation, and maintenance. Site analysis usually entails wetlands vegetation association mapping and analysis, a review of historic conditions, and assessment of the hydrologic, soil, and cultural conditions of the site and any nearby templates of target vegetation association. Goals establish standards for judging success of restoration projects and define monitoring elements and protocols. Wetlands construction design strives to produce a vegetation association that will evolve in a natural fashion and hence will achieve ecological function. Implementation of construction follows a sequence of operations, including the installation of protective flagging for areas to be preserved, removing invasive species, salvaging desired species, grading, planting, fencing, and, in some cases, generating a water supply. Maintenance is vital to ensure that the vegetation association develops as a native vegetation association, and generally is intense during the first 3 years after the cessation of planting. This part of the construction process also focuses on precluding the invasion by harmful exotic species. Effective wetlands construction, therefore, consists of a good design to ensure a match between the selected species and a site, manipulation of the site to improve planting and growth conditions, and maintenance to increase the probability of success.

Seagrass meadows are less amenable to direct improvement techniques than saltmarsh and mangrove habitats. Seagrass restoration projects generally have five specific goals: (1) vegetative cover should develop and persist; (2) equivalent acreage should be attained; (3) increased acreage should be achieved where possible; (4) the species lost should eventually be replaced by the same species; and (5) there should be development of a faunal population equivalent to reference wetlands.[46,47]

According to Alongi,[11] the three most successful methods for restoring seagrasses are

- Use of transplanting entire 10- to 20-cm wide plugs of sediment, roots, rhizomes, and blades from a natural site to the damaged bed
- Hand planting of seeds and seedlings
- Transplanting turfs, sods, and grids that are shallow plugs from a natural bed and do not necessarily include the entire root rhizome system

In a review of 165 seagrass restoration projects that used these methods, Thorhaug[63] reported that 75 were successful. Most of these projects were conducted in the Gulf of Mexico and southeast Florida. The success rate was greatest for the plugging technique. Although the planting method has been effective in some regions (e.g., the tropics), it appears to be less successful than plugging because seeds and seedlings are more susceptible to shifting and resuspended sediments associated with wave action, storms, and other disturbance. Lewis[46] noted that the enhancement, restoration, and creation of seagrass beds require intensive effort and are often unsuccessful regardless of whether seedlings, long shoots, or mature plants are used.

For mangrove forests, restoration can be accomplished by transplanting propagules, young trees, or more mature trees. Survival rates are much lower after planting propagules than after planting young trees. Mortality from disease and disturbance declines as mangrove trees survive past the young-shoot stage, and they attain their fastest rate of growth. The restoration of mangrove systems has been practiced most successfully in the Far East (e.g., Burma, Malaysia, Thailand, and Vietnam).

In the U.S., restoration measures have generally been most successful for estuarine saltmarsh habitats.[64] In many cases, success can be achieved relatively inexpensively and rapidly. In some cases, the re-establishment of a tidal connection is all that is required to achieve successful marsh restoration.[46] Usually, however, restoration efforts on transplanted marshes have involved comprehensive analyses of the effects of hydrology, nutrient additions, tidal elevation, geomorphology, and their synergistic/antagonistic effects.[11,65-67] Despite the greater success of saltmarsh restoration projects, constructed saltmarsh habitats may take many years to approach conditions observed in natural marshes. At some sites, they may never achieve these conditions.

Zedler[61] stressed that two criteria must be considered in assessing the success of a coastal saltmarsh restoration project: (1) comparison of what actually developed with what was promised and (2) comparison of what developed with what was and is still needed in the region. Evaluating a project's success is a long-term process, requiring several years to a decade or more. Detailed and frequent sampling is required to accurately assess ecological communities on a restoration site. Sampling programs must be flexible to delineate community changes related to the seasonal and year-to-year dynamic of a marsh system. As a result, these programs usually entail a multidisciplinary effort conducted over a protracted period of time.

VI. Watersheds

An estuarine watershed is an area of land bounded by ridge lines that catches precipitation and drains into an adjacent estuary. The characterization and assessment of human activities and pollution sources in developed portions of watersheds enable stakeholders, researchers, and government agencies to target and prioritize key watershed pollution issues that must be addressed through a carefully planned management process. Ideally, characterization and assessment reports incorporate descriptions of resources, infrastructure, demographics, and principal human activities within the watershed management area, past and present trends in water quality, and the current state and future projections of the watershed. It is essential to develop environmental criteria to help identify priority water resource concerns. Clearly, both point and nonpoint sources of pollution must be considered. Water quality problems experienced in estuaries usually reflect the combined impacts of multiple land and water uses, taking into account the importance of both point and nonpoint sources of pollution.

The Office of Water at the USEPA has developed a national Index of Watershed Indicators designed to collect, organize, and evaluate multiple sources of environmental information on a watershed basis.[68] This index characterizes the condition and vulnerability of aquatic systems in each of 2111 watersheds in the continental U.S. Many water quality problems in estuaries can be addressed effectively along watershed lines. The primary objectives of the Index of Watershed Indicators are: (1) to characterize the condition and vulnerability to pollution of the watersheds in the U.S.; (2) to provide the basis for dialogue among water quality managers; (3) to measure progress toward the goal that all watersheds will be healthy and productive; and (4) to empower citizens to learn more about their watersheds and to protect them.

The Estuarine Pollution Susceptibility Index provides a tracking method for the susceptibility to pollution of coastal watersheds. Susceptibility is defined here as the relative vulnerability of an estuary to concentrations of dissolved and particulate substances. High susceptibility coastal watersheds predominate along the Texas and Louisiana Coasts, Atlantic Coast of Florida, and the coastal zone of Mid-Atlantic and New England states.

The USEPA[68] has also generated an Index of Agricultural Runoff to show which watersheds have the greatest potential for possible water quality problems from combinations of nitrogen, pesticides, and sediment. Watersheds with the highest composite values have the greatest risk of water quality impairment from agricultural sources. The composite indicator has been constructed by ranking watersheds for each of the three aforementioned components — potential nitrogen runoff, potential pesticide runoff, and potential in-stream sediment loads — and then summing the rankings for each watershed. Results of this effort indicate that coastal watersheds along the Atlantic and Gulf Coasts generally have the highest levels of agricultural runoff potential impact on estuarine systems.

In terms of urban runoff potential, the USEPA used imperviousness as an indicator to predict impacts of land development on estuarine and other aquatic ecosystems. Studies have coupled the amount of imperviousness to changes in the hydrology, habitat structure, water quality, and biodiversity of these ecosystems. Increased imperviousness can alter the hydrology of a receiving stream, increase the rate and volume of runoff, and decrease the capacity of receiving streams to handle floods.[68] Based on this assessment, urban runoff potential is greatest for major metropolitan centers (e.g., Boston, New York, and San Francisco). Hence, estuaries nearby these centers are likely to be most susceptible to pollution impacts from urban runoff.

The National Oceanic and Atmospheric Administration (NOAA) has developed the National Estuarine Inventory to establish a national estuarine assessment capability. The National Estuarine Inventory is included in a larger Coastal Assessment Framework, which identifies all watersheds associated with the coast. In this program, NOAA assesses pollution susceptibility of estuaries by quantifying the dissolved concentration potential, the particle retention efficiency, and the estimated loadings and predicted concentrations of nutrients (i.e., nitrogen and phosphorus). To date, NOAA has assessed over 130 estuaries within the Coastal Assessment Framework.

A. Watershed pollution control

Over the past 25 years, efforts to mitigate the adverse effects of human activities in watersheds on water quality in nearby estuaries and their tributaries have focused principally on reducing point source pollution. Since the implementation of the Federal Water Pollution Control Act and the Clean Water Act, great strides have been made in controlling point source pollution from municipal and industrial dischargers. Despite these advances, it has become clear that point source pollution control alone will not ensure long-term protection and improvement of water quality conditions in estuaries. This is so because pollutants from nonpoint sources in watersheds account for a significant portion of the

Table 1.12 General Comparisons of Point Source and Nonpoint Source Characteristics Impacting Control Strategies and Toxicity to Aquatic Life

Factors	Point Source	Nonpoint Source
Input	Discrete	Nondiscrete (ubiquitous)
Duration	Long term	Short term
Toxicity	Acute/chronic	Acute (usually)
Pollutant(s)	Defined (usually)	Not defined (usually)
Discharge frequency	Continuous	Intermittent (episodic)
Suspended solids	Regulated	High (frequently)
Control	NPDES permit	BMPs

Source: Davies, P. H., Factors in controlling nonpoint source impacts, in *Stormwater Runoff and Receiving Systems: Impact, Monitoring, and Assessment*, Herricks, E. E., Ed., Lewis Publishers, Boca Raton, FL, 1995.

water quality problems encountered in these critically important coastal systems. Davies[50] estimates that two thirds of all pollution sources are of nonpoint origin. Thus, management programs to ameliorate pollution problems in estuarine systems must include strategies to control both point and nonpoint sources of pollution.

1. *Point source pollution control*

Point source and nonpoint source pollution characteristics are fundamentally different as noted previously (Table 1.12). For example, pollutants are usually well defined for point source discharges but poorly defined for nonpoint source discharges. In addition, the input of point source discharges and accompanying pollutants tends to be discrete, whereas the input of nonpoint discharges and accompanying pollutants is diffuse (ubiquitous). The discharge of pollutants through any discernible, confined, and discrete conveyance, including but not limited to any pipe, ditch, channel, tunnel, conduit, well, discrete fissure, container, rolling stock, concentrated animal feeding operation, vessel, or other floating craft, is prohibited in the U.S. in the absence of an NPDES permit as specified by requirements of the Clean Water Act. The conveyance need not be man-made to qualify as a point source.

Strategies for controlling point source pollution discharges differ markedly from those for controlling nonpoint source pollution. For example, controls of point source pollution discharges generally require active treatment processes to meet water quality standards. In contrast, best management practices are typically implemented to reduce or control pollution from nonpoint source discharges. In many respects, point source pollution problems are less complex than nonpoint source pollution and are often resolved by adopting simple treatment and standard compliance approaches.

The NPDES permit process represents an important regulatory mechanism for controlling point source discharges from municipal and industrial facilities into waterways. Discharge permits specify effluent limitations on pollutants present in the effluent as well as monitoring and reporting requirements to characterize the discharge. Effluent limitation standards are either water quality based, as set forth in Section 302 of the Clean Water Act, or technology based, as set forth in Sections 301 and 304 of the Clean Water Act. Estimates of the removal of pollutants that could be achieved by the application of best available technology, best practicable technology, or best conventional technology are used to establish technology-based standards.

To comply with water quality standards and effluent limitations under the NPDES program, permittees have traditionally focused on modifying operations to reduce pollutant levels in receiving waters. These modifications typically involve applying wastewater treatment processes that improve the quality of the effluent, decreasing the rate or volume of effluent discharges, or implementing more advanced technologies to increase performance levels to comply with applicable water quality standards. Significant costs are often associated with efforts to comply with NPDES permits.

Section 402 of the Clean Water Act also requires regulation of certain stormwater point source discharges. As a result, the USEPA has set forth permit requirements for stormwater discharges from medium- and large-sized municipal separate storm sewer systems — known as the final stormwater rule. In regard to stormwater discharges coupled to industrial activity, the USEPA regulations apply to the discharge from any conveyance that is used for collecting and conveying stormwater and that is directly related to manufacturing, processing, or raw materials storage areas at an industrial plant. The rule provides examples of stormwater discharges from facilities to which the permitting requirements apply; these include but are not limited to

> ...stormwater discharges from industrial plant yards; immediate access roads and rail lines used or traveled by carriers of raw materials, manufactured products, waste material, or by-products used or created by the facility; material handling sites; refuse sites; sites used for the application or disposal of process waste waters (as defined at 40 CFR 401); sites used for the storage and maintenance of material handling equipment; sites used for residual treatment, storage, or disposal; shipping and receiving areas; manufacturing buildings; storage areas (including tank farms) for raw materials, and intermediate and finished products; and areas where industrial activity has taken place in the past and significant materials remain and are exposed to stormwater (55 Fed. Reg. 48065).

However, the rule excludes "areas located on plant lands separate from the plant's industrial activities, such as office buildings and accompanying parking lots, so long as the drainage from the excluded areas is not mixed with stormwater drained from [industrial] areas" (55 Fed. Reg. 48065).[69]

In regard to stormwater discharges from municipal sewer systems, USEPA regulations specifically apply to discharges from municipal separate storm sewer systems located in an incorporated place, or a county with a population of at least 100,000 but less than 250,000 (for medium-size municipal separate storm sewer systems), or a population of 250,000 or more (for large municipal separate storm sewer systems). As noted by Koorse,[69] municipal separate storm sewer systems located in any smaller incorporated place or county that the USEPA or the relevant state official determines to be a part of the medium- or large-size municipal separate storm sewer system must also comply with federal regulations (55 Fed. Reg. 48,065). The interrelationship between the smaller and larger systems takes into account physical interconnections between the sewer systems and the location of the system discharges. As a consequence of the final stormwater rule, municipalities will likely be subjected to new regulatory requirements during the next decade.

2. *Nonpoint source pollution control*

Nonpoint source pollution problems cannot be resolved easily and usually require the application of innovative and imaginative control strategies. These problems typically arise from combined impacts of multiple land and water uses, and their resolution depends in part on an understanding of the natural dynamics and uses of watershed lands that drain into adjacent estuaries. Clearly, the most cost-effective strategy in nonpoint source pollution control involves effective planning and site design techniques in the watershed that focus on limiting the generation of waste and chemical toxicants. This strategy, therefore, emphasizes pollution prevention and innovative source controls rather than dealing with wastes after their release to receiving waters.

Watershed management plans must incorporate sound land-use and site planning to control water pollution while providing for economic growth. Nonregulatory measures are important components of these plans, especially in regard to nonpoint source pollution control. Development should proceed in a manner compatible with the natural features of an area and may require municipalities to revise local master plans, zoning ordinances, and subdivision and site review procedures. Sound site-planning procedures are essential, as are land-use design and allocation that need to be consistent with water quality goals.

Best management practices in the watershed may include replacing lawns with natural vegetation (i.e., ground covers, shrubs, and trees) to reduce the use of fertilizers, herbicides, and pesticides, modifying farm practices to mollify erosion, and instituting alternative landscaping and street sweeping to ameliorate pollutant accumulation in stormwater runoff. One of the major impediments to widespread acceptance of best management practices in many communities is the lack of understanding and awareness of nonpoint source pollution by the general public. Many individuals only associate water pollution with point source discharges, such as from large outfalls of municipal and industrial facilities. They have limited understanding of atmospheric deposition and other sources of nonpoint source pollution. Education and lifestyle changes are essential if best management practices are to become effective as part of an overall program that truly controls pollution at its source. Watershed management plans must therefore include an education component that will instruct all individuals on how they contribute to nonpoint source pollution problems and how they can help to remedy these problems. People must also become cognizant of the economic investment that often must be made to minimize nonpoint source impacts. Hence, key components of a watershed management framework are innovative solutions to nonpoint source pollution problems, education programs, and voluntary compliance of the general public to accept management plans and measures to control the pollution.

The success of the NEP hinges strongly on the formation of watershed partnerships where government agencies, nonprofit and volunteer organizations, businesses, educational institutions, and private citizens work together to achieve a common goal. Stakeholder involvement through the formation of partnerships for watershed management will be integral to the development and successful implementation of watershed goals and objectives, including measures to control nonpoint source pollution. The interaction of these watershed partners will be facilitated by an effective outreach program that communicates management plans and decisions as well as the benefits and cost-effectiveness of the programs to the community at large. Well-informed and active participation by a diverse group of partners is crucial to the success of the watershed management framework.

Baker[70] stresses that nonpoint source pollution is the major cause of impairment of U.S. surface waters. Urban and nonurban runoff, mining, land disposal, malfunctioning septic systems, combined sewer overflows, and atmospheric deposition account for a large proportion of the total nonpoint pollution to estuaries. Different strategies are employed to control these various pollution sources. They are examined in detail by Davies[50] and briefly reviewed here.

a. Agriculture. The wide array of practices used to increase agricultural land production can contribute significantly to nonpoint source pollution to streams, rivers, and estuaries. For example, excessive cultivation, overgrazing, and improper irrigation practices promote erosion of tons of soil, with substantial concentrations of fertilizers (nitrates and phosphates), herbicides and pesticides (organochlorine compounds), trace metals (copper, lead, selenium, zinc, etc.), and other chemical contaminants washed into waterbodies from irrigation return flows and direct precipitation and runoff. High BOD and ammonia levels may arise from the runoff of feed lots. The combined effect of elevated sediment and nutrient inputs as well as occasional toxic chemical influx to receiving waters associated with agricultural runoff pose potentially serious pollution problems for many U.S. estuaries.

Strategies employed to control runoff from farm lands include, but are not limited to, the following best management practices: (1) modifying agricultural practices to mitigate erosion and sedimentation; (2) constructing retention basins to trap sediments; and (3) creating wetlands to filter nutrients and reduce concentrations of chemical contaminants. Reduction in the use of fertilizers, herbicides, and pesticides would likely decrease the propensity of some estuaries to support undesirable algal blooms and to suffer from periodic episodes of eutrophication. It would also reduce the risk of toxic chemical contaminant accumulation in estuarine sediments and organisms.

b. Urban runoff. Urban runoff is responsible for a variety of water quality impacts. The severity of these impacts depends largely on the magnitude of urbanization in watersheds and the conditions of receiving waters. Water quality problems associated with urban runoff are often manifested by elevated nutrient levels, suspended solids, chemical toxicants, and pathogens most conspicuously developed during periods of precipitation. Low dissolved oxygen concentrations may also occur at these times.

Large quantities of polluted sediments and other particulate matter derived from urban areas contribute to degraded water quality in receiving waters.[71] Fine sediments, in particular, readily sorb organochlorine compounds, PAHs, heavy metals, and other contaminants that are delivered in urban runoff to streams, rivers, and estuaries. Soluble pollutants present in the water column often precipitate on sediments in suspension and at the sediment–water interface, ultimately contaminating bottom habitats. Benthic organisms exposed to contaminants in bottom sediments exhibit variable effects that are very site specific and contingent upon the bioavailability of the contaminants rather than their absolute concentrations.

Urban areas are major sources of nonpoint pollution because of the large populations inhabiting cities and surrounding areas. Urbanization increases runoff to waterways due in large part to the replacement of porous land surfaces by impervious surfaces. Building and highway construction, degraded man-made drainageways, and aging infrastructure facilitate urban runoff, erosion problems, and heavy sedimentation in receiving waters. Effective best management practices for erosion control consist of sediment dams, filter fences, check dams, and the use of excelsior mats on slopes.

c. Stormwater runoff. All watershed management plans should include some form of stormwater management control dealing with both water quality (chemical contaminants, pathogens, suspended solids, etc.) and quantity. Serious nonpoint source pollution is often linked to stormwater runoff from streets, parking lots, business and residential developments, and municipal and industrial facilities (Table 1.13). Various factors influence the volume of stormwater runoff, such as the watershed and land use characteristics of an area (e.g., amount of farm land, impervious cover, soil type, and slope), storm intensity and duration, and the frequency of antecedent storms. Surface runoff occurs when the capacity of the land to retain precipitation is exceeded. Impervious surfaces increase pollutant loadings in runoff, leading to degraded water quality in receiving waters. For example, runoff from driveways, parking lots, service areas, and highways contains aliphatic and aromatic hydrocarbons from waste oil and automobile emissions that collect in storm sewers and ultimately discharge to waterways. Lawns and gardens of housing developments and some commercial establishments are significant sources of nitrogen and phosphorus as well as herbicides and pesticides, from the application of fertilizers and other chemical treatments. Improperly stored chemicals and spills at industrial sites can be a source of toxic pollutants (e.g., metals and organic chemicals) in stormwater runoff. Other pollutants of concern entering streams and other receiving waters from stormwater runoff include bacterial and viral pathogens, oxygen-demanding substances (organic enrichment), and salt (from road salting) (Table 1.14).

Increasing the area of impervious surfaces in watersheds accelerates the rate of removal of rainfall and other precipitation via stormwater runoff to receiving waters. It also increases runoff volumes while decreasing infiltration and storage. As the area of impervious cover increases, the frequency and severity of flooding events rise, as do greater runoff and stream velocity during storms. In contrast, streamflow declines during periods of prolonged dry weather (loss of base flow). Higher runoff volume results in greater pollutant transport and loading to receiving waters.

Pulsing of stormwater runoff increases peak discharge rates, fostering greater erosion and channel scouring problems. Periods of heavy precipitation increase sediment loading from soil erosion and runoff from farm land, construction sites, and urban areas. Accelerated stormwater runoff from these developed areas raises the concentrations of suspended solids and their associated pollutants in receiving waters. Hence, it is important to formulate a stormwater management plan that will effectively deal with these problems.

Remedial programs to address stormwater runoff problems must consider the entire watershed. These programs should include measures such as regional stormwater management facilities, flood control works, stream restoration and erosion control actions, and the implementation of best management practices that could abate flooding, decrease stormwater runoff, and ameliorate nonpoint source pollution in receiving waters. Best management practices for stormwater runoff consist of improved source controls of pollutants, regulatory controls, and structural controls (e.g., constructed wetlands, detention facilities, filtration practices, water quality inlets, infiltration facilities, and vegetation practices) (Table 1.15). Educational programs that inform the public of the proper uses of fertilizers and pesticides as well as the correct disposal methods for automobile oil and hazardous chemicals are critically important to the long-term success of stormwater runoff control. To improve water quality of receiving waters, particularly in urban areas, local governments must formulate effective land-use and watershed management plans while implementing best management practices.

One of the most widely used best management practices for mitigating impacts of stormwater runoff on streams and other receiving waters is the construction of stormwater ponds (Figure 1.24). These shallow features (0 to 3 m deep) typically are located in first- and second-order headwater streams. To achieve maximum stormwater benefits, many communities have adopted regional stormwater pond policies at the watershed scale, with individual pond sites serving areas ranging in size from about 20 to 200 ha.[72] Schueler and Galli[72] review a regional stormwater pond approach for dealing with stormwater problems. Although stormwater ponds remain the preferred and practical option for dealing with stormwater runoff in many developed areas, they can cause negative impacts on receiving waters themselves when poorly designed and sited with little or no regard to the immediate environment. Schueler and Galli[72] cite the following negative impacts of stormwater ponds: (1) poor quality of pond effluent; (2) wetland disturbance; (3) sacrifice of upstream reaches; (4) loss of stream valley forests; (5) barriers to fish passage; (6) downstream warming; and (7) downstream shifts in trophic status. However, these negative impacts may be offset by the positive effects of the ponds, such as wetland creation, retention of open space, pollutant removal, and flood attenuation.

Table 1.13 Expected Effects of Stormwater Contaminants or Condition on Integrative Measures of Ecosystem Response

Input	Expected Environmental Consequence	Energy Dynamics	Food Web Structure and Complexity	Biodiversity	Critical Species	Genetic Diversity	Dispersal and Migration	Ecosystem Development
Suspended solids (organics)	DO depletion enrichment blanketing	1. Alter carbon flux 2. Reduce complexity of energy pathways and control functions	1. Reduce complexity in food web 2. Eliminate sensitive organisms 3. Favor species tolerant to low DO and high organic enrichment 4. Change connection between organisms and trophic levels	1. Minimal effect if source areas of colonizing species exist 2. Urbanization as a process reduces diversity so change must be viewed holistically	1. Direct effect on species intolerant of low DO and high organic enrichment 2. Indirect effects possible through food web alteration	1. Minimal short-term effect unless species permanently lost due to urbanization 2. Continuous discharges and highly variable discharges may lead to local loss of gene pool	1. Local effects on colonization in altered habitat 2. Potential interference with movement by blocking zones of passage	1. Alteration of successional processes may reset community to an earlier successional state 2. Enrichment leading to alternate stable state 3. Transient effects associated with frequency and magnitude 4. Continuous discharges block advanced successional states in zone of influence

Table 1.13 (continued) Expected Effects of Stormwater Contaminants or Condition on Integrative Measures of Ecosystem Response

Input	Expected Environmental Consequence	Energy Dynamics	Food Web Structure and Complexity	Biodiversity	Critical Species	Genetic Diversity	Dispersal and Migration	Ecosystem Development
Suspended solids (inorganics)	Blanketing morphological change Light reduction — 1° productivity effect(s) Physical damage to organisms	1. Reduce complexity of energy pathways and control functions 2. Reduce primary production 3. Reduce access to buried carbon sources	1. Favor species tolerant of high SS 2. Reduce complexity of food web 3. Eliminate sensitive organisms	1. Minimal effect if source areas of colonizing species exist 2. Urbanization as a process reduces diversity so change must be viewed holistically	1. Direct effect on species intolerant of high SS 2. Indirect effects possible through food web alteration	1. Minimal short-term effect unless species permanently lost due to urbanization 2. Continued high variability may lead to local loss of gene pool	1. Local effects on colonization in altered habitat 2. Potential interference with movement by blocking zones of passage	1. High, continuous SS loading may retard succession of limit processes within a narrow range 2. Low, transient SS loading may alter successional processes usually resetting to an earlier successional state

BOD/COD (oxygen demanding wastes)	DO depletion enrichment	1. Alter carbon flux magnitude 2. Alter carbon flux pathways 3. Reduce complexity of energy pathways and control functions	1. Reduce complexity of food web 2. Eliminate sensitive organisms 3. Favor species tolerant to low DO 4. Change connection between organisms and trophic levels	1. Continuous discharge can create zones of low diversity 2. Minimal effect of transient discharge if source areas of colonizing species exist 3. Urbanization as a process reduces diversity so change must be viewed holistically	1. Direct effect on species intolerant of low DO and high organic enrichment 2. Indirect effects possible through food web alteration	1. Minimal short-term effect unless species permanently lost due to urbanization 2. Continuous discharges and highly variable discharges may lead to local loss of gene pool	1. Local effects on colonization in altered habitat 2. Potential interference with movement by blocking zones of passage	1. Alteration of successional processes may reset community to an earlier successional state 2. Enrichment leading to alternate stable state 3. Transient effects associated with frequency and magnitude 4. Continuous discharges block advanced successional states in zone of influence

Table 1.13 (continued) Expected Effects of Stormwater Contaminants or Condition on Integrative Measures of Ecosystem Response

Input	Expected Environmental Consequence	Energy Dynamics	Food Web Structure and Complexity	Biodiversity	Critical Species	Genetic Diversity	Dispersal and Migration	Ecosystem Development
Nutrients, N and P	1° productivity increase enrichment DO variability	1. Increased nutrient pool leads to higher 1° productivity and increased base of fixed carbon 2. Instability in energy processing possible as enrichment reduces functional redundancy 3. Effects of increased nutrients accumulate over space and time	1. Initial effects may be increased diversity with increased carbon-based food resources 2. Altered connectivity between organisms and trophic levels 3. Cumulative effect may be lower diversity	1. Expected increase or decrease through space or time dependent on loading 2. Effect location will not be adjacent discharge locations	1. Direct effect minimal 2. Indirect effects may alter role and capacity of keystone or resource species in ecosystem	1. Minimal short-term effect unless species permanently lost due to urbanization 2. Continuous loading may lead to local loss of gene pool	1. Minimal short-term effect 2. Continuous loading may lead to local change in habitat type or quality affecting dispersal success	1. Nutrient accumulation will establish new base for successional processes 2. Acceleration of processes possible

Metals	Toxicity (lethal effects, sublethal effects, bioaccumulation) Micronutrients — promote enrichment	1. Direct toxicity will reduce functional redundancy leading to greater variability in energy dynamics 2. Micronutrients may lead to system enrichment 3. Accumulation of toxic metals may reduce top-down control functions	1. Direct toxicity will simplify food web 2. Connection may be altered by elimination of sensitive species 3. Transient discharge effects limited to temporal alterations in web components and connections 4. Continuous discharges produce changes in components and connections through both time and space	1. Elimination of sensitive species, reducing biodiversity 2. Bioaccumulation or bio-concentration may produce effects through time or space that eliminate sensitive species and select for tolerant species producing permanent alteration of biodiversity	1. Direct toxic effect on sensitive species 2. Bioaccumulation and bio-concentration may raise contaminant levels to effect concentrations through time or space	1. Direct, short-term and indirect, long-term effects may limit genetic diversity to species capable of tempering toxic effect through transfer, chelation, or binding	1. Localized effects where exposure concentration exceeds tolerance of colonizing species, limiting dispersal 2. Shock loading may induce migration or movement 3. Continuous discharges may create zones of contamination that affect dispersal	1. Shock loading may eliminate species, initiate successional processes, and lead to alternate stable state 2. Long-term maintenance of elevated concentration will select tolerant species and alter natural successional processes 3. Accumulation or concentration in environment or organisms will limit diversity and alter ecosystem development

Table 1.13 (continued) Expected Effects of Stormwater Contaminants or Condition on Integrative Measures of Ecosystem Response

Input	Expected Environmental Consequence	Energy Dynamics	Food Web Structure and Complexity	Biodiversity	Critical Species	Genetic Diversity	Dispersal and Migration	Ecosystem Development
Toxic organics	Toxicity (sublethal effects, bioaccumulation) Transformation Product toxicity Environmental accumulation	1. Direct toxicity will reduce functional redundancy leading to greater variability in energy dynamics 2. Accumulation of contaminants may reduce top down control functions 3. Detoxication mechanisms have energetic cost to individuals that may affect energy dynamics	1. Direct toxicity will simplify food web 2. Connection may be altered by elimination of sensitive species 3. Transient discharge effects limited to temporal alterations in web components and connections 4. Continuous discharges produce changes in components and connections through both time and space	1. Elimination of sensitive species, reducing biodiversity 2. Bioaccumulation, bioconcentration, or sublethal effects may produce changes through time or space that eliminate sensitive species and select for tolerant species producing permanent alteration of biodiversity	1. Direct toxic effect on sensitive species 2. Bioaccumulation and bioconcentration may raise contaminant levels to concentrations that will affect critical species through time or space	1. Direct, short-term and indirect, long-term effects may limit genetic diversity to species capable of tempering toxic effect through decomposition or transfer	1. Localized effects where exposure concentration exceeds tolerance of colonizing species 2. Shock loading may induce migration or movement 3. Continuous discharges may create zones of contamination that affect dispersal	1. Shock loading may eliminate species, initiate successional processes, and lead to alternate stable state 2. Long-term maintenance of elevated concentration will select tolerant species and alter natural successional processes 3. Accumulation or concentration in environment or organisms will limit diversity and alter ecosystem development

Ammonia	Toxicity DO effect 1° productivity increase Enrichment	1. Toxic effects and DO depletion may reduce complexity and control functions 2. Increased nutrient pool leading to higher 1° productivity and increased base of fixed carbon	1. Reduce complexity of the food web 2. Eliminate sensitive organisms 3. Favor species tolerant of toxic effects or DO depletion changing components and connectance	1. Expected increase or decrease through space or time dependent on loading 2. Effect may be local or at some distance from discharge location	1. Direct effect on species intolerant to toxicant concentrations or DO depletion 2. Indirect effects possible through food web alteration	1. Minimal short-term effect unless species permanently lost due to urbanization 2. Continuous discharges and highly variable discharges may lead to local loss of gene pool	1. Shock loading may induce migration or movement 2. Continuous discharges may create zones of contamination that affect dispersal 3. Potential interference with movement by blocking zones of passage	1. Shock loading may eliminate species, initiate successional processes, and lead to alternate stable state 2. Long-term maintenance of elevated ammonia concentration or DO depletion will select tolerant species and alter natural successional processes
Hydrogen sulfide	Toxicity	1. Toxic effects may alter complexity of energy pathways and control functions	1. Reduce complexity of the food web 2. Eliminate sensitive organisms 3. Favor species tolerant of toxic effects changing components and connection	1. Expected decrease through toxic effect	1. Direct effect on sensitive species	1. Minimal short-term effect unless species permanently lost due to urbanization 2. Continuous discharge of highly variable discharges may lead to local loss of gene pool	1. Local effects on colonization in altered habitat 2. Transient concentration may stimulate movement	1. Shock loading may eliminate species, initiate successional processes, and lead to alternate stable state

Table 1.13 (continued) Expected Effects of Stormwater Contaminants or Condition on Integrative Measures of Ecosystem Response

Input	Expected Environmental Consequence	Energy Dynamics	Food Web Structure and Complexity	Biodiversity	Critical Species	Genetic Diversity	Dispersal and Migration	Ecosystem Development
Bacteria	Public health risk Associated solids effects	1. Presence of bacteria often related to enrichment and corresponding change in energy dynamics	1. Direct effects through enrichment 2. Indirect effects may change components and connections	1. Minimal effect if source areas of colonizing species	1. Indirect effects possible through food web alteration 2. Disease induction possible	1. Minimal short-term effect unless species permanently lost due to urbanization 2. Long-term effects possible through alteration of environmental or biological conditions	1. Indirect effects possible	1. Alteration of successional processes possible
Hydraulic dynamics	Washout of organisms Channel morphology and habitat change Erosion/ sedimentation	1. Removal or dislocation of organisms may alter control functions 2. Habitat alteration may exert control on energy dynamics	1. Organism removal or habitat change may alter both structure and connections	1. Minimal effect if source areas of colonizing species exist 2. Changed hydraulic dynamics may alter diversity — permanence of change dependent on permanence of habitat alteration	1. Direct effect on species affected by changed dynamics 2. Indirect effects possible through food web alteration	1. Minimal short-term effect unless species permanently lost due to urbanization 2. Altered dynamics may lead to local loss of gene pool	1. Altered dynamics may induce migration or movement 2. Scour and removal possible when hydrograph changed 3. Local effects on colonization in altered habitat	1. Alteration of successional processes usually resetting to an earlier successional state 2. Species dominance may change due to disturbance frequency and founder species effect

Temperature	Stress Environmental cues for biological activity Synergism with toxicity	1. Speed of energy transfer functions altered 2. Change in temperature may produce indirect alteration of carbon flux	1. Alteration of components and connections 2. Altered environmental state may affect toxicity leading to changes in components and connections 3. Changed cues may alter biological activity leading to changes in components and connections	1. Change in diversity may be due to direct or indirect causes by ecosystem properties, or processes affected by temperature	1. Temperature variability may alter species capacity to maintain viable populations 2. Temperature cues may affect viability of populations	1. Gene pool may be altered by both direct and indirect effects	1. Variability may lead to increased migration 2. Altered cues may induce migration	1. Successional process speed and duration may be altered by long-term temperature change 2. Succession may be reset by temperature fluctuation
pH	Synergism with metal toxicity Stress	1. Change in species composition may produce indirect effects that alter carbon flux	1. Alteration of components and connections 2. Altered environmental state may affect toxicity leading to changes in components and connections	1. Change in diversity may be due to direct or indirect causes by properties or processes affected by temperature	1. Direct effect on species intolerant of transient or permanent changes in pH 2. Indirect effects through alteration of toxicpotential of contaminants to critical species	1. Gene pool may be altered by both direct and indirect effects	1. Variability may lead to increased migration 2. Transient changes may induce migration	1. Transient changes may reset successional processes 2. Permanent alteration may lead to blocked successional development

Source: Herricks, E. E., Section summary, in *Stormwater Runoff and Receiving Systems: Impact, Monitoring, and Assessment*, Herricks, E. E., Ed., Lewis Publishers, Boca Raton, FL, 1995, 27.

Table 1.14 Pollutants Associated with Urban Runoff, Monitoring Parameters, Pollutant Sources, and Potential Effects

Category	Parameters	Possible Sources	Effects
Suspended solids	Organic and inorganic TSS, turbidity, dissolved solids	Urban/agricultural runoff, combined sewer overflows (CSOs)	Turbidity, habitat alteration, recreational and aesthetic loss, contaminant transport
Nutrients	Nitrate, nitrite, ammonia, organic nitrogen, phosphorus	Urban/agricultural runoff, landfills, atmospheric degradation, erosion, septic systems	Surface waters, algal blooms, ammonia toxicity, groundwater, nitrate toxicity
Pathogens	Indicator bacteria (fecal coliforms and streptococci, enterococci), viruses	Urban/agricultural runoff septic systems, CSOs, boat discharges, domestic/wild animals	Ear/intestinal infections, shellfish bed closure, recreational/ aesthetic loss
Organic enrichment	BOD, COD, TOC	Urban/agricultural runoff, CSOs, landfills, septic systems	DO depletion, odors, fishkills
Toxic pollutants	Trace metals, organics	Pesticides/herbicides, underground tanks, hazardous waste sites, landfills, illegal oil disposal, industrial discharges, urban runoff, CSOs	Lethal and sublethal to humans and other organisms
Salt	Sodium chloride	Road salting	Drinking water contamination, impact on non-salt-tolerant species

Source: Bingham, D. R., Wetlands for stormwater treatment, in *Applied Wetlands Science and Technology*, Kent, D. M., Ed., Lewis Publishers, Boca Raton, FL, 1994, 248.

d. Combined sewer overflows. In many older cities, particularly in the northeastern U.S., combined sewers collect and transport sewage and stormwater through the same conveyance system. During periods of heavy precipitation, the storage and treatment capacities of publicly owned treatment works (POTW) become overloaded with stormwater, resulting in the discharge of urban runoff, industrial wastewater, and untreated sewage to receiving waters through combined sewer overflow (CSO) pipes. Streams, rivers, and estuaries may then be impacted by high biochemical oxygen demand and elevated levels of chemical contaminants. Hence, CSOs are potentially significant contributors to water quality degradation in coastal urban areas and, because of the presence of untreated sewage and associated pathogens, pose a potential human health threat.

CSOs are characterized by intermittent flows and pulsed pollutant loads. Their impacts principally depend on the existing treatment system capacity as well as the regional and local hydrology.[16] The irregularity of CSOs is related to the variable intensity and occurrence of rainfall in an area. In many regions of the U.S., precipitation is seasonal, and CSOs mainly occur during certain months of the year. As a consequence, CSO inputs are discontinuous and differ markedly from continuous, point source loadings. In addition, CSO pollutant inputs even vary over the course of a single storm, with the highest pollutant concentrations observed early in the event and levels decreasing greatly thereafter.

It is extremely costly to reduce pollutant loads from urban runoff and CSOs due largely to the pulsed nature of flows.[50] The development of specific stormwater and CSO treatment and control technologies is currently lacking. Best management practices designed to mitigate

Table 1.15 Options for Treatment of Pollution from Stormwater

Source Controls
Animal waste removal
Catch basin cleaning
Cross connection identification and removal
Proper construction activities
Reduced fertilizer, pesticide, and herbicide use
Reduced roadway sanding and salting
Solid waste management
Street sweeping
Toxic and hazardous pollution prevention
Regulatory Controls
Land acquisition
Land-use regulations
Protection of natural resources
Structural Controls
Constructed wetlands
Detention facilities
Extended detention dry ponds
Wet ponds
Filtration practices
Filtration basins
Sand filters
Water quality inlets
Infiltration Facilities
Dry wells
Infiltration basins
Infiltration trenches
Porous pavement
Vegetative practices
Grassed swales
Filter strips

Source: Bingham, D. R., Wetlands for stormwater treatment, in *Applied Wetlands Science and Technology*, Kent, D. M., Ed., Lewis Publishers, Boca Raton, FL, 1994, 245.

CSO problems must consider the entire watershed and concentrate on ameliorating pollution at its source. For example, pollution control strategies involving both stormwater and CSOs should focus on public education programs dealing with the dangers of illegal dumping and the hazards associated with excessive waste generation and disposal. Runoff source control strategies should also be formulated for watersheds. Various management plans can be effective in this regard, such as the development of household hazardous waste collection programs, the implementation of spill response and containment programs, the preparation of procedures that increase surveillance of chemical storage areas, and the adoption of effective work practices (e.g., the removal of floor drains from storm drains, street sweeping, etc.).[16] The use of retention and detention ponds may help to control CSO discharges without extensive reconstruction of existing pipework systems. As noted by Clifforde,[73] new CSOs constructed in England today incorporate some degree of storage, with a principal objective of limiting peak continuation flows.

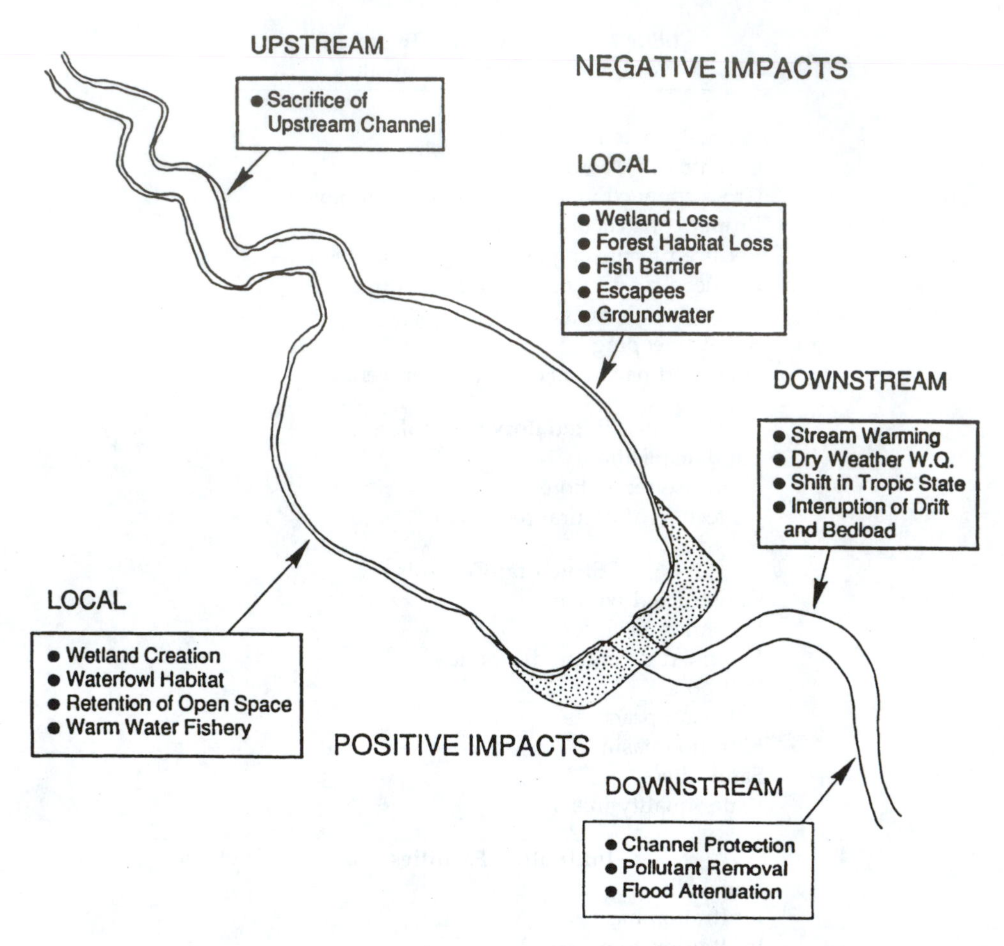

Figure 1.24 Schematic of the positive and negative impacts of ponds used in stormwater runoff control. (From Schueler, T.R. and Galli, J., The environmental impact of stormwater ponds, in *Stormwater Runoff and Receiving Systems: Impacts, Monitoring, and Assessments*, Herricks, E.E., Ed., Lewis Publishers, Boca Raton, FL, 1995, 177.)

e. Silviculture. Erosion and sedimentation problems are often closely coupled to timber harvesting. Clear-cutting practices that strip timber over extensive areas of mountain slopes promote the eventual loss of massive sediment loads that can clog nearby streams. These problems are exacerbated by road construction in remote areas and heavy truck traffic over dirt surfaces.

Nonpoint source pollution control programs designed to mollify the impacts of careless timbering practices on receiving waters usually focus on deforestation, erosion, and sedimentation problems. Best management practices to control erosion and sedimentation associated with timber harvesting are essentially the same as those used for construction projects described previously. Included here are the use of sediment traps, filter fences, check dams, and other remedial efforts to control erosion and subsequent impacts on receiving waters.

f. Mining. Inactive and/or abandoned mines frequently are significant sources of acid waters, metals, and other contaminants to receiving waters. The weathering, leaking, and transport of toxic metals from tailing piles and slag or rock piles are facilitated by the percolation of acidic waters. For example, mine adit water and water percolating through rock and tailing piles with a pH of 3 or less can carry very high concentrations of highly toxic, free, or biologically available metals that may overwhelm the assimilative and buffering capacity of adjacent streams. As a result, aquatic life may be threatened by toxic pollutants far downstream from the mine site and in some cases in nearby estuaries (e.g., San Francisco Bay).

VII. Freshwater inflow alteration

Modification of the natural flow regimes in streams and rivers can have significant environmental and ecological consequences for estuarine receiving waters. Many changes in natural flow regimes derive from attempts by humans to stabilize waterways for protection of infrastructure and property. Other changes result from human needs for recreation and commerce. In many cases, water quality and biotic communities of estuaries are affected by the alteration of influent systems.

Although the causes of natural variations in freshwater inflow to estuaries are well chronicled (i.e., climate cycles, catastrophic storms, floods, etc.), those associated with anthropogenic activities commonly have more dramatic long-term effects. Some human impacts significantly increase freshwater inflow to receiving waters. For instance, as impervious cover increases with urbanization, surface runoff also accelerates, leading to elevated riverine discharges during storm events. The destruction of wetlands not only increases stream flows but also contributes to the loss of natural water storage capacity in river basins. These changes frequently culminate in flooding downstream.[50] Channelization projects intended as flood control measures enhance water movement through watersheds.

In some regions of the U.S., freshwater diversions for agricultural, municipal, and industrial purposes severely restrict freshwater inflow to estuaries (e.g., San Francisco Bay). During the dry season, agricultural and municipal demands for freshwater soar in watersheds, severely curtailing discharges and resulting in near-marine conditions in certain estuaries (e.g., Charlotte Harbor, Florida). Reduced freshwater inflows, together with excessive groundwater pumping for agricultural and other uses, exacerbate saltwater intrusion problems (e.g., Sarasota Bay, Florida). Similarly, water storage reservoirs and dams reduce flows downstream and can be responsible for substantial stream bed degradation. Hydrologic modification of numerous tidal creeks and the damming of major rivers for flood control and water supply developments have altered freshwater inflow in some impounded estuarine tributaries (e.g., Tampa Bay). Poorly controlled, in-river sand or gravel mining also facilitates river bottom degradation downstream and, in some cases, promotes headcutting upstream of the mine. Bed load is removed, and river flow is altered.

The water quality and living resources of an estuary are closely linked to the volume of freshwater inflow. For example, large increases or decreases of freshwater inflow, which shift the salinity gradient in an estuary, directly affect the structure of plant and animal communities. In addition, an estuary's capacity to dilute, transform, or flush contaminants varies with the flux of freshwater inflow. Higher stormwater and surface runoff commonly contributes to greater concentrations of chemical contaminants (i.e., petroleum hydrocarbons, organochlorine contaminants, heavy metals, etc.) in tributary systems. Changes in nutrient and sediment inputs from watersheds likewise arise from modifications of natural flow regimes.

The quantity and quality of river flow are important factors influencing the abundance, distribution, and reproductive success of many estuarine organisms. Species composition and abundance vary with salinity changes along a gradient. In a given area, some species will adapt to the salinity changes, and others will be eliminated. The resultant community composition depends on the characteristics of the particular system and the individual species tolerances.

San Francisco Bay, a large urbanized estuary located at the mouth of the Sacramento and San Joaquin Rivers, provides an excellent example of a system impacted by variable freshwater inflow associated with human activities (see Chapter 5). About 90% of the freshwater inflow to San Francisco Bay originates from the discharge of the Sacramento–San Joaquin Delta into Suisun Bay.[74] However, more than 50% of the freshwater inflow is diverted for agricultural, municipal, and industrial purposes.[75] Most of the diverted water (~85%) is used for agricultural needs (i.e., irrigation), with the remainder utilized by municipal and industrial interests. Among the largest water diversion efforts in the world are the State Water Project and the federal Central Valley Project of California.

Because of the aforementioned diversion projects, water flow in channels within the delta changes direction, thereby reducing freshwater input to the bay. Several biotic impacts arise from the altered natural hydrologic regime in the region. For instance, operation of powerful water project pumps entrains millions of eggs, larvae, and juvenile stages of fish and invertebrates, and increases their mortality. Water pumping not only reroutes the flow but also transports estuarine species into new and potentially hostile areas. Migration of spawning adults or outmigrating young is disrupted in some areas. The decrease of nutrient and organic carbon loading can reduce biological productivity in the bay and, along with changes in species distributions and the basic structure of invertebrate and fish populations related to highly variable freshwater inflow, may lead indirectly to significant shifts in trophic interactions. In other systems where considerable long-term, natural variability of freshwater inflow has been documented, changes in estuarine habitat conditions and trophic organization are also significant.[76,77]

VIII. Shoreline development

The coastal zone incurs multiple anthropogenic impacts associated with steady population growth, development, and recreational pursuits. The shoreline is highly attractive real estate for human habitation — the seven largest metropolitan centers in the U.S. lie in the coastal zone[78] — and the coastal fringe is increasingly valuable for various industrial and commercial endeavors. Approximately 45% of the U.S. population lives in coastal counties, although coastal regions comprise only about 10% of U.S. land area (excluding Alaska).[79] Demographic statistical trends indicate that the total coastal U.S. population may now exceed 125 million.

The degradation of estuarine waters and bordering landmasses has accompanied this domestic and industrial expansion. For example, about 40% of the nation's shellfish beds are closed because of sewage discharges and nonpoint source runoff from coastal towns and cities.[80] In addition, nearly 10,000 ha of coastal wetlands are lost each year, primarily due to development.

Human activities alter coastal landforms in numerous ways. The construction of shorefront homes, industrial installations, and shore-protection structures is evident to even the most casual observer. More subtle alteration, although less conspicuous, also contributes to significant long-term impacts. For example, damage to dunes by careless beach enthusiasts often hastens beach erosion and reduces shore protection. Beach nourishment programs may be undertaken not only to augment the beach sediment supply for recreational purposes but also to protect nature reserves, buildings, and infrastructure.[81]

The literature is replete with investigations of shore-protection structures and their effects on beach and nearshore processes as well as inlets, harbors, and estuaries.[82] These coastal engineering features may be classified as shore-perpendicular and shore-parallel forms. Examples of shore-perpendicular structures are groins and jetties. Constructed of quarrystone, timber, or a combination of stone and timber, groins usually extend seaward for up to several hundred meters from the beach, although they vary greatly in length and height. These structures trap sediment moving alongshore and entering from updrift areas, but they increase erosion downdrift. Differences in sediment characteristics develop on the updrift and downdrift sides.[83] Although groin fields tend to lower the rate of beach retreat and mollify the seasonal fluctuations in beach profiles, they also interfere with wave refraction and breaking as well as surf-zone circulation. Additional rip currents arise, and sediment is redirected, thereby creating more complex topography.[84,85] Two of the most pervasive impacts of groins are sediment starvation and truncation of beach profiles downdrift of the structures.[86,87]

Jetties constrain and direct the flow of seawater at inlets and act as a barrier to longshore transport, which prevents shoaling of the inlet channel. Hence, they can have a profound effect on the sediment movement and hydrology in some estuaries.[84] Similar to groins, jetties generally induce sand starvation in downdrift areas, often contributing to serious erosion problems.[86] In addition to increasing the rate of erosion on the downdrift shoreline, jetties provide a conduit for the transport of sediments by ebb flows to deeper water. As a result, they displace the ebb tidal delta and associated bars farther seaward.[88,89] Both groins and jetties can have a significant impact on local process conditions and beach morphology.

Bulkheads, seawalls, and sloped revetments are types of shore-parallel structures that act as barriers to approaching waves, which tend to break or reflect off them. Although these structures protect bordering land areas, their detrimental geomorphic effects include scour and increased erosion seaward as well as a reduction of beach width. For example, the massive seawall at Sea Bright in northern New Jersey has caused rapid erosion of the shoreline fronting the structure.[90]

Beach revitalization programs are commonly employed in areas impacted by severe erosion. Because coastal facilities and infrastructure require protection, the upper beach generally receives most, if not all, of the fill material, which frequently differs considerably in grain size and other properties from the natural beach sediments. Nourished beaches are readily distinguished from natural beaches by their great width, oversteepened upper beach (with a shape and composition out of equilibrium with natural processes), and prominent vertical scarp (often marking the upper limit of wave reworking at high tide).[84]

Coastal engineers have applied various types of mathematical and computer models to predict the effects of engineering structures on the coastal zone. In some cases, hydraulic models have proven to be extremely useful in understanding potential impacts of these structures on nearshore and beach dynamics. Hydraulic modeling has been especially valuable in predicting the effects of breakwaters on waves in harbors, dredging on sediment movement in estuaries, and current patterns on waste dispersal in coastal regions.[91-93]

IX. Sea level rise

Effective coastal management must consider the vulnerability of beaches, estuaries, and infrastructure to the threat of rising sea level. This requires projected responses of coastal zones to short- and long-term climate variations. Sea levels are rising as a result of both natural processes and human activities that affect regional and global climate in a manner that is not completely understood. Thus, modeling of climate and sea level changes has generated projections with large uncertainties.

Sea level along the Atlantic Coast has risen by ~30 cm during the last century. Much of this sea level rise is ascribed to global climate changes. Measurements indicate that the global mean surface temperature has increased by 5 to 10°C since the Wisconsin glaciation 18,000 year ago.[94] In the past century alone, it has risen ~0.5°C.

Table 1.16 Description of Major Greenhouse Gases[a]

Carbon Dioxide (CO_2)	Currently responsible for approximately one half to two thirds of human contributions to global warming, the concentration of atmospheric CO_2 has risen 25% since the beginning of the Industrial Revolution. Combustion of fossil fuels and deforestation are the main sources of this increase. Because society's basic energy sources produce CO_2, its atmospheric concentration is projected to continue to increase unless strong measures are taken to encourage energy conservation, alternative energy sources, and forest preservation.
Methane (CH_4)	Although methane has a much lower atmospheric concentration than CO_2, it is a more potent greenhouse gas and its concentration is increasing at a rate of 1% per year. Sources include rice paddies, cows, termites, natural gas leakage, biomass burning, landfills, and wetlands. Of the major greenhouse gases, CH_4 concentrations may be the easiest to stabilize with modest cuts in emissions.
Nitrous Oxide (N_2O)	The concentration of nitrous oxide in the atmosphere is increasing at a rate of 0.25% per year. Although nitrous oxide is a more potent greenhouse gas than CO_2, its contribution to global warming is lower because of its low concentration. Anthropogenic sources include fossil fuel and biomass combustion, agricultural fertilizers, and land disturbances. The relative contribution of natural and anthropogenic sources of N_2O is not well understood.
Chlorofluorocarbons (CFCs)	Invented in this century, CFCs have been implicated not only in chemical destruction of stratospheric zone, but also in greenhouse warming. Each CFC molecule has a direct warming effect several thousand times that of a CO_2 molecule. However, CFCs cause an indirect cooling effect by destroying ozone, another greenhouse gas. The extent to which the direct warming effect is offset by the indirect cooling effect has not yet been determined. CFCs are used in refrigerants, aerosol propellants, foam-blowing agents, and solvents. Their atmospheric concentration is increasing at a rate of 4% per year. Substitutes are being developed that are not as chemically stable and therefore will not accumulate as fast in the atmosphere. The Montreal Protocol, an international agreement, and recent U.S. measures currently limit production of these gases, but further limits may be necessary.

[a] *Source*: EPA, Climate Change, EPA 230-F-95-002, U.S. Environmental Protection Agency, Washington, D.C., 1995.

Coastal subsidence increases the level of the sea relative to the elevation of the adjacent shore. Regional downwarping of sediment, sediment compaction, and subsurface fluid withdrawal are the major processes affecting land subsidence from the mid-Atlantic region to the Gulf Coast.[95] In many areas, dramatic increases in groundwater withdrawal have been responsible for much of the sea level rise.

The increase in concentrations of greenhouse gases in the atmosphere due to anthropogenic activities may be significantly accelerating sea level rise. The greenhouse gases of note are primarily carbon dioxide (CO_2) and other radiatively active trace gases, such as methane (CH_4), nitrous oxide (N_2O), and chlorofluorocarbons (CFCs) (Table 1.16). These gases act to warm the earth's surface by trapping heat from the atmosphere, thereby influencing global climate and fostering eustatic sea level changes. Greenhouse climate warming can cause sea level to rise by the combined effect of thermal expansion of the oceans and enhanced melting of mountain glaciers and polar ice sheets.[96,97]

The global atmospheric concentration of CO_2 today amounts to ~350 ppm, up from ~270 ppm before the Industrial Revolution, and it is increasing at a rate of ~1.5 ppm per year.[98] Estimates suggest that CO_2 emissions are responsible for about half of all global warming. The principal sources of CO_2 emissions include electric power generation, manufacturing facilities, automobiles, and trucks. Since the beginning of the Industrial Revolution in the late 1700s, the level of atmospheric CO_2 has increased by ~30%.

Global climate change and associated sea level rise pose risks to coastal ecosystems, resources, and communities. If eustatic sea level continues to rise as projections indicate, shorelines will be subjected to increased rates of erosion. In addition, shore communities will be more susceptible to property damage from storms and episodic flooding. As a result, property values will decline, as will recreational opportunities. The costs for shore protection will escalate, leading to considerable controversy regarding future investment in coastal development. Other potential consequences for coastal communities include upstream penetration of saltwater and intrusion of saltwater into groundwater supplies. Agriculture operations also will be adversely affected.

Coastlines at greatest risk to the damaging effects of rising sea level are those characterized by an erodible substrate (i.e., unconsolidated sediment), high wave/tidal energies, a retreating shoreline, subsidence, high probabilities of major storms (i.e., hurricanes, tropical storms, and extratropical cyclones), and low coastal elevations. Long-term planning and coastal management will be most critical for these high-risk areas. Three principal management strategies for dealing with rising sea levels in the coastal zone are: (1) hardening and protecting the shoreline (i.e., with seawalls, dikes, and beach nourishment) along the entire coast; (2) protecting heavily populated areas only; and (3) abandoning the current development of the coastline and retreating inland.[97]

Rising sea level will also directly impact estuarine systems. Aside from causing shoreline damage and displacement, rising sea level will likely alter tidal prisms and salinities, thus affecting estuarine circulation and biotic communities. Depending on the magnitude of sea level rise, the loss of wetlands can be substantial, resulting in potentially serious impacts on wildlife and fisheries. The establishment of buffer zones to allow for inland migration of wetlands may mitigate some of these impacts.

X. Estuarine biota

Most of the 28 NEPs have identified various biological problems, such as introduced species and declines of fish and wildlife populations, as either high or medium program priorities. Watershed development and accompanying human activities on land and in estuaries threaten biological diversity, productivity, and integrity through decreased water quality, eutrophication, the loss or alteration of habitats, overexploitation of resources, the introduction of invasive exotic species, and other impacts. Although some NEPs face similar environmental problems that threaten living resources, no two estuaries are exactly alike, and hence each requires careful assessment and the development of a CCMP that will restore, maintain, and protect environmental quality. A major goal of all programs is to achieve meaningful and measurable improvement in the quality of aquatic life and resources in the estuarine system.

A. Introduced species

Nine of the 28 NEPs have identified invasion by introduced (non-native) species as a high or medium priority. These exotic species often disrupt the food web structure of estuaries and cause multiple impacts — ecological, economic, and social — on these systems. Intentional or accidental introductions of exotic species can fundamentally alter the trophic organization of estuaries through predation and competition. In extreme cases, the invasive species totally displace or drastically reduce native forms. Aside from food web changes, other biotic impacts of introduced species include reduced species diversity in estuarine communities, shifts in recreational and commercial fisheries, and the possible introduction of detrimental pathogens. By altering wetlands habitat, non-native species may lead to significant secondary impacts, such as increased runoff and greater inputs of chemical contaminants. Other potential effects are the disturbance and destruction of native, non-wetland habitats, increased erosion, and modification of nutrient cycles.

A primary danger of introduced species is that, lacking natural controls in their adopted estuarine habitats, they will outcompete native populations and modify faunal and floral communities.[45] In several NEPs where invasive species have been investigated, the findings are alarming. For example, the introduction of the parasitic protozoan *Haplosporidium nelsoni* in the Delaware Estuary in the 1950s resulted in a dramatic decrease in the abundance of American oyster (*Crassostrea virginica*) populations, which had catastrophic impacts on the multimillion dollar commercial shellfishery of the area. Also known as MSX, *H. nelsoni* rapidly infested oyster beds after being introduced to the estuary, presumably as spores that were transferred from the West Coast of the U.S. or from Asian ballast water. The Delaware Bay oyster industry continues to be severely decimated by the effects of this parasite.

The degradation of native habitats in some Gulf Coast estuaries by invasions of non-native species is a serious problem. Numerous introduced plant species in the Barataria-Terrebonne estuary in Louisiana, for example, have reduced biodiversity in the system and caused significant habitat loss and alteration. Introduced species are also a potential problem in Galveston Bay and Corpus Christi Bay in Texas.

The invasion of the edible brown mussel (*Perna perna*) in the Corpus Christi Bay system in 1990 has caused significant fouling of natural and man-made structures. The brown mussel has proliferated quickly in South Texas waters, establishing healthy populations over a distance of ~1300 km. Its reproductive success has raised concern that even more serious fouling problems may develop in the near future.

The impacts of introduced species appear to be most pronounced in San Francisco Bay. More than 200 nonindigenous populations inhabit bay waters and wetlands, and many of the dominant species in the estuary's communities are introduced forms. Exotic species comprise most of the benthos of the estuary.[99,100] On the inner shallows of the bay, nearly all macroinvertebrates are introduced forms. Among the invertebrates identified in the estuary, about 100 species are introduced forms, including the only two mollusks that support the bay's sport fishery, notably the soft-shelled clam (*Mya arenaria*) and the Japanese little-neck clam (*Tapes japonica*). Some non-native species in the estuary are serious pests that are highly destructive and, therefore, of economic significance (e.g., the oyster drill, *Urosalpinx cinerea*; Asian clam, *Potamocorbula amurensis*; and shipworm, *Teredo navalis*).[75] Nonindigenous species now account for 40 to 100% of the fouling communities as well as the benthic infaunal and epifaunal communities, in different areas of the bay. Nichols and Thompson[101] ascribe the widespread success of many introduced species in the estuary to their opportunistic characteristics that enable them to rapidly colonize underexploited habitats in the system. Others (see Conomos[102]) attribute the success of the introduced forms to the low diversity of native fauna resulting from the geographic isolation and geologic youth of the estuary. The rate of invasion of introduced species has steadily increased during the past several decades.

For more than a century, numerous finfish species have been introduced into San Francisco Bay either deliberately to enhance the recreational and commercial fisheries or accidentally with organisms from other areas of the country. As a consequence, non-native fish species constitute more than half of the fishes in the delta.[103] Most of the common fish have been introduced from the East Coast of the U.S., including the striped bass (*Morone saxatilis*), which remains the primary sports fish of the bay.[75,104]

Consumer demands for frog legs led to the introduction of bullfrogs (*Rana catesbiana*) after overharvesting and natural predation decimated the red-legged frog (*R. aurora*) population.[105,106] Three species of crayfish were also introduced for their potential value as a food source: the signal crayfish, *Pacifasticus leniusculus* var. *leniusculus*; the red swamp crayfish, *Procambarus clarki*; and a species with no common name, *Orconectes virilis*. The signal crayfish now supports a major fishery in the delta.[107]

In regard to floral communities, one of the most aggressive non-native plant species introduced into San Francisco Bay is the smooth cordgrass *Spartina alterniflora*.[108] This vascular plant has successfully invaded saltmarsh habitat in the bay area (i.e., South Bay) and appears to be spreading quickly. There is growing concern that *S. alterniflora* will outcompete the native cordgrass *S. foliosa* as well as other plant species in the marsh. Because *S. alterniflora* grows to a lower elevation on mudflats and differs in stem size and density from *S. foliosa*, possible impacts arising from the displacement of the native form in saltmarshes include changes in associated plant and animal communities, shifts in sedimentation rates, and alterations in habitat value. Several other species of non-native cordgrass have also been introduced to the bay. The implications of their invasion into saltmarsh areas are not completely understood at present.

B. Commercial and recreational fisheries

A major concern of all 28 NEPs is the potential decline in living resources due to deteriorating water quality, habitat destruction, overfishing and overharvesting, and other anthropogenic impacts that threaten the stability and diversity of the ecosystem. Estuaries typically support a large and diverse array of aquatic and wildlife resources. The causes of declining species or resources in these systems often remain unclear, even after detailed investigations are conducted. Natural variations in abundances of populations may be greater than variations resulting from human impacts, which can obfuscate analyses of anthropogenic effects. Hence, additional study is frequently needed, and programs involving both short- and long-term monitoring of biotic communities, habitats, and water quality are usually required to determine the principal factors responsible for diminishing resources. Once the actual causes of declining species or resources are determined, management plans must be formulated and enacted to provide sound remedial solutions to acute and insidious environmental impacts.

An estuary's health is measured in part by the state of its resources. In NEPs, action plans are prepared to address living resource problems, such as declining oyster, clam, fish, and waterfowl populations, damaged habitats, or altered spawning, nursery, and forage grounds. These action plans not only propose immediate protective measures but also consider long-term enhancement strategies. Protective measures may involve tighter regulatory enforcement to preclude additional loss of aquatic habitats, wetlands, and wildlife. They may also recommend more stringent limits on commercial and recreational catches. Enhancement strategies usually deal with the implementation of restoration, creation, and enhancement programs. The planting of wetlands vegetation, creation of aquaculture facilities, and the development of stocking hatcheries provide examples.

Antidegradation policies can be formulated for estuarine waters with exceptionally significant recreational or ecological features. For these valuable waters, it may be necessary to limit shoreline development, severely restrict waste inputs, or designate no discharge areas. The setting of higher standards should be considered. Waters of such great value are commonly designated as high-priority areas for additional study and monitoring.

Assessment of estuarine communities requires at least a fundamental knowledge of the biology of constituent populations. It is important to understand the population dynamics of the biota to accurately delineate the effects of pollutants and other anthropogenic impacts on their life history. To ameliorate or solve these ecological problems and to rehabilitate or protect estuaries, it is necessary to compile data on the structure of estuarine communities and the biotic interactions and responses of the organisms within the systems under investigation. An integral part of evaluating the health of these waters entails the determination of organism–habitat interactions, which, in many respects, is much more problematic. Understanding the effects of abiotic and biotic factors on estuarine organisms is critical to the maintenance and preservation of the aesthetic, commercial, and recreational resources of these invaluable systems.

In each NEP, a critical first step for both technical characterization and the CCMP is to identify and describe valued resources and uses of the estuary, including its living resources.[7] Once the resources are identified, their status and trends must be analyzed, and, if possible, their future conditions must be predicted. This information is important when conducting valuation studies, assessing anthropogenic impacts, and recommending corrective actions in the CCMP.

Living resources of the NEPs are valued for their economic, recreational, aesthetic, and ecological importance. These systems support diverse plant and animal communities integrated through a complex web of ecological relationships. The food web includes submerged aquatic vegetation, phytoplankton, zooplankton, and benthic invertebrates, which, in turn, support fish, amphibians, reptiles, birds, and mammals. The determination of these trophic relationships is necessary to fully understand how the ecosystem functions.

Some fish and shellfish populations support multimillion dollar commercial and recreational fisheries and hence are significant in resource valuation studies. The economic value of fisheries typically plays a major role in gaining the support of local citizens groups, industry, and government agencies in the maintenance and preservation of estuarine habitats. It demonstrates the benefits of public investment in management actions to sustain or improve the health of these ecosystems.

Estuarine environments have long been known for their exceptional recreational and commercial fisheries. Much of the U.S. commercial catch and marine recreational catch embodies species that inhabit estuaries at least part of their lives. The bulk of the fisheries landings in the Gulf of Mexico (>95%) has been attributed to estuarine species. Similar landings data exist for the mid-Atlantic states. The economic value of estuarine-dependent fisheries raises concern over the potential impact of anthropogenic stresses on natural populations of fish.

Twenty-two of the 28 NEPs have identified declines in fish and wildlife populations as either a high- or medium-priority problem.[109] Some of these population declines are attributed to overexploitation and others to decreased water quality, habitat loss and degradation, and introduced species. For instance, in Maryland coastal bays, oyster and scallop shellfisheries largely disappeared during the 20th century. Overfishing as well as disease and predation appear to be contributing factors. Similarly, in the Delaware Estuary, the oyster harvest decreased dramatically from 1 to 3 million bushels per year during the early 1900s to near zero in the 1990s. Disease (i.e., MSX and dermo infestation) is primarily responsible for the decimated shellfishery. The Barnegat Bay estuarine system in New Jersey once supported major hard clam, oyster, and scallop shellfisheries, but overharvest, predation, and changes in water quality resulted in the disappearance of oysters and scallops during the 20th century. Only the hard clam (*Mercenaria mercenaria*) remains a viable commercial species in this system.

In regard to finfish, total fisheries landings in the Peconic estuary dropped from a high of 1.3×10^6 kg in 1980 to less than 1.54×10^5 kg in 1989, although slight recovery is apparent in the 1990s. The causes of the declining fisheries landings have not been unequivocally established. Increased fishing pressure in Albemarle-Pamlico Sounds has led to decreased commercial landings. Eight species of recreationally and commercially important finfish and shellfish appear to be overfished or severely depleted. Commercial landings of sea trout (*Cynoscion arenarius* and *C. nebulosus*) in Sarasota Bay, Florida, have dropped by 50% from historic levels during the 1950s and 1960s. However, sea trout landings have remained relatively stable since the mid-1970s. The decrease in the landings since 1950 has been ascribed to the alteration and degradation of juvenile fish habitats. About 30% of the area of seagrass meadows and 39% of the area of predevelopment tidal wetlands have been lost, and dredge and fill projects have destroyed many hectares of natural habitat, exacerbating the commercial fisheries problems. Between 1966 and 1990, landings of 11 commercially important finfish species in Tampa Bay, Florida, dropped by 24%. Extreme losses were documented for the spotted sea trout (*C. nebulosus*) and red drum (*Sciaenops ocellatus*), whose harvests decreased more than 80% between 1950 and 1990 and 1950 and 1986, respectively. In San Francisco Bay, there have been major declines in fisheries since the 1970s; chinook salmon (*Oncorhynchus tshawytscha*), delta smelt (*Hypomesus transpacificus*), and striped bass (*Morone saxatilis*) provide examples.

Commercial and recreational fisheries are extremely valuable to the NEPs. For instance, the economic value of commercial and recreational fisheries in Long Island Sound exceeds more than $1.2 billion annually. Albemarle-Pamlico Sounds account for much of the total annual revenue of North Carolina's coastal commercial and recreational fishing industry, which is estimated to be about $1 billion. Commercial fishing alone in San Francisco Bay generates more than $25 million annually.

Management plans are being developed and implemented by the NEPs to support and protect commercial and recreational fisheries and other living resources. These plans are incorporating the input of government agencies, industry, educational institutions, and the general public. Major management goals include the restoration of depleted stocks, enhancement and creation of habitat, and the improvement of water quality. The ultimate goal is to establish healthy environments that will support diverse fish and wildlife populations. The sustainability of living resources is a major concern of all NEPs.

References

1. Tammi, C. E., Offsite identification of wetlands, in *Applied Wetlands Science and Technology,* Kent, D. M., Ed., Lewis Publishers, Boca Raton, FL, 1994, 13.
2. Fields, S., Regulations and policies relating to the use of wetlands for nonpoint source pollution control, in *Created and Natural Wetlands for Controlling Nonpoint Source Pollution,* Olson, R. K., Ed., C. K. Smoley, Boca Raton, FL, 1993, 151.
3. Olson, R. K., Ed., *Created and Natural Wetlands for Controlling Nonpoint Source Pollution*, C. K. Smoley, Boca Raton, FL, 1993.
4. USEPA, *Managing Nonpoint Source Pollution: Final Report to Congress on Section 319 of the Clean Water Act (1989),* EPA/506/9-90, U.S. Environmental Protection Agency, Washington, D.C., 1990.
5. Office of Technology Assessment, *Wastes in Marine Environments*, Office of Technology Assessment, OTA-O-334, U.S. Government Printing Office, Washington, D.C., 1987.
6. Robb, D. M., The role of wetland water quality standards in nonpoint source pollution strategies, in *Created and Natural Wetlands for Controlling Nonpoint Pollution*, Olson, R. K. Ed., C. K. Smoley, Boca Raton, FL, 1993, 159.
7. USEPA, *A National Estuary Program Guidance: Technical Characterization in the National Estuary Program*, EPA 842-B-94-006, U.S. Environmental Protection Agency, Washington, D.C., 1994.
8. Chandler, G. T., Coull, B. C., and Davis, J. C., Sediment- and aqueous-phase fenvalerate effects on meiobenthos: implications for sediment quality criteria development, *Mar. Environ. Res.*, 37, 313, 1994.

9. Haebler, R. and Moeller, R. B., Jr., Pathobiology of selected marine mammal diseases, in *Pathobiology of Marine and Estuarine Organisms*, Couch, J. A. and Fournie, J. W., Eds., CRC Press, Boca Raton, FL, 1993, 217.
10. Office of Technology Assessment, *Pollutant Discharges to Surface Waters in Coastal Regions*, U.S. Congress, Office of Technology Assessment, Washington, D.C., 1986.
11. Alongi, D. M., *Coastal Ecosystem Processes*, CRC Press, Boca Raton, FL, 1998.
12. Kennish, M. J., Ed., *Practical Handbook of Marine Science*, 2nd ed., CRC Press, Boca Raton, FL, 1994.
13. Kennish, M. J., Ed., *Practical Handbook of Estuarine and Marine Pollution*, CRC Press, Boca Raton, FL, 1997.
14. Valiela, I., *Marine Ecological Processes*, 2nd ed., Springer-Verlag, New York, 1995.
15. Weinstein, J. E., Anthropogenic impacts on salt marshes — a review, in *Sustainable Development in the Southeastern Coastal Zone*, Vernberg, F. J., Vernberg, W. B., and Siewicki, T., Eds., University of South Carolina Press, Columbia, SC, 1996, 135.
16. National Research Council, *Managing Wastewater in Coastal Urban Areas*, National Academy Press, Washington, D.C., 1993.
17. Dardeau, M. R., Modlin, R. F., Schroeder, W. W., and Stout, J. P., Estuaries, in *Biodiversity of the Southeastern United States: Aquatic Communities*, Hackney, C. T., Adams, S. M., and Martin, W. H., Eds., John Wiley & Sons, New York, 1992, 615.
18. Bricker, S. B. and Stevenson, J. C., Nutrients in coastal waters: a chronology and synopsis of research, *Estuaries*, 19, 337, 1996.
19. USEPA, *Long Island Sound Study: Status Report and Interim Actions for Hypoxia Management*, Tech. Rept., U.S. Environmental Protection Agency, New York, 1990.
20. Sindermann, C. J., *Ocean Pollution: Effects on Living Resources*, CRC Press, Boca Raton, FL, 1996.
21. James, M. O. and Kleinow, K. M., Trophic transfer of chemicals in the aquatic environment, in *Aquatic Toxicology: Molecular, Biochemical, and Cellular Perspectives*, Malins, D. C. and Ostrander, G. K., Eds., Lewis Publishers, Boca Raton, FL, 1994, 1.
22. Rattner, B. A. and Heath, A. G., Environmental factors affecting contaminant toxicity in aquatic and terrestrial vertebrates, in *Handbook of Ecotoxicology*, Hoffman, D. J., Rattner, B. A., Burton, G. A., Jr., and Cairns, J., Jr., Eds., CRC Press, Boca Raton, FL, 1995, 519.
23. Clark, R. B., *Marine Pollution*, 3rd ed., Clarendon Press, Oxford, England, 1992.
24. Kennish, M. J., *Ecology of Estuaries: Anthropogenic Effects*, CRC Press, Boca Raton, FL, 1992.
25. Albers, P. H., Petroleum and individual polycyclic aromatic hydrocarbons, in *Handbook of Ecotoxicology*, Hoffman, D. J., Rattner, B. A., Burton, G. A., Jr., and Cairns, J., Jr., Eds., CRC Press, Boca Raton, FL, 1995, 330.
26. Cormack, D., *Response to Oil and Chemical Marine Pollution*, Applied Science, London, 1983.
27. National Research Council, *Oil in the Sea: Inputs, Fates, and Effects*, National Academy Press, Washington, D.C., 1985.
28. Doerffer, J. W., *Oil Spill Response in the Marine Environment*, Pergamon Press, New York, 1992.
29. Eisler, R., *Polycyclic Aromatic Hydrocarbon Hazards to Fish, Wildlife, and Invertebrates: A Synoptic Review*, Biol. Rept. 85(1.11), U.S. Fish and Wildlife Service, Washington, D.C., 1987.
30. McLeese, D. W., Ray, S., and Burridge, L. E., Accumulation of polynuclear aromatic hydrocarbons by the clam *Mya arenaria*, in *Wastes in the Ocean*, Vol. 6, *Nearshore Waste Disposal*, Ketchum, B. H., Capuzzo, J. M., Burt, W. V., Duedall, I. W., Park, P. K., and Kester, D. R., Eds., John Wiley & Sons, New York, 1985, 81.
31. Barrick, R. C. and Prahl, F. G., Hydrocarbon geochemistry of the Puget Sound region. III. Polycyclic aromatic hydrocarbons in sediments, *Est. Coastal Shelf Sci.*, 25, 175, 1987.
32. Rainbow, P. S. and Dallinger, R., Metal uptake, regulation, and excretion in freshwater invertebrates, in *Ecotoxicology of Metals in Invertebrates*, Rainbow, P. S. and Dallinger, R., Eds., Lewis Publishers, Boca Raton, FL, 1993, 119.
33. Rainbow, P. S., The significance of trace metal concentrations in marine invertebrates, in *Ecotoxicology of Metals in Invertebrates*, Rainbow, P. S. and Dallinger, R., Eds., Lewis Publishers, Boca Raton, FL, 1993, 3.

34. Capuzzo, J. M., Burt, W. V., Duedall, I. W., Park, P. K., and Kester, D. R., The impact of waste disposal in nearshore environments, in *Wastes in the Ocean*, Vol. 6, *Nearshore Waste Disposal*, Ketchum, B. H., Capuzzo, J. M., Burt, W. V., Duedall, I. W., Park, P. K., and Kester, D. R., Eds., John Wiley & Sons, New York, 1985, 3.
35. Bryan, G. W. and Langston, W. J., Bioavailability, accumulation, and effects of heavy metals in sediments with special reference to United Kingdom estuaries: a review, *Environ. Pollut.*, 76, 89, 1992.
36. Kennish, M. J., *Ecology of Estuaries*, Vol. 2, *Biological Aspects*, CRC Press, Boca Raton, FL, 1990.
37. van den Berg, C. M. G., Complex formation and the chemistry of selected trace elements in estuaries, *Estuaries*, 16, 512, 1993.
38. Turekian, K. K., The fate of metals in the ocean, *Geochim. Cosmochim. Acta*, 41, 1139, 1977.
39. Kennish, M. J., Trace metal-sediment dynamics in estuaries: pollution assessment, *Rev. Environ. Contam. Toxicol.*, 155, 73, 1998.
40. Beekman, R. L., Vernberg, F. J., and Vernberg, W. B., Sustainable development, in *Sustainable Development in the Southeastern Coastal Zone*, Vernberg, F. J., Vernberg, W. B., and Swiewicki, T., Eds., University of South Carolina Press, Columbia, SC, 1996, 1.
41. Shabman, L., Land settlement, public policy, and the environmental future of the southeast coast, in *Sustainable Development in the Southeastern Coastal Zone*, Vernberg, F. J., Vernberg, W. B., and Siewicki, T., Eds., University of South Carolina Press, Columbia, SC, 1996, 7.
42. National Research Council, *Restoration of Aquatic Ecosystems*, National Academy Press, Washington, D.C., 1992.
43. Kent, D. M., Ed., *Applied Wetlands Science and Technology*, Lewis Publishers, Boca Raton, FL, 1994.
44. Dahl, T. E., *Wetland Losses in the United States 1970s to 1980s*, Tech. Rept., U.S. Department of Interior, Fish and Wildlife Service, Washington, D.C., 1990.
45. Buchsbaum, R., Coastal marsh management, in *Applied Wetlands Science and Technology*, Kent, D. M., Ed., Lewis Publishers, Boca Raton, FL, 1994, 331.
46. Lewis, R., Enhancement, restoration, and creation of coastal wetlands, in *Applied Wetlands Science and Technology*, Kent, D. M., Ed., Lewis Publishers, Boca Raton, FL, 1994, 167.
47. Fonseca, M. S., Regional analysis of the creation and restoration of seagrass systems, in *Wetland Creation and Restoration: the Status of the Science*, Kusler, J. A. and Kentula, M. E., Eds., Island Press, Washington, D.C., 1990, 171.
48. Reimold, R. J., Wetlands functions and values, in *Applied Wetlands Science and Technology*, Kent, D. M., Ed., Lewis Publishers, Boca Raton, FL, 1994, 55.
49. Darnell, R. M., Impact of human modification on the dynamics of wetland systems, in *Wetland Functions and Values: the State of Our Understanding*, Greeson, P. E., Clark, J. R., and Clark, J. E., Eds., American Water Resources Association, Minneapolis, MN, 1979, 200.
50. Davies, P. H., Factors in controlling nonpoint source impacts, in *Stormwater Runoff and Receiving Systems*, Herricks, E. E., Ed., Lewis Publishers, Boca Raton, FL, 1995, 53.
51. Roberts, T. H., O'Neil, L. J., and Jabour, W. E., *Status and Source of Habitat Models and Literature Reviews*, Miscellaneous Paper EL-85-1, U.S. Army Corps of Engineers, Waterways Experiment Station, Vicksburg, MS, 1987.
52. Schroeder, R. L., *Community Models for Wildlife Impact Assessment: A Review of Concepts and Approaches*, Biol. Rept. 87(2), National Ecology Center, U.S. Fish and Wildlife Service, Washington, D.C., 1987.
53. USFWS, *Habitat Evaluation Procedures*, Tech. Rept., U.S. Fish and Wildlife Service, Washington, D.C., 1980.
54. Short, H. L., Development and use of a habitat gradient model to evaluate wildlife habitat, in *Transactions of the Forty-Seventh North American Wildlife and Natural Resources Conference*, Wildlife Management Institute, Washington, D.C., 1982, 57.
55. Adamus, P. R., Stockwell, L. T., Clairain, W. J., Jr., Morrow, M. E., Rozas, L. P., and Smith, R. D., *Wetland Evaluation Technique (WET), Volume 1: Literature review and Evaluation Rationale, Wetlands Research Program*, Tech. Rept. WRP-DE-2, U.S. Army Corps of Engineers, Waterways Experiment Station, Vicksburg, MO, 1991.

56. Johnson, A. R., Diagnostic variables as predictors of ecological risk, *Environ. Manage.*, 12, 515, 1988.
57. Costanza, R., Natural resource valuation and management: toward an ecological economics, in *Integration of Economy and Ecology: An Outlook for the Eighties*, Jansson, A. M., Ed., University of Stockholm Press, Stockholm, Sweden, 1984, 7.
58. Costanza, R. and Farber, S. C., *The Economic Value of Coastal Wetlands in Louisiana*, Tech. Rept., Louisiana State University, Baton Rouge, LA, 1985.
59. Raphael, C. N. and Jaworski, E., Economic value of fish, wildlife, and recreation in Michigan's coastal wetlands, *Coastal Zone Manage. J.*, 5, 181, 1979.
60. Mitchell, T. C. and Carson, R. T., *Using Surveys to Value Public Goods: The Contingent Valuation Method*, Resources for the Future, Washington, D.C., 1989.
61. Zedler, J. B., Salt marsh restoration: lessons from California, in *Rehabilitating Damaged Ecosystems*, Vol. 1, Cairns, J., Jr., Ed., CRC Press, Boca Raton, FL, 1988, 123.
62. Zentner, J., Enhancement, restoration, and creation of freshwater wetlands, in *Applied Wetlands Science and Technology*, Kent, D. M., Ed., Lewis Publishers, Boca Raton, FL, 1994, 127.
63. Thorhaug, A., Restoration of mangroves and seagrasses — economic benefits for fishes and mariculture, in *Environmental Restoration: Science and Strategies for Restoring the Earth*, Berger, J. J., Ed., Island Press, Washington, D.C., 1990, 265.
64. Kusler, J. A. and Kentula, M. E., Eds., *Wetland Creation and Restoration: The Status of the Science*, Island Press, Washington, D.C., 1990.
65. Gibson, K. D., Zedler, J. B., and Langis, R., Limited response of cordgrass (*Spartina foliosa*) to soil amendments in a constructed marsh, *Ecol. Appl.*, 4, 757, 1994.
66. Wilsey, B. J., McKee, K. L., and Mendelssohn, I. A., Effects of increased elevation and macro- and micronutrient additions on *Spartina alterniflora* transplant success in saltmarsh dieback areas in Louisiana, *Environ. Manage.*, 16, 505, 1992.
67. Havens, K. J., Varnell, L. M., and Bradshaw, J. G., An assessment of ecological conditions in a constructed tidal marsh and two natural reference tidal marshes in coastal Virginia, *Ecol. Eng.*, 4, 117, 1995.
68. USEPA, *The Index of Watershed Indicators*, EPA-841-R-97- 010, U.S. Environmental Protection Agency, Office of Water, Washington, D.C., 1997.
69. Koorse, S. J., The uncertainties of urban stormwater regulation, in *Stormwater Runoff and Receiving Systems: Impacts, Monitoring, and Assessment*, Herricks, E. E., Ed., Lewis Publishers, Boca Raton, FL, 1995, 245.
70. Baker, L. A., Introduction to nonpoint source pollution and wetland mitigation, in *Created and Natural Wetlands for Controlling Nonpoint Source Pollution*, Olson, R. K., Ed., C. K. Smoley, Boca Raton, FL, 1993, 7.
71. Pitt, R. E., Biological effects of urban runoff discharges, in *Stormwater Runoff and Receiving Systems: Impact, Monitoring, and Assessment*, Herricks, E. E., Ed., Lewis Publishers, Boca Raton, FL, 1995, 127.
72. Schueler, T. R. and Galli, J., The environmental impact of stormwater ponds, in *Stormwater Runoff and Receiving Systems: Impact, Monitoring, and Assessment*, Herricks, E. E., Ed., Lewis Publishers, Boca Raton, FL, 1995, 177.
73. Clifforde, I. T., Permitting of combined sewer overflows (CSO) — best management practices in the United Kingdom, in *Stormwater Runoff and Receiving Systems: Impact, Monitoring, and Assessment*, Herricks, E. E., Ed., Lewis Publishers, Boca Raton, FL, 1995, 425.
74. Pereira, W. E., Hostettler, F. D., and Rapp, J. B., Bioaccumulation of hydrocarbons derived from terrestrial and anthropogenic sources in the Asian clam, *Potamocorbula amurensis*, in San Francisco Bay estuary, *Mar. Pollut. Bull.*, 24, 103, 1992.
75. Nichols, F. H., Cloern, J. E., Luoma, S. N., and Peterson, D. H., The modification of an estuary, 231, 567, 1986.
76. Livingston, R. J., Trophic response of estuarine fishes to long-term changes of river runoff, *Bull. Mar. Sci.*, 60, 984, 1997.
77. Livingston, R. J., Niu, X., Lewis, F. G., III, and Woodsum, G. C., Freshwater input to a Gulf estuary: long-term control of trophic organization, *Ecol. Appl.*, 7, 277, 1997.

78. Hall, C. A. S., Howarth, R., Moore, B., III, and Vorosmarty, C. J., Environmental impacts of industrial energy systems in the coastal zone, *Annu. Rev. Energy*, 3, 395, 1978.
79. Wenzel, L. and Scavia, D., NOAA's coastal ocean program: science and solution, *Oceanus*, 36, 85, 1993.
80. Millemann, B., The national flood insurance program, *Oceanus*, 36, 6, 1993.
81. Klomp, W. H., Het Zwanenwater: a Dutch dune wetland reserve, in *Perspectives in Coastal Dune Management*, van der Meulen, F., Jungerius, P. D., and Visser, J. H., Eds., SPB Academic Publishing, The Hague, Netherlands, 1989, 305.
82. Silvester, R. and Hsu, J. R. C., *Coastal Stabilization: Innovative Concepts*, Prentice-Hall, Englewood Cliffs, NJ, 1993.
83. Orme, A. R., Energy-sediment interaction around a groin, *Zeit. Geomorph.*, 34 (Suppl.), 111, 1980.
84. Nordstrom, K. F., Beaches and dunes of human-altered coasts, *Prog. Geogr.*, 18, 497, 1994.
85. Nordstrom, K. F., Developed coasts, in *Coastal Evolution*, Carter, R. W. G. and Woodroffe, C. D., Eds., Cambridge University Press, Cambridge, UK, 1994, 477.
86. Everts, C. H., Beach behavior in the vicinity of groins — two New Jersey field examples, in *Coastal Structures 79*, American Society of Civil Engineers, New York, 1979, 853.
87. Nordstrom, K. F., Effects of shore protection and dredging projects on beach configuration near tidal inlets in New Jersey, in *Hydrodynamics and Sediment Dynamics of Tidal Inlets*, Aubrey, D. G. and Weishar, L., Eds., Springer-Verlag, New York, 1988, 440.
88. Dean, R. G., Sediment interaction at modified coastal inlets: processes and policies, in *Hydrodynamics and Sediment Dynamics of Tidal Inlets*, Aubrey, D. G. and Weishar, L., Eds., Springer-Verlag, New York, 1988, 412.
89. Hansen, M. and Knowles, S. C., Ebb-tidal delta response to jetty construction at three South Carolina inlets, in *Hydrodynamics and Sediment Dynamics of Tidal Inlets*, Aubrey, D. G. and Weishar, L., Eds., Springer-Verlag, New York, 1988, 364.
90. Nordstrom, K. F., Shoreline changes on developed coastal barriers, in *Cities on the Beach*, Research Paper 224, Platt, R. H., Pelczarski, S. G., and Burbank, B. K. R., Eds., University of Chicago, Department of Geography, Chicago, IL, 1987, 65.
91. Fischer, H. B., Ed., *Transport Models for Inland and Coastal Waters*, Academic Press, New York, 1981.
92. Dyke, P. P. G., Moscardini, A. O., and Robson, E. H., Eds., *Offshore and Coastal Modelling*, Springer-Verlag, New York, 1985.
93. Gross, M. G., *Oceanography: A View of the Earth*, 5th ed., Prentice-Hall, Englewood Cliffs, NJ, 1990.
94. Jouzel, J., Raisbeck, J. P., Benoist, F. Y., Lorius, C., Raynaud, D., Petit, J. R., Barkov, N. I., Korotkevitch, Y. S., and Kotlyakov, V. M., A comparison of deep Antarctic cores and their implications for climate between 65,000 and 15,000 years ago, *Quat. Res.*, 31, 135, 1989.
95. Penland, S. and Ramsey, K. E., Relative sea level rise in Louisiana and the Gulf of Mexico: 1908-1988, *J. Coastal Res.*, 6, 323, 1990.
96. Warrick, R. A. and Oberlemans, J., Sea level rise, in *Climate Change, the IPCC Scientific Assessment*, Houghton, J. T., Jenkins, G. J., and Ephraums, J. J., Eds., Cambridge University Press, Cambridge, UK, 1990, 257.
97. Gornitz, V. M., Daniels, R. C., White, T. W., and Birdwell, K. R., The development of a coastal risk assessment database: vulnerability of sea level rise in the U.S. southeast, in *Coastal Hazards: Perception, Susceptibility, and Mitigation*, Finkl, C. W., Jr., Ed., *J. Coastal Res.*, Special Issue 12, The Coastal Education and Research Foundation, Fort Lauderdale, FL, 1994, 327.
98. National Academy of Sciences, *CO_2 and Climate: A Scientific Assessment*, National Academy Press, Washington, D.C., 1979.
99. Carlton, J. T., Introduced invertebrates of San Francisco Bay, in *San Francisco Bay: The Urbanized Estuary*, Conomos, T. J., Ed., Pacific Division, American Association for the Advancement of Science, San Francisco, CA, 1979, 427.
100. Cohen, A. N. and Carlton, J. T., Accelerating invasion rate in a highly invaded estuary, *Science*, 279, 555, 1998.

101. Nichols, F. H. and Thompson, J. K., Persistence of an introduced mudflat community in South San Francisco Bay, California, *Mar. Ecol. Prog. Ser.*, 24, 83, 1985.
102. Conomos, T. J., Ed., *San Francisco Bay: The Urbanized Estuary,* Pacific Division, American Association for the Advancement of Science, San Francisco, CA, 1979.
103. Herbold, B. and Moyle, P. B., *The Ecology of the Sacramento-San Joaquin Delta: A Community Profile,* Biol. Rept. No. 85(7.22), U.S. Fish and Wildlife Service, Slidell, LA, 1989.
104. Skinner, J. E., *An Historical Review of the Fish and Wildlife Resources of the San Francisco Bay Area,* Water Projects Branch Rept. No. 1, California Department of Fish and Game, Sacramento, CA, 1962.
105. Jennings, M. R. and Hayes, M. P., Pre-1900 overharvest of California red-legged frogs (*Rana aurora draytonii*): the indictment for bullfrog (*Rana catesbiana*) introduction, *Herpetologica,* 41, 94, 1985.
106. Hayes, M. P. and Jennings, M. R., Decline of ranid frog species in western North America: are bullfrogs (*Rana catesbiana*) responsible?, *J. Herp.*, 20, 490, 1986.
107. Kimsey, J. B., Fisk, L. O., and McGriff, D., *The Crayfish of California,* Inland Fish Inf. Pamphl. No. 1, California Department of Fish and Game, Sacramento, CA, 1982.
108. Spicher, D. and Josselyn, M. N., *Spartina* (Gramineae) in northern California: distribution and taxonomic notes, *Madrono,* 32, 158, 1985.
109. National Estuary Program, *Background Papers and Management Approaches Summary,* Tech. Rept., San Francisco Bay Workshop, U.S. Environmental Protection Agency, San Francisco, CA, 1997.

chapter two

Case study 1: Long Island Sound Estuary Program

I. Introduction

Prior to the establishment of the National Estuary Program (NEP), Congress appropriated funds in 1985 for the states of Connecticut and New York as well as the U.S. Environmental Protection Agency (USEPA) to research, assess, and monitor the water quality of Long Island Sound. Soon after the enactment of the amended Clean Water Act of 1987, Long Island Sound was designated an Estuary of National Significance and a Management Conference was convened on the sound in March 1988. This conference identified six priority areas of environmental concern that merited further investigation: (1) low dissolved oxygen (hypoxia); (2) toxic contamination; (3) pathogen contamination; (4) floatable debris; (5) the impact of these water quality problems as well as habitat loss and degradation on the health of living resources; and (6) land use and development resulting in habitat loss and degradation of water quality. Low dissolved oxygen was deemed to be the most pressing problem in Long Island Sound.[1]

Focusing on the aforementioned priority issues, a Comprehensive Conservation and Management Plan (CCMP) was developed to protect and improve the health of the sound while ensuring compatible human uses within the system. The CCMP includes specific commitments and recommendations for actions to enhance water quality and protect habitat and living resources. It also proposes actions to educate and involve the public, to improve management of the system, and to monitor its progress.

A. Goals

The Long Island Sound Estuary Program (LISEP) has several well-defined goals:

- To protect and improve the water quality of Long Island Sound and its coves and embayments to ensure that a healthy and diverse living resource community is maintained
- To ensure that health risks associated with human consumption of shellfish and finfish are minimized
- To ensure that opportunities for water-dependent recreational activities are maximized without conflict with ecosystem management
- To ensure that social and economic benefits associated with the use of Long Island Sound are realized to the fullest extent possible, consistent with social and economic costs

- To preserve and enhance the physical, chemical, and biological integrity of Long Island Sound and the interdependence of its ecosystems
- To establish a water quality policy that supports both the health and habitats of the living resources of Long Island Sound and the active and passive recreational and commercial activities of people[1]

The realization of these goals will require the cooperation and concerted effort of government agencies, nongovernment organizations (e.g., civic, conservation, and environmental groups), educational institutions, various stakeholders (e.g., commercial and recreational fishermen, developers, commercial establishments, and industries), and the general public.

II. Physical description

Long Island Sound is a large embayment extending from Hell Gate in the East River at New York City to The Race marked by a chain of islands (e.g., Plum and Fishers Islands) located ~204 km east of the city (Figure 2.1). Long Island borders Long Island Sound on the south, and Westchester County, New York, and the Connecticut shoreline bound it on the north. The mean width of the sound is ~28 km; the maximum width approaches 40 km.[2] Its mean and maximum depths are 20 m and 90 m, respectively.[3]

With a surface area of 3284 km^2, Long Island Sound is the sixth largest estuary in the U.S. It ranks third in terms of total volume (6.2×10^{10} m^3). The land drainage area, amounting to 44,652 km^2, ranks 19th. More than 70% of the total area (i.e., 28,749 km^2) lies in the drainage basin of the Connecticut River.[4] However, the entire watershed of Long Island Sound encompasses portions of several states, including Connecticut (entire state) and parts of New York, Massachusetts, New Hampshire, Vermont, and Canada.

Freshwater enters Long Island Sound from runoff and drainage along coastal areas of Long Island and Connecticut. The discharge of four major tributaries in Connecticut (i.e., Housatonic, Quinnipiac, Connecticut, and Thames Rivers) comprises most of the freshwater input, with the Connecticut River alone being responsible for ~70% of the total inflow. Saline water enters at the western end of the sound from New York Harbor via a tidal strait, the East River. The inflow of deeper oceanic water occurs at the eastern margin of the embayment through Block Island Sound. The Eastern Sill, located at ~72°30′W longitude, rises to a depth of only 21 m where it modulates the volume of deep oceanic seawater input to the western sound. While deeper, more saline waters exhibit a net westerly flow, fresher surface waters move generally eastward. The circulation and mixing patterns throughout the system are complex and not completely understood.

The Long Island Sound watershed is heavily populated. More than 8 million people live in the watershed year round, although the total population expands significantly during the summer tourist season, especially in coastal counties around the sound. The population is not evenly distributed. New York City, which accounts for 42% of the population, covers only 0.4% of the land area. Connecticut, with 33% of the land area, has 37% of the population. New Hampshire, Massachusetts, and Vermont contribute 12.5% of the population, and Westchester, Nassau, and Suffolk Counties in New York contribute an additional 8.3%.[1] The population is projected to exceed 9.4 million by the year 2010.[5] More than 75% of the population increase is expected to occur in Suffolk County on Long Island and the coastal counties of Connecticut and Rhode Island.[3]

The population of the coastal counties around Long Island Sound experienced a dramatic increase from 1950 to 1980. Between 1960 and 1970 alone, the population in these counties increased by nearly 16%. Concomitantly, residential and commercial development expanded rapidly. Table 2.1 summarizes land use in the estuarine drainage area.[6]

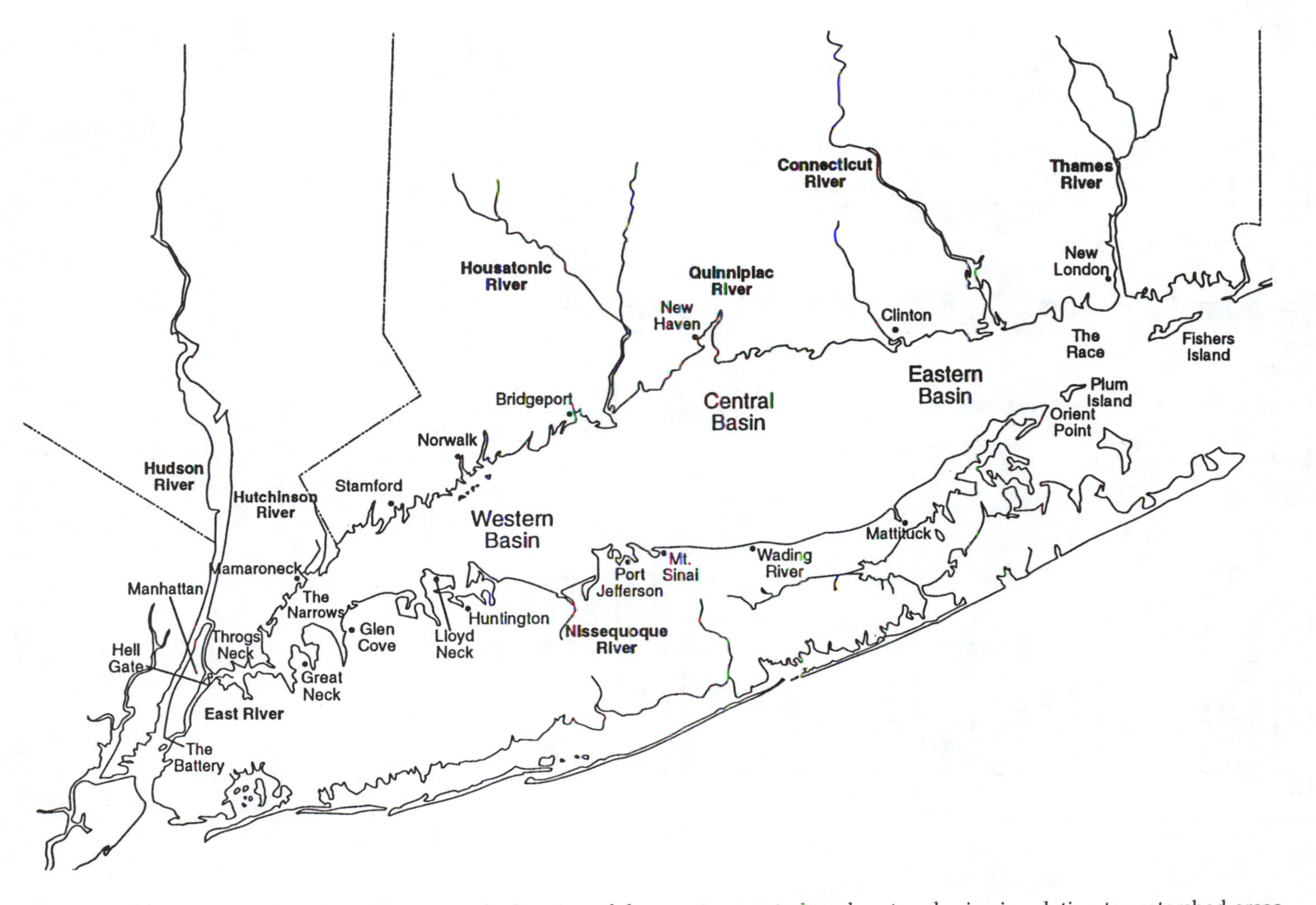

Figure 2.1 Map of Long Island Sound showing the location of the western, central, and eastern basins in relation to watershed areas. (From Strieb, M. and the Living Marine Resources Work Group, *Assessment of Living Marine Resources*, Comprehensive Conservation and Management Plan Supporting Document, U.S. Environmental Protection Agency, New York, 1993.)

Table 2.1 Land Use in the Estuarine Drainage Area (EDA) of Long Island Sound

Land-Use Categories	Area (km²)	Percent
Urban and built up	3,698	25.09
Residential	2,558	17.35
Commercial services	443	3.01
Industrial	140	0.95
Transportation/communication	161	1.09
Industrial/commercial complex	13	0.09
Mixed urban	96	0.65
Other urban	287	1.95
Agriculture	2,128	14.44
Cropland/pasture	2,074	14.07
Other	54	0.37
Range (shrub/brushland)	135	0.92
Forest	8,134	55.18
Deciduous	7,060	47.90
Evergreen	582	3.95
Mixed	492	3.34
Wetland	435	2.95
Forested	295	2.00
Nonforested	140	0.95
Barren	211	1.43
Beaches	2.6	0.02
Other sandy areas	<2.6	<0.02
Mines/quarries	127	0.86
Transitional areas	80	0.54
Total	14,740	100.00

Source: Strategic Assessments Branch, *National Estuarine Inventory Data Atlas, Vol. 2: Land Use Characteristics*, Ocean Assessments Division, National Oceanic and Atmospheric Administration, Rockville, MD, 1987.

Long Island Sound generates ~$5 billion annually for the regional economy from commercial and recreational fishing, boating, swimming, and various tourist activities. Cargo shipping, ferry transportation, and electric power generation add to this total. The sound also has considerable aesthetic value that cannot be assessed purely in terms of economics but nevertheless is of great importance to residents and tourists who have access to the system.

III. Priority problems

Long Island Sound is a highly productive ecosystem characterized by a range of distinctive habitats that support numerous aquatic and wildlife populations. Wetlands and flats, beaches and dunes, bluffs and rocky intertidal zones, natural and artificial reefs, and submerged aquatic vegetation and unvegetated estuarine seafloor provide habitat for plankton, benthic flora and fauna, shellfish, finfish, amphibians, reptiles, mammals, and birds. These biotic groups use the habitats for feeding, nesting, reproduction, and shelter. Habitat loss and alteration together with water quality degradation not only threaten living resources of the sound but also the productivity of the entire system.

Long Island Sound has a long history of pollution problems. Not until the early 1970s, however, was a concerted effort mounted to address water quality issues. At that time, the Nassau-Suffolk Regional Planning Board formulated a wastewater management program that was designed to identify and manage water quality problems. As recounted by Koppelman et al.,[6] the following impacts on water quality were identified for study: (1) extensive and uncontrolled development in watershed areas leading to increased non-point source pollution and impaired water quality in Long Island Sound; (2) urban storm-water runoff; (3) disposal of inadequately treated wastewater by New York City resulting in surface water degradation in the sound; (4) high coliform bacteria counts causing the closure of bathing beaches and the prohibition of shellfish harvesting; (5) nutrient loadings contributing to adverse ecological effects, notably eutrophication of estuarine waters; and (6) alteration of salinity regimes due to decreased freshwater inflow.

Water pollution control programs implemented after enactment of the Clean Water Act measurably improved water quality. Major point sources of pollution to Long Island Sound in the early 1980s included 86 wastewater treatment plants, 255 industrial facilities discharging directly into the estuarine drainage area, 16 electric generating stations, and 14 water treatment plants.[3] However, these sources of pollution are now more tightly regulated and controlled through permit programs, which have helped to restore degraded waters in the sound. Despite this progress, many water quality and habitat problems persist; most can be traced to residential, commercial, and recreational development in the watershed and to human activities along the shoreline and in the estuarine basin.

Development and pollution have caused the alteration and loss of habitat, resulting in diminished productivity of many wetland areas. The input of inadequately treated human sewage as well as domestic and wild animal wastes has increased pathogenic contamination. Low dissolved oxygen levels have likewise impaired water quality, most significantly from mid- to late summer. Toxic substances have contaminated certain bottom habitats. Greater recreational and commercial activities have generated more floatable debris along the shoreline and in open waters. Management strategies have been developed to address all of these problems.

A. Hypoxia

1. Spatial distribution

Oxygen depletion is an ongoing problem in Long Island Sound. Analysis of historical data reveals a general east-to-west gradient of decreasing bottom dissolved oxygen concentrations.[7] Dissolved oxygen levels in bottom waters often fall below 5 mg/l during the mid-July through September period; they typically range from 5 to 9 mg/l in surface waters but may drop to 3 to 4 mg/l in the western narrows and the East River. The current water quality standard for dissolved oxygen is 5 mg/l in New York State and 6 mg/l in Connecticut. Hypoxia, defined by the LISEP as dissolved oxygen concentrations $\leq$3 mg/l, has been a recurrent problem in the deeper waters of the central and western basins as well as in some shallow embayments. It has been observed in bottom waters as far east as Mattituck, New York, and New Haven, Connecticut.

Mild hypoxic conditions were documented in the western basin during the 1950s. However, the severity and widespread nature of the problem were not confirmed until intensive monitoring was conducted from 1986 through 1993. Results of this monitoring effort showed that 50 to 66% of Long Island Sound bottom waters experienced dissolved oxygen levels <5 mg/l during summer from 1987 through 1993 (Figure 2.2). Conditions appeared to be most severe during the summer of 1989 when more than 1300 km^2 (40%) of bottom waters in the sound had dissolved oxygen concentrations of <3 mg/l, and up to 130 km^2 had concentrations of <1 mg/l. The lowest dissolved oxygen levels (0 mg/l), however, were recorded near Hempstead Harbor during late July and August in 1987.[1,8]

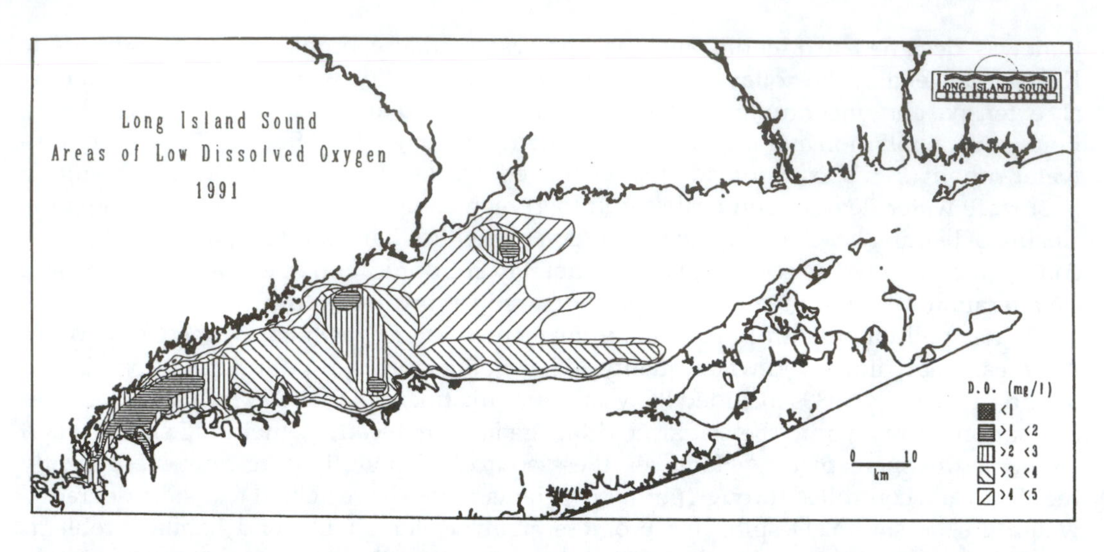

Figure 2.2 Minimum dissolved oxygen levels (<5 mg/l) in the bottom waters of Long Island Sound during the summer of 1991. (From U.S. Environmental Protection Agency, *The Long Island Sound Study: Summary of the Comprehensive Conservation and Management Plan*, Tech. Rept., U.S. Environmental Protection Agency, Stony Brook, NY, 1994.)

2. *Biotic impacts*

Hypoxia is detrimental to biotic communities. When dissolved oxygen levels chronically fall below 3 mg/l, estuarine and marine organisms usually incur physiological and behavioral problems. Anoxic conditions (dissolved oxygen ~0 mg/l) nearly always culminate in massive resource mortalities and consequently have significant adverse ecological effects on bottom communities and habitats. Persistently low dissolved oxygen concentrations in the bottom waters of the central and western basins during late summer may result in the following biotic impacts: (1) an increase in mortality of early life stages of some fish populations (e.g., bay anchovy, *Anchoa mitchilli*; Atlantic menhaden, *Brevoortia tyrannus*; cunner, *Tautogolabrus adspersus*; tautog, *Tautoga onitis*; and sea robin, *Prionotus evolans*), lobster (*Homarus americanus*), starfish (*Asterias forbesi*), and various benthic infaunal species; (2) a reduction in the abundance and diversity of adult fish; (3) a decline in the growth rate of newly settled lobsters (*H. americanus*) and possibly juvenile winter flounder (*Pseudopleuronectes americanus*); (4) a decrease in disease resistance of exposed organisms; and (5) a diminution in habitat value.[8]

A number of specific biotic impacts of oxygen depletion have been chronicled. The Connecticut Department of Environmental Protection, conducting trawl sampling in the sound, found significantly reduced catches of fish in waters severely depleted in dissolved oxygen. For example, the total mean catch per tow of fish increased from 38 at sites with dissolved oxygen concentrations of 1 to 2 mg/l to 300 at sites with dissolved oxygen concentrations of >2 mg/l. The number of fish species per tow also increased from 1.6 at sites with dissolved oxygen levels of <1 mg/l to 11 at sites with dissolved oxygen levels of >3 mg/l. In addition, the relative abundance of fish also declined by 40% in a 780 km^2 area where dissolved oxygen levels were <4 mg/l. The greatest impacts were evident in demersal species (e.g., winter flounder, *Pseudopleuronectes americanus*; sea robins, *Prionatus evolans*; and skates, *Raja* spp.), which had biomass values (kg/tow) reduced by more than 50% in hypoxic areas. Trawl catches in the Hempstead area (both total catch and number of species) were ~50% lower than those in less impacted waters of central Long Island Sound. Similar reductions in abundance and diversity of benthic invertebrates were noted in oxygen-depleted waters.

Laboratory studies conducted by the Office of Research and Development of the USEPA on various life stages of fish and crustaceans from the sound support the findings of the Connecticut field surveys. These studies, which focused on the sensitivity of eggs, larvae, and juveniles of target organisms (e.g., mud crabs, *Rithropanopeus harrisi*; lobster, *Homarus americanus;* and grass shrimp, *Palaemontes vulgaris*) to low dissolved oxygen concentrations, revealed that impacts are significant when dissolved oxygen levels drop below 3.5 mg/l. The most severe effects were manifested at dissolved oxygen concentrations of <2 mg/l; however, some effects were also observed at levels between 3.5 and 5.0 mg/l.

Fish kills associated with low dissolved oxygen levels are relatively common in the harbors and bays of Long Island Sound. For example, Greenwich, Stamford, Norwalk, Bridgeport, New Haven, Hempstead, and Cold Spring Harbors, Manhassett and Oyster Bays, and some East River tributaries have been sites of fish kills involving Atlantic silversides (*Menidia menidia*), Atlantic menhaden (*Brevoortia tyrannus*), and winter flounder (*Pseudopleuronectes americanus*). Hypoxia in the main body of the sound has also caused fish kills. Dead crabs, lobsters, starfish, and other invertebrates have been reported during hypoxic events in the western sound. A crash of the benthic community in this area between the summer of 1972 and spring of 1973 may have been due to an hypoxic event.[1]

Hypoxia may be responsible for major changes in benthic community structure. Bivalves, which tend to be less sensitive to low dissolved oxygen than crustaceans, echinoderms, and worms, are the most abundant members of the benthic community in the western sound, where hypoxia is relatively severe. Crustaceans and worms, in contrast, are more abundant in the central and eastern basins.

Hypoxia can change benthic community structure not only by modulating the abundance and distribution of constituent populations but also by altering trophic interactions. Mobile species typically move away from oxygen-deficient areas. Nonmotile forms, however, remain essentially in place, commonly incurring serious physiological impacts and impaired reproduction. Increased mortality of forage-based species, which serve as important food sources for organisms higher in the food chain, can be detrimental to the entire community structure.

3. *Causes of hypoxia*

Nutrient enrichment in Long Island Sound is closely coupled to the occurrence of hypoxic events. High nitrogen inputs spur rapid phytoplankton growth and periodic blooms that contribute to large organic loads. Bacterial decomposition of organic matter below the pycnocline, which forms in summer, promotes oxygen depletion in bottom waters. Hypoxia development is a cumulative process through the summer months corresponding to a period of thermally controlled stratification of estuarine waters. A highly stratified water column associated with the formation of the pycnocline exacerbates oxygen depletion because it prevents the mixing of surface and bottom waters. Hence, although the surface waters remain well oxygenated as a result of atmospheric and photosynthetic inputs of oxygen, bottom waters become progressively depleted in oxygen. This condition often spreads over extensive areas of the bottom until surface waters cool in the early fall, the density gradient declines, and the pycnocline breaks down.

Deep-water oxygen depletion in summer is well known in fjord-type estuaries and in some larger coastal plain systems. In the deeper waters of central Chesapeake Bay (below 10 m), for example, oxygen levels decrease to near zero in summer. Increased stratification of the water column in February and March induced by the spring freshet from the Susquehanna River minimizes vertical oxygen transfer that may lead to hypoxia or anoxia during the period from May to September.[9,10] As in the case of Long Island Sound, severe oxygen depletion in bottom waters of Chesapeake Bay commonly persists until water column mixing increases in late summer and early fall.

The genesis of hypoxia in Long Island Sound can be traced to nitrogen inputs from both natural and anthropogenic sources. Natural influx of nitrogen to the sound mainly occurs via groundwater transport, riverine inflow, and atmospheric deposition. Major anthropogenic sources of nitrogen are sewage treatment plants, faulty septic systems, agricultural and lawn fertilizers, and automobile emissions. The total nitrogen load to the sound in 1992 amounted to 84,914 metric tons (mt), with 36,197 mt (43%) derived from natural sources and the remaining load of 48,717 mt (57%) from anthropogenic sources. Natural sources of nitrogen are not subject to management controls; consequently, the LISEP has targeted those human activities most amenable to management actions for priority attention.

Of the estimated 90,629 mt/yr of nitrogen currently entering Long Island Sound, 37,558 mt/yr are from natural sources and 53,071 mt/yr are from anthropogenic sources. About 35,381 mt/yr of nitrogen originate from human activities in the drainage basin. Point source discharges (primarily sewage treatment plants) deliver 33,566 mt/yr of nitrogen to the sound, and nonpoint source discharges, such as agricultural and stormwater runoff, contribute another 1814 mt/yr. Atmospheric deposition, in turn, yields 5897 mt/yr of nitrogen; 3357 mt/yr of this total falls directly onto the estuarine surface and 2540 mt/yr onto the watershed. About 9707 mt/yr of nitrogen enter the sound across its boundaries through the East River and The Race. This represents the combined input from natural sources as well as from human activities in areas that drain to New York-New Jersey Harbor and Block Island Sound. Connecticut tributaries transport 2087 mt/yr of nitrogen from activities north of the state line. A breakdown of nitrogen loads from anthropogenic sources is as follows: 63.2% from point sources, 18.3% across boundaries, 6.3% from direct atmospheric deposition, 4.8% from indirect atmospheric inputs, 3.8% from tributary import, and 3.4% from nonpoint sources.

Nitrogen occurs in both inorganic (i.e., nitrate, nitrite, and ammonia) and organic forms, although phytoplankton preferentially use inorganic nitrogen for their growth. Organic nitrogen bound in live and dead plant and animal tissues is not available for plant growth until decomposition processes release a usable, inorganic form of the element. Thus, nitrogen cycled through the food chain and from bottom sediments is important in the generation of nutritionally useful nitrogen for the sound.

4. *Management strategies*

Water quality models predict that reductions of nitrogen inputs to Long Island Sound will decrease phytoplankton production, increase dissolved oxygen levels in bottom waters, and decrease the probability of hypoxic events. As a result, the management program has implemented a three-phase approach to reduce nitrogen loads to the sound from point and nonpoint source discharges in the watershed. The long-term goal is to substantially reduce the duration of hypoxic events and the overall area in the sound affected by hypoxia.

The Phase 1 actions to improve dissolved oxygen levels in Long Island Sound were approved by the Policy Committee of LISEP in December 1990. Phase 1 established a *no-net-increase* policy for controlling nitrogen discharges by *freezing* nitrogen loadings in critical areas at 1990 levels to preclude a worsening of hypoxia. Several major agreements and objectives to control point and nonpoint sources were achieved during Phase 1:

- Permits regulating critical point source dischargers were modified to cap nitrogen loads at 1990 levels.
- New York City and the New York State Department of Environmental Conservation reached full agreement on sewage treatment plant limits, which froze total nitrogen loadings at 1990 levels.

- Existing nonpoint source and stormwater management programs focused on nitrogen control with the objective of freezing the loads.
- Highest priorities for management control were assigned to coastal sub-basins where nitrogen loads from nonpoint sources were estimated to be the highest.
- It was agreed that tributary loads to the Long Island Sound must be assessed to begin planning for their control.

One approach to assess nutrient loads is the application of water quality models. LIS 2.0, a two-dimensional water quality model that approximates the biological and chemical processes of the sound, shows that the minimum dissolved oxygen levels in the bottom waters of the system during late summer can be increased to an average of ~3.5 mg/l by substantially reducing nitrogen loadings from sources within the drainage basin.[8] Such an increase will significantly reduce hypoxia in the sound.

The Management Conference of the LISEP has established interim dissolved oxygen target levels. In waters that currently meet state standards, the objective is to maintain existing dissolved oxygen concentrations. In waters that do not meet state standards but where dissolved oxygen levels exceed 3.5 mg/l, the objective is to increase the levels to meet the standards. In waters with dissolved oxygen levels of <3.5 mg/l, the objective is to ensure that dissolved oxygen concentrations do not fall below 1.5 mg/l while attempting to raise short-term dissolved oxygen levels to 3.5 mg/l.

Phase 2 management actions focused on the process of reducing nitrogen discharges to Long Island Sound. An initial step was to adopt the aforementioned interim targets for improving dissolved oxygen levels. To achieve these targets, Connecticut and New York made commitments to begin low-cost reduction of nitrogen discharges to the sound. The two states agreed to remove ~6865 mt/yr of the total in-basin, human-induced nitrogen load, which in 1992 amounted to 37,014 mt. Management strategies of point source reductions have concentrated on the application of retrofits and a variety of other low-cost nitrogen removal technologies at selected sewage treatment plants. Connecticut agreed to spend about $18.1 million to reduce the aggregate annual nitrogen load from 15 sewage treatment plants by nearly 816 mt (~25%) by 1995. New York State agreed to spend $103.1 million to reduce the aggregate annual nitrogen load from 11 sewage treatment plants by 6048 mt (~25%). As a result of the rigorous measures undertaken by both Connecticut and New York, the total nitrogen load to Long Island Sound from point and nonpoint sources within the watershed decreased from 39,000 mt in 1992 to 34,100 mt in 1996, representing a 12.6% decline.

Phase 2 actions for controlling nonpoint sources of nitrogen are directed at priority coastal sub-basins. Coastal nonpoint pollution control programs have been developed to address this issue. Both Connecticut and New York have agreed to implement the actions necessary to achieve *no net increase* of nitrogen loads from nonpoint sources. They are also utilizing existing programs to meet the broader pollution control objective of reducing nitrogen loads from nonpoint sources (Table 2.2).

Phase 2 nitrogen load reductions will reap two kinds of benefits for the living resources of Long Island Sound. As demonstrated by the LIS 2.0 simulation, the minimum dissolved oxygen concentrations in the bottom waters of the sound in summer will be raised on average 0.9 mg/l (from 1.5 to 2.4 mg/l) by applying the Phase 2 management actions. These improved oxygen conditions will decrease the amount of degradable estuarine habitat by ~10%. In addition, the actions will yield more than a 30% reduction in the area of the sound most severely affected by hypoxia.[1,8]

Results of Phase 1 and 2 agreements, although significantly lowering nitrogen loads, will neither meet the interim dissolved oxygen targets nor achieve the long-term dissolved oxygen goal. Therefore, additional nitrogen reduction is required to attain the ultimate

Table 2.2 Actions to Reduce Nitrogen Loads in Long Island Sound from Nonpoint Sources

Ongoing Programs	Responsible Parties/Status
The states of Connecticut and New York will continue to use their existing authority to manage nonpoint source pollution and appropriate federal grants such as Clean Water Act Section 319, 604(b), and 104(b) to carry out projects that will help prevent increases and, to the extent possible, achieve rductions in the nonpoint source nitrogen loads from high priority drainages identified in the Connecticut and New York portions of the Long Island Sound watershed.	Both the CTDEP and the NYSDEC will use these programs to continue to manage nonpoint sources of nitrogen. Nonpoint source management annual program costs, statewide, are $2.5 million in Connecticut.
The states of Connecticut and New York are developing their coastal nonpoint source control programs, as required by Section 6217 of the Coastal Zone Management Act.	These efforts were initiated in 1992 by the CTDEP and the NYDOS to implement requirements of Section 6217. The effort is funded at about $250,000 per year for both states combined. The states are using their programs to address nonpoint nitrogen control.
The states of Connecticut and New York will continue to implement general stormwater permit progams to control the discharge of stormwater from industrial, construction, and municipal activities, in accordance with the EPA's national program regulations. These permits will regulate discharges from construction activity greater than five acres and from eleven industrial categories.	These base programs run by the CTDEP and the NYSDEC, at a staff commitment cost of about $300,000 per year, provide a mechanism for controlling nonpoint sources of nitrogen from key urban sources.
The states of Connecticut and New York will continue to implement their existing permitting programs, such as the inland and tidal wetlands programs, to address nonpoint nutrient control with respect to Long Island Sound management needs, as appropriate.	General permitting programs for tidal and inland wetlands, run by the CTDEP and the NYSDEC, protect vital natural functions of nitrogen and other pollutant removal that wetlands afford. The CTDEP spends about $7 million per year on nonpoint source and wetland management.
The states of Connecticut and New York will implement the requirements of the reauthorized Clean Air Act to achieve additional nitrogen emission controls. Major actions include reduction of nitrous oxide emissions through adoption of statewide enhanced vehicle inspection and maintenance programs and stricter emission controls for stationary sources such as power plants.	Both the CTDEP and the NYSDEC are implementing aggressive emission control programs as part of the federal Clean Air Act that will reduce atmospheric loadings of nitrogen to the sound. The cost of these new initiatives specific to nitrogen control has not been estimated.

Commitments	Responsible Parties	Time Frame	Estimated Cost
The EPA will make nonpoint source management of nitrogen and other pollutants identified by the LISS, through wetlands and riparian zone protection as well as best management practices implementation, high priorities for fiscal year 1994 funding under Sections 319, 104(b), and 604(b) of the Clean Water Act.	EPA	1993–1994	Redirection of base program
Investigate expansion of stormwater permitting programs to regulate communities with populations fewer than 100,000 that border Long Island Sound within high priority management zones.	CTDEP, NYSDEC	1994	Redirection of base program
In cooperation with New York State, Westchester County is developing a nonpoint source management plan that will include implementing best management practices for nonpoint source nitrogen control, monitoring their effectiveness and establishing a Westchester County management zone (or bubble) for assessing compliance with the nitrogen load freeze. The LISS will explore extending the bubble concept to other management zones throughout Connecticut and New York State portions of the Long Island Sound drainage.	NYSDEC, Westchester County	1993–1996	$500,000 one-time cost
Westchester County will implement the recommendations of the County Executive's Citizen Committee on Nonpoint Source Pollution in Long Island Sound.	Westchester County, Local Government	1993 initiation and continuing	$500,000 per year, $200,000 per year for the first 3 years, $600,000 for implementation
Point and nonpoint nitrogen load estimates will be made in the City of Stamford to assess feasibility of a point/nonpoint source *trading* program. A cost-effective mix of management options will be proposed that may be used to help decide how nitrogen reduction targets can be met once they are established.	CTDEP, City of Stamford	1992–1994	$87,000 one-time planning effort
New York State will pursue the expansion of the State Building Code to include provisions for erosion and sediment control and stormwater practices for all construction activities in order to prevent increases in nonpoint nitrogen runoff.	NYDEC, NYSDOS	1993–1994	Redirection of base program
Provide technical assistance to coastal municipalities to address impact of hypoxia in their municipal regulations and plans of development, as required by state law.	CTDEP	1993 and continuing	Redirection of base program
Advocate the use of the June nitrate test on agricultural lands to ensure that fertilizer applications to crops do not exceed crop needs.	CTDEP, NYSDEC	1993 and continuing	Redirection of base program

Table 2.2 (continued) Actions to Reduce Nitrogen Loads in Long Island Sound from Nonpoint Sources

Recommendations	Responsible Parties	Time Frame	Estimated Cost
In addition to continuing general stormwater permitting programs, the state of New York should determine if the general permit adequately regulates nitrogen from activities subject to national stormwater regulations.	NYSDEC	—	$50,000
Explore the expansion of current requirements for federally licensed or permitted projects to obtain a water quality certification in New York to protect water quality from sources of pollution to include all projects adjacent to wetlands and other sensitive areas (e.g., adjacent to wetlands) or those that exceed a minimum size (e.g., greater than one acre).	NYSDEC	1994–1995	$50,000
The states of Connecticut and New York should develop a habitat restoration plan that includes a list of potential project sites and priorities. Wetland projects that are in close proximity to priority nitrogen management areas should be highlighted.	CTDEP, NYSDEC, NYSDOS	—	See Chapter VII, "Management and Conservation of Living Resources and Their Habitats."
Evaluate Maryland's Critical Areas regulations and the reported nutrient reduction benefits and make recommendations of the potential value of a similar program for Long Island Sound.	LISS	1993–1995	$50,000

Source: U.S. Environmental Protection Agency, Long Island Sound Study Comprehensive Conservation and Management Plan, Final Report, U.S. Environmental Protection Agency, Stony Brook, NY, 1994.

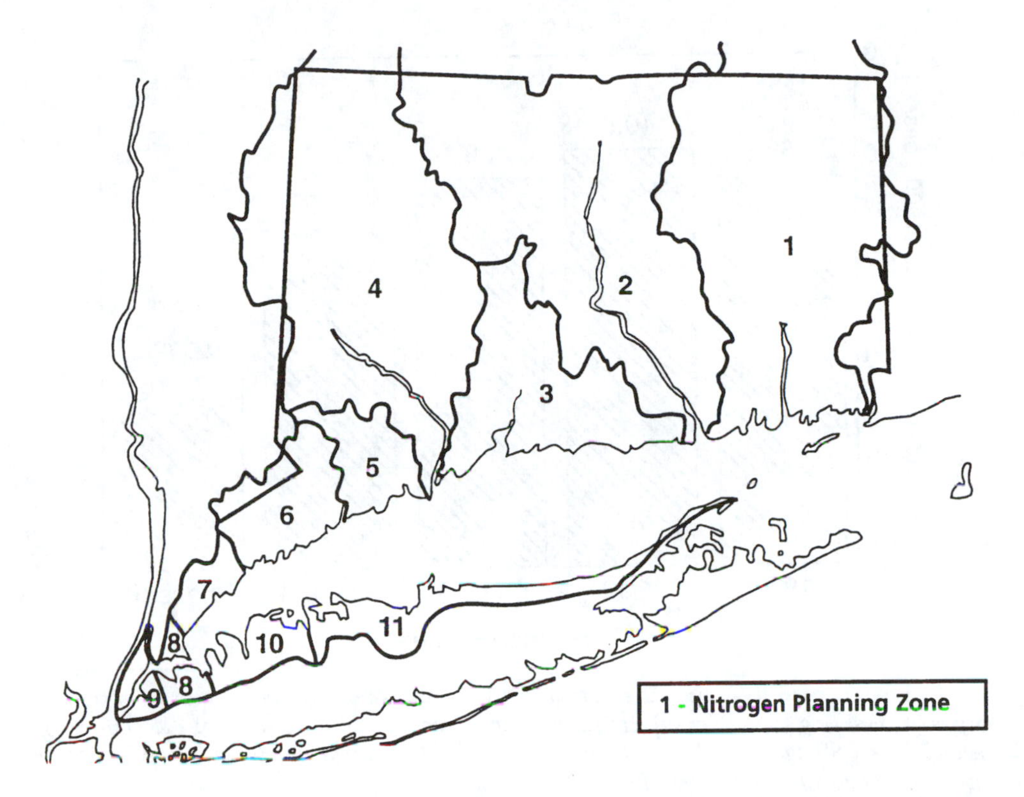

Figure 2.3 Watershed geographic zones established by the Long Island Sound Estuary Program for managing total nitrogen loads to Long Island Sound: (1) Thames; (2) Connecticut; (3) Quinnipiac; (4) Housatonic; (5) Saugatuck; (6) Norwalk; (7) Westchester; (8) Bronx/Queens; (9) Manhattan/Queens; (10) Nassau; (11) Suffolk. (From U.S. Environmental Protection Agency, *Long Island Sound Study: Proposal for Phase III Actions for Hypoxia Management*, EPA 840-R-001, U.S. Environmental Protection Agency, Stony Brook, NY, 1997.)

goal of eliminating hypoxic impacts caused by anthropogenic activity. This reduction will be accomplished during Phase 3, which has set specific nitrogen targets for each of 11 watershed management zones (Figure 2.3). These zones have been established to facilitate the identification of nitrogen sources and the development of comprehensive watershed plans.

The LIS 3.0 model is a key element of Phase 3 actions to further control nitrogen loads. It was developed by coupling a hydrodynamic model, which was completed by the National Oceanic and Atmospheric Administration (NOAA) in 1993, with the LIS 2.0 water quality model. The impacts of nitrogen loads from watershed areas can be delineated using LIS 3.0. The model provides a means of relating nitrogen sources from specific geographic areas in the drainage basin to the genesis of hypoxia in portions of the sound. Its application enables the estuary program to accurately determine the most beneficial and cost-effective nitrogen load reduction targets for different geographic management zones established in the system.[1]

Specific commitments of the Phase 3 plans are as follows:

- The completion of the LIS 3.0 model
- The development of Long Island Sound-based nitrogen reduction and dissolved oxygen targets for the management zones
- The formulation of zone-by-zone plans to achieve load reduction targets

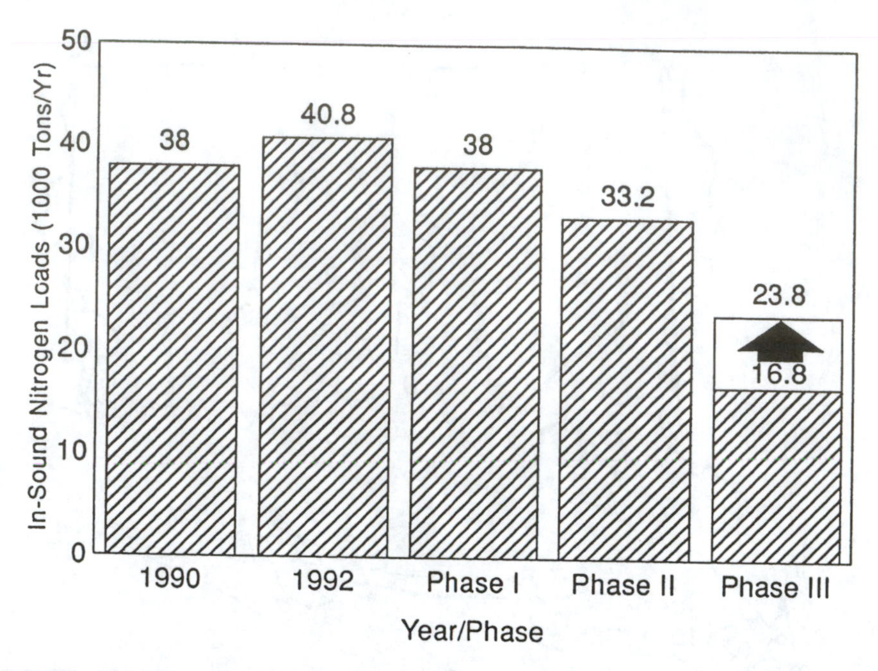

Figure 2.4 Phased plan to reduce the annual load of anthropogenic point and nonpoint source discharges in the Long Island Sound drainage basin. (From U.S. Environmental Protection Agency, *Long Island Sound Study Comprehensive Conservation and Management Plan,* Final Report, U.S. Environmental Protection Agency, Stony Brook, NY, 1994.)

- The support of innovative, feasible, and cost-effective technologies to reduce point and nonpoint sources of nitrogen
- The establishment of a firm timetable for accomplishing the load reduction targets by zone within 15 years, with progress measured in 5-year increments
- The implementation of long-term nitrogen control plans to ensure steady increases in dissolved oxygen and reductions in the area impacted by hypoxia

Figure 2.4 illustrates the phased plan to reduce the annual nitrogen load from anthropogenic point and nonpoint sources in the Long Island Sound drainage basin. The recommended 58.5% reduction in nitrogen load is to be phased in according to the following schedule: 40% in 5 years, 75% in 10 years, and 100% in 15 years. This phased plan is designed to assure steady progress in achieving nitrogen reduction targets commencing after completion of zone-by-zone plans. The total estimated cost to achieve maximum nitrogen reduction from all point sources is $6 to $8 billion.

Although management plans to control nitrogen loads from point sources have focused on nitrogen reduction options for the 70 sewage treatment plants in the 11 management zones, controls on nitrogen loads from nonpoint sources have been implemented under the broader context of watershed management. Habitat protection and restoration activities (e.g., stormwater detention basins, artificial wetlands, cleaning catch basins, and streetsweeping) have been proposed to limit nonpoint source inputs of nitrogen to the sound. The effectiveness of these approaches must be coupled to concerted education and outreach programs.

To determine if the aforementioned management actions are benefiting the system, a hypoxia monitoring and assessment program has been implemented. This program provides ongoing characterization of the physical and chemical conditions of the sound, the sources of nutrients, and the biological effects of hypoxia. Monitoring of dissolved oxygen, nutrients, and physical–chemical conditions is conducted on a regular schedule at a series of stations along an axial (east–west) transect in the sound. This effort is necessary to assess progress in meeting the nitrogen reduction and dissolved oxygen targets as well as to evaluate ecosystem responses.

Model analyses indicate that the maximum area of Long Island Sound that is unhealthy for marine life will be reduced by 75% if the nitrogen reduction targets of the program noted previously are realized. Other positive biological effects are also projected for the sound. For example, off Stony Brook, New York, and New Haven, Connecticut, larval mortality will decline by ~84 and 65%, respectively, with adverse impacts on fish abundance being eliminated in both areas. In the western narrows, mortality of sensitive larvae will decrease by 67%. There will also be a 97% reduction of adverse effects on fish abundance in this area. Of particular note, adverse impacts on winter flounder abundance will drop by 99%. In addition, abundance of lobsters will rebound as water quality improves.[11]

B. Pathogens

A serious concern of the LISEP is the occurrence of pathogens. These disease-causing microorganisms, which include certain bacteria, viruses, protozoa, and fungi pose a potential health threat to humans who either ingest contaminated shellfish or come in direct contact with contaminated waters in Long Island Sound. A number of acute illnesses (e.g., cholera, dysentery, gastroenteritis, hepatitis A, salmonellosis, and typhoid fever) are caused by human exposure to pathogens from contaminated sites. Pathogen contamination not only impacts human health but also adversely affects the local and regional economy by causing the closure of beaches and the restriction of shellfish harvesting. Commercial and recreational fishing activity can be greatly impacted by severe outbreaks of shellfish and finfish diseases.

1. Sources

Pathogens in the sound originate from both point and nonpoint sources of contamination and are concentrated in untreated or inadequately treated human sewage and the wastes of wildlife populations and domestic animals. Human pathogen sources are significant. Sewage treatment plants with capacity limitations, plant design flaws, inadequate maintenance or system operation, combined sewer systems, or disrepaired sewage conduits commonly discharge inadequately treated sewage and associated pathogens. Failing septic systems and illegal sanitary connections between sewer and stormwater systems yield additional loadings of pathogens. The largest sewage treatment plants in New York City and Westchester County have been historically responsible for the greatest point source loadings of pathogens to the western sound, where pathogen contamination is chronically most severe. More than 100 different human enteric bacteria may be found in treated municipal wastewater and urban stormwater runoff.[12]

Combined sewers, which exist in many older cities in the region (e.g., New York, Bridgeport, New Haven, Norwalk, and Norwich), carry both stormwater and sewage to municipal facilities for treatment. During rainfall events, however, the sewage treatment plants often cannot handle the capacity of waste, and the systems commonly overflow at designated points, releasing untreated sewage, wastewaters, urban runoff, and associated pathogens. Coliforms are commonly detected in local receiving waters at levels exceeding health standards for 2 or 3 days following storm events. The impacts of stormwater and combined sewer overflows are primarily contingent upon the regional and local hydrology and the existing system capacity.[12] Because precipitation is seasonally variable, combined sewer overflows will affect the sound more greatly during certain periods (e.g., spring) than others (e.g., summer and fall).

Aside from sewage treatment plants and combined sewer overflows, urban and non-urban surface runoff transports considerable quantities of pathogens to the sound. Many of these pathogens derive from nonhuman sources, including wastes of wildlife and domestic animals. The frequency and intensity of precipitation largely control these non-point source loadings to the system.

According to the USEPA,[8] the relative importance of different sources of fecal coliforms (an indicator of pathogen contamination) to Long Island Sound on an annual average basis is as follows:

- 51.6% from river influx, originating from sewage treatment plants, failing septic systems, and other sources
- 47.3% from urban runoff, including combined sewer overflows
- 1.1% from direct discharges of sewage treatment plants and industrial facilities

Vessel sewage discharges, although contributing a small percentage of pathogens to the sound, may create localized water quality problems, particularly in those embayments where large numbers of vessels congregate and circulation is greatly limited. Discharges from vessels in the vicinity of beaches and shellfish beds can exacerbate poor water quality conditions. Strategies formulated to mitigate vessel discharges include an increase in the number of marine pumpout facilities on the sound, the implementation of best management practices at marinas, and the creation of vessel no discharge zones.

2. *Impacts*

Pathogen contamination directly impacts human uses of the system, particularly bathing beaches and shellfish beds as noted previously. When they are suspected of being contaminated by pathogens, bathing beaches and shellfish grounds are closed or restricted to protect public health. For example, over the 5-year period between 1986 and 1990, various beaches along the sound were closed for a total of 1440 days. Among the chronically closed beaches were Scudder Park, Gold Star Battalion, the Mamaroneck Area, Huntington Beach Community, the Hempstead Harbor area, Centerport Yacht Club, Fleets Cove, Mamaroneck Beach Cabana and Yacht Club in New York, and the beaches in the Norwalk and Milford areas of Connecticut.[8] Most beach closings occurred in the western sound, especially in heavily populated areas and/or in enclosed embayments subject to poor tidal flushing. The closings were due to elevated coliform numbers observed during routine sampling, elevated coliform numbers anticipated subsequent to rainfall events, and sewage treatment plant operations and maintenance problems.[13] Most of the beach closings in Connecticut, which resulted in 196 lost beach days, were attributable to sewage treatment plant malfunctions, whereas those in New York were ascribable to rainfall-associated events.[1]

In 1990, 7442 ha of productive shellfish beds were classified as restricted/prohibited in Connecticut, compared to 19,513 ha in New York. In the period from 1986 to 1990, the number of harvest-limited hectares increased by 9% (13,981) in Connecticut waters and only 0.1% (145) in New York waters. The majority of harvest-limited hectares were in embayments and tributaries close to suspected pathogen sources (sewage treatment plants, septic systems, stormwater outfalls, and marinas). During 1990, only 10% of the hectares approved for shellfish harvest in New York and 19% in Connecticut were located in embayments and tributaries, the remainder being in the open waters of the sound.

Although economic losses have been incurred from the closure of productive shellfish beds, a relay or transplant program has mitigated some of the impact. For instance, the relay or transplantation of oysters and hard clams from restricted and prohibited areas in Connecticut has accounted for ~85 to 90% of the harvest from approved waters. The relay program allows shellfish to be moved from restricted areas to certified areas for depuration and subsequent harvest.[1]

Shellfish-growing areas may be affected by various sources of pathogenic contamination that increase from east to west in Long Island Sound. The impact of sewage treatment plants, once of overwhelming importance to degraded water quality in the western narrows, has decreased substantially in recent years concomitant with upgrades in the level of sewage treatment and disinfection together with improvements in maintenance, monitoring, and enforcement. Nevertheless, malfunctions of these systems can alter water quality. For example, a failed sewer line in Greenwich, Connecticut, on May 12, 1998, resulted in the release of more than 2 million liters of raw sewage into the sound. Although upgrades in the technical components and operation of sewage treatment plants have led to measurable improvements in water quality, nonpoint source pollution — notably surface runoff, faulty septic systems, and boats — remains a principal source of pathogenic contamination in many areas of the sound.

3. Monitoring and assessment

Both Connecticut and New York monitor shellfish grounds for the presence of pathogen indicators. State shellfish monitoring programs incorporate stringent sampling protocols to protect the public from contaminated shellfish. The programs require ambient water quality monitoring of shellfish-harvesting grounds for bacteria indicative of contamination. They also conduct detailed pollution source surveys to identify potential sources of pathogens, to restrict shellfish harvests consistent with the results of field monitoring and surveying, and to enforce the restrictions.[13]

Connecticut and New York classify shellfish growing areas according to the guidelines of the National Shellfish Sanitation Program.[14] These guidelines specify various shellfish-harvesting restrictions. For example, waters within the confines of marinas or overnight anchorage areas are automatically classified as prohibited to shellfish harvesting due to the potential for discharge from on-board marine sanitation devices.

Sanitary surveys of shellfish-growing areas in Connecticut waters are the responsibility of the Connecticut Department of Agriculture-Aquaculture Division (DA/AD) and the Connecticut Department of Environmental Protection, with the assistance of municipal health agencies and town shellfish commissions. The shellfish sanitation program of the New York State Department of Environmental Conservation is authorized to collect and analyze bacteriological water quality samples and to assess conditions of shellfish-growing areas in New York waters. The efforts of these state shellfish agencies have effectively controlled the outbreaks of shellfish-related foodborne disease in the Long Island Sound region, thereby protecting the health of the general public.[13]

Connecticut and New York likewise monitor the water quality of bathing beaches, focusing on the concentrations of pathogen indicators. In Connecticut, local health departments monitor town beaches, and the Connecticut Department of Environmental Protection monitors state beaches. In contrast, county governments monitor bathing beaches in New York, with each county having slightly different sampling strategies.

New York bathing beach monitoring programs utilize either total or fecal coliform bacterial counts as a water quality indicator. Because pathogen inputs rise appreciably during periods of rainfall, water samples are specified as either taken in "wet" or "dry" periods. The sample is considered wet when rainfall occurs within 48 hours prior to sampling, more than 1.02 cm of rain accumulates within a 24-hour period, or more than 0.51 cm of rain falls in a 2-hour period. Beaches are closed when the measured indicator levels exceed New York's coliform standards for beach water quality as specified in Sub Part 6-2 of the New York State Sanitary Code. In addition, administrative closure automatically closes beaches in some areas following rainfall events or when sewage treatment plants malfunction. Because of elevated coliform counts and pathogen-associated human health risks predicted for these events, administrative closures are used as precautionary measures. New York government agencies generally recommend that people refrain from swimming at bathing beaches after significant rainfall.

Connecticut uses enterococcal organisms as its human pathogen indicator. Prior to 1989, total coliform counts were employed in this capacity. When the geometric mean for a sampling station exceeds 33 enterococcal organisms per 100 ml or a single sample exceeds 61 coliform/100 ml, bathing conditions are deemed unsatisfactory, and beaches are subject to closure. Administrative beach closures in Connecticut can be based on historical rainfall data that determine the amount of rainfall required to cause bacterial counts to exceed acceptable levels for bathing.

4. *Management action plans*

There are two primary goals for managing pathogen contamination in Long Island Sound: (1) to increase the amount of area certified/approved for shellfish harvesting while adequately protecting the public health and (2) to eliminate public bathing beach closures while adequately protecting the public health. Management strategies to achieve these goals include actions to control the major sources of pathogen contamination (e.g., combined sewer overflows and stormwater runoff) and site-specific plans for each harbor, embayment, or discrete shellfish bed area. Other actions address sewage treatment plant malfunctions, vessel discharges, on-site (septic) systems, monitoring and assessment, and public education.[1]

a. Combined sewer overflows. Both Connecticut and New York are implementing long-term combined sewer overflow abatement programs to control pathogen contaminant inputs to the sound. Abatement programs are being implemented in New York City and several larger Connecticut cities (i.e., New Haven, Norwalk, Jewett City, Norwich, Derby, Shelton, and Bridgeport). The western sound is particularly susceptible to combined sewer overflow problems, and this area is receiving much of the attention.

The corrective measures being implemented are costly. For example, in Connecticut ~$27 million is required to remedy combined sewer overflow problems by the year 2000 in the cities of Norwalk, Jewett City, Derby, Norwich, and Shelton. For the larger cities of Bridgeport and New Haven, corrective actions will be phased in over a 20-year period at an estimated cost of $91 million and $125 million, respectively. The combined sewer overflow abatement program for New York City meanwhile will cost ~$1.5 billion, with enforceable completion dates for various aspects of the program occurring during the years 2001 to 2006.

b. Nonpoint sources. The diffuse nature of nonpoint source inputs to Long Island Sound is problematical, requiring the implementation of various tools of control, notably best management practices (both structural and nonstructural permits), changes in building codes, consent agreements, and public education. To better manage stormwater from industrial and construction activities, Connecticut and New York are implementing general statewide stormwater permit programs in accordance with the national program regulations of the USEPA.[8] New York City has initiated a pilot program to control stormwater discharges using enforceable instruments (i.e., permits or consent agreements), and their effectiveness is now being assessed in controlling and managing stormwater that causes the closure of bathing beaches and shellfish grounds. Most of the extensive programs established in Connecticut and New York to control stormwater runoff problems are based on nonpoint source control programs under Section 319 of the Clean Water Act and Section 6217 of the Coastal Zone Act Reauthorization Amendments. A primary focus of both states is to develop coastal nonpoint pollution control programs to reduce pathogen discharges. Table 2.3 provides a list of these programs.

Table 2.3 Actions to Control Pathogen Contamination in Long Island Sound from Nonpoint Sources

Ongoing Programs	Responsible Parties/Status		
Implement the state nonpoint source management initiatives supported with funding from Section 319 of the Clean Water Act.	The CTDEP and the NYSDEC administer programs to reduce loadings from nonpoint sources of pathogens, with federal financing at 50% of authorized levels.		
Develop state coastal nonpoint source control programs, as per Section 6217 of the Coastal Zone Management Act to address nonpoint source pathogen load from the Long Island Sound coastal zone.	The CTDEP, the NYSDEC, and the NYSDOS are responsible for developing the program at the state level, while the EPA and the NOAA have oversight responsibilities at the federal level.		
Implement general stormwater permit programs to control the discharge of stormwater from industrial, construction, and municipal activities, as per EPA regulations.	The CTDEP and the NYSDEC are responsible for implementing and managing their permit programs. New York State has initiated its statewide stormwater permitting efforts by focusing on the Long Island Sound watershed, while Connecticut's stormwater permitting program considers regional benefits for Long Island Sound. Both states have issued two general permits each, one for construction activities and one for all industrial activities, as per definitions in federal stormwater regulations. This requires applicants to develop and implement comprehensive stormwater pollution prevention plans and controls.		
Provide technical assistance to coastal municipalities to address impacts of pathogens in their municipal regulations and plans of development, as required by state law.	The CTDEP assists local municipal managers to reduce inputs, using existing staff.		
Commitments	**Responsible Parties**	**Time Frame**	**Estimated Cost**
Pursue changes of the State Building Code to include provisions for stormwater management.	NYSDEC, NYSDOS	1994/1995	Redirection of base program
Initiate a pilot program to control stormwater discharges using enforceable instruments (i.e., permits or consent agreements). Connecticut and New York will evaluate the effectiveness of the pilot program for more widespread implementation.	NYSDEC	Ongoing/ continuous	$100,000
Recommendations	**Responsible Parties**	**Time Frame**	**Estimated Cost**
Expand current requirements for federally licensed or permitted projects to obtain a water quality certification to include all projects in sensitive areas or where a contaminant or parameter is found to exist at or exceeding threshold value.	NYSDEC	1994/1995	See Table 5 of Chapter III, "Hypoxia," for details.

Source: U.S. Environmental Protection Agency, Long Island Sound Study Comprehensive Conservation and Management Plan, Final Report, U.S. Environmental Protection Agency, Stony Brook, NY, 1994.

Table 2.4 Actions to Control Pathogen Contamination in Long Island Sound from Sewage Treatment Plants

Ongoing Programs	Responsible Parties/Status		
Minimize malfunctions of treatment systems and eliminate dry weather overflows and illegal hook-ups to storm sewers through aggressive management programs. Ensure prompt notification and response and take quick enforcement action.	The CTDEP and the NYSDEC, using existing enforcement programs, will take administrative actions in cases where the closure of beaches or shellfish beds could have been prevented by proper operation and maintenance of STPs.		
Identify and take priority enforcement actions to control wet weather overflows from sewers caused by excessive infiltration and inflow.	The CTDEP and the NYSDEC, in coordination with local municipalities, administer programs to detect and correct illegal sewer hook-ups and control dry weather overflows from sanitary sewers.		
Commitment	Responsible Parties	Time Frame	Estimated Cost
Implement a beach and shellfish closure action plan to take immediate corrective and priority enforcement actions addressing improperly treated municipal discharges. Preventable incidents involving beaches and shellfish areas will be emphasized.	CTDEP, NYSDEC, EPA	Ongoing/ continuous	Redirection of base program

Source: U.S. Environmental Protection Agency, Long Island Sound Study Comprehensive Conservation and Management Plan, Final Report, U.S. Environmental Protection Agency, Stony Brook, NY, 1994.

c. Point sources. Sewage treatment plants are responsible for only ~1% of the fecal coliform loadings to the sound, but because the plants discharge large volumes, their pathogen inputs can be locally significant. Management objectives focus on minimizing malfunctions at the plants and responding quickly when malfunctions occur. Both Connecticut and New York have committed to identifying and controlling problems associated with dry and wet weather overflows from sanitary sewers, illegal hook-ups to storm sewers, and incidents of raw sewage waste discharge or improperly treated municipal discharge that culminate in the closure of bathing beaches and shellfish beds. A beach and shellfish closure action plan will enable officials to initiate immediate enforcement and seek corrective actions and penalties when sewage treatment plant malfunctions impact beach and shellfish grounds. If sewage treatment plants malfunction and emergency conditions arise, prompt notification and response are essential to minimize beach and shellfish bed closures (Table 2.4).

d. On-site systems. Faulty septic systems are a source of pathogens to surface waters. Because more than half of all homes and businesses in the watershed have septic tanks, it is important to ensure that they are properly sited and maintained. Connecticut and New York are coordinating management actions with local governments when on-site septic systems are determined to be failing and impacting bathing beaches and shellfish-growing areas. New York is evaluating existing septic system controls, including system monitoring, required maintenance, and repair and replacement of failing systems, to determine if they are sufficient to protect coastal ecosystems.[1]

e. Vessel discharges. Local areas of Long Island Sound that are heavily utilized by boats, such as harbors and some embayments, can be impacted by vessel sewage discharges.

One management strategy is to designate specific harbors and embayments as vessel "No Discharge Areas." This designation will provide additional protection from vessel sewage discharges beyond the protection afforded by the federal marine sanitation device standards. Sites that would potentially benefit from the No Discharge Area designation in Connecticut are Greenwich, Stamford, Norwalk, Westport, Fairfield, Bridgeport, Stratford, Milford, New Haven, West Haven, Branford, Madison, Clinton, Westbrook, Groton, and the Connecticut River from its mouth to Windsor. Those identified in New York are Lloyd Harbor, Huntington Harbor, Port Jefferson Harbor, and Oyster Bay.[13]

It is imperative that adequate pumpout and treatment facilities exist in areas heavily used by boats. Connecticut and New York are assessing pumpout facilities in the sound and the strategies to increase these facilities. Both states have received Clean Vessel Act grants to conduct pumpout need surveys, to determine the effectiveness of existing facilities, and to install additional vessel sewage pumpout facilities in the system. Table 2.5 summarizes the management actions in place to control pathogen contamination from vessel discharges.

f. Monitoring and assessment. Monitoring and assessment of pathogen contamination are needed to ensure proper management of bathing areas and shellfish-harvesting areas. Monitoring programs have been implemented to determine the concentrations of pathogen indicators for the protection of beachgoers and shellfish consumers. Assessment of the monitoring data is required to document the areas that are fit for recreational activities and harvesting of shellfish as well as to evaluate the success of pathogen abatement actions.

Additional and more intensive surveys are necessary. New monitoring, assessment, and research programs must be developed to clearly delineate the pathogen sources affecting the uses of Long Island Sound and its resources and the geographical extent, temporal duration, and frequency of contamination impacting the bathing beaches and shellfish beds. These programs will also identify bathing areas and shellfishing grounds in need of further assessment.

Management programs will be enhanced by annual meetings of coastal municipal health directors who examine and refine monitoring and bathing beach closure protocols and information sharing. Workshops will be held to update appropriate and consistent methods for bathing beach monitoring and laboratory analysis. An appropriate management strategy is to assess impacts and assign priorities to areas where management actions are most likely to be beneficial.

g. Education. A key element to improving overall water quality conditions in the sound is to provide educational opportunities for the general public, municipal officials, stakeholders, and various recreational user groups. Education will not only facilitate implementation of the management actions but also understanding of pathogen issues to ensure reduced risk of contamination and exposure. The recommendation is for the development and implementation of a public education plan that will target specific audiences. The plan will be implemented in cooperation with federal, state, and local public outreach and environmental education programs.

C. Toxic substances

Many estuaries, especially those near urban centers, are burdened with high concentrations of chemical contaminants that can cause adverse ecosystem or human health risks. Toxic contaminants include a wide array of naturally occurring and anthropogenic substances, such as halogenated hydrocarbons (e.g., DDTs, PCBs, chlordane, dieldrin, etc.),

Table 2.5 Actions to Control Pathogen Contamination in Long Island Sound from Vessel Discharges

Ongoing Programs	Responsible Parties/Status
During the permitting process, minimize the impacts of boat dockage facilities and temporary live-aboard anchorages by considering their proximity to productive and certified shellfish waters, existing boat channels, wetlands, and critical habitat areas, and tidal flushing in the waterway.	The CTDEP, the NYSDEC, and the NYSDOS, through existing regulations such as the Tidal Wetland Act, Protection of Waters, Water Quality Certification, and Coastal Nonpoint Source Program.
Consider the impacts of vessel discharges through appropriate resource management and recovery programs and will limit or condition the siting or operation of boating facilities as necessary to minimize such impacts.	The CTDEP and the NYSDEC administer these existing programs. Siting of facilities is already considered in the permitting process.

Commitments	Responsible Parties	Time Frame	Estimated Cost
New York State and Connecticut will apply to the EPA to create vessel *No Discharge* areas in specific embayments and harbors after ensuring the sufficient availability of pump-out stations and treatment facilities.	CTDEP, NYSDEC, EPA, Local municipalities	Ongoing/continuous	Redirection of base program
New York State has identified Huntington and Lloyd Harbors as areas requiring additional protection, and the EPA has Public Noticed its tentative determination that there are adequate pump-out facilities in these areas.	NYSDEC, EPA	1993/1994	Redirection of base program
Connecticut, through a 319 grant, will ensure completion of a marina and mooring area water quality assessment guidance document. Connecticut has also completed a marinas *best management practices* project report for nonpoint sources of pollution, which may be used to develop requirements for use of certain best management practices at marinas. New York State will review these documents for potential incorporation into state management programs.	CTDEP, NYSDEC	Ongoing/continuous	Redirection of base program
Complete regulations to require pump-out facilities as required by, and in accordancce with, state law.	CTDEP	Ongoing/continuous	Redirection of base program
The states of Connecticut and New York have received funding from the Federal Clean Vessel Act to conduct a pump-out needs survey, determine the effectiveness of existing facilties, develop and implement plans for construction of additional pump-out stations by marinas, and prepare education/information plans.	CTDEP, NYSDEC	Initiated 1993/ completion 1995	$1 million for NY, $120,000 for CT
Collect information on sewage discharge controls in Long Island Sound, disinfection chemicals used, and boater education and sewage treatment plant acceptance of pump-out wastes. Evaluate availability of treatment capacity of pump-out wastes and secure commitments from municipalities to accept these wastes.	NYSDEC, Municipalities	Initiated 1994/ completion 1994	$42,000

Source: U.S. Environmental Protection Agency, Long Island Sound Study Comprehensive Conservation and Management Plan, Final Report, U.S. Environmental Protection Agency, Stony Brook, NY, 1994.

Table 2.6 List of Toxic Contaminants of Concern in the Long Island Sound National Estuary Program

Metals
Cadmium
Chromium
Copper
Lead
Mercury
Zinc
Chlorinated Hydrocarbons
Polychlorinated biphenyls
Pesticides
Chlordane
DDT, DDD, DDE
Dieldrin
Heptachlor
Lindane
trans-Nonachlor
Polynuclear Aromatic Hydrocarbons

Source: U.S. Environmental Protection Agency, Long Island Sound Study Comprehensive Conservation and Management Plan, Final Report, U.S. Environmental Protection Agency, Stony Brook, NY, 1994.

polycyclic aromatic hydrocarbons (e.g., chrysene, fluorene, naphthalene, perylene, and pyrene), and heavy metals (e.g., copper, lead, mercury, and zinc). Although the USEPA has identified 129 substances nationwide as priority pollutants, the LISEP has compiled a shorter list of toxic substances of concern (Table 2.6). The concentrations of some of these contaminants may be sufficiently high to be a threat to organisms in localized areas of the sound. In addition, certain contaminants are not only toxic but also extremely persistent, resisting breakdown and creating acute as well as insidious pollution problems. Furthermore, the database on various contaminants in the sound is too sparse to draw meaningful conclusions. Future monitoring and assessment programs must address this deficiency.

1. Sources

In 1985, NOAA conducted a National Coastal Pollutant Inventory for Long Island Sound, which examined the relative importance of different sources of toxic substances to the sound. Efforts by the LISEP later supplemented this information. Four source categories of contaminants are recognized. Listed in decreasing order of significance, these are: (1) river inflow; (2) sewage treatment plants; (3) urban runoff; and (4) other sources. Numerous human activities in the watershed promote toxic contaminant inputs to the sound. For example, pesticide use in agricultural, residential, and urban areas releases chlorinated hydrocarbon compounds that wash into the tributaries and storm sewers. Metal-finishing industries are sources of heavy metals. Marinas and boats also contribute heavy metals as well as aliphatic and polycyclic aromatic hydrocarbon compounds. Atmospheric deposition appears to be another important pathway of toxic contaminant input.

a. River inflow. Because of its large discharge volume, the Connecticut River is the most significant source of toxic contaminants to Long Island Sound on a mass loading scale. The Naugatuck, Quinnipiac, and Thames Rivers in Connecticut are also major sources. All of these rivers discharge considerable amounts of heavy metals, but it is unclear what fraction originates from human activity as opposed to natural processes. In addition, atmospheric deposition directly augments the concentrations of heavy metals and organic contaminants in the river systems.

b. Sewage treatment plants. A large fraction of heavy metals and organic contaminants also derives from sewage treatment plants, which represent the second largest source of toxic substances to the sound. The bulk of these contaminant loads originates from high-volume sewage treatment plants in New York City. For example, sewage treatment plants located in the Bronx and Queens Boroughs account for a substantial portion of the total heavy metal loads discharged by sewage treatment plants in watershed areas.

c. Urban runoff. The third most significant source of toxic contaminants in Long Island Sound is the input from urban runoff, combined sewer overflows, and stormwater. This combined source, which releases polycyclic aromatic hydrocarbons, chlorinated hydrocarbons (e.g., PCBs), heavy metals (e.g., lead), and other substances, has its greatest impact on waters in the western sound. Approximately 80% of the annual urban runoff in this area originates from a part of the watershed extending from western Suffolk County through New Haven County.

d. Other sources. Many other human activities contribute toxic substances to the sound, albeit in relatively minor concentrations. Electric generating stations, for example, release heavy metals such as copper and zinc, but this accounts for <10% of the total load of these metals. Chemical and oil spills also add toxic substances to the system, although these events take place only sporadically and hence do not create chronic problems. Landfills may be potential sources of various chemical contaminants. However, they have not been studied in sufficient detail to determine their potential impact.

2. *Contaminant concentrations*

Quantitative data have been collected on toxic substances in the water column, bottom sediments, and biota. The database is most extensive for contaminants in bottom sediments and in tissues of organisms, with reliable and detailed data on dissolved contaminants in the water column being relatively scarce.[15] The National Status and Trends Program of NOAA monitors trace metal and organic contaminant concentrations in surface sediments and biota (mussels and flounder) at multiple sampling sites in the sound. The Environmental Monitoring and Assessment Program of the USEPA also samples stations for analysis of chemical contamination. The LISEP has recommended the implementation of a comprehensive, coordinated monitoring program to fully evaluate toxic contamination problems in the system.

a. Water column. Data are deficient on water column toxic organic contaminants. The most useful quantitative data exist on trace metal contamination in waters of the western sound and the East River collected in 1991 as part of the New York-New Jersey Harbor Estuary Program. These data show that the mean concentrations of trace metals in the water column of the western sound did not exceed the state standards of either Connecticut or New York. However, the state water quality standards were exceeded for mercury in the East River. Dissolved trace metal concentrations appear to be relatively consistent in plankton throughout the sound (Table 2.7).

b. Sediments. The concentrations of trace metals and toxic organic compounds are elevated in bottom sediments at a number of localized sites in Long Island Sound, indicating that the contaminants are not uniformily distributed in the system. For example, trace metal concentrations in bottom sediments of the western sound are significantly

Table 2.7 Concentrations of Metals (μg/1) in the Dissolved Phase of Long Island Sound Waters Calculated Using Mean and Maximum Plankton Levels

Metal	Mean	Maximum	Mean Ocean[a]
Silver	0.0038	0.0044	0.0027
Cadmium	0.074	0.120	0.078
Chromium	0.290	0.380	0.208
Copper	0.072	0.440	0.256
Nickel	0.520	2.000	0.472
Lead	0.056	0.160	0.0021
Zinc	0.315	1.000	0.390

[a] From Bruland, K. W., Trace elements in seawater, in Chemical Oceanography, Vol. 8, Riley, J. P. and Chester, R., eds., Academic Press, London, 1983, 158.

Source: U.S. Environmental Protection Agency, Long Island Sound Study Comprehensive Conservation and Management Plan, Final Report, U.S. Environmental Protection Agency, Stony Brook, NY, 1994.

Table 2.8 Mean Total Concentrations (mg/kg dry wt) of Metals in Sediments from Three Areas of Long Island Sound and Their Relationship to Criteria and Nationally and Locally "High" Concentrations

	Area			NYSDEC Guidelines		High Levels	
Metal	WLIS	CLIS	ELIS	Low Effect Level	Severe Effect Level	NOAA[a]	Harbors[b]
Arsenic	9.0	5.6	6.2	6	33	—	50–60
Silver	3.0	0.6	0.39	—	—	0.74	—
Cadmium	1.4	0.4	0.16	0.6	10	0.72	24–35
Chromium	138	79	37	26	110	135	510–570
Copper	121	57	9.5	16	110	55	2000–7700
Mercury	0.7	0.21	0.1	0.2	2	0.30	7–17
Nickel	25	16	8	16	75	—	90–665
Lead	89	31	13	31	250	52	1150–1960
Zinc	198	99	35	120	820	172	1000–4800

[a] NOAA Status and Trends nationally high sites.

[b] Highest levels observed in the Army Corps of Engineers data.

Source: U.S. Environmental Protection Agency, Long Island Sound Study Comprehensive Conservation and Management Plan, Final Report, U.S. Environmental Protection Agency, Stony Brook, NY, 1994.

greater than those of central and eastern basins (Table 2.8). The LISEP has also documented high levels of trace metals and some toxic organic contaminants at 12 urbanized harbors, rivers, and embayments (Table 2.9). These elevated contaminant levels may reflect a remnant effect of historical discharges that occurred prior to the implementation of state and federal Clean Water Act requirements. The sediment load of toxic substances at some of these sites has not decreased appreciably over time. Although high levels of the aforementioned contaminants have been observed at several urbanized harbors and embayments, the available data on nearshore sediments are not comprehensive enough to characterize the entire shoreline area.

Higher contaminant levels occur in bottom sediments of the western sound than the eastern sound due to the closer proximity to contaminant sources and the greater enrichment of organic carbon in fine grained sediments in the western reach. The mean concentrations of trace metals in the western sound typically exceed the high values defined by

Table 2.9 Long Island Sound Harbors Identified as having Highly Contaminated Sediments (Normalized for Carbon) and the Contaminants Which Were at Least Locally Elevated

Harbor or River	Contaminant							
	Pb	Cd	Cu	Ni	Hg	Zn	Cr	PCB
Bridgeport Harbor	X							X
Milford Harbor	X			X				X
Stamford Harbor	X	X				X		X
Connecticut River		X			X			
Housatonic River			X	X			X	
New Haven Harbor			X	X		X		X
New Rochelle Creek			X	X		X		
New London Harbor				X		X		
Norwalk Harbor				X	X			
Northport Harbor					X			
Hutchinson River						X		
Branford Harbor						X		

Source: U.S. Environmental Protection Agency, Long Island Sound Study Comprehensive Conservation and Management Plan, Final Report, U.S. Environmental Protection Agency, Stony Brook, NY, 1994.

Table 2.10 Concentrations of Organic Contaminants (μg/kg dry wt) in the Fine-Grained Fraction of Long Island Sound Sediments Compared to Nationally "High" Sites

Substance	LIS Average	NOAA "High"
tPCB[a]	249	200
tDDT	36	40
tPAH	7814	3900
tChlordane	7.7	5.5

[a] From NOAA, National Status and Trends Program for Marine Environmental Quality, Progress Report. A Summary of Data on Chemical Contaminants in Sediments from the National Status and Trends Program, NOAA Tech. Memo. NOS OMA 59, National Oceanic and Atmospheric Administration, Rockville, MD, 1991.

Source: U.S. Environmental Protection Agency, Long Island Sound Study Comprehensive Conservation and Management Plan, Final Report, U.S. Environmental Protection Agency, Stony Brook, NY, 1994.

the Status and Trends Program of NOAA based on national surveys. The bottom sediments of some harbors and embayments in other areas of the sound, however, exhibit elevated concentrations of both heavy metals and organic contaminants (e.g., PCBs) that are comparable to those of the western reach. The mean concentrations of PCBs in bottom sediments generally do not exceed 1 ppm, although locally higher levels have been recorded at a number of harbor sites (e.g., Bridgeport, Milford, New Haven, Norwalk, and Stamford Harbors). The Status and Trends Program also reported that the sediment concentrations of PCBs and a few other organic contaminants (i.e., DDT, chlordane, and polycyclic aromatic hydrocarbons) at a number of locations, primarily in the western sound, exceed NOAA's nationally high values for these contaminants (Table 2.10).

Trace metal and toxic organic contaminant concentrations in sediments of the western sound are relatively high on a national scale.[16] Comparing the *lowest-effect-level* guidelines of the New York State Department of Environmental Conservation with the sediment toxicity levels in the sound, some regional differences are apparent (Table 2.8). For instance, the mean sediment concentrations of arsenic and chromium in the eastern sound exceed the guidelines. In the central basin, the mean sediment concentrations of copper, chromium, lead, mercury, and nickel equal or exceed the guidelines. The greatest effects appear in the western sound, where the mean sediment concentrations exceed the guidelines for all metals examined (i.e., arsenic, cadmium, chromium, copper, lead, mercury, nickel, silver, and zinc). Of particular importance are copper and chromium concentrations that exceed the *severe-effect-level* guidelines. The levels of copper are high enough to be potentially toxic to benthic organisms. Enrichment of copper, lead, and zinc in surface sediments are 3 to 10 times above background.

In summary, elevated levels of metals and organic compounds are site specific and depend in large part on the proximity of the contaminant sources in the watershed as well as the sediment properties and distribution dynamics in the sound. The highest contaminant levels generally occur in nearshore sediments, mostly urbanized harbors, rivers, and embayments. In offshore sediments, higher toxic levels are recorded in the western reach than in the central and eastern basins. Sediment toxicity tests completed by the Environmental Monitoring and Assessment Program of the USEPA on samples from the eastern sound near Mattituck, in Black Rock Harbor (a highly contaminated part of Bridgeport Harbor), in the Housatonic River near Devon, and in the area around Throgs Neck have revealed significant toxicity to amphipods (i.e., the tube-dwelling forage species, *Ampelisca abdita*). However, other sediment samples tested by the Status and Trends Program of NOAA found no toxicity despite elevated toxic contaminant levels.

c. Organisms. Toxic contaminants have also been monitored in the tissues of organisms, principally bivalves (i.e., blue mussels, *Mytilus edulis* and oysters, *Crassostrea virginica*) and finfish (i.e., winter flounder, *Pseudopleuronectes americanus*). As mentioned previously, mussels and oysters have been successfully employed as sentinel organisms to monitor contaminants in coastal waters because they have low or undetectable (MFO) enzyme activity for metabolizing the substances. Therefore, these organisms typically bioaccumulate the contaminants with little alteration and can be used to identify "hot spot" locations in the environment.

The tissue contaminant database for Long Island Sound is most comprehensive for the blue mussel (*Mytilus edulis*).[16-21] Turgeon and O'Connor[16] compared tissue contaminant data collected by the Mussel Watch Program of the USEPA from 1976 to 1978 to that collected by the National Status and Trends Mussel Watch Program of NOAA from 1986 to 1988. Results of this comparative study indicate a decadal decrease in lead concentrations and an increase in copper levels in the sound. There was also possibly a decreasing trend in cadmium and chlordane levels based on declining concentrations observed in mussel tissues during the 1986 to 1988 period. However, later investigations revealed no statistically significant trends in blue mussel trace metal concentrations.[15] In addition, no hot spots of contamination were evident based on these investigations.

Examining NOAA Status and Trends data for 1986 to 1988, trace metal concentrations in blue mussels at most Long Island Sound locations did not rank high on a national level. The highest rankings (in the top 20 sites) were for chromium in the Connecticut and Housatonic Rivers, copper at Throgs Neck, Mamaroneck Harbor, Hempstead Harbor, and in the Connecticut and Housatonic Rivers, and lead at Mamaroneck Harbor and Throgs Neck. Toxic organic compounds (i.e., polycyclic aromatic hydrocarbons, chlordane, dieldrin, DDT, and PCBs), however, exhibited relatively high levels in the mussel tissues.[4] Few trends were ascertained from the oyster tissue data. Although the mussel and oyster data are spatially limited, several areas of the sound clearly show high contamination levels based on tissue analyses. Most notable are the areas around Throgs Neck, the lower Housatonic River near Devon, and the urban harbors of Bridgeport, Mamaroneck, and Hempstead.[8] Minor east–west gradients in contaminant concentrations of bivalve tissues are evident, and few hot spots of contamination have been ascertained based on bivalve body burden levels.

Of all the tissue contaminants examined, PCBs are of greatest concern. For example, the PCB levels in American eel (*Anguilla rostrata*), bluefish (*Pomatomus saltatrix*), and striped bass (*Morone saxatilis*) have exceeded the U.S. Food and Drug Administration (USFDA) action level of 2.0 ppm, and as a result both Connecticut and New York have issued consumptive advisories on these finfish species. However, low concentrations of PCBs have been noted in winter flounder (*Pseudopleuronectes americanus*), summer flounder (*Paralichthys dentatus*), blackfish (*Centropristis striata*), and lobster (*Homarus americanus*) tail and claw. PCB levels often exceed 2.0 ppm in lobster hepatopancreas, and the concentrations of cadmium and copper are also elevated in this digestive gland. Consequently, consumptive advisories have likewise been issued on this shellfish species. Finally, consumption advisories have been issued on some waterfowl (e.g., mergansers) that have accumulated elevated tissue contaminant levels.

Sublethal toxic effects on resident organisms of the sound have also been investigated, with the focus being on the pathology and reproductive processes of winter flounder (*Pseudopleuronectes americanus*) and hard clams (*Mercenaria mercenaria*). More specifically, studies have been conducted on abnormalities in embryos and histopathological condition of liver, blood, and kidney tissues of winter flounder as well as embryo and larval abnormalities and development success of hard clams.[22-25] In these studies, samples were collected at several harbor and embayment sites, such as Bridgeport, New Haven, and Norwalk Harbors and Niantic Bay. Gronlund et al.[22] observed the highest incidence of biochemical and pathological abnormalities in winter flounder from New Haven Harbor. The livers of finfish samples from this site showed high prevalences of histopathological changes and DNA alterations. However, the prevalences of lesions in these samples were not as high as those previously reported at more contaminated East Coast estuaries (e.g., Boston Harbor, Massachusetts, and Raritan Bay, New York). The sediment concentrations of aromatic hydrocarbons and heavy metals are relatively high in New Haven Harbor (Figures 2.5 and 2.6), and the total contaminant levels are sufficiently high to exceed effects thresholds.[15]

Stiles et al.,[23] focusing on hard clams at five locations in Long Island Sound (i.e., Greenwich, Norwalk, Bridgeport, Milford, and New Haven Harbors), found a negative relationship between high contaminant levels and reproductive success of the bivalves. Results of their study suggest that environmental quality in some areas may be detrimental to hard clam reproduction and survival. Although adults are hardy and quite tolerant of chemical contamination, their gonads may be impacted during gametogenesis, especially at the elevated contaminant levels observed near industrialized sites (e.g., Bridgeport Harbor). Hard clam embryos of specimens from the relatively contaminated sediments of Black Rock Harbor displayed higher frequencies of abnormal chromosome numbers than those of specimens from the other four sampling locations. In addition, abnormal larvae were more prevalent in specimens from Black Rock Harbor. This may be indicative of the long-term sublethal effects of chemical contamination. Elsewhere, hard clams from Norwalk Harbor also exhibited reduced normal fertilization and development success.

In a study of the reproductive success of winter flounder at six sites in the sound (i.e., Norwalk, Milford, New Haven, Madison, Shoreham, and Hempstead waters), Perry et al.[24] demonstrated that fish embryos from New Haven Harbor were usually the most aberrant. The embryos consistently showed higher rates of mitotic abnormalities, lower mitotic counts, and high percentages of abnormally differentiated cells. The observed chromosome

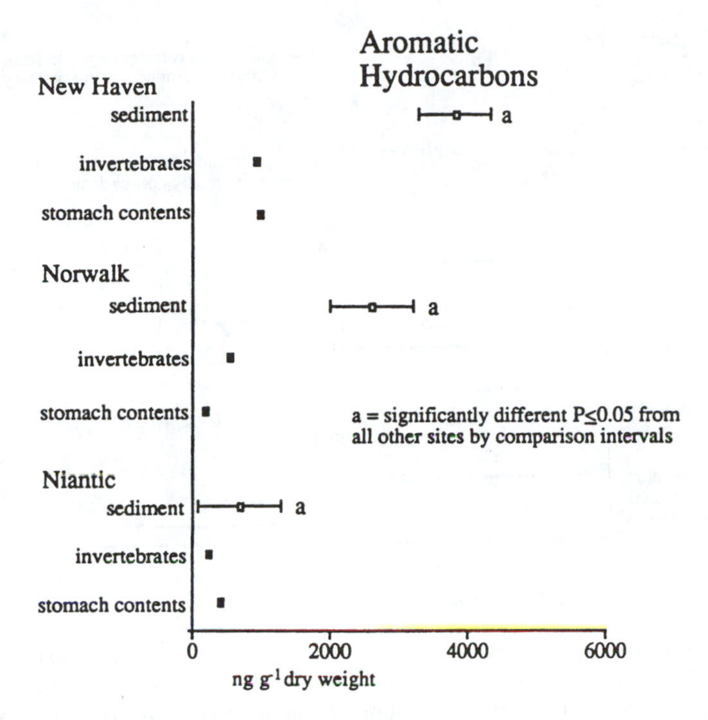

Figure 2.5 Concentrations (ng/g dry wt) of aromatic hydrocarbons in sediments, infaunal invertebrates, and stomach contents of winter flounder from New Haven, Norwalk, and Niantic Harbors. Open squares indicate ± comparison intervals. Solid squares indicate results of analysis of composites. Lowercase letters denote results of statistical analysis. (From Gronlund, W.D., Chan, S.-L., McCain, B.B., Clark, R.C., Jr., Myers, M.S., Stein, J.E., Brown, D.W., Landahl, J.T., Krahn, M.M., and Varanasi, U., Multidisciplinary assessment of pollution at three sites in Long Island Sound, *Estuaries*, 14, 299, 1991.)

abnormalities may be the result of sublethal effects of mutagenic chemicals. Subtle indications of embryo abnormalities were also noted in fish from Hempstead Harbor and Shoreham. The reproductive success of winter flounder from New Haven Harbor was reduced, as reflected by the lower percentage of viable hatch and the small larvae produced relative to other sampling sites.[25]

d. Human Health Risk. To protect public health, Connecticut and New York have issued consumption advisories for Long Island Sound finfish (i.e., American eel, *A. rostrata*; bluefish, *Pomatomus saltatrix*; and striped bass, *Morone saxatilis*) and shellfish (i.e., crabs, *Callinectes sapidus* and lobsters, *Homarus americanus*) based on contaminant levels in their tissues (Table 2.11). These advisories, which have no regulatory weight, are designed to ensure that a person's exposure to a contaminant does not exceed acceptable risk levels. Hence, they specify the recommended maximum consumption levels for the seafood-consuming public.

The U.S. Food and Drug Administration (USFDA) has developed two other levels of criteria (i.e., action levels and tolerance levels) that provide national protection and thus have a bearing on seafood consumption from the sound. Action levels refer to the minimum concentrations of chemical substances in food that may cause the USFDA to take enforcement action. Tolerance levels carry the most regulatory weight. They represent the maximum permissible concentrations of contaminants established by the USFDA. Both human health risks of toxicity and the economic impacts of seafood restrictions are considered by the USFDA in developing action levels and tolerances.

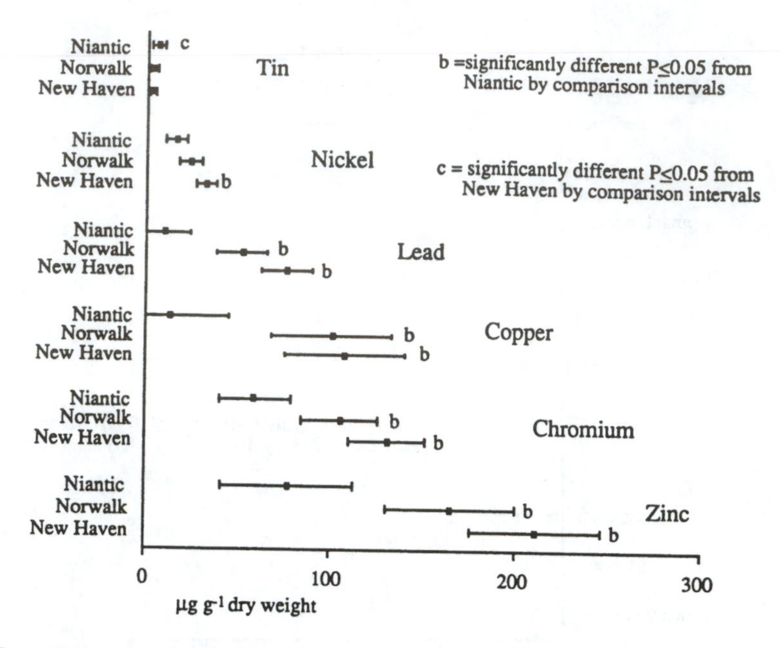

Figure 2.6 Concentrations (µg/g dry wt) of zinc, chromium, copper, lead, nickel, and tin in sediments from New Haven, Norwalk, and Niantic Harbors. Squares represent means ± comparison intervals. Lowercase letters indicate results of statistical analysis. (From Gronlund, W.D., Chan, S.-L., McCain, B.B., Clark, R.C., Jr., Myers, M.S., Stein, J.E., Brown, D.W., Landahl, J.T., Krahn, M.M., and Varanasi, U., Multidisciplinary assessment of pollution at three sites in Long Island Sound, *Estuaries*, 14, 299, 1991.)

Of all contaminants in Long Island Sound seafood, PCBs are of greatest concern. These toxic chemicals occur in elevated concentrations in some edible finfish and shellfish. Recreationally important, lipid-rich species, which tend to accumulate PCBs and other contaminants (e.g., bluefish, *P. saltatrix 121*and striped bass, *M. saxatilis*), are noteworthy examples.

3. *Management actions*

The Long Island Sound Toxic Substance Management Plan incorporates a number of actions to protect and restore the system from the adverse effects of toxic chemical contamination. The following actions are recommended for four priority areas of documented impact or characterization/problem identification needs:

- Continuing and, where appropriate, enhancing existing regulatory and pollution prevention programs, which have already greatly reduced toxic substance inputs to the sound
- Further evaluating sediments where toxic contamination problems exist to determine the feasibility of remediation
- Improving communication to the public of any legitimate health risks from consumption of seafood species from the sound
- Coordinating and strengthening monitoring activity for toxic substances to improve understanding and management of toxic contamination problems[1]

a. Existing pollution prevention and regulatory programs. Various ongoing state and federal permit programs and enforcement activities as well as an array of existing nonregulatory management initiatives have greatly reduced and minimized toxic contaminant loads to Long Island Sound. The CCMP[1] for the sound provides several examples:

Table 2.11 Consumption Advisories Issued by New York and Connecticut for Fish and Seafood Products Taken from Long Island Sound

Consumption Advisories
New York
To minimize potential adverse health impacts, the New York State Department of Health recommends:
Eat no more than one meal (1/2 lb) per week of fish from the East River to the Throgs Neck Bridge.
Eat no more than one meal per week of bluefish and American eel.
American eels from the East River should not be eaten.
Eat no more than one meal per month of striped bass taken from the marine waters of western Long Island Sound, which includes that portion of the Island west of a line between Wading River and the terminus of Route 46 near Mastic Beach.
Eat no more than one meal (1/2 lb) per week of striped bass taken from eastern Long Island marine waters.
Women of childbearing age, infants, and children under 15 should not eat striped bass taken from Long Island marine waters.
It is recommended that the hepatopancreas (liver, mustard, or tomalley) of crabs and lobsters not be eaten because this organ has high contaminant levels.
The health implications of eating deformed or cancerous fish are unknown. Any grossly diseased fish should probably be discarded. Levels of PCB, mirex, and possibly other contaminants of concern (except mercury) can be reduced by removing the skin and fatty portions along the back, sides, and belly of striped bass and bluefish.
Connecticut
Sensitive groups (pregnant women, nursing mothers, children under 15, and women who plan to become pregnant soon) are warned against eating bluefish larger than 25 inches. If consumers in this group choose to eat bluefish larger than 25 inches, consumption should be limited to no more than a few meals per year.
Sensitive groups are warned against eating any striped bass from Long Island Sound and nearby waters. If consumers in this group choose to eat striped bass, consumption should be limited to no more than a few meals per year.
The general population is advised to eat only a moderate number of striped bass and bluefish meals (18 per year or fewer), eat smaller fish when possible, remove fatty portions of the fish when cleaning them and broil them so that fat drips away.
It is suggested that the hepatopancreas (tomalley or mustard) of lobster should be eaten only in moderation.

Source: U.S. Environmental Protection Agency, Long Island Sound Study Comprehensive Conservation and Management Plan, Final Report, U.S. Environmental Protection Agency, Stony Brook, NY, 1994.

- The development of standards and criteria for toxic discharges
- Pollution prevention, pretreatment, and waste reduction programs
- Water quality-based effluent limits for point sources
- Toxic substance bans or use limitation such as those imposed on PCBs, DDT, and leaded gasoline
- Remediation of inactive hazardous waste sites
- Domestic waste management, including recycling programs and hazardous waste collection days, developed by state and local agencies
- Oil and chemical spill response programs

- Long Island Sound Research Fund studies on toxic source, fate, and ecological impact
- Agricultural management practices such as integrated pest management and runoff controls
- State and federal coastal dredging permitting programs
- Seafood consumption advisories

These efforts are funded under current programs.

Connecticut and New York continue to support pretreatment programs that ensure control of toxic discharges to sewage treatment plants. Both states are also ensuring that facilities comply with their National Pollution Discharge Elimination System Programs. To reduce the allowable concentrations of toxic pollutants from the previous permitted values, Connecticut and New York are reviewing municipal and industrial discharges. These programs, if successful, should substantially lower the discharge of toxic materials to meet adopted criteria for toxic substances.There is also a commitment of the states to encourage adequate funding for the continuance and expansion of pollution prevention site visit programs by targeting industrial discharges. In addition, toxic substance management programs of the New York-New Jersey Harbor Estuary Program will enhance toxic substance control in Long Island Sound. All of these programs will be facilitated by improved education strategies. Table 2.12 contains specific actions relevant to source control and pollution prevention of toxic contamination in the sound.

Nonpoint source pollution control remains a primary issue of Connecticut and New York environmental programs. These states have completed Section 319 (Clean Water Act Amendments of 1987) assessments and are currently implementing management plans that address nonpoint problems. Efforts are under way to develop improved stormwater management, to adopt best management practices, and to investigate new nonpoint control technologies in watershed areas that will reduce inputs of toxic substances to the sound.

b. Sediment contamination. As noted previously, bottom sediments in the western sound as well as in more localized areas (e.g., urbanized harbors, rivers, and embayments) have elevated levels of toxic substances. Contaminant data from these areas are being reviewed by the LISEP on a site-by-site basis. Additional assessments of toxic contaminant distribution in the bottom sediments will be recommended to fully characterize the problem and ascertain the need for remediation. The feasibility, technical approach, cost, and value of remediating contaminated sediments with heavy burdens will be based on these assessments.

Remediation plans will be prepared for target areas. To this end, the estuary program is reviewing and evaluating the remediation approaches formulated by the Great Lakes and New York-New Jersey Harbor Estuary Programs (Table 2.13). The objective is to transfer the most appropriate and cost-effective technology to the contamination problems encountered in bottom sediments of Long Island Sound.

c. Risk communication. Consumption advisories have been issued for selected finfish and shellfish of the sound due, in particular, to PCB contamination. Actions to improve human health risk management entail coordinated efforts by the states of Connecticut and New York to review health risk and advisory recommendations and formulate consistent plans designed to minimize human exposure to toxic substances. This uniform approach will facilitate risk communication and understanding of contamination problems in seafood products. The actions will also promote common approaches to releasing and

Table 2.12 Toxic Contaminant Source Control and Pollution Prevention Actions for Long Island Sound

Ongoing Programs	Responsible Parties/Status
The states of Connecticut and New York, and the Army Corps of Engineers will continue to regulate dredging and the disposal of dredged sediments through existing permit programs.	CTDEP, NYSDEC, NYSDOS, EPA, USACOE
The states of Connecticut and New York and the EPA will continue their pretreatment programs to ensure that toxic discharges to sewage treatment plants are controlled. The states of Connecticut and New York, through their Pollution Discharge Elimination System Programs, will continue to ensure that facilities comply with their permit limits.	CTDEP, NYSDEC
The states of Connecticut and New York and the EPA will apply pollution-prevention techniques, as appropriate, to both direct and indirect discharges of toxic substances by emphasizing wastewater minimization, recycling of wastewater, and alternative processes and chemicals to reduce toxicity and toxic loads and to minimize effects on all environmental media.	CTDEP, NYSDEC, EPA. Both states and the EPA have established policies on pollution prevention to highlight the importance and benefits of controlling pollution before it enters the waste stream and potentially impacts the environment. Connecticut has established pollution prevention as a public policy by statute and has begun a program to institutionalize multimedia pollution prevention in regulatory programs, eliminate barriers to pollution-prevention initiatives, and identify targets for an outreach program. New York's policy is to reduce the generation and discharge of pollutants to all environmental media consistent with sound facility management and economic practices.
The states of Connecticut and New York will review municipal and industrial discharge permits to surface waters to reduce the allowable concentrations of toxic pollutants from the previous permitted values.	CTDEP, NYSDEC. The net result will be a substantial reduction in the discharge of toxic materials over the next few years to meet adopted criteria for toxic substances in the states' waters.

Commitments	Responsible Parties	Time Frame	Estimated Cost
The LISS will encourage adequate funding to continue and expand pollution prevention site visit programs targeting industrial dischargers to the sound and its tributaries.	LISS	Initiated 1993/continuing	Minimal staff time
As part of the NY-NJ Harbor Estuary Program, total maximum daily loads, wasteload allocations for point sources, and load allocations for nonpoint sources will be developed to ensure that water quality standards for mercury are met in NY-NJ Harbor, the East River, and Long Island Sound.	HEP NJDEPE NYSDEC EPA	1994	Redirection of base program
As part of the New York-New Jersey Harbor Estuary Program, the states of New York and New Jersey will establish water quality-based effluent limits for copper, mercury, and six other toxic metals, as necessary. Permits will be subsequently modified.	NJDEPE NYSDEC	Completed December 1994	Redirection of base program

Table 2.12 (continued) Toxic Contaminant Source Control and Pollution Prevention Actions for Long Island Sound

Recommendations	Responsible Parties	Time Frame	Estimated Cost
Support education on the environmental impact of using home, garden, and commercial hazardous chemicals and pesticides and will continue to provide guidance on how to minimize use of these chemicals and properly dispose of them through household hazardous waste collection.	LISS	Initiated 1993/continuing	$20,000; see Chapter X, "Public Involvement and Education," for details
Evalute mass loadings of toxic contaminants and determine their relationship to ambient water and sediment quality.	LISS CTDEP NYSDEC	—	$200,000 per year
Identify and assign priorities to toxic substances which should be banned from use and for which *virtual elimination of discharge* should be the goal.	LISS CTDEP NYSDEC	—	$200,000 per year

Source: U.S. Environmental Protection Agency, Long Island Sound Study Comprehensive Conservation and Management Plan, Final Report, U.S. Environmental Protection Agency, Stony Brook, NY, 1994.

Table 2.13 Actions Addressing Sediment Contamination in Long Island Sound

Commitments	Responsible Parties	Time Frame	Estimated Cost
The LISS will review the National Oceanic and Atmospheric Administration (NOAA) 1991 sediment chemistry and toxicity survey results of harbors and embayments, when available in the spring 1994. This will supplement the available data.	LISS NOAA	Completed 1994	Existing staff to be used
The LISS will provide a preliminary review of the data on sediment contamination on a site-by-site basis. State and federal experts will evaluate the problem at each site and recommend additional assessments needed to fully characterize the problem, ascertain the need for and feasibility of remediation, and prepare a remediation plan.	LISS	Ongoing	Existing staff to be used
The City of Glen Cove plus its review committee will evaluate the contamination of Glen Cove Creek.	NYSDEC, City of Glen Cove	1994/1995	$250,000
The LISS will review and evaluate sediment remediation approaches developed in the Great Lakes ARCS Program and HEP.	LISS	1994/1995	Existing staff to be used
Recommendations	**Responsible Parties**	**Time Frame**	**Estimated Cost**
Conduct further assessments and develop site plans addressing the feasibility, technical approach, cost and value of conducting remediation activities for Black Rock Harbor and Glen Cove Creek, where data may be sufficient to conduct case study analyses. Recommend other harbors for characterization and feasibility studies to be conducted at a rate of two harbors per year.	LISS	Ongoing	$250,000 per harbor or $500,000 per year

Source: U.S. Environmental Protection Agency, Long Island Sound Study Comprehensive Conservation and Management Plan, Final Report, U.S. Environmental Protection Agency, Stony Brook, NY, 1994.

publicizing advisories for target species, thereby leading to effective risk communication and reduced human exposure to toxic substances.

d. Monitoring. A detailed monitoring plan is required to identify toxic contamination problems, causes, and trends in Long Island Sound. The LISEP recommends the development of a coordinated monitoring program consisting of multiple elements that will focus on toxic contamination of water, sediment, and tissue media (Table 2.14). Two federal monitoring programs will be continued. These include the Mussel Watch and Benthic Surveillance components of NOAA's Status and Trends Program and the USEPA's Environmental Monitoring and Assessment Program. These monitoring programs survey toxic substance levels in bottom sediments and tissue samples of mussels and finfish, providing information on the status and trends of toxic contaminants in tissues and sediments as well as geographic problem areas.

Table 2.14 Actions for Monitoring and Assessing Toxic Contaminants in Long Island Sound

Ongoing Program	Responsible Parties/Status
The Mussel Watch and Benthic Surveillance components of NOAA's Status and Trends Program and the EPA's Environmental Monitoring and Assessment Program provide regular and systematic sampling of contaminant levels in the sound.	EPA, NOAA. NOAA's Mussel Watch and Benthic Surveillance components of the National Status and Trends Program have been ongoing since 1984 in Long Island Sound. Annual samples of mussels, sediments, and fish tissues are taken and analyzed for several toxic substances, providing a continuing monitoring base to identify trends in Long Island Sound water quality. Similarly, EPA's Environmental Monitoring and Assessment Program has looked at toxic impacts and toxic substance levels in tissue samples from the sound since 1990.

Commitments	Responsible Parties	Time Frame	Estimated Cost
A monitoring workshop was held to integrate findings of the LISS and develop a comprehensive, sound-wide monitoring plan for toxic substances.	LISS	Initiated 1993/ completed 1994	See Chapter IX, "Continuing the Management Conference"
Under the auspices of the New York-New Jersey Harbor Estuary Program (HEP), the U.S. Army Corps of Engineers has agreed to develop a work plan and budget to develop system-wide models for PCBs, mercury, and other toxic pollutants that will provide the technical foundation for comprehensive efforts to eliminate these contamination problems in the Sound-Harbor-Bight system. The Corps of Engineers and other participants have agreed to seek the funding necessary to complete these models. Special attention will be directed to fully account for nonpoint sources of mercury.	HEP USACOE	1994	Existing staff to be used
Monitoring initiatives will be coordinated with the EPA Regional-Environmental Monitoring Assessment Program (R-EMAP) to further the understanding of sediment toxicity and benthic community structure gradients in western Long Island Sound.	CTDEP NYSDEC EPA	Initiated 1993/ completed 1994	$200,000

Recommendations	Responsible Parties	Time Frame	Estimated Cost
Conduct site-specific characterization surveys of water, sediment, and biota in harbors where active sources of toxic substances are believed to persist. Conduct at a rate of two harbors per year.	CTDEP NYSDEC	—	$200,000 per harbor or $400,000 per year
Identify sources and sites of PCB loadings to the sound ecosystem from in-sound and NY-NJ Harbor Estuary sources. Focus on reducing and eliminating PCB loadings on a priority basis, concentrating on areas of known contamination such as Black Rock Harbor.	CTDEP NYSDEC EPA	—	$200,000 per year
Monitor contaminant levels in selected estuarine organisms to ascertain their effects on the biology of the species and their effects on the edibility of the species.	LISS CTDEP NYSDEC EPA NMFS USFWS	—	$300,000 per year
Implement the recommendations from the LISS Monitoring Plan to improve contaminant monitoring.	LISS	—	$15,000

Source: U.S. Environmental Protection Agency, Long Island Sound Study Comprehensive Conservation and Management Plan, Final Report, U.S. Environmental Protection Agency, Stony Brook, NY, 1994.

Aside from continuing federal monitoring programs, the LISEP recommends the monitoring of contaminant levels in edible fish and shellfish to ensure compliance with the newly proposed fish safety initiative of the USFDA. Recommendations are also proposed to improve monitoring of toxic substances. In addition, recommendations are made to identify the sources and sites of PCB loadings to the sound and to reduce and eliminate the loadings on a priority basis. Finally, the New York-New Jersey Harbor Estuary Program and U.S. Army Corps of Engineers have agreed to develop systemwide models for PCBs, mercury, and other toxic pollutants that will provide the technical foundation for comprehensive efforts to eliminate these contamination problems from the Sound-Harbor-Bight system.

One of the most cost-effective means of preventing future degradation of the Long Island Sound system is to preclude the input of toxic substances via pollution abatement and regulatory programs. The reduction of toxic contaminant loads will benefit the system in two important ways. First, sediment quality will improve, thereby mollifying contaminant risks to estuarine life. Second, it will also lower health risks to the seafood-consuming public. The success of this effort will be contingent upon the aquisition of consistent funding for existing programs associated with the control, monitoring, and assessment of toxic substances.[8]

e. Research. Effective management of toxic contamination requires greater understanding of complex processes that can only be attained through basic research initiatives. For instance, more research is needed on the relationship between organism body burdens and toxic responses. Additional quantitative data must also be obtained on the trophic level transfer and bioaccumulation effects of contaminants up the food chain. The relationship between toxicity and specific causative agents is largely unresolved and should be investigated in future research projects as well. The estuary program recommends an evaluation of the use of an ecological risk assessment approach for more widespread application to identify toxicity and its sources in embayments and harbors.[1]

D. Floatable debris

One of the most visible impacts of human activity in the coastal zone is trash floating on estuarine and nearshore waters and accumulating on beaches and along shorelines. Improperly discarded waste and litter represents the principal source of floatable debris in estuaries. "Floatables" impact environmental quality and human activities in several ways. Although rarely dangerous to humans, floatable debris is potentially hazardous to estuarine animals when ingested or when the organisms become entangled in it. Subsequent to ingesting debris, animals frequently suffocate or starve. Others entangled in the debris usually become immobilized and trapped, rendering them vulnerable to predation and drowning.

Floatable debris is aesthetically displeasing and often offensive, which can adversely affect human uses of coastal resources. In extreme cases, the public abandons littered beaches and fishing grounds. Boating activity may decline substantially as engine propellers and cooling water intakes become increasingly fouled by debris, leading to failed systems. As a consequence, recreational businesses (e.g., marinas, bait and tackle shops, and seafood markets) and the tourism industry may suffer significant economic losses. For example, excessive debris floating on the open waters of the sound and stranded on beaches and shorelines during the summer of 1988 resulted in an estimated loss of $1 to $2 million to the Long Island economy. Numerous local and regional businesses (e.g., seafood retailers and restaurants) suffered dramatic economic hardships at this time due to recurring floatable debris events. In addition, many social activities were canceled or curtailed in response to public hysteria regarding the perceived pollution problems.

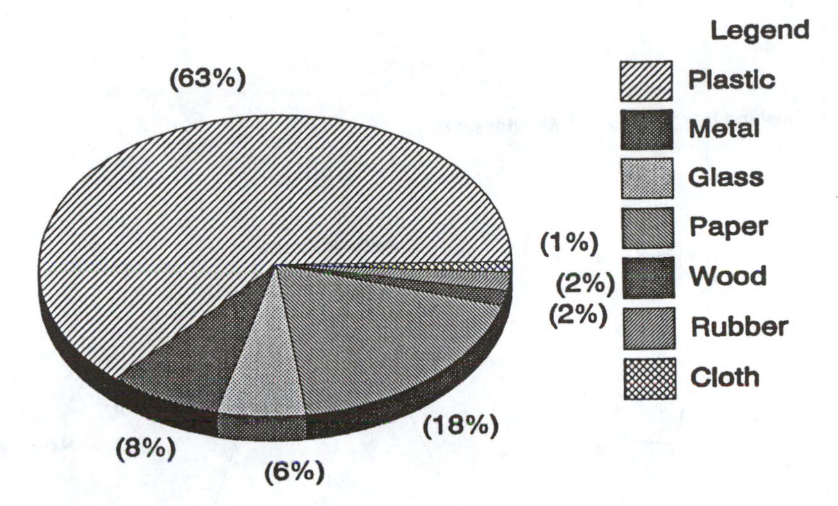

Figure 2.7 Composition of debris collected on Connecticut beaches during the 1990 National Beach Cleanup based on the number of items in each category. (From Masters, M.H. and Freeman, D., *Floatable Debris — Assessment of Conditions and Management Recommendations*, Comprehensive Conservation and Management Plan Supporting Document, U.S. Environmental Protection Agency, Stony Brook, NY, 1993.)

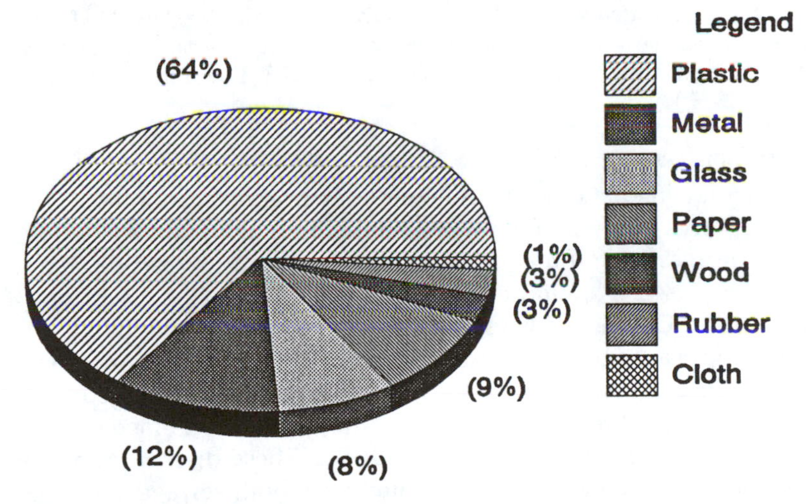

Figure 2.8 Composition of debris collected on New York beaches during the 1990 National Beach Cleanup based on the number of items in each category. (From Masters, M.H. and Freeman, D., *Floatable Debris — Assessment of Conditions and Management Recommendations*, Comprehensive Conservation and Management Plan Supporting Document, U.S. Environmental Protection Agency, Stony Brook, NY, 1993.)

1. *Debris composition*

Floatable debris consists mainly of waste material or litter from products most commonly used by people in their daily lives. Included here are items such as plastic cups and bottles, styrofoam, glass bottles, aluminum cans, paper, and cigarette filters. During the 1990 National Beach Cleanup, plastics accounted for more than 60% of the total debris collected on both Connecticut and New York beaches (Figures 2.7 and 2.8). Metal, glass, paper, wood, rubber, and cloth products comprised the remainder. The percent composition of debris was similar to that of the national totals.[26]

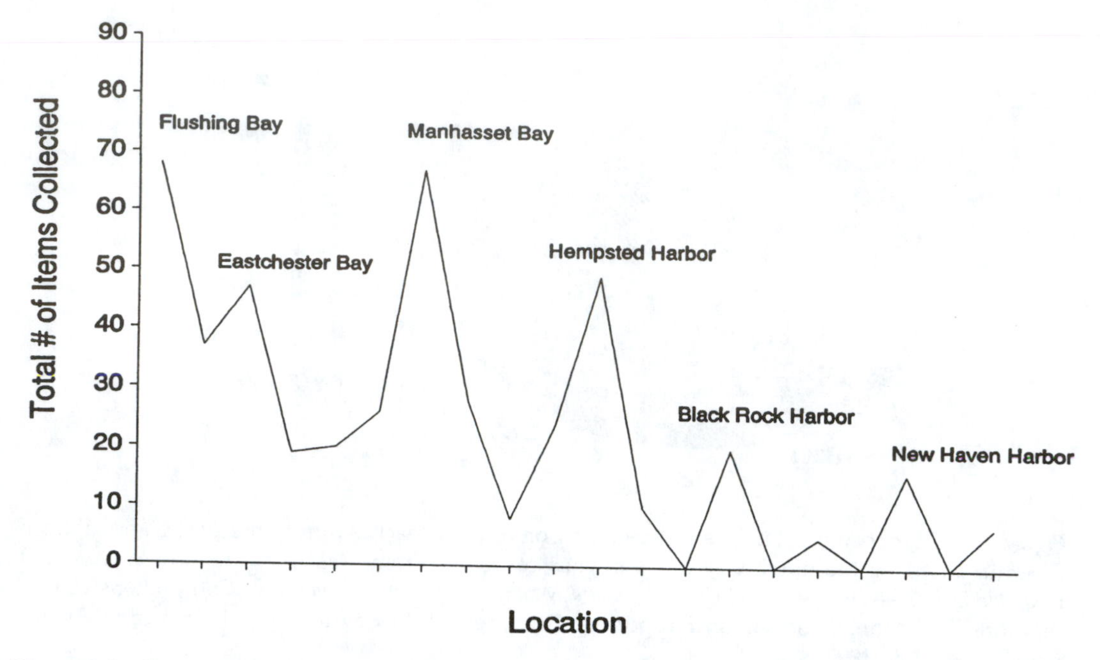

Figure 2.9 Occurrence of floatables in Long Island Sound July 24–26, 1989, showing the highest occurrence of debris associated with harbor/bay entrances. (From Masters, M.H. and Freeman, D., *Floatable Debris — Assessment of Conditions and Management Recommendations*, Comprehensive Conservation and Management Plan Supporting Document, U.S. Environmental Protection Agency, Stony Brook, NY, 1993.)

2. *Debris sources*

Floatable debris that enters Long Island Sound may originate from litter and other waste materials improperly disposed of anywhere in the drainage basin. However, most of the debris actually derives from three primary sources: (1) stormwater discharges and combined sewer overflows; (2) New York Harbor and tributaries to the sound; and (3) shoreline visitors and boaters.[8] Of these sources, inputs from stormwater discharges and combined sewer overflows appear to be quantitatively the most significant. More than 90 combined sewer overflows in Connecticut (i.e., coastal municipalities of Bridgeport, New Haven, and Norwalk) discharge directly to the sound. Another 200 combined sewer overflows in New York City discharge directly to the East River. These discharges are problematical because they contain floatable debris associated with both sewage and street litter from stormwater runoff that enters the western sound. Conditions worsen during periods of heavy precipitation. The influx of floatables from New York Harbor, which can be considerable, exacerbates the problem.

The concentration of floatables decreases from west to east in the open channels of Long Island Sound. A detailed survey of floatables conducted in the system July 24–26, 1989, found the highest occurrence of debris associated with harbor/bay entrances both in Connecticut and New York, indicating that these areas are important sources of the waste.[26] Debris concentrations decreased eastward (Figure 2.9). The floatables were concentrated in surface debris slicks by convergent surface currents or frontal lines.

3. *Management actions*

There are three principal strategies for controlling floatable debris in the sound: (1) combined sewer overflow abatement and stormwater management; (2) physical removal of debris from influent systems, beaches, and open waters; and (3) education.[1] New York City is implementing a comprehensive combined sewer overflow abatement program that includes plans for the areas of Newtown Creek, the East River, and Flushing Bay. This program involves a citywide planning effort to control floatable debris discharges from combined sewer overflows by removing the debris from the waste stream before it can enter the sound. The underground construction of multimillion-liter retention basins is also being planned by the city. Enforceable deadlines for construction of the retention basins range from the year 2001 to 2006. In addition, abatement alternatives (e.g., street cleaning, catch basin maintenance and replacement, and booming and skimming) are being evaluated.

Connecticut is also implementing a long-term abatement strategy for combined sewer overflows. Actions under way to reduce debris loadings to the sound include improved management efforts to control combined sewer overflows and stormwater discharges. New engineering designs are being developed to minimize the release of floatable debris from these systems. Enforceable administrative orders, which address the separation of sewerage and stormwater drainage systems, are in place for Bridgeport, New Haven, and Norwalk. Municipalities with combined sewer overflows on major tributaries (i.e., Norwich, Jewett City, Derby, and Shelton) are also subject to these orders.

Connecticut and New York are continuing their statewide stormwater permit programs to manage stormwater from industrial and construction activities. Both states issue permits for contruction and industrial activities identified by federal stormwater regulations, with applicants required to develop and implement comprehensive stormwater pollution prevention plans and controls that minimize the potential for pollution runoff from storms. Sewage treatment plants have been outfitted with devices such as screens that filter out floatable debris so that it will not reach the sound. The LISEP has recommended the continual maintenance of these devices to ensure that floatable debris is continually and effectively removed from the waste stream.

The estuary program supports the National Beach Cleanup Program. Volunteers have conducted annual cleanups of Long Island Sound shorelines in the fall season since 1988. The Connecticut Sea Grant Program and the New York Department of Environmental Conservation coordinate this effort. During the summer season, routine beach cleanups at Connecticut and New York State beaches are conducted once and twice per day, respectively. Floatable debris action plans developed by the New York-New Jersey Harbor Estuary Program should significantly reduce the amount of floatable debris entering the sound from the New York-New Jersey Harbor.

Other cleanup measures in place to reduce marine debris involve a network of more than 100 volunteer groups that work on recycling projects, trash removal from beaches, and litter reduction of tributaries and other areas of the watershed. Litter reduction from boats is also a high priority because uncontrolled vessels are a substantial source of debris in open waterways during the summer season. Marinas have initiated programs in recent years that encourage boaters to use proper solid waste handling and recycling methods.

The expansion of floatable debris education programs is likewise recommended by the estuary program. A public education campaign (*Clean Steets/Clean Beaches*) implemented by a coalition of public and private groups in New York and New Jersey during 1992 has proven to be an effective anti-litter program. It has made many people aware that much street debris ultimately becomes stranded on Long Island Sound beaches. The goal of the campaign is to modify the behavior of people to prevent littering in the watershed and sound. To this end, education videos, stencil programs, and public signs emphasizing the problems caused by littering have been effectively utilized. The USEPA has awarded $100,000 in grants for the anti-litter campaign.[1,8]

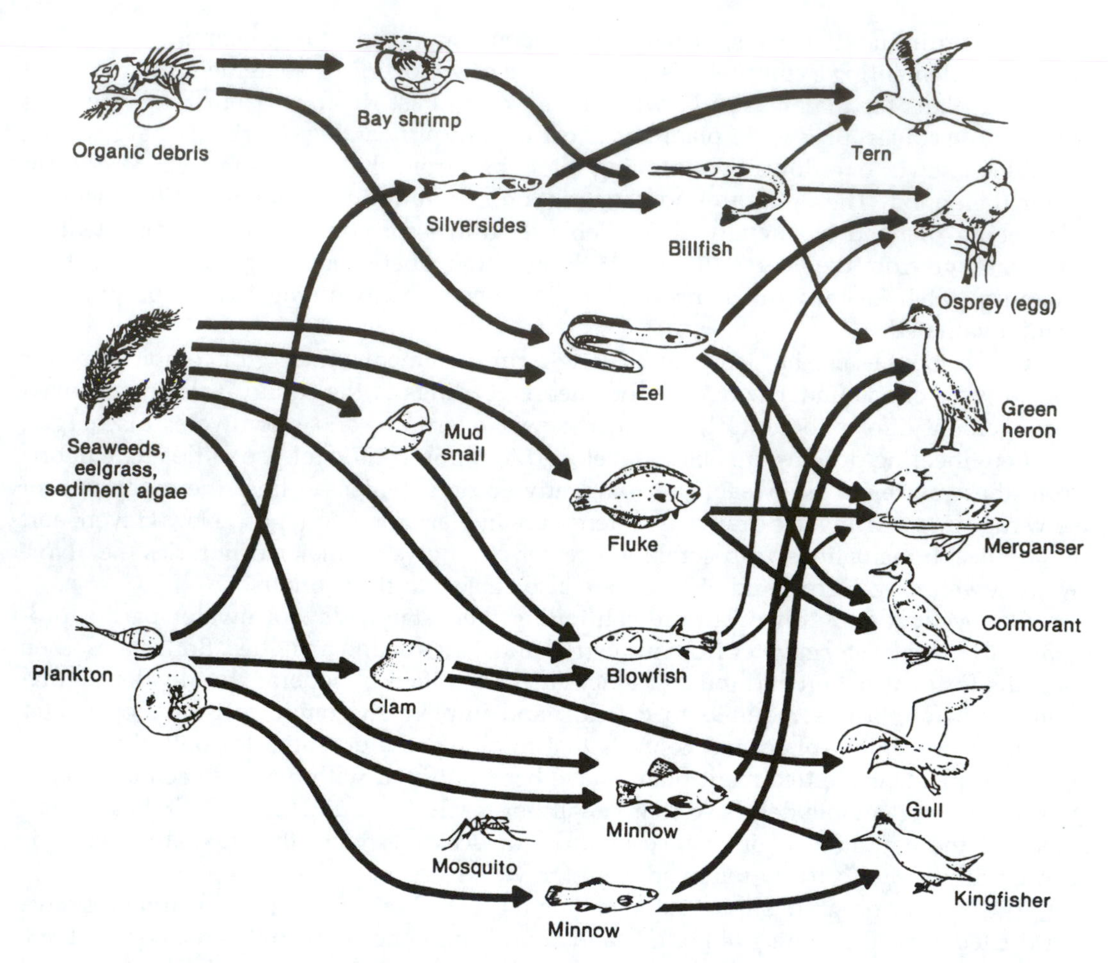

Figure 2.10 Generalized trophic relationships in Long Island Sound. (From U.S. Congress, Office of Technology Assessment, *Wastes in Marine Environments*, OTA-O-334, U.S. Government Printing Office, Washington, D.C., 1987.)

IV. Living resources

Long Island Sound is a highly productive ecosystem characterized by rich and diverse biotic communities inhabiting a wide array of distinctive habitats. Numerous species of plankton, benthos, and nekton reside in the sound and its tributaries. Reptile, mammal, and bird populations occur in nearby wetlands and other areas of the watershed. All of the organisms in the sound — producers, consumers, and decomposers — form a complex food web in which individuals are interconnected by their trophic relationships (Figure 2.10).

A. Phytoplankton

A taxonomic list of phytoplankton identified in Long Island Sound contains about 200 species, 40 of which are designated as major constituents.[27-29] Diatoms and dinoflagellates predominate. Some of the abundant diatom species include *Skeletonema costatum*, *Thalassiosira nordenskioldii*, *T. nitzschoides*, *Rhizosolenia delicatulum*, and *Ditylum brightwellii*. Commonly observed dinoflagellates are *Prorocentrum* spp., *Peridinium* spp., *Gyrodinium* spp., and *Exuviella* spp.

A seasonal periodicity of phytoplankton taxa is conspicuous in the sound. Diatoms and silicoflagellates attain peak abundance in the winter, whereas dinoflagellates reach numerical dominance in the summer. Total phytoplankton productivity is highest during the summer months. At times of maximum phytoplankton numbers, cell counts exceed 10^6 cells/l.

A winter–spring diatom bloom develops between late January and March, and it is dominated by two species, *Thalassiosira nordenskioldii* and *Skeletonema costatum*. Early in the season when water temperature and light intensity are reduced, *T. nordenskioldii* predominates and is superseded by *S. costatum* as the primary form at higher temperatures and light intensities. *Asterionella japonica*, *Chaetoceros* spp., *Lauderia borealis*, *Leptocylindrus danicus*, *P. trochoideum*, and other species occur somewhat later. The community composition changes from April to August with the appearance and growth of many dinoflagellate and microflagellate species, particularly in the summer. Diatoms may remain dominant until the end of May, followed by dinoflagellates, in a classic diatom to dinoflagellate succession. *Skeletonema costatum* dominates phytoplankton blooms in the early fall,[30] when *Chaetoceros* spp. also increase in abundance.[27]

The species composition of phytoplankton varies spatially in Long Island Sound. Species that tolerate lower salinity waters occur in the western sound, and those forms adapted to more oceanic conditions are found in the eastern sound. Phytoplankton abundance also varies across the sound, with low numbers recorded in the East River, highest concentrations between the Throgs Neck and Lloyd Neck areas, and intermediate numbers eastward from Lloyd Neck.[29]

B. Zooplankton

Copepods (i.e., *Acartia hudsonica*, *A. tonsa*, and *Temora longicornis*) dominate the zooplankton community. Tiselius and Peterson[31] described two copepod assemblages in the sound: (1) a boreal assemblage characterized by *A. hudsonica*, *T. longcornis*, and *Pseudocalanus* spp. and (2) a subtropical assemblage dominated by *A. tonsa*, *Paracalanus crassirostris*, *Oithona similis*, and *Labidocera aestiva*. Peak abundance of the boreal copepods takes place in April or May. Their numbers diminish in June, and they generally disappear by the end of July. The subtropical assemblage begins to increase in abundance in the summer and commonly peaks during August and September when food availability and temperature are optimum.[29] Although the boreal copepod populations vanish during the summer and fall, the subtropical forms persist in the plankton year round.[32] The boreal and subtropical assemblages coexist in the sound during the winter and spring. Pulsing of the boreal forms in the spring contributes to their ecological dominance early in the year.

The number of chaetognaths (*Sagitta elegans*) and cladocerans rises rapidly during June and July, subsequent to the diminution of boreal copepods but prior to the peak development of subtropical species. Meroplankton comprises a significant fraction of the zooplankton community in late spring and summer.[33] However, meroplankton abundance decreases in the fall. Low abundances of meroplankton are recorded between November and April, although barnacle larvae remain relatively common from January to March.

C. Benthic communities

1. Benthic flora

Eelgrass (*Zostera marina*) and kelp (*Laminaria digitata*, *L. longicruris*, and *L. saccharina*) are the dominant benthic macroflora in Long Island Sound, and diatoms are the dominant benthic microflora. Widgeon grass (*Ruppia maritima*), another seagrass species, also exists in the sound but in small amounts. Eelgrass beds are not widely distributed in the system, being confined to areas along the Connecticut shore and Fishers Island, New York. The densest concentrations of eelgrass occur in small bays at the mouths of rivers (e.g., Niantic, Thames, and Mystic Rivers). The north shore of Long Island is devoid of eelgrass.[34]

Among kelp beds, *Laminaria saccharina* has the widest distribution; it grows throughout sound. *Laminaria digitata* and *L. longicruris* have a more restricted distribution. They are limited to the area east of New Haven Harbor along the Connecticut coast. Substantial kelp beds cover rocky substrate at Stamford Harbor and Norwalk Islands, Connecticut, and rocky headlands on the southern shore (e.g., Crane Neck Point, Long Island).[35]

2. *Benthic fauna*

Bottom sediment composition strongly influences the structure of benthic faunal communities. Sediment size decreases westward in Long Island Sound with the finest grain sizes recorded near New York City. The larger-sized particles are concentrated in bays and harbors and in the eastern sound. The diversity of benthic infauna is greater in fine sands than muds. Filter feeders (e.g., sponges, bryozoans, tunicates, bay scallops, oysters, and soft- and hard-shelled clams) occupy sandy bottoms. Deposit feeders (e.g., capitellids, glycerids, and tellinids) predominate in finer, muddy sediments with higher concentrations of organic matter. Among important benthic carnivores in the sound are gastropods (e.g., oyster drills, *Urosalpinx cinerea* and whelks, *Busycon* spp.), crustaceans (e.g., blue crabs, *Callinectes sapidus* and lobsters, *Homarus americanus*), echinoderms (e.g., starfish, *Asterias forbesi*), and finfish (e.g., tautog, *Tautoga onitis*; winter flounder, *Pseudopleuronectes americanus*; and summer flounder, *Paralichthys dentatus*). There are also various scavengers that consume large amounts of detrital material on the estuarine floor (e.g., amphipods, crabs, and mud dog whelks, *Ilyanassa obsoleta*).

Several benthic species support recreational and commercial fisheries. For example, surf clams (*Spisula solidissima*), hard clams (*Mercenaria mercenaria*), eastern oysters (*Crassostrea virginica*), and lobsters (*Homarus americanus)* provide millions of dollars of revenue for the regional economy. The long-term viability of these fisheries is dependent on the quality of estuarine waters and benthic habitats.

D. *Finfish*

Only a few species numerically dominate the finfish assemblages. For example, the most abundant finfish populations in the sound are small forage species, notably the bay anchovy (*Anchoa mitchilli*) and Atlantic silverside (*Menidia menidia*). These smaller forms are particularly important because they support recreationally and commercially important species (e.g., bluefish, *Pomatomus saltatrix*; striped bass, *Morone saxatilis*; and weakfish, *Cynoscion regalis*) as well as other larger populations that help to maintain a healthy ecosystem.

The finfish of the sound can be grouped into several assemblages: (1) residents present in the sound year round; (2) warm-water migrants abundant primarily from spring to fall; (3) cool-water migrants inhabiting the system from fall to spring; (4) marine strays; and (5) freshwater strays. Peak species diversity and abundance of individuals occur during the warmer months of the year. Warm-water migrants account for a significant increase in species diversity at this time.

E. *Birds*

Long Island Sound is an important staging and overwintering area for thousands of waterfowl, shorebirds, and seabirds. Many rocky islands located along the northern shoreline and in the central and western basins are ideal breeding grounds for birds because they remain largely devoid of human impacts that destroy nesting habitat and interfere with breeding behavior. In addition, many natural predators (e.g., rats, *Rattus norvegicus* and raccoons, *Procyon lotor*), which prey on eggs and young birds, often do not inhabit these islands. A variety of avifauna populations nest on the islands, such as gulls (e.g., herring gull, *Larus argentus*; laughing gull, *L. atricilla*; and great black-backed gull, *L. marinus*), terns (e.g., common tern, *Sterna hirundo*; least tern, *S. antillarum*; and roseate tern, *S. dougalli*), black skimmers (*Rhynchops niger*), and osprey (*Pandion haliaetus*).

Wading birds (e.g., great blue heron, *Ardea herodias*; great egret, *Casmerodius albus*; and glossy ibis, *Plegadis falcinellus)* and waterfowl (e.g., black ducks, *Anas rubripes*; wood ducks, *Aix sponsa*; Canada geese, *Branta condenses*; scaup, *Aythya* spp.; and mergansers, *Mergus* spp.) also utilize habitat in the Long Island Sound system. Waterfowl hunting on the sound provides recreational opportunities for many people, and it generates significant revenue for the region. The long-term success of avian populations in the Long Island Sound region is contingent upon the conservation of habitats, protection of sites for breeding and other functions, and the maintenance of untainted food supplies and good water quality.

F. Other fauna

Long Island Sound also supports numerous reptilian and mammal populations, some of which are rare, endangered, or threatened. For example, loggerhead turtles (*Caretta caretta*) and green turtles (*Chelonia mydas*) are on the Connecticut, New York, and federal threatened species lists. Kemp's Ridley turtle (*Lepidochelys kempi*), also an inhabitant of this system, is on the Connecticut, New York, and federal endangered species lists.

Harbor seals (*Phoca vitulina*), gray seals (*Halichoerus grypus*), harbor porpoises (*Phocoena phocoena*), and bottle-nosed dolphins (*Tursiops truncatus*) likewise occur here. These marine mammals are important predators, consuming fish and invertebrates. There is some evidence that the harbor seals have increased in abundance in recent years. However, the porpoises and dolphins are less abundant today, perhaps due to increased boating activity and degraded water quality.

Tidal wetlands are important habitat for many reptiles, mammals, and birds. Diamondback terrapins (*Malaclemys terrapin*) inhabit marshes of the area and are occasionally seen along the shorelines of the sound. Many herbivorous mammals also utilize marsh habitats; for example, muskrats (*Ondatra zibethicus*) and several other rodents consume marsh plants. Some are prey of hawks and other raptors. Various mammals in the wetlands are carnivores feeding at the top of the food chain.

V. Habitats

Long Island Sound organisms not only depend on each other through the food web but also on an array of habitats that are used as feeding, nesting, and nursery areas. Important habitats include eelgrass and kelp beds (and other types of submerged aquatic vegetation), unvegetated bottom sediments, natural and artificial reefs, dredged material disposal sites, the water column, rocky intertidal areas, sand and mud flats, and beaches. Critical habitats in nearby watershed areas are tidal wetlands (saltmarshes, brackish marshes, and tidal freshwater marshes) and transitional and upland vegetation.

A. Eelgrass and kelp beds

Both eelgrass and kelp beds are valuable habitats. In the case of eelgrass (*Zostera marina*), several faunal groups live on the leaves: (1) epiphytic felt flora and micro- and meiofauna; (2) sessile fauna attached to the leaves; (3) mobile epifauna moving on the leaf surfaces;

and (4) swimming fauna attached to the leaves. Stems and rhizomes harbor a second category of fauna (e.g., amphipods, bivalves, and nest-building polychaetes). Nekton (e.g., decapod crustaceans and fishes), which swim under the leafage and commonly seek the shelter and protection of this habitat, comprise the third faunal group. The fauna inhabiting bottom sediments (i.e., epibenthic and infaunal invertebrates) constitute the fourth group.[36]

Eelgrass beds in Long Island Sound also serve as significant nursery habitat for a wide diversity of organisms. For instance, juvenile grass shrimp (*Palaemonetes* spp.), blue crabs (*Callinectes sapidus*), and Atlantic silversides (*Menidia menidia*) frequently inhabit these beds. Eelgrass leaves commonly act as substrate for the attachment of eggs from species living in the beds.

Similar to eelgrass, kelp beds provide food and habitat for estuarine organisms. However, these beds are concentrated along rocky intertidal areas rather than on soft bottom sediments in subtidal waters where the eelgrass beds exist. Few herbivores graze directly on kelp, an exception being sea urchins (*Arbacia punctulata*). Some gastropods, crabs, and finfish also feed on kelp.[37] Many organisms use kelp beds as shelter from predators rather than as a source of food.[38]

B. *Tidal wetlands*

Four different types of tidal wetlands occur in the Long Island Sound system: tidal flats (sand and mud), saltmarshes, brackish marshes, and freshwater marshes. The extent of submersion of the habitat strongly controls the composition of the wetlands. The duration of tidal inundation is a key factor in the development of plant zonation in the marsh habitats.

Tidal flats, which occupy the area between open estuarine waters and saltmarshes, are relatively common along the shoreline. Along the north shore, tidal flats are most conspicuous in Greenwich Cove, outside the Saugatuck River, Bridgeport Harbor, and the Housatonic River. On the southern shore, they are well developed within and outside the Nissequogue River, Stony Brook Harbor, and in Conscience Bay, Mt. Sinai Harbor, Northport Bay, Mill Neck Creek, and Little Neck Bay.[34] Benthic invertebrates are well represented in this habitat, particularly polychaetes, bivalves, gastropods, and crustaceans. These organisms provide forage for many crabs, finfish, shrimp, and other animals that are not permanent residents but frequently visit the habitat to feed.

The smooth cordgrass (*Spartina alterniflora*) dominates the lower saltmarsh habitat. Fiddler crabs (*Uca* spp.), grass shrimp (*Palaemonetes* spp.), ribbed mussels (*Geukensia* [*Modiolus*] *demissa*), and certain fish species (e.g., Atlantic silverside, *Menidia menidia*) are commonly found in this area. The saltmeadow grass (*Spartina patens*) is abundant on high marshes and is associated with black grass (*Juncus gerardii*), glassworts (*Salicornia* spp.), sea lavender (*Limonium carolinianum*), spike grass (*Distichlis spicata*), prairie cordgrass (*Spartina pectinata*), marsh elder (*Iva frutescens*), and groundsel bush (*Baccharis halimifolia*). Various herbivores inhabit this area of the saltmarsh, and some species of birds (e.g., rails, *Rallus* spp. and willets, *Catoptrophorus semipalmatus*) nest here.

Brackish marshes represent a transition habitat between saltmarshes and tidal freshwater marshes. They primarily grow along rivers (e.g., Connecticut, Housatonic, Nissequogue, and Quinnipiac Rivers). As an ecotone, the brackish marshes contain a mixture of saltmarsh and tidal freshwater marsh vegetation. Although a variety of plant species occupies this habitat, tall graminoids predominate. A wide variety of birds, mammals, and reptiles utilize the brackish marshes, and many are important herbivores, predators, and scavengers.

Tidal freshwater marshes cover a smaller area of the Long Island Sound system than saltmarshes, but they display a greater species diversity. Emergent freshwater plants predominate in these marshes, including herbaceous and broad-leafed plants, grasses, shrubs, and rushes. Short, broad-leafed emergents dominate the lower elevations, and taller grasses (graminoids) dominate the higher elevations. Phytoplankton and rooted or attached vegetation occur subtidally, and several fish species are common in tidal fresh waters, such as the American eel (*Anguilla rostrata*), bay anchovy, (*Anchoa mitchilli*), mummichog (*Fundulus heteroclitus*), and juvenile Atlantic menhaden (*Brevoortia tyrannus*). There are fewer aquatic invertebrates (e.g., crustaceans, bivalves, and polychaetes) here than in the saltmarshes. However, amphibians, reptiles, and mammals are often observed. Shorebirds, wading birds, waterfowl, and raptors either nest or frequently visit this habitat.

C. Other habitats

Several other habitats in the system support productive biotic communities. For example, there are four dredged material disposal sites that are in use in the sound (i.e., the New London dumping grounds south of the Thames River, the Cornfield Shoals Dredged Material Disposal Area south and west of the Connecticut River, the Central Long Island Sound Dredged Material Disposal Area south of New Haven, and the Western Long Island Sound Dredged Material Disposal Site off Darien). When dredged material is dumped at these sites, new habitat is created and biological recolonization takes place within 12 to 18 months. Benthic recolonization of these disturbed habitats follows a successional pattern, with opportunistic, pioneering species settling initially followed by equilibrium forms.[39]

Natural and artificial reefs serve as sites of attachment for a multitude of organisms. Bartlett (Waterford) and Penfield (Fairfield) Reefs as well as the Stratford Shoal are natural reefs consisting of rocky substrates. Smithtown Bay is the site of the only intentional artificial reef in the sound. These habitats attract considerable numbers of fish and lobsters that are attracted to the reefs for shelter and the rich food supply.

Among the physically harshest habitats are rocky intertidal areas, beaches, dunes, and bluffs. There is generally less biological activity in these habitats than in the subtidal areas of the sound. The communities of plants and animals found here have adapted to unusually difficult physical conditions, such as shifting sediments, scouring, large temperature fluctuations, dessication, salt spray, and breaking waves. Nevertheless, some populations rely on these habitats for survival. For example, horseshoe crabs (*Limulus polyphemus*) breed on sandy beaches and deposit their eggs there. Common (*Sterno hirundo*), least (*S. antillarum*), and roseate (*S. dougallii*) terns and the piping plover (*Charadrius melodus*) nest on the beaches, as do diamondback terrapins (*Malaclemys terrapin*). The box turtle (*Terrapene carolina*) often visits dunes in the area in search of berries and other food.

VI. Anthropogenic impacts

Human activities in Long Island Sound and its watershed have directly impacted the system, as manifested by a reduction in the overall abundance and diversity of living resources. Water pollution is a primary cause of the diminished resources. In addition, the degradation and loss of habitat have contributed to the negative effects. Overfishing likewise has adversely affected some fisheries, such as the surf clam (*Spisula solidissima*), American lobster (*Homarus americanus*), and winter flounder (*Pseudopleuronectes americanus*). Despite these problems, abundant living resources still exist in the sound and its watershed, and they require careful management for their long-term protection and maintenance.[1,34]

A. Living resources

There are six principal areas of management concern that deal with anthropogenic impacts on the living resources and habitats of Long Island Sound. These include adverse effects of hypoxia, toxic contamination, pathogens, floatable debris, overharvesting, and physical alteration of the environment. These problem areas collectively threaten the health, diversity, and distribution of plants and animals or their habitats.

Depending on its severity, hypoxia elicits a variety of responses by marine organisms. For example, the physiological functioning and behavior of the organisms may be altered by low to moderate levels of hypoxia. Altered physiological responses commonly appear as decreased aerobic metabolism and metabolic rate, increased ventilation, and decreased growth and reproduction.[40] Aberrant behavior may be manifested by the emergence of benthic infauna from burrows, increased swimming activity of fish, and the skimming of the air–water interface by organisms to increase oxygen intake. Animals subjected to hypoxia often are more vulnerable to predation and disease. In addition, there is usually an effective loss of valuable habitat area because the organisms can no longer utilize it. In extreme cases, hypoxic events can alter the structure of entire communities through massive mortalities and changes in trophic relationships. As a result, the productivity of plant and animal communities may be significantly reduced.

Many organisms in the sound are exposed to low levels of toxic contaminants (e.g., chlorinated hydrocarbons, polycyclic aromatic hydrocarbons, and trace metals) in bottom sediments, overlying waters, and biotic media. At lower concentrations, these contaminants can impact growth and reproduction or cause other sublethal effects. At high concentrations, they may be lethal. Because the highest concentrations of toxic substances occur in bottom sediments, notably in the western sound and industrial embayments, benthic organisms are generally at greatest risk to exposure. However, bioaccumulation and biomagnification of some contaminants (e.g., DDTs and PCBs) threaten certain upper-trophic-level organisms (e.g., American eel, *Anguilla rostrata*; bluefish, *Pomatomus saltatrix*; and striped bass, *Morone saxatilis*) and, hence, pose a human health concern.

Most pathogens that cause diseases in the resident plant and animal populations are naturally occurring, in contrast to human pathogens that mainly derive from anthropogenic activities. For example, *Haplosporidium nelsoni* (MSX) is a naturally occurring pathogenic protozoan that causes mass mortalities of oyster (*Crassostrea virginica*) populations. Winter flounder (*Pseudopleuronectes americanus*) are occasionally infected with the intracellular parasite, *Glugea stephani*, which usually causes death. Sewage-associated human pathogens are well documented, such as *Vibrio cholerae*, *Salmonella* spp., and *Shigella* spp. Some sewage that enters the sound is raw or inadequately treated and, thus, may pose a risk to humans who come in contact with it.

Although floatable debris is unsightly and a potential navigation hazard, it is not a major threat to estuarine life in the sound. However, there are documented cases of a few organism deaths due to entanglement or ingestion of debris. Of great concern is the potential losses of endangered species from floatable debris.

Overfishing in Long Island Sound has become a serious issue. The harvesting pressure on certain species (e.g., surf clams, *Spisula solidissima*; lobster, *Homarus americanus*; bluefish, *Pomatomus saltatrix*; and winter flounder, *Pseudopleuronects americanus*) has elicited calls for stricter management of the fisheries. Overharvesting of surf clams in Long Island Sound occurred during the mid-1980s and early 1990s. The winter flounder population decreased markedly in 1992. The American lobster continues to be the target of intense fishing pressure. Effective fishery management of commercially and recreationally important finfish and shellfish populations is essential to maintaining consistent levels of seafood production from the system. It is also vital to restoring diminished stocks from overharvesting impacts.

B. Habitats

Apart from direct impacts on living resources, physical disturbances of habitats associated with human activities have been detrimental to numerous plant and animal communities. Approximately 25 to 35% of the tidal wetlands of Long Island Sound have been eliminated during the past century by dredging for navigation purposes and filling for landfills, roads, and commercial, industrial, and residential development. Legislative action during the past few decades, however, has virtually arrested the loss of tidal wetlands habitat. Strict federal and state regulations are now protecting tidal wetlands against human destruction, but they cannot reverse some of the impacts caused by historic degradation.[34]

Submerged aquatic vegetation, particularly eelgrass beds, has also experienced a significant decline in abundance and distribution in the sound. Impacts on this habitat have been largely attributed to increased nitrogen loadings, phytoplankton blooms, elevated turbidity levels, and reduced light transmission that have rendered broad areas of the estuarine bottom unsuitable for eelgrass growth. The elevated nitrogen levels may have prompted a shift in vegetation type from eelgrass to certain benthic algae (e.g., *Ulva lactuca*), which are more tolerant of high nitrogen levels.

Human-built structures (e.g., breakwaters, groins, jetties, and seawalls) interrupt sediment transport that naturally replenishes beaches and dunes, leading to increased erosion of these habitats. Pedestrian and vehicular traffic exacerbate degradation of beaches, dunes, and bluffs. Uncontained public access to some areas has resulted in the destruction of much fragile vegetation that naturally trap and retain sediments. The net effect is greater vulnerability of these habitats to additional erosion and degradation.

Other human activities that adversely affect habitats in the sound are dredging and dredged material disposal operations. Both activities temporarily destroy the benthic habitat and benthic communities. The construction of electric generating stations and other coastal installations also physically destroys extensive habitat area. Dams built on tributary systems decrease flow and limit anadromous finfish (e.g., alewife, *Alosa pseudoharengus*; blueback herring, *Alosa aestivalis*; American shad, *A. sapidissima*; and Atlantic salmon, *Salmo salar*) access to critical habitats. Some species unable to reach their breeding habitats may decrease significantly in abundance or, in the most severe case, totally disappear from the region.

VII. Management actions

Several ongoing conservation programs deal specifically with the protection of living resources and the preservation, restoration, and enhancement of habitats in the Long Island Sound system. Included here are

- Fisheries management programs, focusing on species regulation and restoration and population monitoring
- Wildlife management programs, concentrating on the protection and restoration of endangered and threatened species of plants and animals as well as population monitoring and assessment
- State and federal regulatory programs, targeting the protection of tidal wetlands and submerged aquatic vegetation
- Habitat restoration and enhancement programs[8]

Table 2.15 Actions for Managing Harvested Species in Long Island Sound

Ongoing Programs	Responsible Parties/Status
Development and implementation of fishery management plans, including research, monitoring, and conservation law enforcement activities	The CTDEP, the NYSDEC, the NMFS, and the USFWS cooperate under the auspices of the Atlantic States Marine Fisheries Commission and the New England and Mid-Atlantic Fishery Management Councils to develop plans that reduce fishing mortality, prevent overfishing, and increase stock size and yield from Long Island Sound (and all Atlantic coast) fisheries. Research, monitoring, and conservation law enforcement activities are integral components of such activities, costing Connecticut in excess of $1,000,000 per year in state and federal funds.
Management of shellfish aquaculture activities including resource monitoring	In state-managed waters, the Connecticut Department of Agriculture's Aquaculture Division, the NYSDEC, and private shellfish companies engage in practices intended to enhance production of oysters and hard clams as well as manage other available resources (e.g., surf clams) as needed. In waters under municipal jurisdiction, a number of towns have shellfish commissions that manage town shellfish beds for recreational and sometimes joint recreational/commercial harvests. In Connecticut, the state program costs approximately $1,250,000 for staff, base programs, and cultch (shell) acquisition. Municipal programs are often conducted for $5,000 or less.

Source: U.S. Environmental Protection Agency, Long Island Sound Study Comprehensive Conservation and Management Plan, Final Report, U.S. Environmental Protection Agency, Stony Brook, NY, 1994.

Management actions specific to the principal water quality problems of the sound (i.e., hypoxia, toxic substances, pathogens, and floatable debris) have been detailed previously and will not be recounted here. The following discussion addresses living resources and habitat management actions.

A. Species management actions

The LISEP has developed management plans that enhance ongoing efforts to protect and restore endangered and threatened species. Both Connecticut and New York are implementing the requirements of the federal Endangered Species Act. Lists of endangered, threatened, and rare species are periodically revised and published. Recommendations have been proposed to formulate legislation or regulations that will minimize disturbance to essential habitats of rare plants and animals.

Fishery and waterfowl management plans are also being developed and implemented to preclude overharvesting of commercially and recreationally important species, to reduce mortality, and to increase stock size and yield from the sound. Associated research and monitoring programs are being conducted in support of these plans. Various recommendations have been advanced to enhance harvestable species. For example, the construction of artificial reefs and the installation of fishways or fishlifts to bypass obstructions to anadromous fish migrations will ultimately improve productive fishing opportunities. The expansion of shellfish seed stocking, oyster cultch placement, and aquaculture ventures is intended to enhance clam and oyster production (Table 2.15).

For general aquatic life protection, commitments have been made by federal, state, and local agencies to plan dredging operations that will minimize adverse effects on aquatic organisms and their habitats. The mitigation of entrainment and impingement mortality of aquatic organisms at industrial facilities, especially power plants, is another goal of the program. This may be achieved by the application of best available technology at the facilities.

Other measures to protect balanced indigenous communities in the sound entail efforts to prevent the introduction of undesirable exotic and nuisance species. Introduced species (e.g., the common reed, *Phragmites australis*) pose a potential threat to many endemic organisms because they often outcompete these organisms and rapidly spread. Certain habitats can be greatly altered and the ecological balance destroyed. Therefore, it is important to protect native plant and animal populations from these undesirable invasions. Even some native species that attain high abundances can be potentially harmful by preying upon, outcompeting, and overwhelming other species in the system. Examples of such species are the Norway rat (*Rattus norvegicus*), raccoon (*Procyon lotor*), and mute swan (*Cygnus olor*). Domestic animals also can be a threat to endangered species by preying on the eggs and young of these sensitive populations.

B. Habitat management actions

Management actions designed to protect, conserve, and restore habitats in Long Island Sound are multifaceted. They include pollution abatement, land-use regulation, land acquisition, easements, and habitat restoration and enhancement initiatives (Tables 2.16 and 2.17). The LISEP also strongly supports ongoing habitat protection efforts funded outside the purview of the program. For example, Connecticut and New York continue to pursue the restoration of coastal habitats (e.g., submerged aquatic vegetation beds, tidal wetlands, beaches, dunes, and coastal woodlands) through programs administered by both state and federal agencies (e.g., Connecticut Department of Environmental Protection, Connecticut Department of Transportation, New York State Department of Environmental Protection, USEPA, U.S. Fish and Wildlife, and U.S. Army Corps of Engineers). Coastal regulatory programs have also provided significant protection of tidal wetlands and other habitats. They regulate potentially detrimental activities such as dredging, filling, and construction of lagoons, docks, piers, and other structures. However, not all habitats are afforded equal protection.

The LISEP recommends the acquisition of additional funding to enhance the existing state and federal programs that are targeting habitat restoration and maintenance. Prospective funding sources are the National Coastal Wetlands Conservation Grants, the Land and Water Conservation Fund, Long Island Sound Challenge Grant Funds, Long Island Sound Cleanup Funds, and the U.S. Army Corps of Engineers Section 22 Planning Funds, among others. An estimated $1.7 million is needed to implement habitat restoration and enhancement projects.[8]

Land acquisition is an important component of habitat protection in Long Island Sound. Both Connecticut and New York have land acquisition and management programs that include listings that prioritize the most sensitive habitats in need of protection and acquisition. These states acquire habitats by easement or other means. Funds may derive from the National Coastal Wetland Conservation Program, the North American Waterfowl Management Plan, the Land and Water Conservation Fund, or other state and federal sources. Clearly, substantial and consistent funding is required for the ongoing success of the land acquisition program.

C. Education, monitoring, assessment, and research

It is imperative to educate the public about the living resources of the sound. To this end, the LISEP has recommended the development of an outreach program not only to inform individuals of the valuable plant and animal communities occurring in the sound and nearby watershed areas but also of the the critically important habitats that support them. In addition, a citizens monitoring program is recommended as a component of outreach.

Table 2.16 Actions Dealing with the Protection and Acquisition of Habitat in Long Island Sound

Ongoing Programs	Responsible Parties/Status
The states of Connecticut and New York and the USACOE will continue to implement their permit programs and coastal consistency provisions of states' Coastal Management Programs to regulate use and development of aquatic resources and critical habitats such as tidal and freshwater wetlands, intertidal flats, submerged aquatic vegetation beds, beaches, and dunes. These programs also regulate dredging and the disposal of dredged sediments at designated sites in Long Island Sound. Open water disposal is only permitted at the designated open water sites and may only occur if the disposal will not cause adverse impacts to estuarine organisms.	Programs are managed by the NYSDEC, the CTDEP, the USACOE, the EPA, and the NYSDOS that are essential to habitat preservation and conservation. Key permit programs include Tidal Wetlands, Structures and Dredging and Filling, and Coastal Management and Inland Wetlands and Watercourses Programs in Connecticut; Protection of Waters and Freshwater Wetlands program, and Coastal Erosion Hazard Protection in New York; Section 404 of the Clean Water Act, Section 10 of the Rivers and Harbors Act, and Section 103 of the Marine Protection, Research and Sanctuaries Act. These are the primary programs that regulate activities in coastal waters and freshwater wetlands to protect and minimize adverse impacts to aquatic habitats. The states and federal agencies routinely update dredged sediment disposal plans and procedures as new testing and management protocols are developed. Annual program costs are $1.15 million in Connecticut.
Connecticut will continue to reduce habitat degradation caused by stormwater runoff projects (e. g., chronic dilution effects and sedimentation) through the goal of retaining the first one inch of runoff.	The CTDEP and local governments are implementing stormwater management actions in accordance with stormwater general permitting guidelines and the standards in the Coastal Management Act to avoid or minimize habitat degradation caused by stormwater runoff. This is accomplished through the goal or requirement of retention of the first one inch of runoff.
Connecticut and New York have programs to acquire by easement, fee simple acquisition, or other means habitats important for populations of plants and animals. These programs include the development of priority listings for acquisition and protection. Connecticut and New York have land acquisition and management programs that use state funds and federal fund programs such as the Land and Water Conservation Fund, the National Coastal Wetland Conservation Program, and the North American Waterfowl Management Plan to protect and acquire coastal lands and wetlands.	Both states have had a long history of acquiring lands and wetlands along the shoreline and in the Long Island Sound watershed. In Connecticut, the CTDEP is responsible for land acquisition programs and for the management of parks, forests, and wildlife management areas. The CTDEP is responsible for the management of over 114 different management areas, totaling over 11,700 acres of land and wetland, located along its tidal shorelines. The NYSDEC and the CTDEP are the primary parties responsible for initiating acquisition projects. In Connecticut, the Recreation and Natural Heritage Trust Program is the principal state funding program for land acquisition. Examples of coastal habitats that have been acquired with this fund include Cedar Island in Clinton, Davis Farm at Barn Island in Stonington, Beacon Hill in Branford, and Selden Island in Haddam. In 1992, Connecticut established a Migratory Bird Conservation Stamp Program containing a dedicated fund, a portion of which will be used for acquisition related to migratory bird protection and enhancement. In 1992, Connecticut's statewide program costs were $17,000,000.

The USFWS maintains a national system of refuges, which includes the Stewart B. McKinney National Wildlife Refuge in Connecticut (i. e., Salt Meadow, Chimon Island, Sheffield Island, Goose Island, Milford Point, and Falkner Island units) and Long Island National Wildlife Refuge Complex in New York (i.e., Oyster Bay and Target Rock units).	These units in Long Island Sound are owned and managed by the USFWS. Congress has authorized the expansion of the McKinney National Wildlife Refuge and the Service is currently pursuing acquisition of a portion of the Great Meadows complex in Stratford, Menunketesuck Island, and wetlands in Westbrook. Three million dollars have been appropriated for these sites to date, and the remaining acquisition costs are projected at $11 million.
Congress has authorized the creation of the Silvio Conte Connecticut River National Fish and Wildlife Refuge within the Connecticut River watershed for the purpose of conserving, protecting, and enhancing the Connecticut River Valley populations of plants, fish, and wildlife; preserving natural diversity and water quality; fulfilling international treaty obligations relating to fish and wildlife; and providing opportunities for scientific research and education.	The USFWS is responsible for the development of recommendations with respect to defining and designating refuge boundaries, developing a management strategy for the river and identifying lands for acquisition. The Service is working cooperatively with the states and heritage programs to collect information for *Species of Special Emphasis* and significant concentration areas for these species. As part of this analysis, the Service has identified the lower tidal section of the Connecticut River as a nationally significant fish and wildlife habitat complex.
Connecticut has established a Migratory Bird Conservation Stamp Program, the proceeds of which can be used for acquisition and management. The newly created state income tax form check-off for endangered species, natural areas preserves, and watchable wildlife creates a fund that can be used for the identification, protection, conservation, management, and education activities related to the above listed wildlife and habitats.	These programs are statewide programs administered by the CTDEP and a portion of the proceeds are expected to be directed to projects associated with Long Island Sound. Connecticut has completed its first issue duck stamp and prints, and the sale of art products will be an ongoing program. Projects are soon to begin under this program and will include restoration and wildlife conservation. This is the first year for the check-off program.

Recommendations	Responsible Parties	Time Frame	Estimated Cost
Create a Long Island Sound Reserve System consisting of areas of land and water of outstanding or exemplary scientific, educational, or biological value to reflect regional differentiation and variety of ecosystems and to include representatives of all of the significant natural habitats found in the sound. Where appropriate, sites will be selected from existing lands and wetlands held for conservation purposes so that acquisition funds will be directed toward those lands in private ownership that are needed to complete the reserve system. The primary activities in the recommendation include site identification (2 years) and site protection through the development of management plans, acquisition where necessary, and site management.	CTDEP, NYSDEC, New York State Office of Parks and Recreation and Historic Preservation, USFWS, Long Island Sound Bi-State Committee	—	$50,000 per year for each state for staff to identify sites, develop acquisition strategies, and manage the reserve complex. Acquisition costs will depend upon areas identified for protection through purchase.

Table 2.16 (continued) Actions Dealing with the Protection and Acquisition of Habitat in Long Island Sound

Recommendations	Responsible Parties	Time Frame	Estimated Cost
Connecticut and New York should continue to acquire or protect through less than fee simple means, significant coastal habitats through funding sources such as the Land and Water Conservation Fund, the National Coastal Wetland Conservaton Program, the North American Waterfowl Management Plan, Connecticut's Recreation and Natural Heritage Trust Program, Connecticut's Migratory Bird Conservation Stamp Program, New York's Environmental Protection Fund, and, where appropriate, natural resource damages recovered under CERCLA or OPA90.	CTDEP, NYSDEC, Assistance of local governments, environmental groups, and federal granting agencies	—	$50,000 per year for each state for staff
Acquire and protect those sites that are considered priorities for acquisition in the New York State Open Space Conservation Plan. Sites include Oyster Bay Harbor ($5 million), Porpoise Channel ($2 million), Plum Point ($1 million), and Udall's Cove ($8 million). Other sites on Long Island Sound that among the state's highest priority acquisition sites include: Bronx River Trailway, Udall's Ravine, Alley Creek ($750,000), Long Creek and Mattituck Creek ($340,000), Premium River ($750,000), and Cedar Beach Creek ($186,000).	NYSDEC, New York State Office of Parks and Recreation and Historic Preservation	—	Priority sites for acquisition total $16 million
Acquire and protect those sites that are considered priorities for acquisition in Connecticut. The Great Meadows site is the highest priority. (See also Ongoing Programs, previous page.)	CTDEP, USFWS	—	$14 million
Encourage activities of existing Long Island Sound-specific land trusts and encourage formation of new trusts, to seek donations and easements of localized habitat areas for the plants and animals of Long Island Sound.	NYSDEC, EPA-LIS Office	—	Redirect base program

Source: U.S. Environmental Protection Agency, Long Island Sound Study Comprehensive Conservation and Management Plan, Final Report, U.S. Environmental Protection Agency, Stony Brook, NY, 1994.

Table 2.17 Actions Focusing on the Restoration and Enhancement of Aquatic and Terrestrial Habitats in the Long Island Sound System

Ongoing Programs	Responsible Parties/Status
Connecticut, New York, and federal agencies will continue to pursue the restoration of degraded tidal wetlands.	These programs are administered by the NYSDEC, the NYSDOS, the CTDEP, the CTDOT, the U.S. Fish and Wildlife Service, the USACOE, and the EPA.
	Since 1980, the CTDEP has, in cooperation with many partners, restored over 1000 acres of degraded tidal wetlands, The CTDEP uses the Long Island Sound Cleanup Account to fund the restoration of degraded tidal wetlands. The CTDEP has created a tidal wetland restoration program with staff and specialized equipment with annual operating costs of $350,000. The CTDEP receives commitments of approximately $800,000 per year from the CTDOT's Intermodal Surface Transportation Efficiency Act (ISTEA) program to fund wetland restoration projects associated with transportation facilities.
	The USFWS provides, on average, $45,000 of Partners in Wildlife Funds to Connecticut to conduct wetland restoration and also provides staff and equipment to assist in tidal wetland restoration. It also provides challenge grant monies to conduct tidal pool and pan restoration activities in its Connecticut refuges.
Through Connecticut's coastal permit programs and consistency with the Connecticut Coastal Management Act, applicants may be required to protect, restore, or enhance aquatic resources.	These programs are managed by the CTDEP. Retrofits or removal of tide gates have been required to increase tidal flows to tidal wetlands and embayments and offsetting of unavoidable wetland losses for public benefit projects such as bridge replacements through wetland restoration.
Connecticut is preparing a tidal wetland management plan that includes an identification of potential wetland restoration sites.	The responsible party is the CTDEP. This project has been funded by NOAA's Office of Ocean and Coastal Resources Management.
Connecticut will continue the Coves and Embayment Restoration Program to restore degraded tidal and coastal embayments and coves.	Since 1982, the CTDEP has sponsored, in cooperation with coastal municipalities, the restoration of 20 sites. In 1989, the Connecticut legislature amended the Clean Water Fund to create the Long Island Sound Cleanup Account, which has provided increased funding to this program. Annual restoration costs average $500,000 per year. The Department will continue to request appropriations for this account as needed.

Table 2.17 (continued) Actions Focusing on the Restoration and Enhancement of Aquatic and Terrestrial Habitats in the Long Island Sound System

Ongoing Programs	Responsible Parties/Status
Connecticut, New York, and federal agencies currently administer programs for the restoration of habitats other than tidal wetlands such as dunes, submerged aquatic vegetation beds, and coastal woodlands.	The NYSDEC, the CTDEP, and the USFWS are the responsible parties. The CTDEP continues to conduct dune restoration activities on state lands and assists municipalities and private citizens with their restoration projects. The CTDEP created the Long Island Sound License Plate Fund, which provides funding for restoration projects. In 1993, $25,000 was specifically set aside for municipal dune restoration projects. Management of coastal upland habitats is conducted chiefly on Connecticut Wildlife Management Areas. The USFWS has begun to manage coastal uplands in the McKinney National Wildlife Refuge units.
New York is phasing out, and Connecticut prohibits, maintenance ditching of mosquito ditches in favor of selective use of open marsh water management techniques to control mosquitos and restore pools and ponds on tidal wetlands.	The responsible parties are the CTDEP and the NYSDEC in cooperation with mosquito control agencies and federal agencies. The CTDEP, the USACOE, the USFWS, NOAA's National Marine Fisheries Service (NMFS), and the EPA agreed to discontinue maintenance of mosquito ditches in Connecticut's tidal wetlands since 1985 and to allow the selective use of the open marsh water management as a mechanism to restore the natural character and habitat diversity of tidal wetlands.

Commitments	Responsible Parties	Time Frame	Estimated Cost
Coastal America, a cooperative effort of several federal agencies, is conducting a study in Connecticut to evaluate the impacts of transportation facilities upon ten tidal wetland sites. This study is being sponsored by the CTDEP and undertaken by the USACOE. When the study is completed, restoration plans will be developed for those sites where a transportation facility is shown to be the cause of the degradation. Restoration is expected to be implemented through a combination of ISTEA, Water Resources Development Act, Long Island Sound Cleanup Account funds, New York's Environmental Protection Fund, and, where appropriate, natural resource damages recovered under CERCLA or OPA90.	CTDEP CTDOT Coastal America Partners	Restoration projects will proceed as funding is approved.	$100,000 for the initial study; restoration costs will vary for each project site.
Connecticut's Coves and Embayments Program will complete nine restoration projects in progress and commitments to begin three new projects.	CTDEP in cooperation with the municipality sponsor	Varies depending on project	$263,625 for projects in progress and $123,475 for projects to commence.

Recommendation	Responsible Parties	Time Frame	Estimated Cost
Connecticut and New York should continue to pursue the use of funds from the following programs, and explore additional funding sources, to support restoration and enhancement activities described in the previous recommendation: The Land and Water Conservation Fund, the Intermodal Surface Transportation Efficiency Act (ISTEA) Enhancement Program, the Partners in Wildlife Program, Section 319 of the Clean Water Act, Army Corps of Engineers Section 22 Planning Funds, the Water Resources Development Act, National Coastal Wetlands Conservation Grants, the North American Waterfowl Management Plan, Connecticut's Long Island Sound Cleanup Funds, and the Coastal Zone Management Act.	CTDEP CTDOT NYDOT NYSDEC NYSDOS EPA USACOE USFWS	Ongoing	Existing staff will be used; project costs vary from site to site.
The rapid displacement of native brackish and fresh tidal plant communities on the Connecticut River has been identified as the single most significant habitat problem in this estuary. A specific restoration program for the control of common reed in these tidal wetlands needs to be implemented to check and reverse the spread of common reed and develop the most efficient means of effecting this restoration. Control techniques need to be evaluated for the full range of wetland habitat types on the river. Baseline surveys will be established and post-control monitoring over multiple years will be conducted.	CTDEP USFWS	3 years	$130,000 for amphibious mulching machine and $100,000 for staff, supplies, and monitoring.
New York should continue to phase out maintenance ditching for mosquito control. These programs should receive additional support for selective use of open marsh water management techniques to control mosquitos and restore pools and ponds on tidal wetlands.	NYSDEC in cooperation with mosquito control agencies	—	$1,000 per acres for open marsh water management.
Obtain long-term funding for Connecticut wetland restoration staff.	CTDEP	Upon approval of funding	$250,000 per year for staff.

Table 2.17 (continued) Actions Focusing on the Restoration and Enhancement of Aquatic and Terrestrial Habitats in the Long Island Sound System

Recommendations	Responsible Parties	Time Frame	Estimated Cost
Connecticut and New York should develop a restoration plan for the full range of coastal terrestrial and estuarine aquatic habitats adjacent to and in Long Island Sound. The restoration plan will include a list of potential restoration projects and a priority listing of projects to be implemented. Preliminary sites identified for future restoration in New York include: City Island ($300,000), Pelham Bay Park ($400,000), Wading River ($50,000), Sunken Meadow Creek ($50,000), Crab Meadow ($50,000), and Mattituck Creek ($100,000). Other sites in New York where costs have not been estimated include Pugsley Creek, Udall's Cove, Oak Neck Creek, Frost Creek, and East Creek. Connecticut has estimated that ten priority sites could be restored for $750,000, or approximately $75,000 per site.	CTDEP NYSDEC NYSDOS EPA NOAA USACOE USFWS	3 years	$50,000 per year for each state for three years; restoration costs will vary depending upon project type.
New York should strengthen its capabilities for implementing programs that restore degraded habitats. This should be undertaken in cooperation with the implementation of the Long Island Sound Regional Coastal Management Plan.	NYSDEC NYSDOS	—	$250,000 per year.

Source: U.S. Environmental Protection Agency, Long Island Sound Study Comprehensive Conservation and Management Plan, Final Report, U.S. Environmental Protection Agency, Stony Brook, NY, 1994.

Individuals engaged in this monitoring program will make observations of the living resources in the system to aid managers in identifying problems and assessing the effects of their management actions.

Scientific monitoring of the system will be continued in an effort to determine the condition of finfish, shellfish, and wildlife resources as well as valuable habitat. An integral component of the Long Island Sound monitoring and management programs is a sound-wide open water fishery survey. Connecticut and New York conduct various surveys of commercially important finfish and shellfish populations as well as surveys of colonial waterbirds. The strategy is to computerize data collected from these monitoring efforts and other elements of the program into a Geographic Information System for rapid access and assessment of living resources.

Detailed examination of the database on living resources of the estuary has identified a number of data gap areas. For example, commitments have been made to evaluate the causes of the decline of eelgrass in the sound, changes in the phytoplankton community resulting from nitrogen enrichment, effects of hypoxia on bottom feeding finfish, and vegetation changes in a restoring tidal wetland. The Long Island Sound Research Fund will be an important source of funding for research of these problems. A stable source of funding will be required to address future data gap issues.

VIII. Land use and development

The myriad of human activities in watershed areas greatly affects water quality and habitat in the Long Island Sound system. With more than 8 million people inhabiting the drainage area of the sound, water quality is closely coupled to land use and development in the 4.14×10^6 ha watershed area. Land-use policies often have not considered water quality protection. Hence, water quality and habitats in the system remain vulnerable to development in the watershed despite the generally improved conditions realized during the past 2 decades with the implementation of more stringent regulatory programs. This is due in large part to varied and dispersed nonpoint sources of pollution and ever-increasing human activities along the shoreline and in open waters of the sound.

The LISEP has presented five major recommendations to improve land-use planning and management of the entire watershed.[8]

- The impacts from existing development are significant, particularly in urbanized areas, and must be reduced to improve coastal water quality. These areas should be targeted for nonpoint source management, including public education, infrastructure upgrades, spill prevention and response, and flood and erosion control. Also, abandoned or underutilized sites should be a high priority for remediation and reuse.
- The impacts from new development are also significant and must be minimized to prevent further degradation of water quality. Progressive planning and management should ensure the application of best management practices, protection of wetlands, minimization of land disturbances, improvement of access, and maintenance of appropriate water-dependent uses.
- To improve land-use decision-making that incorporates effective water quality and habitat protection, better information, training, and technical assistance must be available. Training, technical assistance, and financing should be made available to local governments as well as education for the public, professionals, and trade organizations. This will help develop consistent land-use and natural resource information and management practices in the region.

- Conservation of natural resources and open space is vital to the long-term protection of Long Island Sound. Open space preservation and conservation practices must be aggressively pursued. This might be accomplished through a watershed-based planning approach that integrates protection of surface waters with programs and plans that guide growth and development.
- Public access is essential to public use and enjoyment of the sound, especially because improvements to water quality involve public costs. Public access improvements should be aggressively pursued throughout the watershed using a combination of traditional techniques, such as fee-simple acquisition, and innovative techniques, such as transfer of development rights and tax credits.

Managing the impacts of land use and development and formulating remedial actions to protect and enhance water quality and habitat in Long Island Sound are formidable tasks. Coastal management programs in Connecticut and New York have provisions for guiding land use and development in the watershed. However, specific actions addressing improvements in water and habitat quality in light of current land-use and development policies are incomplete. New actions must be advanced and the necessary funding must be obtained to successfully effect long-term change in the system.

IX. Program implementation

There are three principal elements essential to the implementation of the management plan for Long Island Sound: (1) continuation of the Management Conference; (2) public involvement and education; and (3) adequate funding. Continuation of the Management Conference will provide the following:[1]

- Tracking, monitoring, and reporting of program implementation
- Enhancing implementation of actions
- Obtaining commitments from participating agencies for implementation
- Seeking adequate program funding
- Continuing public involvement and education

Public involvement and education are vital to protecting, restoring, and improving water quality, living resources, and habitats in the Long Island Sound system. This is accomplished by fostering a greater understanding and appreciation of the sound and its watershed through various education initiatives. For example, information on protecting and restoring the sound can be effectively exhibited to the public at local fairs and conventions. It can likewise be forwarded to all municipalities for general public distribution. A public outreach network will increase community awareness and stewardship regarding the system. Primary and secondary schools serve as excellent outlets of information on living resources for children. Training and education on environmental issues germane to Long Island Sound must also be extended to the environmental decision-making community in Connecticut and New York. Finally, it is imperative to encourage public participation in activities relating to the cleanup and protection of the sound and its watershed.

Additional state, federal, and private funding must be secured to meet the needs of existing, enhanced, and implementation programs. Existing program funding relies heavily on commitments from the Connecticut Department of Environmental Protection, New York State Department of Environmental Conservation, the USEPA, local governments, and other federal, state, and local agencies. For enhanced program support, the Management Conference has recommended funding under the Long Island Sound Improvement Act, other available federal Coastal Zone Management Act and Clean Water

Act funds as well as additional state and local funding sources. The Management Conference also has recommended that Congress appropriate $50 million to fund a Long Island Sound Challenge Grant program in support of management plan implementation.[1,8]

The costs of cleanup and restoration of Long Island Sound are significant. Many millions of dollars are required for adequate funding of existing, enhanced, and implemention programs as documented in the CCMP. Such funding is necessary for the revitalization and enhancement of this critically important system.

References

1. USEPA, *Long Island Sound Study: Comprehensive Conservation and Management Plan*, Final Report, U.S. Environmental Protection Agency, Stony Brook, NY, 1994.
2. Schubel, J. R., Long Island Sound in time and space, in *Long Island Sound: Issues, Resources, Status, and Management*, Gibson, V. R. and Connor, M. S., Eds., NOAA Estuary-of-the-Month Seminar Series, No. 3, National Oceanic and Atmospheric Administration, Estuarine Programs Office, Washington, D.C., 1987, 1.
3. Wolfe, D. A., Monahan, R., Stacey, P. E., Farrow, D. R. G., and Robertson, A., Environmental quality of Long Island Sound: assessment and management issues, *Estuaries*, 14, 224, 1991.
4. Robertson, A., Gottholm, B. W., Turgeon, D. D., and Wolfe, D. A., A comparative study of contaminant levels in Long Island Sound, *Estuaries*, 14, 290, 1991.
5. Terleckyj, N. E. and Coleman, C. D., Data and methods, in *Regional Growth in the United States: Projections for 1989–2010*, Summary Volume 1, National Planning Association Data Services, Washington, D.C., 1989, 31.
6. Koppelman, L. E., Weyl, P. K., Gross, M. G., and Davies, D. S., *The Urban Sea: Long Island Sound*, Praeger Publishers, New York, 1976.
7. Parker, C. A. and O'Reilly, J. E., Oxygen depletion in Long Island Sound: a historical perspective, *Estuaries*, 14, 248, 1991.
8. USEPA, *The Long Island Sound Study: Summary of the Comprehensive Conservation and Management Plan*, Tech. Rept., U.S. Environmental Protection Agency, Stony Brook, NY, 1994.
9. Taft, J. L., Taylor, W. R., Hartwig, E. O., and Loftus, R., Seasonal oxygen depletion in Chesapeake Bay, *Estuaries*, 3, 242, 1980.
10. Officer, C. B., Biggs, R. B., Taft, J. L., Cronin, L. E., Tyler, M. A., and Boynton, W. R., Chesapeake Bay anoxia: origin, development, and significance, *Science*, 223, 22, 1984.
11. USEPA, *Long Island Sound Study: Proposal for Phase III Actions for Hypoxia Management*, EPA 840-R-97-001, U.S. Environmental Protection Agency, Stony Brook, NY, 1997.
12. National Research Council, *Managing Wastewater in Coastal Urban Areas*, National Academy Press, Washington, D.C., 1993.
13. Chiarella, L. A., *Pathogen Contamination — Assessment of Conditions and Management Recommendations*, Tech. Rept., New York State Department of Environmental Conservation, Stony Brook, NY, 1993.
14. National Shellfish Sanitation Program, *Manual of Operations: Part I and Part II*, U.S. Department of Health and Human Services, Public Health Service, U.S. Food and Drug Administration, Washington, D.C., 1990.
15. USEPA, *Toxic Substance Contamination — Assessment of Conditions and Management Recommendations*, Tech. Rept., U.S. Environmental Protection Agency, Stony Brook, NY, 1993.
16. Turgeon, D. D. and O'Connor, T. P., Long Island Sound: distributions, trends, and effects of chemical contamination, *Estuaries*, 14, 279, 1991.
17. Goldberg, E. D., Bowen, V. T., Farrington, J. W., Harvey, G., Martin, J. H., Parker, P. L., Risebrough, R. W., Robertson, W., Schneider, E., and Gamble, E., The Mussel Watch, *Environ. Cons.*, 5, 101, 1978.
18. Goldberg, E. D., Koide, M., Hodge, V., Flegal, A. R., and Martin, J., U.S. Mussel Watch: 1977-1978 results on trace metals and radionuclides, *Est. Coastal Shelf Sci.*, 16, 69, 1983.
19. Greig, R. A. and Sennefelder, G., Metals and PCB concentrations in mussels from Long Island Sound, *Bull. Environ. Contam. Toxicol.*, 35, 331, 1985.

20. Nelson, W. G., Phelps, D. K., Galloway, W. B., Rogerson, P. F., and Pruell, R. J., *Effects of Black Rock Harbor Dredged Material on the Scope for Growth of the Blue Mussel, Mytilus edulis, after Laboratory and Field Exposures*, Tech. Rept., U.S. Environmental Protection Agency, New York, 1987.
21. NOAA, *National Status and Trends Program for Marine Environmental Quality: A Summary of Data on Tissue Contamination from the First Three Years (1986–1988) on the Mussel Watch Project*, NOAA Tech. Memo. NOS OMA 49, National Oceanic and Atmospheric Administration, National Ocean Service, Rockville, MD, 1989.
22. Gronlund, W. D., Chan, S.-L., McCain, B. B., Clark, R. C., Jr., Myers, M. S., Stein, J. E., Brown, D. W., Landahl, J. T., Krahn, M. M., and Varanasi, U., Multidisciplinary assessment of pollution at three sites in Long Island Sound, *Estuaries*, 14, 299, 1991.
23. Stiles, S., Choromanski, J., Nelson, D., Miller, J., Greig, R., and Sennefelder, G., Early reproductive success of the hard clam (*Mercenaria mercenaria*) from five sites in Long Island Sound, *Estuaries*, 14, 332, 1991.
24. Perry, D. M., Hughes, J. B., and Hebert, A. T., Sublethal abnormalities in embryos of winter flounder, *Pseudopleuronectes americanus*, from Long Island Sound, *Estuaries*, 14, 306, 1991.
25. Nelson, D. A., Miller, J. E., Rusanowsky, D., Greig, R. A., Sennefelder, G. R., Mercaldo-Allen, R., Kuropat, C., Gould, E., Thurberg, F. P., and Calabrese, A., Comparative reproductive success of winter flounder in Long Island Sound: a three-year study (biology, biochemistry, and chemistry), *Estuaries*, 14, 318, 1991.
26. Masters, M. H. and Freeman, D., *Floatable Debris — Assessment of Conditions and Management Recommendations*, Comprehensive Conservation and Management Plan Supporting Document, U.S. Environmental Protection Agency, Stony Brook, NY, 1993.
27. Conover, S. M., Oceanography of Long Island Sound, 1952–1954. IV. Phytoplankton, *Bull. Bingham Oceanogr. Coll.*, 15, 62, 1956.
28. Smayda, T. J., A survey of phytoplankton dynamics in the coastal waters from Cape Hatteras to Nantucket, in *Coastal and Offshore Environmental Inventory, Cape Hatteras to Nantucket Shoals*, Mar. Publ. Ser. No. 2, University of Rhode Island, Kingston, RI, 1973.
29. Monteleone, D. M., Cerrato R. M., Lonsdale, D. J., and Peterson, W. J., Abundance and seasonality of key forage species, in *Characterization and Assessment of Potential Impacts of Hypoxia on Forage Species in Long Island Sound*, Tech. Rept., Marine Sciences Research Center, State University of New York, Stony Brook, NY, 1992.
30. Riley, G. A., The plankton of estuaries, in *Estuaries*, Publ. 83, Lauff, G. H., Ed., American Association for the Advancement of Science, Washington, D.C., 1967, 316.
31. Tiselius, P. T. and Peterson, W. T., Life history and population dynamics of the chaetognath *Sagitta elegans* in central Long Island Sound, *J. Plankton Res.*, 8, 183, 1986.
32. Peterson, W. T., Abundance, age structure and *in situ* egg production rates of the copepod *Temora longicornis* in Long Island Sound, New York, *Bull. Mar. Sci.*, 37, 726, 1985.
33. Levinton, J. S., *Marine Ecology*, Prentice-Hall, Englewood Cliffs, NJ, 1982.
34. Strieb, M. and the Living Marine Resources Work Group, *Assessment of Living Marine Resources*, Comprehensive Conservation and Management Plan Supporting Document, U.S. Environmental Protection Agency, Stony Brook, NY, 1993.
35. Egan, B. and Yarish, C., The distribution of the genus *Laminaria* (Phaeophyta) at its southern limit in the western Atlantic Ocean, *Bot. Mar.*, 31, 155, 1988.
36. Kikuchi, T., Faunal relationships in temperate seagrass beds, in *Handbook of Seagrass Biology: An Ecosystem Perspective*, Phillips, R. C. and McRoy, C. P., Eds., Garland STPM Press, New York, 1980.
37. Dayton, P. K., Ecology of kelp communities, *Annu. Rev. Ecol. Syst.*, 16, 215, 1985.
38. Carter, R. W. G., *Coastal Embayments: An Introduction to the Physical, Ecological, and Cultural Systems of Coastlines*, Academic Press, New York, 1988.
39. Rhoads, D. C., McCall, P. L., and Yingst, J. Y., Disturbance and production on the estuarine seafloor, *Am. Sci.*, 66, 577, 1978.
40. McEnroe, M., Review of Physiological Effects of Hypoxia on Forage Base Organisms, in *Characterization and Assessment of Potential Impacts of Hypoxia on Forage Species in Long Island Sound*, Tech. Rept., Division of Natural Sciences, State University of New York, Purchase, NY, 1992.

chapter three

Case study 2: Delaware Estuary Program

I. Introduction

The Delaware Estuary is a large, heavily industrialized system that receives drainage from parts of Delaware, New Jersey, Pennsylvania, and New York (Figure 3.1). The 217-km-long body of water is a coastal plain estuary — a drowned river valley of the Delaware River — that stretches from the mouth of Delaware Bay at a line drawn between Cape May Point, New Jersey, and Cape Henlopen, Delaware (38°50′32″ N, 75°03′15″W; river kilometer 0) to the falls at Trenton, New Jersey (river kilometer 217). It represents one of the largest estuaries in the U.S., with its surface area of 1773 km^2 being exceeded only by that of the Chesapeake Bay and Long Island Sound.

Based on differences in salinity, turbidity, and biological productivity, three major ecological zones are delineated in the Delaware Estuary (Table 3.1). The lower zone, which extends from the mouth of Delaware Bay up to river kilometer 79 near Artificial Island, consists primarily of Delaware Bay. It is characterized by high salinity, low turbidity, and the highest primary production of the three estuarine zones (>90%).[1] Surrounded by areas that are largely agricultural, the lower zone supports most of the living resources of the estuary and includes 80 to 95% of the estuarine surface area and water volume. It exhibits the least human disturbance of the three estuarine zones. The middle or transition zone, extending from the lower zone boundary to river kilometer 127 (Artificial Island to Marcus Hook), has low to moderate salinity (0 to 15‰), high turbidity, and low biological production. This zone is an area with moderate anthropogenic influence, and it marks the gradual beginning of the salinity gradient. The upper zone of tidal freshwater, which stretches upstream from the middle zone boundary to the falls at Trenton, is typified by low biological production, elevated coliform counts, and high turbidity. The most severe impacts from development and industrialization occur here.

The Delaware Estuary has historically experienced substantial water pollution problems, although improvements have occurred during the past 2 decades (Table 3.2). More than 150 industries and municipalities discharge wastewaters into the estuary. In addition, about 300 combined sewer overflows periodically release sewage wastes and other contaminants to the system. Together with ongoing human development, these effects continually threaten estuarine water quality and sensitive habitats.

More than 6 million people reside in areas surrounding the Delaware Estuary. The watershed supports great concentrations of heavy industry. For example, the second largest national oil refining petrochemical center in the U.S. is located in this region. The largest freshwater port in the world is also found along the Delaware River. As a result

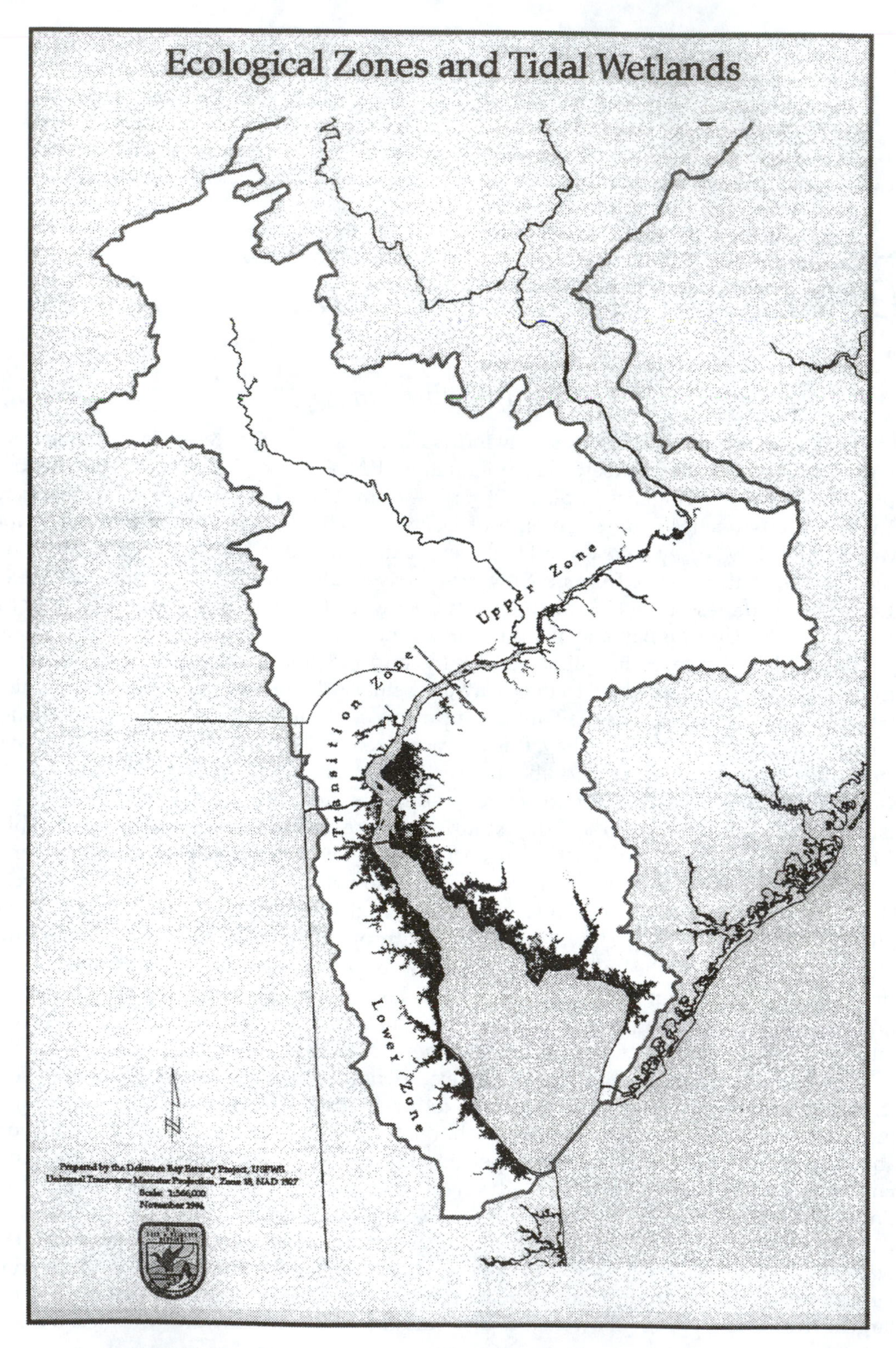

Figure 3.1 Map of the Delaware Estuary and watershed showing three ecological zones (lower, transition, and upper zones) of study south of Trenton. Dark areas depict tidally influenced wetlands. (From Breese, G., Delaware Bay Estuary Project, U.S. Fish and Wildlife Services, 1994.)

Table 3.1 Ecological Zones of the Delaware Estuary Designated by the Delaware Estuary Program

	Geographical range	River kilometer	Area[a]
Zone I (tidal fresh)	Marcus Hook, PA, to Trenton, NJ	130–217	55
Zone II (transition)	Artificial Island to Marcus Hook, PA	80–130	118
Zone III (Delaware Bay)	Marcus Hook, PA to the bay mouth (Cape May and Cape Henlopen transect)	0–80	1600

[a] Kilometers.

Source: Sutton, C. C. et al., The Scientific Characterization of the Delaware Estuary, Tech. Rept., Delaware Estuary Program, U.S. Environmental Protection Agency, New York, 1996.

Table 3.2 Water Quality History of the Delaware Estuary

Approximate Period	Water Quality Description
1800–1860	Local water supplies experience increased incidence of pollution; water quality of Philadelphia Harbor becomes a concern; the first pollution survey is conducted in 1799.
1880–1910	Increased water pollution leads to an acceleration of waterborne diseases associated with the consumption of river water.
1936–1960	Population and industrial growth, together with an expansion of urban water and sewer systems, results in the rapid degradation of water quality in the estuary.
1960–1980	The newly formed Delaware River Basin Commission adopts higher water quality standards; federal construction funds are made available for wastewater treatment plants; water quality improves throughout the Delaware River.
1980–Present	Several environmental problems persist; for example, along with the presence of toxic chemical contaminants in some areas, dissolved oxygen conditions remain a concern; the principal goal is to make the entire estuary fishable and swimmable.

Source: Sutton, C. C. et al., The Scientific Characterization of the Delaware Estuary, Tech. Rept., Delaware Estuary Program, U.S. Environmental Protection Agency, New York, 1996.

of the high-density population and heavy industry in the Delaware watershed, the estuary has been subject to a long history of environmental impacts, which have degraded water quality, impacted living resources, and accelerated habitat loss and alteration.

The Delaware River watershed drains an area of 33,000 km^2 in Delaware, New Jersey, Pennsylvania, and New York. The landscape varies considerably over the length of the estuary. For example, tidal wetlands surround much of Delaware Bay, flanking the lower end of the watershed. Appalachian mountain plateaus occur in the upper watershed. Between these topographic extremes are four heavily populated urban centers: Wilmington, Camden, Philadelphia, and Trenton.[2]

A great diversity of flora and fauna inhabits the Delaware Estuary, and its wetlands (~164,000 ha) are of national significance. Living resources of the estuary include rich communities of plankton, benthos, finfish, and mammals. Delaware River wetlands provide breeding, nesting, staging, and feeding grounds for finfish, amphibians, reptiles, mammals, waterfowl, and shorebirds.

II. Delaware Estuary Program goals and objectives

A. Goals

The general goals of the Delaware Estuary Program (DELEP) are as follows:

- Provide for the restoration of living resources of the Delaware Estuary and protect their habitats and ecological relations for future generations
- Reduce and control point and nonpoint sources of pollution, particularly toxic pollution and nutrient enrichment, to attain the water quality conditions necessary to support abundant and diverse living resources in the Delaware Estuary
- Manage water allocations within the estuary to protect public water supplies and maintain ecological conditions in the estuary for living resources
- Manage the economic growth of the estuary in accordance with the goals of restoring and protecting the living resources of the estuary
- Promote greater public understanding of the Delaware Estuary and greater participation in decisions and programs affecting the estuary

B. Objectives

Several specific objectives of the DELEP also apply:

- Restore population levels of harvestable species of finfish and invertebrate species to levels that will support sustainable recreational and commercial fisheries
- Restore or maintain populations of birds dependent on the Delaware Estuary to levels deemed attainable by comprehensive analysis
- Restore or maintain populations of estuarine-dependent amphibians, reptiles, and mammals to levels deemed attainable by comprehensive analysis of natural populations
- Maintain or restore an assemblage of organisms and their habitat throughout the Delaware Estuary and tidal wetlands that contribute to the ecological diversity, stability, productivity, and aesthetic appeal of the region
- Ensure an adequate supply of freshwater to the estuary to maintain habitats, distribution of salinity, and human population in 2020
- Preserve acreage and enhance quality of shoreline and littoral habitat to sustain a balanced natural system
- Restore and maintain the physical and environmental conditions necessary to achieve target levels of estuarine species
- Optimize sediment quantity and quality in a manner that maintains or enhances a balanced indigenous estuarine biota and habitat
- Promote and enhance ample and high quality water-based and associated terrestrial-based recreational opportunities with sustained availability for public use
- Develop programs and actions that will be mutually beneficial to both the economy and environment of the estuary by forging a partnership with industry, commerce, and local governments all jointly in pursuit of continued economic vitality of the region while enhancing and preserving its living and natural resources
- Preserve and enhance cultural resources and traditions in the estuary region and promote their accessibility to the public
- Promote pollution prevention technologies and strategies that protect estuarine resources from point and nonpoint pollution sources and, if possible, from all types of catastrophic oil or chemical spills[2]

III. Physical oceanography

Delaware Bay has a mean depth of 9.7 m; however, most of the bay (>80%) is less than 9 m deep. The deepest areas occur along the western part of the bay, where the maximum

Table 3.3 Water Inputs to the Delaware Estuary Shown with Distance (Kilometers) Upstream from the Mouth of the Delaware Bay

Source	Distance (km)	Drainage Area (km^2)	Average Annual Flow (m^3/s)(ft^3/s)
Delaware River at Trenton	210	17,560	319 (11,280)
Intermediate small tributaries	No data	13,367	51 (1,800)
Schuylkill River at Philadelphia	150	4,944	78 (2,750)
Intermediate small tributaries	No data	1,202	18 (650)
Christina-Brandywine near Wilmington	110	1,475	21 (750)
Intermediate small tributaries	No data	4,514	63 (2,240)
Total at mouth	0	33,062	550 (19,470)

Source: Sutton, C. C. et al., The Scientific Characterization of the Delaware Estuary, Tech. Rept., Delaware Estuary Program, U.S. Environmental Protection Agency, New York, 1996.

depth amounts to 46 m. Bottom sediments in the bay, which are affected by bottom shear stresses, consist primarily of sands, but they grade to greater concentrations of silts, clays, and organic matter in the upper estuary.[3]

The Delaware Estuary is a tidally dominated system, although freshwater inflow and the wind also influence hydrodynamic conditions. The nontidal Delaware River accounts for 60% of the freshwater flow into the estuary. The Schuylkill River contributes an additional 10% of the flow, with the remainder derived from the Chesapeake and Delaware Canal, small rivers, and nonpoint source runoff.[4,5] The tidal range increases from about 1.3 m at the mouth of Delaware Bay to 2.5 m at the head of tide at Trenton. The tidal excursion is approximately 10 km, and the average flushing time is about 90 days.[6]

The mean rate of freshwater inflow to the Delaware Estuary is ~550 m^3/s (Table 3.3). The Delaware River accounts for most of this input. The mean rate of flow of the Delaware River (at Trenton) amounts to 319 m^3/s, and that of the Schuylkill River equals 78 m^3/s.[7]

The estuary is well mixed vertically and only partially stratified because of strong tidal flow. At the mouth of Delaware Bay, the ratio of tidal to freshwater flow is 300:1. Some stratification of the water column occurs during periods of high freshwater input, most notably from March to early May when spring runoff is high. However, this stratification does not persist, and the system returns to a well-mixed condition in summer, fall, and winter, which precludes the development of anoxia or hypoxia in bottom waters.[8] Although the estuary is vertically homogeneous, a conspicuous horizontal gradient exists along the north–south axis of the system.

Hires et al.,[9] reviewing the work of Pape and Garvine,[10] reported the following findings on current velocities in Delaware Bay and adjacent shelf areas:

1. Surface velocities in Delaware Bay are generally directed seaward. There is a persistent deviation in the direction of current toward the Delaware side of the bay. This deflection could be caused by the Coriolis Force.
2. Surface current velocities in the bay increase with distance downstream, which is the usual pattern for estuarine circulation in partially mixed estuaries. Mean velocities near the bay mouth are about 10 cm/s.
3. At the bay mouth and on the continental shelf, mean surface currents are generally directed to the south. Surface current velocities at shelf stations are consistently greater than at bay stations.
4. The near-bottom mean currents at all shelf stations are directed onshore.

5. The magnitude of the near-bottom mean velocities is generally less than 10% of the surface speeds at all stations.
6. Within the bay, the mean bottom currents exhibit a marked tendency to be directed toward the nearest shoreline. For stations on the Delaware side of the deep channels, the bottom currents are directed toward the Delaware shoreline. At stations on the New Jersey side of the ship channel, bottom currents are directed toward the New Jersey shore.
7. A tentative conclusion is that the effects of wind forcing on subtidal circulation is considerably more important than variability in freshwater discharge.

IV. Land and water use

A. Land use

As people settle in an area, they build houses, municipal and industrial facilities, shopping centers, highways, parks, and other infrastructure components. The development of land for agriculture as opposed to heavy industry typically results in different ecological effects on watersheds and receiving waters. "Land use" refers to the way land is developed, and it has a major influence on water quality in an area, the quantity of runoff as well as groundwater flow, all of which can affect the hydrodynamics and biotic communities in nearby streams, rivers, and estuaries. Development also fosters various human activities that directly degrade or alter valuable wetlands and estuarine habitats.

Trenton, Philadelphia, Camden, and Wilmington are the major urban centers along the Delaware River as noted previously. Suburban regions essentially connect these centers in a corridor along the river. Chemical, oil, steel, and other heavy industries are concentrated at the urban centers.[11] From the upper Delaware River to Trenton, forested and rural areas give way to suburban development and finally to urban coverage. A suburban/agricultural mix is evident south of Trenton, but the Philadelphia/Camden area as well as the riverfront south to Wilmington are heavily urbanized. Undeveloped open space predominates south of the Delaware Memorial Bridge. Salem, Cumberland, and Cape May Counties in New Jersey have vast expanses of saltmarshes bordering Delaware Bay, and rural and agricultural areas are evident inland. The cape or lower peninsula of Cape May County has recently become suburbanized. Much of the Delaware Estuary coast is undeveloped. Developed land from Wilmington south to Delaware City grades from urban/industrial to rural and agricultural areas.

In many areas, the saltmarshes along Delaware Bay have been extensively altered by parallel-grid ditching for mosquito control, construction of impoundments, salt hay farming, and drainage for agricultural use. They continue to be the focus of wetlands management interests. Their preservation is essential to support a wide diversity of living resources.

Population growth is closely coupled to development. In the Delaware River watershed, 10 of 22 counties experienced more than a 20% increase in population growth between 1970 and 1990, thereby creating greater demand for land development and resource use. The Delaware Valley Regional Planning Commission noted that 63% of the area in its nine-county region was undeveloped, and 37% was developed. By the year 2020, 51% of the area is expected to be developed. Population growth and development along the estuarine shoreline are also expected to increase concomitant with reconstruction and denser housing. This additional shoreline development will invariably lead to greater nonpoint source pollution, the potential for additional point source pollution, and direct physical impacts on estuarine habitats.

B. Water use

Because more than 6 million people inhabit the estuary watershed, the Delaware River is one of the most intensely used rivers in the U.S. It provides water for municipal, industrial, and other needs. Population growth and development in watershed areas of the Delaware River and Estuary led to the construction of 23 reservoirs in the Delaware River Basin during the past century to ensure adequate water supplies. With a combined storage capacity of 1.5×10^{12} l, the reservoirs not only provide water for municipal supplies, but also for the generation of hydroelectric power, river flow augmentation and/or maintenance, and other purposes. About half of the reservoir water supply is reserved for New York City, which owns three large reservoirs in the upper parts of the basin.[12] The city has the legal right to divert an average of 3.04×10^9 l/d of water from the Delaware River Basin, but to do this it must sustain a flow in the Delaware River of 52.5 m^3/s at Montague, New Jersey.

The Delaware River Basin is a source of drinking water for approximately 20 million people. Power generation, industry, and agriculture are other large consumptive users. Six electric generating plants were among the 10 largest users of Delaware River water in 1990. These plants required large volumes of water to cool condensers during the electric generation process. The remaining principal water users were two oil refineries, a chemical manufacturer, and the City of Philadelphia.

In 1991, total water withdrawals from the Delaware River Basin averaged more than 27.7×10^9 l/d. Power generation accounted for 68% of this total (19.2×10^9 l/d), and industry and the public water supply sectors accounted for 15% each. The average consumptive water use for 1991, in turn, totaled 1.18×10^9 l/d, with the public water supply, power generation, agricultural, and industry sectors being responsible for 42, 22, 17, and 14% of the total, respectively. Consumptive water use refers to water that is not available for reuse or instream flow protection; consequently, it is of particular concern to water managers. By the year 2020, the consumptive water use in the Delaware River Basin is expected to exceed 1.52×10^9 l/d.

Rapid population growth increases the demand for water from aquifer systems, which can depress water levels in the aquifers and reduce the discharge of groundwater to streams, wetlands, the Delaware River, and Delaware Bay. In some areas where natural land cover has been replaced by impervious surfaces, the net rate of groundwater recharge may also decline significantly, leading to additional decreases in groundwater levels and reductions in the baseflow of streams discharging to the bay. In severe cases of groundwater withdrawal, the probability of salt water intrusion into area aquifers increases, as is evident in Cape May, New Jersey.

Water use in the Delaware Estuary is considerable. The total freshwater inflow to the estuarine system amounts to 21.3×10^{12} l/yr. The total surface withdrawal in the estuary is 8.36×10^{12} l/yr; power generation removes 77% of the water (6.46×10^{12} l), and the domestic and industrial sectors remove 22% (1.9×10^{12} l).[13] Much of the water removed from the estuary is ultimately returned, albeit often altered in physical or chemical properties (e.g., temperature, pH, nutrients, and toxic chemical concentrations).

V. Water quality

A. Dissolved oxygen and bacteria

During the 1940s and 1950s, water quality in the Delaware Estuary was severely degraded by industrialization of the Delaware River Basin, the growth of major cities, and the expansion of urban water and sewer systems. This degradation was manifested by anoxia

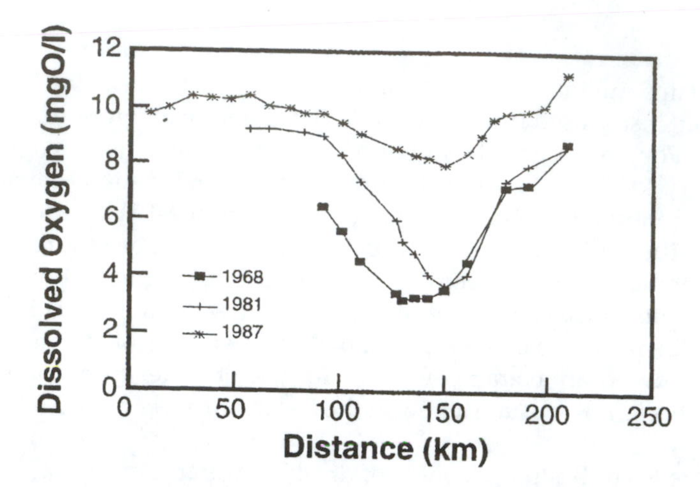

Figure 3.2 Dissolved oxygen concentrations in the Delaware Estuary. Concentrations represent annual averages from weighted monthly intervals. (From Sutton et al., The Scientific Characterization of the Delaware Estuary, Tech. Rept., Delaware Estuary Program, U.S. Environmental Protection Agency, New York, 1996.)

and hypoxia of waters extending from Wilmington to Trenton. Depleted dissolved oxygen levels culminated in massive fish kills and the obstruction of andromous fish migration, most notably between Marcus Hook and Philadelphia.[14] However, since the 1960s, water quality has improved significantly concomitant with the enactment of more stringent state and federal environmental legislation, the establishment of the Delaware River Basin Commission (DRBC) (a four-state and federal government regulatory and management agency for water resources), and the development of environmental education programs. The DRBC adopted much higher water quality standards and set waste load allocations for municipal and industrial discharges. With a focus on improved sewage and industrial wastewater treatment, fecal coliform bacteria levels decreased substantially, as did biochemical oxygen demand and the concentration of nutrient inputs (nitrogen and phosphorus).

Dissolved oxygen levels and pH increased during the 1970s and 1980s (Figure 3.2). As acid waste influx declined, waters became more alkaline. Improved pH and alkalinity measures were ascribed to industrial waste cleanup. Despite these improvements, however, water quality in the Camden/Philadelphia metropolitan area remained inadequate to support fishable/swimmable classifications of the DRBC.[5,15] Areas rated as swimmable/nonswimmable are so designated largely based on fecal coliform numbers and the risk of associated pathogens.[2] The impact of combined sewer overflows appears to be at least partially responsible for the degraded water quality in this area.

Even though nutrient inputs have generally dropped in the Delaware Estuary during the past several decades, they still rank among the highest of all major estuaries in the U.S. Nevertheless, these high nutrient levels have not resulted in eutrophic conditions in the estuary. The observed algal biomass and production values are not excessive, and massive phytoplankton blooms rarely occur.

In 1990, the DRBC assessed the compliance of the Delaware Estuary with the goals of the Clean Water Act using four primary parameters (i.e., dissolved oxygen, bacteria concentrations, temperature, and pH). Of the 2.24×10^5-ha area of the estuary and bay investigated by the DRBC, 96% had good water quality, 1% had good to fair water quality, 2% had fair water quality, and 1% had poor water quality. In addition, 93% of the estuary and bay met the fishable goal of the Clean Water Act, and 99% met the swimmable goal. By comparison, 100% of the free-flowing river met both the fishable and swimmable goals. Even higher levels of water quality were delineated in 1993.[16]

Table 3.4 Cause and Source of Impaired Uses on the Delaware River

Zone	Location	Uses Impaired	Cause	Source
River	Upper nontidal (few miles)	Fish, aquatic life; recreation; public water supply	High pH	Aquatic plants
	Upper nontidal (few miles)	Recreation; public water supply	High bacteria levels	Point and nonpoint sources
	Lower nontidal (5 miles)	Fish consumption	Chlordane, PCBs	Nonpoint sources
1	Tidal river (Zones 2–4)	Fish, aquatic life	Low dissolved oxygen	Point sources of BOD
	Tidal river (Zones 2–4)	Fish and shellfish consumption	Chlordane, PCBs	Point and nonpoint sources
2	Estuary (lower Zone 5)	Fish, aquatic life	Low dissolved oxygen	Point and nonpoint sources
3	Estuary/bay (Zone 6)	Shellfish consumption	Bacterial infestations	Point and nonpoint sources locally

Source: Sutton, C. C. et al., The Scientific Characterization of the Delaware Estuary, Tech. Rept., Delaware Estuary Program, U.S. Environmental Protection Agency, New York, 1996.

Although ambient monitoring data and evaluative assessments indicate that water quality in the Delaware Estuary has generally improved since 1990, some problems remain. The DRBC has documented the principal causes and sources of impaired uses in the system (Table 3.4). In the upper, nontidal river, bacteria numbers are the main cause of impairment. In the tidal river and estuary, low dissolved oxygen levels are of more paramount importance, but toxic chemicals and bacterial infestation of shellfish also appear to be a major concern in some areas.

The DRBC continues to oversee and assess water quality conditions in the estuary. As a consequence, dissolved oxygen levels have gradually increased during the critical summer period, reaching levels greater than 4 mg/l in areas where they once approached 0.1 mg/l. The improved water quality appears to have had a positive effect on organisms inhabiting the estuary. The greater abundance of American shad (*Alosa sapidissima*) and striped bass (*Morone saxatilis*) in recent years, for example, may be attributable to the enhanced water quality.

B. Nutrients

Nitrogen and phosphorus loads in the Delaware Estuary amount to ~7500 mmol N/m^2/yr and 600 mmol P/m^2/yr, respectively. These values are much higher than those of other major estuaries in the U.S. For example, the loading of nitrogen and phosphorus is 10 times less in Chesapeake Bay than in the Delaware Estuary. Northern San Francisco Bay, which also experiences high nutrient loading, has nitrogen and phosphorus inputs about four times and two times less, respectively. The total concentration of nitrogen in the estuary currently ranges from 1.5 to 3 mg N/l, down from a historic range of 2 to 4 mg N/l, but it is still excessive. Ammonium levels along the length of the estuary amount to ~0.05 to 0.4 mg N/l; those of nitrate are ~0.1 to 2 mg N/l. Phosphate levels along the length of the estuary, in turn, range from ~0.02 to 0.12 mg P/l.[2]

Nutrients enter the estuary from both point and nonpoint sources, although comprehensive data on nonpoint source inputs to surface and groundwater bodies in watershed areas are largely unavailable. The highest nutrient inputs originate from five municipal sewage treatment plants in the Philadelphia area.[17] Nitrate and phosphate concentrations progressively decrease from about Marcus Hook to the mouth of the estuary where the lowest concentrations are found.[2]

Total phosphorus levels peak in areas near points of maximum inputs in the tidal river and in areas of low salinity and high turbidity. Minimum concentrations occur near the mouth of Delaware Bay. As conveyed by Lebo and Sharp,[18] the distribution patterns of dissolved reactive phosphorus, dissolved organic phosphorus, and particulate phosphorus along the estuary are the result of spatial and temporal patterns in phosphorus inputs, turbidity, river flow, and biological production. Statistically significant decreasing trends of total phosphorus as well as total nitrogen and ammonia have been recorded throughout the estuary. In the tidal river of the Delaware Estuary, total phosphorus levels have declined fourfold.

The export of nutrients from the estuary to the coastal ocean precludes eutrophication problems in Delaware Bay.[17] In the middle estuary, high turbidity depresses phytoplankton production, despite an adequate nutrient supply. Phytoplankton production may also be depressed in some reaches of the estuary because of growth inhibitation by toxic substances.[19]

VI. Toxic substances

Various toxic substances have been detected in the water column, sediments, and biota of the Delaware Estuary. Most notable in this regard are heavy metals and organic contaminants that enter the estuary by both point sources (industrial and municipal effluents) and nonpoint sources (surface runoff, groundwater input, and atmospheric deposition). Highest concentrations of chemical contaminants occur in urbanized areas, especially near Philadelphia.[20-23] They are of concern because of their potential harmful effects (both sublethal and lethal) to estuarine organisms and to humans who consume contaminated seafood products. Some toxic substances found in the estuary are carcinogens, mutagens, and/or teratogens.

Municipal and industrial facilities discharge a large number of toxic substances to the estuary. The DRBC[15] identified 12 metals and 115 chemical compounds released to the estuary from these facilities. Many toxic contaminants also derive from nonpoint sources of pollution, although these later inputs have been more difficult to track.

Table 3.5 provides a list of representative toxic substances recorded in the Delaware Estuary. Frithsen et al.[23] estimate that a minimum of 1×10^6 kg of toxic substances enter the estuary each year. Approximately 62% originates from point source discharges, and 38% originates from nonpoint sources (Table 3.6). In regard to nonpoint sources, loadings from urban runoff appear to be much more significant than those from agricultural runoff and atmospheric deposition (Table 3.7).

A. Trace metals

The concentrations of trace metals in the Delaware River rank among the highest observed in rivers on the east coast (Table 3.8). Along the length of the estuary, trace metal concentrations tend to decrease from the transition zone to the coastal ocean due to fewer dischargers in the lower zone and increasing dilution by seawater. Total loadings of arsenic, chromium, copper, and lead to the estuary are about 1×10^5 kg/yr, and those of mercury are approximately 1×10^4 kg/yr.[2]

Water column monitoring between 1970 and 1990 showed that the concentrations of all trace metals except arsenic and nickel frequently exceeded U.S. Environmental Protection Agency (USEPA) water quality criteria for protection of aquatic life. Over the years, there has been a pattern of reduced maximum concentrations of some of the metals (e.g., cadmium, chromium, cobalt, lead, mercury, selenium, and silver), although the average

Table 3.5 Representative Toxic Substances of Concern in the Delaware Estuary

Metals	
Aluminum	Lead[a]
Arsenic[a]	Mercury[a]
Beryllium	Nickel
Cadmium	Selenium
Chromium[a]	Silver[a]
Copper[a]	Zinc[a]
Volatile Organics	
Acrolein	1,2-Dichloroethane[a]
Acrylonitrile	Tetrachloroethene[a]
Benzene	Toluene
Carbon tetrachloride	Trichloroethene
Chloroform	Vinyl chloride
Nonvolatile Organics: Polycyclic Aromatic Hydrocarbons (PAHs)[a]	
Acenaphthene	Dibenzo [a,h] anthracene
Acenaphthylene	Dibenzothiophene
Anthracene	Fluoranthene
Benzo [a] anthracene	Fluorenes
Benzofluoranthenes	Indeno [1,2,3-c,d] pyrene
Benzo [g,h,i] perylene	Naphthalene
Benzopyrenes	Perylene
Biphenyl	Phenanthrene
Chrysene	Pyrene
Organochlorines: Chlorinated Pesticides	
Aldrin	Heptachlor
Chlordane[a]	Hexachlorobenzene
DDT and its metabolites[a]	Mirex
Dieldrin[a]	Pentachlorophenol
Endosulfan	Toxaphene
Endrin	
Polychlorinated Biphenyls (PCBs)	
Others	
Dinitrophenol	Nitrophenol
Nitrobenzene	Phenol

[a] These substances were named to the preliminary list of toxic pollutants of concern by the Delaware Estuary Programs' Toxics Task Force (USEPA, 1994).

Source: Sutton, C. C. et al., The Scientific Characterization of the Delaware Estuary, Tech. Rept., Delaware Estuary Program, U.S. Environmental Protection Agency, New York, 1996.

concentrations of these contaminants have not changed significantly. However, the average concentrations of other trace metals (e.g., copper and zinc) appear to be decreasing in the water column throughout the estuary. Nickel, in turn, may be increasing in the middle and upper zones.[2] Results of water column studies by Riedel and Sanders[24] indicate that cadmium, copper, lead, and zinc are bioavailable in significant quantities, arsenic and nickel less bioavailable, and chromium and selenium least bioavailable. Table 3.9 provides a budget for the total fluxes of trace metals through the estuary.

Table 3.6 Toxic Pollutants (Substances) of Concern and Rationale for Preliminary Listing by the Delaware Estuary Program's Toxics Task Force

Pollutant	Rationale	Possible Sources
	Pesticides/PCBs/PAHs	
PCBs	Current consumption advisories issued by NJDEP, PADER, and DNREC; sediment contamination	Nonpoint sources including Superfund sites
PAHs	Observed sediment toxicity and excesses of NOAA effects levels (ER-L and ER-M)	Nonpoint sources including atmospheric deposition
DDT, DDE, and DDD	Elevated tissue levels in fish and birds; 1990 data collected by DRBC indicate levels exceeding 10^{-6} risk level; possible excess of chronic aquatic life WQC for DDD; sediment contamination	Nonpoint sources (runoff from existing or abandoned sites); point sources (DDD, 12 discharges)
Dieldrin	1990 data collected by DRBC indicate levels exceeding 10^{-6} risk level	Nonpoint sources including abandoned sites (Cobbs Creek, PA)
Chlordane	Current Consumption advisories issued by NJDEP and PADER; 1990 data collected by DRBC indicate levels exceeding 10^{-6} risk level	Nonpoint sources
	Metals	
Lead	Possible excess of chronic aquatic life WQC; monitoring data for 1992 indicate excess of proposed WQC in the lower Estuary (river mile 60.6 and 66.0)	Point sources (53 discharges); nonpoint sources?
Zinc	Elevated levels in shellfish tissue; Christina River ambient water quality criteria exceeded; excess of ER-M in much of estuary	Point sources (83 discharges)
Copper	Possible excess of chronic aquatic life WQC; monitoring data for 1992 indicate excesses of proposed WQC in the lower estuary (river mile 60.6, 66.0, and 71.0); sediment contamination	Point sources (58 discharges); nonpoint sources?
Mercury	Possible excess of chronic aquatic life WQC; sediment contamination	Point sources (24 discharges); nonpoint sources?
Arsenic	Possible excess of human health WQC for carcinogenic effects	Point sources (16 discharges); nonpoint sources?
Chromium	Possible excess of chronic aquatic life WQC for hexavalent chromium but not for trivalent chromium; sediment contamination	Point sources (39 discharges); nonpoint sources?
Silver	Possible excess of chronic aquatic life WQC	Point sources (22 discharges); nonpoint sources?
	Volatile Organics	
1,2 Dichloroethane	Possible excess of human health WQC for carcinogenic effects; monitoring data for 1990 indicate excess of proposed WQC between river mile 71.0 and 107.1	Point sources (8 discharges); nonpoint sources?
Tetrachloroethene	Possible excess of human health WQC for carcinogenic effects	Point sources (9 discharges); nonpoint sources?

Table 3.6 (continued) Toxic Pollutants (Substances) of Concern and Rationale for Preliminary Listing by the Delaware Estuary Program's Toxics Task Force

Pollutant	Rationale	Possible Sources
	Toxicity	
Chronic toxicity	Possible excess of chronic aquatic life WQC for whole effluent toxicity; study in November 1990 indicated chronic toxicity of ambient water samples collected at river mile 69.0 and between river mile 97.5 and 107.1	Point sources (51 discharges); nonpoint sources?

Source: Sutton, C. C. et al., The Scientific Characterization of the Delaware Estuary, Tech. Rept., Delaware Estuary Program, U.S. Environmental Protection Agency, New York, 1996.

Table 3.7 Toxic Substance Loadings to the Delaware Estuary

	Source				Percent of Total Loading by Substance
	PS(½)	UR(½)	AR(½)	AD(½)	
As	43.8/7.3	8.9/3.2	46.6/92.2	0.7/2.0	10.4
Cr	87.4/20.1	11.6/5.8		1.0/3.8	14.3
Cu	82.1/18.7	15.6/7.7		2.3/9.0	14.2
Pb	70.3/13.2	24.5/10.0		5.2/16.9	11.7
Hg	10.1/0.2	10.1/0.3		79.8/20.2	0.9
Ag	100.0/2.2				1.4
Zn	52.6/33.4	43.5/59.8		4.0/43.6	39.6
PAH		95.1/10.6		4.9/4.4	3.2
Chlor. Pest.	39.5/0.4	2.6/0.1	57.9/7.8		0.7
PCBs	66.7/<0.01			33.3/0.1	<0.01
Volatile Organics	79.0/4.5	21.0/2.6			3.5
Percent of Total Loading by Source	62.3	28.8	5.2	3.6	99.9/99.9

Note: 1 = Percent loading of a substance by source; 2 = percent contribution of a substance to loading from a source; PS = point source; UR = urban runoff; AR = agricultural runoff; and AD = atmospheric deposition.

Source: Sutton, C. C. et al., The Scientific Characterization of the Delaware Estuary, Tech. Rept., Delaware Estuary Program, U.S. Environmental Protection Agency, New York, 1996.

Table 3.8 Concentration of Dissolved Trace Metals in Some East Coast Rivers

	Trace Metal (μg/l)							
River	Cd	Co	Cu	Fe	Mn	Ni	Pb	Zn
Delaware	0.17	0.42	2.36	32.9	155	3.86	0.27	12.1
Susquehanna	0.089	1.0	1.21	57.3	655	5.75	0.21	2.62
Southeastern U. S. (average)	0.078		0.56	30.7	18	0.26		0.64
Hudson	0.25		3.24	31.9	10.7	2.41		8.83
Connecticut	0.10		4.17	113	45.9			0.98
Potomac								0.55

Source: Sutton, C. C. et al., The Scientific Characterization of the Delaware Estuary, Tech. Rept., Delaware Estuary Program, U.S. Environmental Protection Agency, New York, 1996.

Table 3.9 Budget for the Total Fluxes of Trace Metals through the Delaware Estuary

	Trace Metal (g/s)						
	Fe	Mn	Co	Cu	Ni	Cd	Zn
Input Fluxes							
River	400.0	135.8	0.62	2.41	3.90	0.20	12.87
Marsh	885.0	102.0	0.67	0.96	0.14	0.06	3.80
Atmosphere	1.4	0.2	0.02	0.14	0.10	0.02	0.56
Total	1286.2	238.0	1.31	3.51	4.14	0.28	17.23
Export Fluxes							
Tidal exchange	176.4	19.0	0.172	2.21	2.67	0.191	7.48
Sediments (calculated)	1110.0	219.0	1.14	1.30	1.47	0.09	9.75

Source: Sutton, C. C. et al., The Scientific Characterization of the Delaware Estuary, Tech. Rept., Delaware Estuary Program, U.S. Environmental Protection Agency, New York, 1996.

Bottom sediments are a major repository of trace metals and other chemical contaminants in the estuary. Storms, dredging, and other natural and anthropogenic effects resuspend the sediments and may remobilize the metals. The reintroduction of trace metals to the water column and their release into surficial sediment pore waters increase the bioavailability of the contaminants. Trace metals are potentially toxic above a threshold availability.[25] Metal toxicity appears to be a function of the free metal ionic activity in seawater rather than the total concentration of the metal.[26]

Trace metal concentrations recorded at some locations in the estuary are at levels where toxic effects may be manifested in organisms. Based upon levels (effects range-low; ER-L) recognized by Long and Morgan,[27] the most harmful concentrations of trace metals occur in the greater Philadelphia area of the middle and upper estuary. The following trace metals exist in certain bottom sediments at or above ER-L levels (1) cadmium, chromium, lead, mercury, nickel, silver, and zinc in the middle estuary, and (2) cadmium, chromium, copper, lead, mercury, nickel, and zinc in the upper estuary.[21,22,28,29] Fine-grained, organic-rich sediments, particularly in urban areas, have elevated trace metal concentrations.[30] Costa and Sauer[22] ascertained that sediment copper and zinc concentrations in bottom sediments tend to increase from the lower to the upper estuary. Copper concentrations ranged from <10 to >70 μg/g dry weight and zinc concentrations from <100 to >600 μg/g dry weight, with highest levels recorded in the upper estuary. Other metals (e.g., chromium and mercury) did not exhibit a general trend of increasing concentrations in bottom sediments from the lower to the upper estuary.

A wide array of trace metals has been found in the tissues of finfish and shellfish collected in the estuary, including those of concern to the Delaware Estuary Program's Toxics Task Force (see Table 3.6). Some of the reported concentrations are troubling. For example, the concentrations of chromium in white perch (*Morone americana*) and the concentrations of chromium and lead in channel catfish (*Ictalurus punctatus*) collected from the middle estuary have exceeded the nationwide 90th percentile of fish burdens of these metals. In the upper estuary, the concentrations of chromium in the brown bullhead (*Ameiurus nebulosus*) and white catfish (*A. catus*), chromium and lead in the channel catfish (*I. punctatus*), and copper, cadmium, chromium, and zinc in the white perch (*M. americana*) have exceeded the nationwide 90th percentile for those metal concentrations in finfish.[2] Relatively high levels of copper, lead, and silver tissue burdens have also been documented in blue crab samples from the middle estuary.[31] Elevated levels of tissue metals (i.e., cadmium, copper, and chromium) have likewise been registered in the eastern oyster (*Crassostrea virginica*) from the lower estuary.[28]

B. Volatile organic contaminants

The Delaware Estuary receives an estimated 5.5×10^4 kg/yr of volatile organics from both point and nonpoint sources.[23] Approximately 70% of the total loading is ascribed to the input of 1,2-dichloroethane (ethylene chloride) and tetrachloroethene. Other commonly occurring volatile organics in the upper estuary include chloroform and trichloroethene. These contaminants are carcinogenic and/or mutagenic, and they can be highly toxic. The most persistent volatile organics in estuarine environments are those denser than the water and/or more than slightly soluble. Although very toxic, the volatile organics are readily degraded by environmental (e.g., photooxidation) and biological processes. Between 1970 and 1990, no volatile organic exceedences of acute water quality criteria for protection of aquatic life were reported in the estuary.[20]

C. PAHs and phenolic compounds

Polycyclic aromatic hydrocarbons (PAHs), ubiquitous nonvolatile organic xenobiotics, are of considerable concern in estuarine and marine environments because of their potential carcinogenicity, mutagenicity, and teratogenicity to aquatic organisms and humans. Although not all PAHs are potent carcinogens, mutagens, and teratogens, the environmental impact of most of them remains uncertain, and only a rudimentary knowledge exists with regard to the physical, chemical, and biological processes controlling their behavior in estuarine environments.[32] The highest concentrations of PAHs occur in urbanized estuaries.[33]

PAHs are widespread in estuarine and marine environments because they originate from multiple natural and anthropogenic sources, as recounted previously. Among important natural sources of the contaminants are grass and forest fires, marine seep and volcanic emissions, sedimentary diagenesis, and microbial degradation. Anthropogenic sources of significance include municipal and industrial wastewaters, waste incineration, oil spills, and fossil fuel combustion. Atmospheric deposition of PAH compounds generated from the pyrolysis of fossil fuels is a primary delivery system. However, urban and agricultural runoff as well as groundwater influx account for substantial inputs of PAHs to estuarine environments.

In the Delaware Estuary, PAH loadings amount to 3.28×10^4 kg/yr.[23] These contaminants have a high affinity for particulate matter and readily sorb to sediments and other particulate surfaces. Hence, they typically partition out of the water column and accumulate in bottom sediments. PAHs found in estuarine sediments originate principally from fossil fuel combustion and oil pollution.[22,23] Highest sediment total PAH concentrations, exceeding 15 ng/g dry weight, are observed in the middle and upper estuary (Figure 3.3). In these areas, the sediment concentrations of total PAHs are high enough to be toxic to organisms. However, assessment of the toxic effects of PAHs on organisms in the estuary is lacking.[2]

Phenolic loadings to the estuary exceed 3×10^4 kg/yr.[23] Most of the phenolic compounds derive from urban runoff. During the 1980s, investigators suggested a trend of increasing concentrations of phenolic compounds throughout the system, but this pattern was never confirmed.[20] These compounds are readily degraded in estuarine environments, which decreases the probability of major impacts on biota and habitats in the Delaware Estuary.

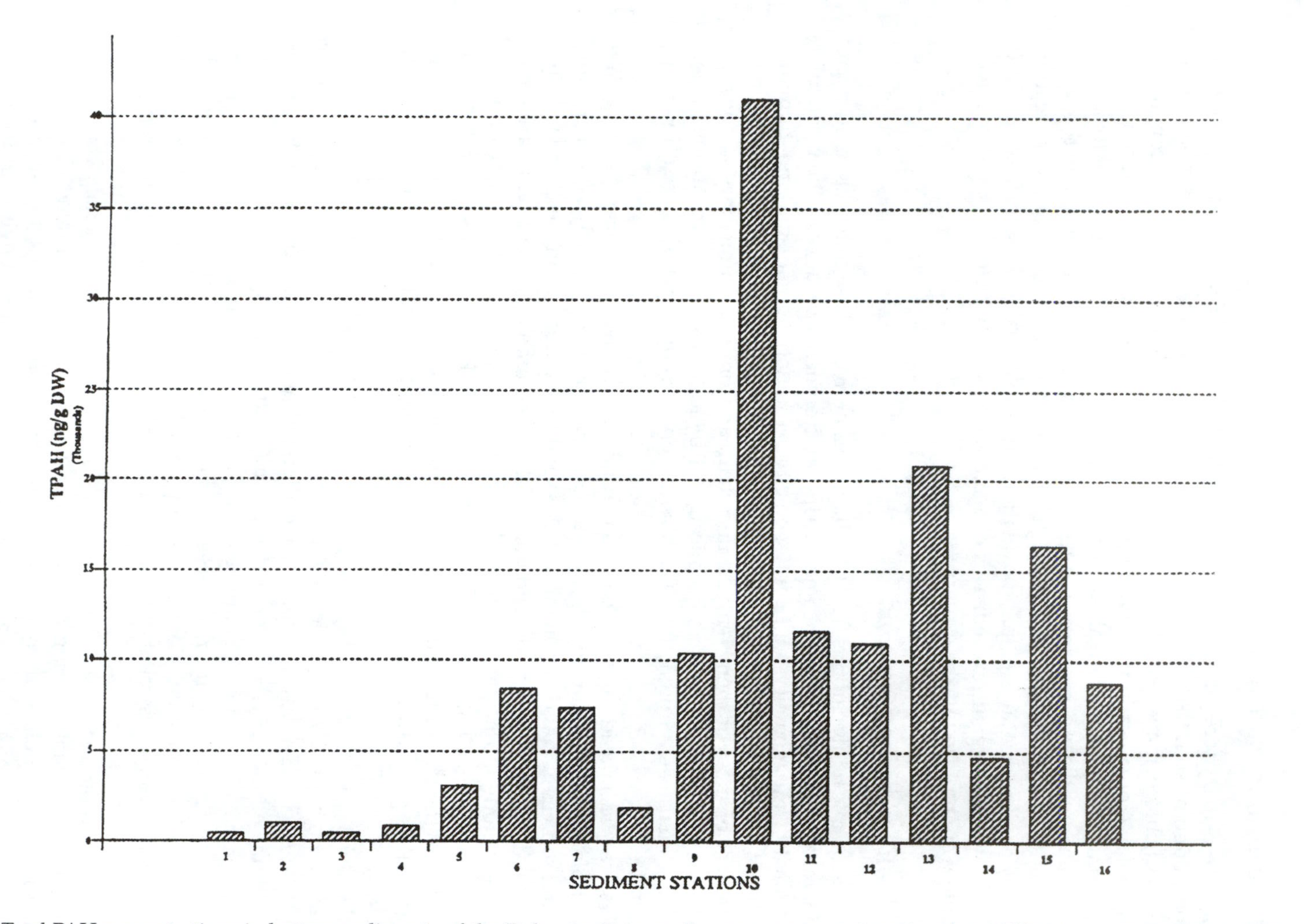

Figure 3.3 Total PAH concentrations in bottom sediments of the Delaware Estuary (lower estuary, stations 1 to 4; middle estuary, stations 5 to 9; and upper estuary, stations 10 to 16). Values in ng/g dry weight. (From Sutton et al., The Scientific Characterization of the Delaware Estuary, Tech. Rept., Delaware Estuary Program, U.S. Environmental Protection Agency, New York, 1996).

D. Chlorinated hydrocarbons

Chlorinated hydrocarbon compounds have been commonly used by humans as biocides (insecticides, herbicides, and fungicides) and as substances with a variety of industrial applications (e.g., PCBs). The synthetic compounds are generally toxic to estuarine organisms. They exhibit great stability, mobility, hydrophobicity, and persistence in the environment. Due to their lipophilicity, chlorinated hydrocarbon compounds tend to accumulate in lipid-rich tissues of organisms, and they undergo biomagnification such that upper-trophic-level consumers harbor the highest concentrations of the contaminants. Thus, adverse effects are often most pronounced among predatory fish and mammal populations.

Much of the focus of the Delaware Estuary Program has been on PCBs and a few chlorinated pesticides (i.e., DDTs, chlordane, and dieldrin).[2] Several properties of these contaminants — great stability, water insolubility, and particle adsorption — result in their accumulation in bottom sediments, which serve as a repository for the particle-sorbed or complexed fractions. Consequently, benthic inhabitants and benthic-feeding finfish are exposed to higher concentrations of the organochlorine compounds. Because of biomagnification, however, piscivorous fishes may be at greater risk to the toxic effects of these substances.

Frithsen et al.[23] estimated a PCB loading figure of 89 kg/yr for the Delaware Estuary. Most of this input (approximately two thirds) is attributed to point sources, and the remainder is attributed to atmospheric deposition. The loading of DDTs, in turn, is estimated to be 7900 kg/yr, with almost 60% delivered by agricultural runoff. Urban runoff is the major agent responsible for chlordane influx to the estuary, yielding an estimated total loading of 100 kg/yr. Data provided by the DELEP[29] indicate that dieldrin concentrations exceed acceptable risk levels. Dieldrin, like chlordane, is extremely toxic. A significant amount of this pesticide appears to be introduced into the estuary from nonpoint sources.

Bottom sediments in the Delaware Estuary are contaminated with the aforementioned organochlorine contaminants at levels that may be toxic to many organisms. The ER-L levels of Long and Morgan[27] are often exceeded by PCBs, DDTs, chlordane, and dieldrin in the sediments. PCB values average in excess of 75% of the estuarine sediment concentrations reported nationwide.[2] Concentrations of DDTs are at toxic levels throughout the estuary, with highest levels (>50 ng/g dry weight) recorded in sediments of the middle and upper estuary.[2,21,22,28] Exceedences of ER-L levels by these pesticides are a concern. Bottom sediments of the estuary likewise contain some of the highest concentrations of chlordane found on a nationwide basis, and dieldrin levels at certain sites in the upper and middle estuary are significantly elevated.

Numerous organisms in the Delaware Estuary, including shellfish (e.g., oysters, *Crassostrea virginica;* blue mussels, *Mytilus edulis;* and blue crabs, *Callinectes sapidus*) and finfish (e.g., striped bass, *Morone saxatilis;* white perch, *M. americana*; and American eel, *Anguilla rostrata*) of recreational and commercial importance, are contaminated with organochlorine compounds.[28,31,34-36] Elevated tissue burdens are common, and levels that exceed U.S. Food & Drug Administration (USFDA) action levels have been frequently reported. For example, in the upper estuary, levels that exceed USFDA action levels have been documented for PCB tissue concentrations in the American eel (*A. rostrata*), common carp (*Cyprinus carpio*), white catfish (*Ameiurus catus*), brown bullhead (*A. nebulosus*), and white perch (*M. americana*). Similarly, in the middle estuary, levels that exceed USFDA action levels have been recorded for PCB tissue concentrations in the American eel (*A. rostrata*), white catfish (*A. catus*), and striped bass (*M. saxatilis*). In the lower estuary, tissue concentrations of PCBs in the white perch (*M. americana*) and windowpane flounder (*Scopthalmus aquosus*) have exceeded the USFDA action level. In addition, chlordane tissue concentrations in channel catfish (*Ictalurus punctatus*) and white perch (*M. americana*) in the upper estuary have exceeded the USFDA action level for this contaminant.

It is unclear to what extent exposure to toxic substances is impacting the health of estuarine organisms. The DRBC[37] reported a coupling among point source discharges, toxic substances in the water column, and the health of organisms in the middle and upper estuary (from Artificial Island to Trenton). Toxicological studies showed that ambient water, particularly from the Philadelphia area, was chronically toxic to test organisms, signaling potentially serious toxic effects in some areas of the estuary. Subsequently, the USEPA[36] examined more than 3000 fish (42 species) throughout the estuary for gross external pathologies (e.g., fin rot, neoplasia, ulcers, etc.). Only six of these fish were observed with gross external pathologies, five from the lower estuary and one from the middle estuary. The DRBC[38] also registered at least 10 incidences of gross external pathologies in 996 fish (4 species). The few incidences of gross external pathologies discerned in these two studies do not support the view of gross damage to fish tissues from toxic substances in the estuary.

Because of the tendency of organochlorine compounds to biomagnify in food chains, organisms at higher trophic levels have been the target of detailed investigations regarding these contaminants. Environmental contaminants (i.e., DDTs, PCBs, chlorinated pesticides, and trace metals) are suspected of impacting predatory birds (raptors). The reproductive success of the osprey (*Pandion haliaetus*), peregrine falcon (*Falco peregrinus*), and bald eagle (*Haliaetus leucocephalus*) has been adversely affected by eggshell thinning and/or embryo toxicosis.[39-41] Although low hatching rates (due to the effects of toxic substances) have been linked to poor production of raptors in the estuary, the losses of eggs and chicks to predators as well as poor brooding associated with food shortages or predator harassment cannot be discounted.[2]

Government agencies have issued fish consumption advisories because of concerns associated with environmental contaminants. Species of fish from the Delaware Estuary with consumption advisories owing to contamination by DDTs, PCBs, chlordane, dioxin, and/or mercury include sharks, American eel (*Anguilla rostrata*), white catfish (*Ameiurus catus*), channel catfish (*Ictalurus punctatus*), chain pickerel (*Esox niger*), white perch (*Morone americana*), striped bass (*M. saxatilis*), and bluefish (*Pomatomus saltatrix*). These advisories reflect the risk to humans of consuming contaminated seafood products from the estuary.

VII. Habitat and living resources

A. Habitat

Ecological investigations of the Delaware estuarine ecosystem have focused on living resources, the interrelationships of key species, and the habitats that support biotic communities.[42] From the estuarine floor to upland areas in watersheds, an array of habitats provide food, water, shelter or cover, and space suitable for numerous organisms. Sneddon et al.[42] have described in detail four major habitat types in the system:

1. Forest and woodlands
   a. Upland deciduous forests and woodlands
   b. Mixed forests and woodlands
   c. Wetland forests and woodlands
2. Shrublands
   a. Freshwater shrub wetlands
   b. Saltwater shrub wetlands

3. Herbaceous vegetation
   a. Upland grasslands
   b. Freshwater herbaceous vegetation of coastal plain ponds
   c. Freshwater nontidal marsh
   d. Freshwater tidal marsh
   e. Submerged freshwater vegetation
   f. Brackish tidal marsh
   g. Submerged brackish marsh
   h. Saltmarsh vegetation
   i. Sparsely vegetated beach
4. Unvegetated aquatic systems
   a. Marine waters
   b. Estuarine waters

1. *Forest and woodlands*

This habitat type is characterized by a diversity of vegetation, notably upland deciduous forests (mainly nonwetland oak-dominated forests), upland mixed forests and woodlands (forests and open-canopy woodlands of mixed deciduous and coniferous species, usually oaks and pines), and wetland forests and woodlands (deciduous, mixed or coniferous forests, and woodlands). Upland deciduous forests provide nesting and foraging habitat for many resident and migratory birds. In addition, numerous invertebrates, reptiles, and mammals utilize this habitat. Human activities have impacted most upland forests of the system. There are few large, mature undisturbed stands; many hectares of forests have been cleared for development and agriculture. Nevertheless, the upland forested habitat plays an important role as a pollutant filter, protecting tidal wetlands from the influx of contaminants in runoff.

As in the case of upland deciduous forests and woodlands, mixed forests and woodlands support numerous assemblages of invertebrates, reptiles, mammals, and birds. Logging, development, and other human-use activities have fragmented these habitats. Upland habitats are largely regulated at the local level, and hence are rather easily exploited for human needs. Consequently, those populations of animals most dependent on unfragmented uplands have decreased significantly in abundance.

Atlantic white cedar swamps, red maple swamps, tidal swamps, and floodplain forests form expanses of wetland forests in the Delaware Estuary area. Various human practices have disrupted many wetland forest stands. For example, residential development, impoundments, land reclamation for aquaculture, and logging have caused a substantial decline in white cedar (*Chamaecyparis thyoides*).

2. *Shrublands*

Shrub wetlands are saturated or seasonally flooded habitats consisting of freshwater and saltwater components. They represent a transition between herbaceous wetlands and adjacent uplands, often occupying a zone between river margins and palustrine forests. As such, shrub wetlands play a significant role in erosion control, pollution filtration, and water quality management. Various wildlife populations use shrub swamps for breeding habitat, e.g., black-crowned night heron (*Nycticorax nycticorax)*, muskrat (*Ondatra zibethicus*), and raccoon (*Procyon lotor*). Avifauna (e.g., willow flycatcher, *Emidonax traillii* and yellow-billed cuckoo, *Coccyzus americanus*) also breed in this habitat and, in some cases, overwinter there.

A number of human activities impact shrub wetlands. Logging, ditching, waste disposal, and water pollution have degraded many hectares of these wetlands over the years.

Losses to filling and diking have been detrimental to both saltwater shrub wetlands and salt marshes.

3. Herbaceous vegetation

The largest group of floral communities in the watershed consists of herbaceous plants, which have been organized into nine broad categories as listed previously. Among the most important herbaceous vegetation are grasses, sedges, rushes, and other plants occurring in extensive tidal wetlands that fringe the lower estuary. Because tidal wetlands are tightly regulated, they are less impacted by human activity than estuarine uplands in the study area. Tidal marshes extend to Trenton, and variations in salinity along the lower and transition zones greatly influence the composition of plant and animal communities. High-salinity marshes at the mouth of Delaware Bay, characterized by salinity values of 28 to 30‰, are primarily dominated by smooth cordgrass (*Spartina alterniflora*) and salt hay (*S. patens*). Smooth cordgrass (*S. alterniflora*) predominates on the lowest sections of the tidal marshes, with salt hay (*S. patens*) being most abundant at higher elevations that are only periodically inundated by tidal flooding (e.g., during spring tides). Near the shrub border at still higher elevations, salt bushes (*Baccharis halimifolia* and *Iva frutescens*) peak in abundance.[42] Other plant species commonly observed in high-salinity marshes include spike grass (*Distichlis spicata*), saltworts or glassworts (*Salicornia* spp.), and marsh lavender (*Limonium carolinium*).[2] The diversity of plant communities in these marshes is relatively low.

As salinity decreases upestuary, the species diversity of the plant communities increases. Other species begin to appear along the salinity gradient; for example, big cordgrass (*Spartina cynosuroides*) and sedges (*Scirpus* spp.) give way to cattails (*Typha* spp.), saltmarsh feabane (*Pluchea purpurascens*), swamp rose mallow (*Hibiscus moscheutos*), and other forms.[2] Brackish tidal marshes are quite extensive along the Delaware Estuary. Four species attain maximum development in brackish tidal marshes: big cordgrass (*S. cynosuroides*), narrow-leaved cattail (*T. angustifolia*), giant reed (*Phragmites australis*), and Olney three-square (*Scirpus americanus*).[42]

Although grasses predominate in both the tidal and brackish marshes, they are nearly absent in freshwater tidal marshes. These marshes are fairly extensive along the upper reaches of the estuary, being dominated by broad-leaved perennials, herbaceous annuals, sedges, reedlike perennials, and some hydrophytic shrubs. In the lower reaches, forbs (e.g., pickerelweed, *Pontederia cordata*; arrow arum, *Peltandra virginica*; and arrowhead, *Sagittaria latifolia*) typically are most abundant. Wild rice (*Zizania aquatica*) is also a dominant or codominant form, most notably in areas behind the forbs. The giant reed and cattails reach peak abundance in the upper reaches of these biotopes.

Nontidal freshwater wetlands occur beyond the head of tide and are composed of characteristic freshwater plants. These wetlands extend a considerable distance inland and are found in large concentrations along many influent systems. They are inundated very infrequently, perhaps seasonally or during severe storm events, in contrast to tidally influenced saltmarsh habitats that are inundated daily.

Marsh habitats in the Delaware Estuary system serve several critical functions. They are not only extremely productive habitats but also provide important habitat for numerous invertebrates, fishes, amphibians, reptiles, mammals, and birds. In addition, they act as filters, removing nutrients and chemical contaminants, and thus are effective agents in pollution control. A significant amount of detritus generated on marsh surfaces is exported to open waters of the river and bay where various organisms may utilize it. The marshes, therefore, appear to be an important component of the total energy flow in the system.

The state of wetlands habitat in the Delaware Estuary system improved during the 1970s after enactment of the Wetlands Act. Between 1953 and 1973, heavy losses of tidal marshes occurred in Cape May, Cumberland, and Salem Counties (New Jersey), with as much as 25% of the marshes being lost in New Jersey alone.[43] Today, the system contains ~164,000 ha of wetlands habitat. However, human impacts have resulted in the loss of ~21 to 24% of the originally existing wetlands area in the estuary region.[44,45] The greatest losses have occurred in nontidal freshwater and forested wetlands, owing to development and associated human activities.[2] The wetlands continue to be altered by the encroachment of undesirable native species (e.g., *Phragmites australis*), the invasion of exotic species, and the input of pollutants.

In past decades, an array of human activities has degraded the wetlands habitat. Notable in this regard are impacts on marshes due to the diking of areas for agriculture conversion, construction of impoundments and parallel grid ditching for mosquito control, point construction for waterfowl, dumping of dredged spoil as fill for development purposes, channelization of tidal creeks for navigation, and salt hay farming. The conversion of habitat for agricultural uses has accounted for large losses of freshwater marshes and forested wetlands. Current management strategies are focused on the preservation of wetlands as open space, the restoration of impounded wetlands, and the implementation of less damaging mosquito control methods.

Economic costs associated with the loss and alteration of wetlands habitat are substantial. As this habitat is degraded, the costs of flood protection and control, erosion, and water treatment in watershed areas escalate. In addition, a reduction in potable water supplies may develop from diminished groundwater retention and salt intrusion. Furthermore, lower abundances of recreationally and commercially important fish and wildlife populations, which depend on these habitats, can have significant social and ecological ramifications.

4. *Unvegetated aquatic systems*

Open waters and benthic habitats of the Delaware River and Bay support diverse assemblages of organisms, including numerous species of phytoplankton, zooplankton, benthos, fish, microbial decomposers, and other thriving populations. The success of these assemblages is closely linked to the occurrence of various habitats in the estuary. Because environmental conditions vary considerably in, on, and along the bottom (e.g., sediment type, salinity, light intensity, currents, and depth) and different water masses exist within the system, distinct assemblages of species occupy different areas of the estuary.

Many commercially and recreationally important species of organisms depend on the Delaware Estuary for survival. Among finfish populations, bluefish (*Pomatomus saltatrix*), weakfish (*Cynoscion regalis*), summer flounder (*Paralichthys dentatus*), and striped bass (*Morone saxatilis*) are recreationally and/or commercially important. Shellfish that continue to be harvested include oysters (*Crassostrea virginica*), hard clams (*Mercenaria mercenaria*), mussels (*Mytilus edulis*), and blue crabs (*Callinectes sapidus*). Oyster production, once extremely profitable, has declined substantially since the mid-1950s due primarily to disease (MSX and dermo parasites) and the low success of seed beds.[46,47] Despite significant reductions in the production of some recreationally and commercially important species during the past several decades, the fisheries of the Delaware Estuary remain a viable and extremely valuable commodity for the local and regional economy.

B. *Living resources*

1. *Introduction*

The Delaware Estuary is a nutrient-rich environment typified by high primary and secondary production. A complex food web exists in the system structured by a high diversity of plants and animals living in, on, and around the estuary, including plankton, benthos, finfish, birds, mammals, and other organisms. Principal energy flow through the aquatic environment occurs through grazing pathways because the estuary is essentially devoid of submerged aquatic vegetation, and hence detritus accumulation in the river and bay is limited. In contrast, shallow coastal bays to the north (e.g., Barnegat Bay, New Jersey), which harbor extensive seagrass beds, exhibit well-developed detritus food webs.

Phytoplankton form the base of the grazing food web of the estuary, supporting a multitude of zooplankton and higher-trophic-level organisms, some of which are of recreational and commercial importance. Grazing food webs are generally most conspicuous in deeper systems, such as Delaware Bay and Long Island Sound. Aquatic communities in the Delaware Estuary appear to be relatively healthy despite high nutrient concentrations and turbidity levels as well as elevated chemical contaminants in some areas.

The Science and Technical Advisory Committee (STAC) of the Delaware Estuary Program used the following characteristics to determine the ecological importance of specific groups (or guilds) of species:

- Produce significant quantities of organic matter for the food web
- Are food for other resources in the estuary
- Significantly control or modify the population levels or seasonal dynamics of other plants and animals within the estuary by grazing, predation, or disturbance
- Control or modify some process (e.g., benthic nutrient regeneration) that, in turn, influences other resources
- Are classified as "endangered," threatened," or "protected" by federal or state agencies
- Are a shared resource with other estuaries or even other hemispheres, as in the case of migratory birds

Employing these characteristics, the STAC developed a priority species list, also referred to as key species, for the Delaware Estuary (Table 3.10). Some of these organisms are examined in greater detail.

2. *Phytoplankton*

The species composition, abundance, and production of the phytoplankton community in the estuary are strongly influenced by light, nutrients, and salinity. Freshwater species dominate the upper sections of the estuary, and marine forms dominate the lower river and bay. Aside from the spatial distribution patterns of the phytoplankton, temporal patterns are conspicuous. For example, diatoms dominate the phytoplankton community in the winter and spring and are the most important constituents of the winter–spring blooms. Phytoplankton diversity increases during the summer and fall when many flagellates appear.[48,49]

Seasonal cycles of phytoplankton biomass and production are similar to those in other mid-Atlantic estuaries.[1] Peak biomass in the lower estuary occurs during March, and in the upper estuary it takes place during July. The mean primary production rate for the entire estuary is estimated at 307 g $C/m^2/yr$, with values ranging from 7 g $C/m^2/yr$ in turbidity maximum zones to 392 g $C/m^2/yr$ in the central region of the lower estuary. Primary production rates are relatively high in the tidal freshwater region.[2] The influx of large amounts of nutrients in the upper estuary associated with sewage discharges, industrial discharges, and agricultural runoff may be largely responsible for this spatial difference. The effect of turbidity (i.e., light limitation) is a major factor controlling primary production in the middle estuary.[50]

More than 250 species and over 100 genera of phytoplankton have been recorded in the Delaware Estuary over an annual cycle.[48] Watling et al.[51] registered 113 phytoplankton species in Delaware Bay; whereas diatoms predominated from fall through spring, small flagellates were most abundant in summer. Principal diatom populations comprising the spring bloom are chain-forming species, notably *Skeletonema costatum*, *Thalassiosira nordenskioldii*, *Asterionella glacialis*, *Chaetoceras* sp., and *Rhizosolenia* sp.[48,52] The present composition of the phytoplankton community, in which diatoms predominate with a minor representation of nuisance species, is a desired condition for the estuary.[49]

Phytoplankton production in Delaware Bay appears to be increasing, with values nearly doubling between 1980 and 1990. Although phytoplankton production is clearly on the rise in the bay, trends are not evident for the transition or tidal river portions of the estuary. Estuarine and marine species dominate the phytoplankton community of the lower bay, and their production may be limited by light and nutrient (i.e., phosphorus) levels.

3. Zooplankton

The zooplankton community is subdivided into two groups of organisms based on the duration of planktonic life: holoplankton and meroplankton. Holoplankton include those forms that spend their entire life in the plankton (e.g., copepods and cladocerans), and meroplankton include those forms that remain planktonic for only a portion of their life cycle (e.g., larvae of benthic invertebrates and fish). Zooplankton represent the principal herbivorous component of the Delaware Estuary, consuming large numbers of phytoplankton and serving as an essential link in the energy flow of the system.

Copepods numerically dominate the zooplankton community of the Delaware estuary, comprising ~85% of the total biomass.[53] Six resident copepod species are most important, namely, *Acartia hudsonica*, *Acartia tonsa*, *Eurytemora affinis*, *Halicyclops fosteri*, *Oithona colcarva*, and *Pseudodiaptomus pelagicus*. *Acartia tonsa* is the most abundant form in the estuary.[54] Although abundant year round, copepods are occasionally eclipsed by pulses of meroplankton. Peak zooplankton densities occur in the lower transitional zone of the estuary.

Seasonal changes in copepod abundance are conspicuous. As stated by Stearns[54] (p. 35), "… During the winter, *Eurytemora affinis* is abundant upestuary, *Acartia hudsonica* and *Pseudodiaptomus pelagicus* occur throughout the system, and *Oithona colcarva* has been reported in Delaware Bay. During the spring, these species continue to be present in the estuary. *Halicyclops fosteri* begins to appear in late spring in the low salinity zones of the estuary proper and its tributaries. During the summer, *A. hudsonica* is replaced by a tremendous production of *A. tonsa* throughout the estuary. *Pseudodiaptomus pelagicus*, *H. fosteri*, and *O. colcarva* also increase in abundance at this time, while *E. affinis* declines." Copepods are most abundant during summer. The mean number of zooplankton at this time commonly exceeds 0.5×10^5 individuals/m^3 in the bay area (Figure 3.4).[55]

Salinity greatly influences the distribution of zooplankton in the system. The aforementioned copepod species, for example, have different salinity tolerances, and hence their spatial distribution patterns vary in the estuary.[54] Marine and estuarine forms (e.g., calanoid copepods) dominate the zooplankton community in the lower bay, whereas cyclopoid copepods, cladocerans, and gammarid amphipods dominate in the tidal portion upestuary. Mysids are widely distributed, providing a significant source of food for fish over extensive areas.[2]

4. Benthos

Benthic organisms — bottom-dwelling plants and animals living attached to, on, in, or near the estuarine bed — constitute a critically important food-chain link between primary producers and higher-trophic-level organisms. Some species (e.g., the eastern oyster, *Crassostrea virginica*) also serve as habitat formers for other organisms in the estuary. The

Table 3.10 Delaware Estuary Program's Priority[a] Species List

Aquatic Invertebrates

Jellyfish
Copepods
Saltmarsh invertebrates
Soft bottom (mud or sand) — oligohaline freshwater community
Soft bottom (mud or sand) polyhaline community
Hard bottom (gravel or oystershell) polyhaline community
American oyster
Mysid shrimp
Horseshoe crab
Blue crab
Dragonflies
Saltmarsh mosquito

Fish

American shad
River herrings
Marine forage fish
Freshwater marsh killifishes
Brackish marsh killifishes
Drums
Structure oriented fishes
Catfishes
Carp
Minnows
Sunfish (*Centrarchids* and *Esocides*)
Sturgeons
American eel
Atlantic menhaden
Weakfish
Sharks
Skates and rays
Perch
Striped bass
Flounder
Bluefish
Important biomass fish

Amphibians and Reptiles

Sea turtles
Diamondback terrapin
Snapping, mud, spotted, and red-bellied turtles (all "marsh" or "pond" turtles)
Marbled and tiger salamanders; New Jersey chorus, coastal plain leopard, and wood frogs; and eastern spadefoot toad (all "vernal pond breeders")

Birds

Migratory and nontidal pond shorebirds
Willet
American woodcock
Northern harrier
Short-eared owl
Barn owl
Bald eagle
Osprey

Table 3.10 (continued) Delaware Estuary Program's Priority[a] Species List

Birds
Herons, egrets, and bitterns
American black duck
Northern pintail, mallard, green-winged teal
Snow and Canada geese
Sea/bay ducks
Swamp/forest nesters
Laughing gull
Marsh wren and coastal plain swamp sparrow
Saltmarsh sparrows
Rails
Migratory passerines
Migratory raptors
Mammals
Bats
Marine mammals
River otter
Muskrat
Marsh rice rat
Meadow vole
Beaver
White-tailed deer

[a] Sometimes referred to as "key species."

Source: Sutton, C. C. et al., The Scientific Characterization of the Delaware Estuary, Tech. Rept., Delaware Estuary Program, U.S. Environmental Protection Agency, New York, 1996.

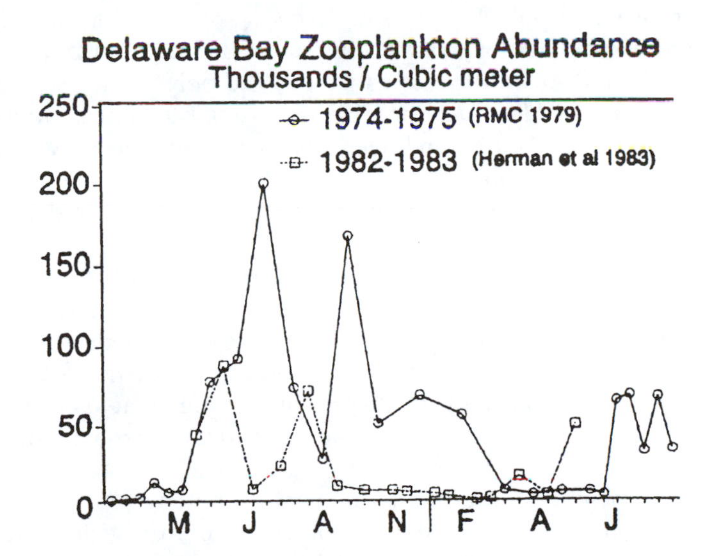

Figure 3.4 Zooplankton abundance in Delaware Bay between 1974 and 1975 and between 1982 and 1983 over a several month period. (From Sutton et al, The Scientific Characterization of the Delaware Estuary, Tech. Rept., Delaware Estuary Program, U.S. Environmental Protection Agency, New York, 1996.)

eastern oyster is a dominant benthic invertebrate in Delaware Bay, building natural beds or reefs (called seed beds) over much of the upper bay and thus creating habitat for numerous invertebrates.[56] Many smaller benthic infauna (e.g., polychaetes) mix bottom sediments via bioturbation. They not only have the potential to structurally alter the bottom habitat but also may regulate biogeochemical cycling of nutrients and other substances.[2] In addition, benthic invertebrates are particularly vulnerable to pollution and other environmental disturbances because of their limited mobility. As a result, they have been employed as indicators of the overall ecological health of estuaries and the focus of many environmental impact studies.

Physical and chemical factors in the estuarine environment clearly influence the functional morphology and behavior of the benthos. Salinity changes along the longitudinal axis of an estuary modulate abundance and diversity of the benthos, although salinity profiles tend to be more stable in interstitial than overlying waters. Consequently, the benthic infauna may be less impacted than the epifauna by salinity variations in the water column. The species composition of benthic communities depends greatly on sediment type, which often varies appreciably within short distances. Fluctuations in other physical and chemical factors (e.g., dissolved oxygen, temperature, turbidity, wave action, and turbulence) can also alter the structure of benthic communities. The availability of organic matter and oxygen below the sediment–water interface likewise has profound effects on the vertical distribution of the infauna.[57]

Biotic factors, such as predation and species competition, cannot be disregarded when investigating the occurrence and distribution of benthic fauna. They also can act as limiting factors; hence, the mere tolerance of a species to physical and chemical conditions often does not provide sufficient explanation for an observed distribution pattern. The occurrence of a species depends on biological adaptation as well. Similarities in reproductive seasons, modes of feeding, size, and other biotic factors, for instance, may enable cohabitation of several species in the same general environment.

In the lower Delaware Estuary, bivalves and polychaetes dominate the soft bottom polyhaline communities. In medium sands near the mouth of Delaware Bay, surf clams (*Spisula solidissima*) and sand dollars (e.g., *Echinarachnius parma*) are most abundant.[58] *Nucula proxima* and several polychaetes (e.g., *Nepthys* spp.) predominate in siltier areas. The most common assemblage of benthic infauna in the subtidal bay is dominated by bivalves (*Tellina agilis* and *Ensis directus*) as well as the polychaetes (*Glycera dibranchiata* and *Heteromastus filiformis*).[59] However, three bivalves (*Gemma gemma*, *Mulinia lateralis*, and *Mya arenaria*) are also quite abundant in lower salinity mesohaline, silty-fine sands. Abundance of benthic infauna in subtidal Delaware Bay appears to be less than in other estuarine systems.[58] The benthic community of the bay has not changed significantly since the 1970s.[60]

In lower salinity waters upestuary, benthic infaunal assemblages are complex and dynamic, varying with salinity and bottom sediment type. An investigation of the benthic macroinvertebrate community in the Delaware River between the C & D Canal and Trenton revealed 129 taxa of 9 phyla.[2] However, few of these taxa appear to be seasonally and regionally abundant. In these waters, oligochaetes dominate the community, with less abundant forms being chironomids, turbularians, amphipods, isopods, bivalves, and polychaetes.

Other important members of the estuarine benthos belong to the hard bottom, polyhaline community. Some common members of this community include blue mussels (*Mytilus edulis*), barnacles (*Balanus* spp.), sea squirts (*Molgula spp.*), northern corals (*Astrangia poculata*), and several species of sponges, hydroids, bryozoans, anemones, and caprellid amphipods. The hard bottom community also provides habitat and shelter for juvenile fish and many other invertebrates. Many organisms in the community also serve as a food source for recreationally or commercially important crustaceans, fish, and waterfowl.[61]

Prominent members of the saltmarsh macroinvertebrate community are fiddler crabs (*Uca* spp.), grass shrimps (*Palaemonetes* spp.), ribbed mussels (*Geukensia* [*Modiolus*] *demissa*), coffee-bean snails (*Melampus bidentatus*), and mud snails (*Ilyanassa obsoleta*). These organisms have several key ecological roles in the estuarine system. They facilitate the processing of organic matter and the cycling of nutrients in marsh habitats. They also serve as important food-web links in the marsh–estuarine system.[62]

Some areas of the Delaware Estuary exhibit benthic degradation. Based on data collected in the estuary by the Environmental Monitoring and Assessment Program (EMAP) of the USEPA, the following benthic community conditions are indicated, as defined by the EMAP benthic index:[63]

- From the vicinity of the C & D Canal northward to Trenton, ~60% of the tidal river has benthic communities classified as either degraded (31% of area) or severely degraded (29% of area).
- Benthic community conditions are less degraded in Delaware Bay. Only 13% of the area south of the C & D Canal is classified as degraded, and 4% is classified as severely degraded. The benthic community is considered to be generally healthy in the bay based on these results. The degraded benthos appear to be associated with the presence of toxic chemical contaminants, notably PCBs, pesticides, PAHs, and trace metals.

Some benthic organisms are described as key species either because they are a vital component of the ecology of the estuary or because of their economic importance. Three examples are the horseshoe crab (*Limulus polyphemus*), eastern oyster (*Crassostrea virginica*), and blue crab (*Callinectes sapidus*). All of these species are of considerable recreational and commercial importance.

a. Limulus polyphemus. The horseshoe crab population of Delaware Bay is the largest in the world. Horseshoe crab eggs provide food for numerous animals, particularly migrating shorebirds. In addition, the species is economically important in the manufacture of a product used to detect pyrogens (bacterial contamination) in injectable drugs and surgical implants. Commercial fishermen also harvest horseshoe crabs for use as bait in eel, conch, and lobster traps. Heavy harvesting has taken place since the late 1980s. Historically, the abundance of horseshoe crabs has fluctuated in response to commercial fishing, being reduced by accelerated fishing pressure. However, it is difficult to obtain accurate absolute abundance figures for this species in Delaware Bay because population surveys do not typically census animals occurring off the beaches in subtidal waters.

Horseshoe crabs are important predators of various benthic invertebrates, particularly clams, small crustaceans, and polychaete worms.[64] They grow slowly, reach sexual maturity in about 9 to 11 years, and may live as long as 15 to 20 years.[2,64,65] As a result of its slow growth and relatively long generation time, this species does not rebound rapidly from heavy exploitation.

Sandy beaches are critical habitat for spawning of horseshoe crabs. The degradation or loss of this type of habitat can severely undermine the long-term viability of the species. During a single spawning event, a female deposits a cluster of about 4000 eggs in beach sands at depths of 10 to 25 cm and may lay as many as 20 egg clusters each season.[64,66,67] Spawning takes place from April to July; it peaks in May and June, and, at this time, substantial harvesting of the resource is evident. Females are preferentially pursued by harvesters because their eggs attract commercially important target species. To obtain their catch, harvesters often search beaches along Delaware Bay where the crabs congregate to spawn.

There is concern that overharvesting of horseshoe crabs, especially during the critical reproductive period, may have significantly reduced population abundance in recent years

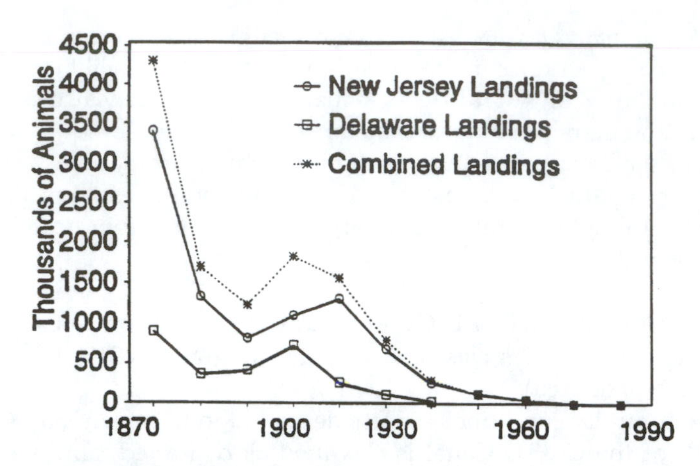

Figure 3.5 Mean annual harvest of horseshoe crabs (*Limulus polyphemus*) in Delaware and New Jersey from 1870 to 1990. (From Sutton et al., The Scientific Characterization of the Delaware Estuary, Tech. Rept., Delaware Estuary Program, U.S. Environmental Protection Agency, New York, 1996).

(Figure 3.5). For example, overharvesting may have been responsible for the declining horseshoe crab numbers reported in 1995.[2] Various horseshoe crab harvest regulations have been implemented during the 1990s to protect the resource as well as to support shorebird populations that depend on the resource for survival. These regulations have imposed limits on the species harvest to ensure the long-term viability of the horseshoe crab population along the estuary.

b. Crassostrea virginica. The eastern oyster (*Crassostrea virginica*) is both ecologically and commercially important in the Delaware Estuary. It occurs in natural beds or reefs that produce seed oysters, which support the oyster industry. These reefs also form valuable habitat for a multitude of other organisms in the estuary. Thus, they play a significant role in the estuarine food web.

Pollution is less of a problem than disease for oysters inhabiting the estuary. Predation by the oyster drill (*Urosalpinx cinerea*) remains another concern. Diseases, in particular, have devastated the oyster industry of Delaware Bay. The most severe diseases are MSX and Dermo caused by waterborne parasitic protozoans (*Haplosporidium nelsoni* and *Perkinsus marinus*, respectively). MSX, which first appeared in 1957, killed 50 to 95% of the bay's oysters over the next 2 years.[56] It has plagued the oyster population ever since. Dermo disease became prevalent in the bay during the 1990s, inflicting heavy losses on both seed and planted oysters. The reduction in the amount of hard substrate in the bay for the settlement, survival, and growth of oyster larvae and postlarvae has also limited production of the fishery.

Oyster landings have decreased substantially over the last century. Highest landings were observed between 1885 and 1890, when a peak annual harvest of more than 3 million bushels was recorded (Figure 3.6). The annual oyster harvest ranged from 1 to 2 million bushels between 1890 and 1930 and then dropped to about 1 million bushels until 1957. By 1960, the harvest had fallen to only 49,000 bushels due to MSX attack. A diminished fishery has persisted since 1960. Seed beds were closed for several years during the past decade. Efforts are under way at Rutgers University and the Virginia Institute of Marine Science to develop disease-resistant oysters to MSX and Dermo so that the fishery can be rejuvenated.

c. Callinectes sapidus. The blue crab (*Callinectes sapidus*) is a commercially and recreationally important species in the Delaware Estuary. In terms of dollar value, the harvest of blue crabs exceeds that of all other shellfish species in the system. The blue crab fishery

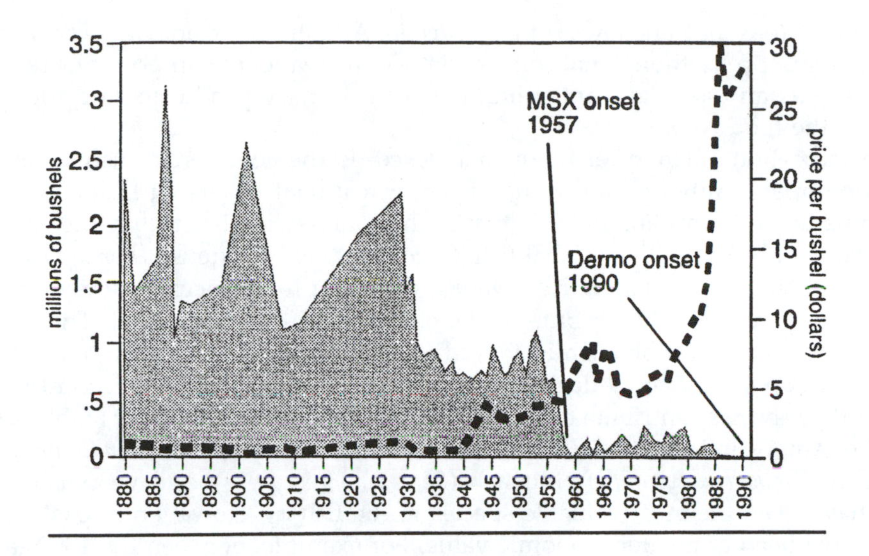

Figure 3.6 Landings data (shaded) and value (dashed line) for the eastern oyster (*Crassostrea virginica*) in the Delaware Estuary from 1880 to 1990. Note acute drop in landings data between 1950 and 1960 due to MSX infestation. (From Sutton et al., The Scientific Characterization of the Delaware Estuary, Tech. Rept., Delaware Estuary Program, U.S. Environmental Protection Agency, New York, 1996).

has increased greatly during the last 30 years concomittant with the decline of the oyster fishery. Commercial fishermen harvest blue crabs primarily via baited pots, although some crabs are also harvested from bottom sediments in winter via dredges.

The entire Delaware Estuary, including tidal freshwater areas, provides habitat for the blue crab. Adult females concentrate in high-salinity waters near the mouth of the estuary, and adult males prefer low-salinity areas. Blue crabs are most common in shallow waters (<4 m) during the warmer months of the year, but they move to deeper waters and burrow into bottom sediments during the winter.[68]

The blue crab is a voracious predator of bivalves, crustaceans, polychaetes, and small fish. However, its diet also includes plant material, and hence it is considered an omnivore.[69] Because the blue crab is widely distributed in the estuary and consumes large concentrations of juvenile bivalves, the species can have a significant impact on the abundance of other commercially or recreationally important populations (e.g., hard clams, *Mercenaria mercenaria*). The blue crab continues to be a major component of Delaware Estuary ecology due to its large numbers and significant trophic interaction with other species in the system.

5. *Finfish*

A rich and diverse assemblage of finfish, both residents and migrants, utilizes the Delaware Estuary as a feeding, spawning, and/or nursery ground. Freshwater, estuarine, and marine forms are represented. More than 200 species of finfish have been recorded in the estuary, with several supporting major fisheries. Priority species include alewife (*Alosa pseudoharengus*), American shad (*A. sapidissima*), blueback herring (*A. aestivalis*), American eel (*Anguilla rostrata*), Atlantic menhaden (*Brevoortia tyrannus*), Atlantic sturgeon (*Acipenser oxyhynchus*), white perch (*Morone americana*), striped bass (*M. saxatilis*), weakfish (*Cynoscion regalis*), bluefish (*Pomatomus saltatrix*), spot (*Leiostomus xanthurus*), scup (*Stentomus versicolor*), Atlantic croaker (*Micropogonias undulatus*), black drum (*Pogonias cromis*), channel catfish (*Ictalurus punctatus*), white catfish (*Ameiurus catus*), summer flounder (*Paralichthys dentatus*), windowpane flounder (*Scophthalmus aquosus*), and common carp (*Cyprinus carpio*).[70]

Other species (e.g., bay anchovy, *Anchoa mitchilli*; Atlantic silverside, *Menidia menidia*; and mullets, *Mugil* spp.), although not commercially or recreationally important, are significant ecologically as an essential trophic link between primary producers and top-level consumers in the food web.[71]

The recreational fishery of the estuary exceeds the commercial fishery in terms of economic impact. Although the value of the recreational fishery in Delaware Bay alone was estimated at $25 million in 1990, that of the commercial fishery for the entire estuary amounted to only $1.4 million.[2] Weakfish (*Cynoscion regalis*), bluefish (*Pomatomus saltatrix*), and summer flounder (*Paralichthys dentatus*) comprise the principal species of the recreational fishery, and they contribute to the commercial fishery as well. The recreational harvest of weakfish and bluefish is about twice that of the commercial harvest. There is some indication of an overall decline in the Atlantic weakfish stock in recent years.

About 30 species contribute to the commercial fishery of the estuary (Table 3.11). Since 1990, the American eel (*Anguilla rostrata*), American shad (*Alosa sapidissima*), Atlantic menhaden (*Brevoortia tyrannus*), bluefish (*Pomatomus saxatilis*), and weakfish (*Cynoscion regalis*) have been the most valuable species in the fishery. However, in past years other species have been of greater economic value. For example, between the 1850s and 1920s alewife (*Alosa pseudoharengus*), blueback herring (*A. aestivalis*), American shad (*A. sapidissima*), and Atlantic sturgeon (*Acipenser oxyhynchus*) were of paramount importance. These populations eventually crashed due to overexploitation and pollution. During the past 50 years, 11 different species were major contributors to the economic value of the estuary fishery (Table 3.11). Commercial landings data reveal that marine (estuarine-dependent) species (i.e., weakfish, *C. regalis*; bluefish, *P. saltatrix*; summer flounder, *Paralichthys dentatus*; Atlantic menhaden, *B. tyrannus*; and spot, *Leiostomus xanthurus*) have largely replaced the prominent upriver forms (e.g., American shad, *Alosa sapidissima*; alewife, *A. pseudoharengus*; blueback herring, *A. aestivalis*; and Atlantic sturgeon, *Acipenser oxyhynchus*) as the primary species of the commercial fishery during the past century.[72]

Finfish populations in the Delaware Estuary typically exhibit large year-to-year variations in abundance due to climatic, environmental, and anthropogenic factors. Natural and anthropogenic factors, in combination or separately, have been coupled to diminishing fish stocks.[73,74] Some of the declines of Delaware Estuary fisheries this century have been attributed to pollution as well as to the loss and degradation of habitat. Two primary nursery areas for fish are (1) low-salinity waters at the head of the estuary and (2) wetlands habitat, notably open water areas at the head of the estuary that receive fish eggs, larvae, and juveniles from freshwater spawners, semi-anadromous and anadromous fish, estuarine spawners, in addition to some larvae of fish from Delaware Bay and even the ocean. Degraded water quality in these areas, therefore, can have devastating effects on many ichthyofauna of the estuary.

The impact of habitat degradation and destruction may be more dramatic. The decrease in available nursery area for juvenile fish development (e.g., wetlands) can contribute to long-term reductions in fisheries landings of some species, as can the decline of available spawning habitat due to obstructions in waterways (e.g., dams and pollution blocks) that prevent access of certain forms (e.g., anadromous species) to spawning grounds.[72] The loss of habitat is an insidious process that has the potential to alter the composition of fish assemblages in the estuary. Population reductions of American shad, alewife, blueback herring, Atlantic sturgeon, and striped bass have been linked to habitat degradation.[2]

Aside from degraded water quality and habitat destruction, overfishing is a major factor affecting living resources. Diminished levels of Delaware River American shad (*Alosa sapidissima*) and Atlantic sturgeon (*Acipenser oxyhynchus*) stocks over the years provide examples. Changes in fish populations due to anthropogenic activities often

Table 3.11 Fishes of the Delaware Estuary Contributing to the Historic and/or Recent Commercial Fishery

Sharks (includes dusky shark, dogfish, and spiny dogfish)
Atlantic sturgeon
American eel[a]
Blueback herring[a]
Alewife[a]
American shad[a]
Atlantic menhaden[a]
Common carp[a]
Catfish (includes white catfish, brown bullhead, and channel catfish)
Silver hake
Red hake
White perch[a]
Striped bass[a]
Black sea bass
Bluefish[a]
Scup
Weakfish[a]
Spot[a]
Atlantic croaker[a]
Black drum
Tautog
Atlantic mackerel
Spanish mackerel
Butterfish
Summer flounder
Windowpane

[a] These species variously were major contributors to the economic value of the estuary fishery for nine years of record, 1947 to 1990.

Source: Sutton, C. C. et al., The Scientific Characterization of the Delaware Estuary, Tech. Rept., Delaware Estuary Program, U.S. Environmental Protection Agency, New York, 1996.

cannot be differentiated from those caused by natural factors. This is so because of the great variation in natural factors that influence the distribution and abundance of fish populations, such as temperature, salinity, turbidity, food availability, and predation pressure.[70] Hence, it is generally difficult to unequivocally establish a link between a specific anthropogenic factor and a change in the structure of fish communities in the estuary.

a. State of fisheries. The following information describes the state of fisheries for several key species of the Delaware Estuary.

- Weakfish (*Cynoscion regalis*). This migrating species is most abundant in the middle and lower estuary at depths of less than 10 m. It usually ranks among the top five most abundant finfish populations and is currently the most economically important form in the estuary.
- Bluefish (*Pomatomus saltatrix*). This species is relatively abundant in the lower and middle estuary, but rare in the upper estuary. It uses the estuary as a nursery and feeding ground. The bluefish has increased in commercial importance since 1990, and it remains a recreationally important species.
- Striped bass (*Morone saxatilis*). This major predator comprises a smaller fraction of the estuarine fish fauna today than in earlier times. The historically significant commercial

fishery for striped bass ended during the 1980s, but the species remains an economically important commodity in the estuary because of its recreational importance.

- Summer flounder (*Paralichthys dentatus*). The Delaware Estuary is an important nursery ground for this species. Although distributed throughout the lower estuary, the summer flounder is not particularly abundant. It has not contributed significantly to the commercial fishery since the mid-1960s. Nevertheless, the flatfish remains a primary target of recreational fishermen.
- White perch (*Morone americana*). This resident species occurs throughout the estuary, utilizing the system as a spawning and nursery ground. It is an important component of the recreational fishery in the middle and upper estuary but not the lower estuary where migrating marine species account for most of the recreational harvest. The commercial catch of white perch in the estuary has decreased substantially from peak levels in the late 1980s.
- Atlantic menhaden (*Brevoortia tyrannus*). The peak commercial harvest of Atlantic menhaden occurred during the 1950s. The stock was depressed in the 1970s but has increased somewhat since that time. Atlantic menhaden migrate into the estuary in the spring to feed; they emigrate to the coastal ocean in the fall.
- American shad (*Alosa sapidissima*). A century ago, the American shad supported an immense commercial fishery, but significant reductions in population abundance during the early 1900s severely limited its viability. However, an important recreational fishery for this anadromous species exists primarily in the nontidal Delaware River and its major tributaries.
- American eel (*Anguilla rostrata*). Ubiquitously distributed through the main body of the Delaware Estuary and its tributaries, the American eel is both a commercially and recreationally important species. The annual commercial harvest of this catadromous form has been relatively constant over the years. The American eel usually is not the primary target species of recreational fishermen, but it contributes to the overall recreational harvest of the estuary.

6. *Birds*

The avifauna of the Delaware Estuary is rich and diverse, with more than 300 species having been documented in the system over an annual period. The seasonal occurrence and abundance of avian populations of the estuary vary considerably. There is a constant flux of populations, as many birds migrate into the system while others emigrate out. Four principal avifauna groups are identified based on their use of the estuary: (1) resident nester-species that nest along the estuary; (2) winter resident species that inhabit the estuary during winter; (3) spring and/or fall migrant species that visit the estuary only during migration; and (4) breeding season foraging species that regularly fly to the estuary to forage while nesting at sites other than the estuary.[2] Avian populations are major components of the estuarine food web, feeding at low, intermediate, and upper trophic levels. Hence, they play a significant role in energy flow through the system. The following discussion examines the five principal avifauna groups of the estuary: waterfowl, shorebirds, wading birds, rails, and raptors.

a. Waterfowl. A diverse assemblage of waterfowl species utilizes the Delaware Estuary system. Among the prominent members of this group are divers (e.g., scoters, *Melanitta* spp.; ruddy ducks, *Oxyura jamaicensis*; and red-breasted mergansers, *Mergus serrator*) and dabblers (e.g., black ducks, *Anas rubripes*; green-winged teal, *A. crecca*; blue-winged teal, *A. discors*; pintail, *A. acuta*; and wood duck, *Aix sponsa*). Some waterfowl have distinct habitat preferences. For example, black ducks (*A. rubripes*) breed in tidal marshes

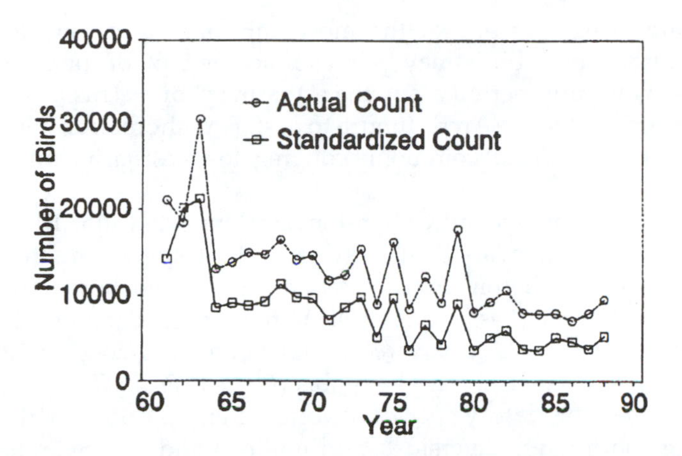

Figure 3.7 Abundance of black ducks (*Anas rubripes*) throughout their range between 1960 and 1990. (From Sutton et al., The Scientific Characterization of the Delaware Estuary, Tech. Rept., Delaware Estuary Program, U.S. Environmental Protection Agency, New York, 1996).

of the estuary. Wood ducks (*A. sponsa*) occur in wooded swamps along tributary systems, whereas mallards (*Anas platyrhynchos*) nest in impoundment areas. Snow geese (*Chen caerulescens*) feed heavily on vegetation in salt and brackish marshes. They arrive in the fall and leave when the marsh surface freezes. They may attain very high numbers (up to ~200,000 birds).

Many waterfowl that nest in prairie potholes and Arctic tundra to the far west and north of the estuary overwinter along both shores of the lower bay. They often reach high abundances there. Tidal wetlands of the estuary are overwintering sites for black ducks (*Anas rubripes*), mallards (*A. platyrhynchos*), pintails (*A. acuta*), and green-winged teal (*A. crecca*). Diving ducks, which commonly winter in the open waters of the estuary, include the common goldeneye (*Bucephala clangula*), greater scaup (*Aythya marila*), and scoters (*Melanitta* spp.).

There is growing concern regarding the drop in the abundance of black ducks (*Anas rubripes*) along the estuary, although the population decrease has been documented throughout its range (Figure 3.7). Several factors may be responsible for this decline, notably overharvesting and reduced habitat for breeding and wintering. However, other factors also may have contributed to the diminishing numbers of these birds.

b. Shorebirds. The Delaware Estuary is a major staging area in the spring for shorebirds that migrate from South America. Horseshoe crab eggs comprise a critically important food source for these birds. Delaware Bay beaches, in particular, represent primary feeding sites. Included among the most important shorebird species using the bay beaches are the red knot (*Calidris canutus*), sanderling (*C. alba*), dunlin (*C. alpina*), semipalmated sandpiper (*C. pusilla*), and ruddy turnstone (*Aernaria interpre*). More than 30 different species of shorebirds have been observed on the beaches and coastal marshes of the estuary at various times, with an estimated 1 to 1.5 million shorebirds using these habitats each spring. Some are extremely abundant. For example, up to 70% of the North American population of red knot (*C. canutus*) may use Delaware Bay at any one time.[75] Shorebirds occupy tidal marshes and mudflats of the estuary year round.[2]

Although the estuary is a very important staging area for shorebirds, few species breed there. One exception is the willet (*Catoptrophorus semipalmatus*), which breeds in saltmarshes along the estuary. This bird also feeds on invertebrates and some plants in the saltmarshes as well as in mudflats, creeks, and ditches.[76]

The Delaware Estuary is one of the most important areas in North America for migrating shorebirds. These birds may gain as much as 50% of their body weight in fat over a 10- to 14-day feeding period along the estuarine shore.[2] This location is vital to the birds, which later migrate to the Arctic tundra to nest. Few shorebirds winter in the estuary. However, dunlin (*Calidris alpina*) commonly congregate on estuarine mudflats at this time.

c. *Wading birds.* Jenkins and Gelvin-Innvaer[77] reported that 10 species of colonial nesting birds inhabit the Delaware Estuary system. These species are the great blue heron (*Ardea herodias*), little blue heron (*Egretta caerulea*), tricolored heron (*E. tricolor*), snowy egret (*E. thula*), great egret (*Casmerodius albus*), green-backed heron (*Butorides striatus*), black-crowned night heron (*Nycticorax nycticorax*), yellow-crowned night heron (*N. violaceus*), cattle egret (*Bubulcus ibis*), and glossy ibis (*Plegadis falcinellus*). Preferring saline or brackish habitats, these predatory species feed near the top of the food chain on various forage fishes (e.g., anchovies, silversides, and mullets) and invertebrates (e.g., bivalves, gastropods, and crabs).

Wading birds use wetlands habitat extensively. Herons, egrets, and ibises feed regularly in tidal wetlands of the estuary. Although the glossy ibis (*P. falcinellus*) consumes large amounts of invertebrates and fish larvae, herons and egrets feed heavily on fish. Shallow quiet waters are primary feeding habitat for wading birds; included here are saltmarsh panes and ponds, brackish and freshwater marshes, and tidal and nontidal creeks and rivers.

Most wading birds inhabit the Delaware Estuary during the spring, summer, and fall, and they emigrate to southern coastal regions for the winter.[77] Important nesting habitat can be found on estuarine islands. For instance, the largest wading nesting colony of mixed species on the Atlantic Coast north of Florida exists on Pea Patch Island in upper Delaware Bay.[78] The wading birds use different habitats for nesting, roosting, and feeding. Physical and chemical alteration of wetlands habitat along the estuary can have a devastating impact on these birds.

d. *Rails.* Six rail species have been documented in the Delaware Estuary region (i.e., king rails, *Rallus elegans*; Virginia rails, *R. limicola*; clapper rails, *R. longirostris crepitans*; black rails, *Laterallus jamaicensis*; yellow rails, *Coturnicops noveboracensis*; and sora rails, *Porzana carolina*). These species inhabit tidal marshes of the estuary. Two are migrating forms (yellow rails, *C. noveboracensis* and sora rails, *P. carolina*), and four are breeding species (king rails, *R. elegans*; Virginia rails, *R. limicola*; clapper rails, *R. longirostris crepitans*; and black rails, *L. jamaicensis*).[79] The migrating yellow rail (*C. noveboracensis*) is a relatively rare inhabitant of Delaware Estuary marshes. It tends to prefer drier, brackish salt-hay marsh areas during its spring (April and May) and fall (September and October) stopover periods. Similar to the yellow rail (*C. noveboracensis*), the black rail (*L. jamaicensis*) inhabits drier, brackish marshes. Along with the king rail (*R. elegans*) and clapper rail (*R. longirostris crepitans*), the black rail (*L. jamaicensis*) nests in the region.

In contrast to the yellow rail (*Coturnicops noveboracensis*) and black rail (*Laterallus jamaicensis*), the clapper rail (*Rallus longirostris crepitans*) prefers wetter marsh habitat dominated by cordgrass (*Spartina alterniflora*). They nest in areas where both cordgrass (*S. alterniflora*) and fiddler crabs (*Uca* spp.) proliferate. Aside from consuming fiddler crabs (*Uca* spp.), clapper rails (*R. longirostris crepitans*) readily ingest snails and other invertebrates. Nesting clapper rails (*R. longirostris crepitans*) arrive in March, and females begin to lay eggs in June. Despite occasional heavy predation on the eggs, the species is prolific and well established in the estuary.

e. Raptors. The Delaware Estuary supports a significant number of predatory bird species, including a variety of hawks, eagles, falcons, and owls. Several habitats are particularly important. For example, hardwood swamps provide nesting and feeding grounds for barred owls (*Strix varia*) and red-shouldered hawks (*Buteo lineatus*). Upland forest, forest edges, and woodlots are havens for Cooper's hawks (*Accipiter cooperii*), red-tailed hawks (*B. jamaicensis*), and great horned owls (*Bubo virginianus*). The shores of the Delaware serve as nesting areas for bald eagles (*Haliaetus leucocephalus*), osprey (*Pandion haliaetus*), and peregrine falcons (*Falco peregrinus*).[2]

In addition to raptors that nest and overwinter in the watershed, many migratory raptors utilize the Delaware Estuary area as a stopover to feed, rest, and roost. Niles and Sutton[80] noted that as many as 80,000 raptors fly through the mouth of the estuary each fall, and many also fly through the area in the spring. These birds include both short-distance migrants as well as long-distance forms (e.g., broad-winged hawk, *Buteo platypterus*), which winter in Central and South America. These migrating raptors are ecologically important to the system because they consume a wide array of animals from insects to mammals. Several migrating raptor species are listed as endangered or threatened — six in New Jersey, two in Delaware, and two at the federal level.

Raptor populations have been historically plagued by toxic chemical contaminants in their tissues, notably halogenated hydrocarbons (i.e., DDTs and PCBs). This is so because these predators feed near the top of the food chain and tend to bioaccumulate the contaminants (from numerous prey organisms) to high concentrations. Eggshell thinning caused by DDT contamination resulted in serious reductions in osprey and bald eagle populations between 1945 and 1975, and they nearly disappeared from the Delaware Estuary system. With the banning of DDT and PCB use in the U.S. during the 1970s, the bald eagle population has recovered. For example, the estuary nesting population of bald eagles increased substantially between 1982 and 1994 from 1 pair to 11 pairs. Osprey recovery has been less dramatic, with only 24 of the 200 osprey nests in New Jersey now occurring in the Delaware Estuary system.[2]

f. Environmental and anthropogenic factors. A number of environmental and anthropogenic factors influence the abundance and distribution of avian populations in the estuary. Chief among these are natural climate variations, hunting pressure, chemical contamination as well as habitat alteration and loss. The fragmentation of upland forested habitat coupled to development pressure remains a serious concern.[74]

During the past century, long-term trends in abundance of several key bird species of the Delaware Estuary have been delineated. Among the species with relatively low abundance at the turn of the century that have rebounded in recent years are the great egret (*Casmerodius albus*), snowy egret (*Egretta thula*), wood duck (*Aix sponsa*), dunlin (*Calidris alpina*), red knot (*C. canutus*), black-bellied plover (*Pluvialis squatarola*), and semipalmated sandpiper (*Charadrius semipalmatus*). Species more common 100 years ago than today are the bald eagle (*Haliaetus leucocephalus*), northern harrier (*Circus cyaneus*), piping plover (*C. melodus*), and peregrine falcon (*Falco peregrinus*).

7. Marine mammals

Whales, dolphins, porpoises, and harbor seals have been observed in the lower estuary. Most whales reported in this area are juvenile humpbacks (*Migaptera novaeangliae*). In 1991, 20 humpback whales were recorded in Delaware Bay. A right whale (*Balaena glacialis*) was discovered in the estuary in 1994, and a finback whale (*Balaenoptera physalus*) was killed there by a passing vessel in 1995.

Both bottlenose dolphins (*Tursiops truncatus*) and harbor porpoises (*Phocoena phocoena*) have appeared in the lower estuary at various times. The bottlenose dolphin is relatively abundant here from spring through fall. The harbor porpoise, in turn, can be found in the lower reaches of the estuary during the winter. At this time, harbor seals (*Phoca vitulina*) also periodically move into lower estuarine waters. Gray seals (*Halichoerus grypus*) and harp seals (*Pagophilus groenlandicus*) occasionally wash ashore each year.[2]

VIII. Management plan

Although the Delaware Estuary has exhibited significant improvement during the past 2 decades in water quality as well as other chemical, physical, and biological conditions, many problems remain that potentially threaten the environmental integrity of the system. Included here are impacts associated with point and nonpoint source runoff, habitat loss and alteration, toxic chemical contaminants, the decline of recreationally and commercially important finfish and shellfish, and a variety of watershed development issues. Specific action plans that have been formulated to protect and enhance the water quality and living resources of the estuary are contained in the CCMP[29] for the estuary. The action plans address the aforementioned problems confronting the estuary, and the CCMP proposes measures to ameliorate or correct them.

The action plans for the estuary cover seven major program areas: land management, water-use management, habitat and living resources, toxics, education and involvement, monitoring, and regional information management service. Table 3.12 provides a list of these action plans, which focus on specific priority problems. They were formulated over a 4-year period in a cooperative effort by numerous individuals representing government agencies, academic institutions, environmental organizations, industry, commercial establishments, and the general public. The following discussion on the action plans derives heavily from the CCMP.[29]

A. Land management

Five categories of action items are germane to land management. These include: (1) Sustainable Development (Action Item L1) designed to promote and facilitate development that provides housing, jobs, and revenue without destroying natural resources of the watershed; (2) Watershed-based Land Planning for Nonpoint Source Pollution Control (Action Items L2 to L5) designed to support the efforts of surrounding states in achieving the program objectives of the Clean Water Act and Coastal Zone Management Act to reduce the input of nonpoint source pollutants, both nutrients and toxic substances, into the tributaries and mainstem of the Delaware Estuary; (3) Increased Regional Coordination, Planning, and Decision-Making (Action Items L6 to L9) designed to provide incentives for regional coordination in planning and infrastructure decision-making; (4) Technical Assistance, Funding, and Streamlining (Action Items L10 to L14) designed to provide local governments with the information, data, and means to use tools to achieve environmentally sound planning and to streamline and better coordinate programs to support these changes; and (5) State and Local Regulatory Changes (Action Items L15 to L18) designed to promote the preservation of natural resources, to reduce pollutant emissions, and to streamline planning and land development processes at all levels of government.

There are several key features of a sustainable development strategy. First, the environment and the economy must be integrated at all levels of decision-making. Second, a shared vision must be developed for the estuary, and the common and conflicting values that affect the region's sustainability must be identified. Third, it is necessary to identify the

Table 3.12 Delaware Estuary Program CCMP Action Items

Action Item	Land Management
L1	Develop a comprehensive sustainable development strategy for the Delaware Estuary
L2	Support watershed-based planning
L3	Support the implementation of coastal zone act management measures
L4	Support the establishment of riparian corridor protection programs
L5	Support the implementation of urban best management practices
L6	Identify and support Greenspace Program plans to protect natural resource areas related to the estuary
L7	Support environmental agreements among municipalities and counties
L8	Develop environmental guidelines for county master plans and encourage and provide incentives for municipal conformance
L9	Expand state and/or regional planning and technical guidance to local governments
L10	Establish a land-use planner circuit rider
L11	Continue or expand municipal planning grants program
L12	Conduct training and workshops
L13	Establish and/or increase support for mapping/GIS activities
L14	Develop sustainable development business/industry incentive programs
L15	Encourage and support compact development as an element of comprehensive planning for communities
L16	Develop policies and incentives to encourage redevelopment in previously developed areas
L17	Develop policy options to address the tax revenue impact of conservation lands on municipalities
L18	Develop self-assessment techniques and an awards program to encourage municipalities to adopt environmentally sensitive planning, zoning, and site development practices
Action Item	**Water-Use Management**
W1	Promote implementation of water conservation rate structures/conservation retrofitting programs by water/wastewater utilities
W2	Conduct studies for tributary watersheds experiencing stream diminution problems
W3	Encourage water utilities to utilize water conservation techniques and conjunctive-use methods to prevent long-term lowering of groundwater levels
W4	Encourage the reuse of wastewater for nonpotable purposes
W5	Encourage water and wastewater utilities to conduct integrated resource plans
W6	Support efforts to ensure freshwater flows to the estuary to meet water supply needs to the year 2020
W7	Encourage coordination of dredging activities and priorities and the management of dredged material within the region
W8	Utilize RIMS for information management that facilitates port operations and safety
W9	Support private sector efforts on oil spill response and pollution prevention
W10	Develop, publish, and implement a comprehensive public access management strategy
W11	Inventory available pump-out stations and address any identified deficiencies
W12	Develop and implement strategies to achieve the "fishable/swimmable" goals of the Clean Water Act

Table 3.12 (continued) Delaware Estuary Program CCMP Action Items

Action Item	Habitat and Living Resources
H1	Assure compliance with existing interstate species management plans and prepare plans for additional appropriate species
H2	Establish a procedure for enhancing compatibility among species management plans
H3	Develop a natural community classification system to assist in the protection of these communities
H4	Coordinate and enhance wetlands management within the estuary
H5	Target habitat enhancement opportunities for present and future action
H6	Develop and implement an estuary-wide policy to evaluate proposed intentional introductions of exotic species and prevent unintentional ones
H7	Develop measures to protect shoreline and littoral habitats that are threatened by sea level change
H8	Facilitate coordination among the states to update and improve environmental sensitivity index mapping for hazardous spill response information
H9	Consider priority species in regulatory reviews and environmental impact statements
H10	Protect rare species through a landscape approach
Action Item	**Toxics**
T1	Implement a toxics management strategy to assist environmental managers in developing regional prevention and control strategies
T2	Assist residents in the proper use and disposal of chemicals
T3	Develop and adopt uniform water quality criteria for toxic pollutants which will be used by regulatory agencies to regulate point and nonpoint sources
T4	Implement phased limits on toxic pollutants using the TMDL concept
T5	Identify the sources of contaminated sediments and identify control strategies and mitigation alternatives
T6	Develop a uniform program for issuing fish/shellfish consumption advisories
Action Item	**Education and Involvement**
E1	Continue existing public participation program
E2	Hold and attend public meetings and workshops
E3	Continue holding annual events to raise public awareness of the estuary
E4	Develop educational initiatives in support of the land management action plan
E5	Develop educational initiatives in support of the water-use action plan
E6	Develop educational initiatives in support of the habitat and living resources action plan
E7	Develop educational initiatives in support of the toxics action plan
E8	Conduct and publish public attitude surveys
E9	Determine priority educational messages and targeted audiences
E10	Promote ecotourism in the estuarine region
E11	Encourage use of citizen monitoring activities and best available technology for monitoring
E12	Promote "hands-on" educational activities and volunteer stewardship opportunities
E13	Support floating classrooms
E14	Develop and publish outreach articles in trade magazines and journals

Table 3.12 (continued) Delaware Estuary Program CCMP Action Items

Action Item	Education and Involvement
E15	Meet the demand for existing and new publications that will increase public awareness
E16	Utilize electronic bulletin boards to disseminate information
E17	Establish estuarine resource sections within existing libraries and environmental centers
E18	Organize and implement storm drain stenciling programs
E19	Urge school administrators to incorporate estuary education in curricula and establish challenge grants
E20	Develop and place permanent estuary displays
E21	Develop a mascot for the estuary
E22	Establish a Delaware Estuary environmental badge
E23	Develop and place watershed signs on roadways and promote watershed education
Action Item	**Monitoring**
M1	Establish an interim monitoring advisory group
M2	Establish a permanent monitoring implementation team
M3	Establish the office of monitoring and mapping coordination
M4	Implement the minimal monitoring program
M5	Implement the expanded monitoring program
M6	Evaluate and report monitoring information
Action Item	**RIMS**
R1	Implement RIMS on a pilot scale for one year
R2	Implement RIMS in expanded form

Source: Delaware Estuary Program, The Delaware Estuary: Discover Its Secrets, Comprehensive Conservation and Management Plan for the Delaware Estuary, Delaware Estuary Program, U.S. Environmental Protection Agency, New York, 1996.

organizations and existing resources participating in and affecting sustainable development in the estuarine region. Fourth, it is imperative to assess institutional patterns and individual behavior in the region and determine where fundamental changes are necessary. Fifth, the formulation of a technical assistance "tool box" (e.g., assessment methods, indicators, etc.) will promote the integration of economic, environmental, and social factors in planning and policies.

To attain the objectives of watershed-based land planning for nonpoint source pollution control, efforts are under way to generate regional satellite and Geographic Information System (GIS) information for the states that will assist them in the identification of critical areas within watersheds. This information will also be useful in developing a ranking of tributary watersheds. In addition, it will be valuable in stormwater management planning to more effectively control nonpoint source pollution.

Increased regional coordination, planning, and decision-making will enable local governments to better plan and manage growth while maintaining economic vitality. Improved planning and zoning of watersheds will lead to improved environmental conditions. Although land-use decision-making takes place principally at the muncipal level, increased regional coordination and planning will facilitate the improvement of water quality, protection of habitat, and preservation of open space.

Technical assistance, funding, and streamlining are essential to effectuate change through the implementation of the CCMP action items. To promote protection of natural resources, funding should be coupled to regional/county watershed plans. Dedicated funding for updating comprehensive plans, zoning ordinances, and subdivision regulations is an important means to encourage improvements in the municipal planning process. The state of New Jersey provides municipalities with grants for projects, such as regional coordination of multimunicipality issues, GIS integration, and natural resource inventories. Comprehensive state initiatives that make grants available to municipalities for watershed planning are integral to the long-term success of the Delaware Estuary Program.

It is important for municipalities and states to review regulations and assess the need for regulatory changes to enhance the ecological integrity of the system. Particularly notable are regulations that affect development in watersheds. Recommendations of specific changes in regulations are targeting redevelopment in previously developed areas to mitigate impacts on the environment.

B. *Water-use management*

Twelve action items (W1 to W12; Table 3.12) have been formulated to address water-use management issues. They focus on three priority areas: water supply, port/navigation activities, and public access and recreational use. Millions of people in the Delaware Estuary watershed utilize the system for recreation, commerce, transportation, and food. More than 20 million people depend on water supplies of the Delaware River Basin for drinking and industrial needs. More than 6 million residents live in the estuarine region alone (from Trenton, New Jersey, to the sea), and they place significant demands on the basin's water resources. Concern exists regarding the long-term water supply for these people. Consumptive water use in the Delaware River Basin is expected to exceed 1.5×10^9 l/d by the year 2020.

Periodic basinwide water supply shortages and regional groundwater overdrafts have occurred in the Delaware River Basin. As the regional population grows, area aquifers exhibit some degree of stress. Two of the most stressed groundwater systems are the Potomac Raritan-Magothy aquifer in the New Jersey Coastal Plain and the Triassic Lowlands in southeastern Pennsylvania and central New Jersey. Water withdrawals associated with increased municipal and industrial development have resulted in the termination of freshwater flows from the Potomac Raritan-Magothy aquifer to the Delaware Estuary. Primary recharge for the aquifer now derives from the estuary and nearby streams. In addition, water levels in the aquifer have dropped below ~30 m, with additional withdrawal rates threatening the aquifer's yield and promoting the inland movement of the saltwater–freshwater transition zone.

Homeowners relying on groundwater in the Triassic Lowlands have experienced increasing episodes of water shortages. Rapid development in the Triassic Lowlands area since 1950 has been responsible for excessive pumping from the aquifer and substantial decreases in water table levels. As a consequence, groundwater withdrawals exceeding an average of more than 38,000 l/d in a groundwater-protected area of Pennsylvania (i.e., all of Montgomery County, 36 municipalities in Bucks County, 25 communities in Chester County, 3 townships in eastern Berks County, and 1 township in southern Lehigh County) require a Delaware River Basin Commission permit. This stringent regulatory program is designed to provide long-term planning and protection of water supplies. Action Items W1 to W6 directly relate to water supply problems.

Water-use management also involves assessment of the economic growth of ports along the estuary, waterborne transportation, and associated effects on living resources. The health of the regional economy is closely linked to routine operations at the ports of

Philadelphia, Camden, Gloucester City, Salem, and Wilmington. The import and export of materials at these ports are responsible for the successful operation of numerous industrial and municipal facilities in the Delaware River Basin. Aside from increasing industrial and municipal productivity and efficiency, the ports are responsible for creating many jobs in the region. It is estimated that the port complex generates an annual income of more than $3 billion and supports 180,000 jobs through waterborne commerce.

Long-term port management is critical for maintaining the economic stability and services of the region. Port activities rely on dredging by the U.S. Army Corps of Engineers to keep shipping channels clear for navigation. The port complex handles millions of metric tons of petroleum products each year, ranking second only to the Gulf region in crude oil imports alone. Consequently, Action Items W7 to W9 have been advanced to deal with dredging as well as oil spill response and pollution prevention to protect habitats and living resources in the system. Management plans proposed by the DELEP seek a common ground to sustain both economic vitality and environmental health of the Delaware River watershed and estuary.

The economic vitality of the region is also tied to the recreational use of the estuary, which depends in large part on public access to the waterfront. The ability of people to move freely to, from, and along the water is integral to fostering public interest in the resources of the system and to stimulating a sense of ownership and stewardship among area residents and visitors. These factors are crucial to protecting the habitats and living resources of the estuary. Action Items W10 to W12 propose management strategies for enhancing public access and improving water quality, both of which are necessary for increasing recreational use of the system. One objective of the DELEP is to ensure that waterfront and public access areas are properly managed for the public enjoyment of future generations.

C. Habitat and living resources

The preservation and protection of habitat and living resources in the Delaware Estuary are vital goals of the CCMP. More than 300,000 ha of wetlands and open water habitats support a multitude of terrestrial, aquatic, and avian species, representing the lowest to the highest trophic levels. Past anthropogenic practices have taken a toll on these habitats. For example, over the 20-year period between 1954 and 1974, as much as 21 to 24% of the originally existing wetlands of the estuary were lost, with freshwater and forested wetlands sustaining the greatest reduction in area.[44,45] Tighter federal regulations and more effective conservation management practices since the early 1970s have dramatically reduced impacts on the remaining wetlands habitat (>164,000 ha) of the estuary. Nevertheless, this habitat is highly vulnerable to catastrophic events, such as oil spills and releases of other toxic substances. Upland habitats are much less regulated and protected and hence continue to be altered at significant rates by development and related human activities.

Ten action items (H1–H10; Table 3.12) deal specifically with habitat and living resources of the estuary. These action items have been formulated with the following strategies in mind: (1) to restore and maintain acreage and quality of the habitats that contribute to the ecological diversity, productivity, and aesthetic appeal of the region and (2) to restore and maintain healthy populations of finfish, invertebrates, amphibians, reptiles, mammals, and birds. They address the following subject areas:[29]

- Coordination and integration of species management plans to ensure more comprehensive conservation (Action Items H1 and H2)
- Identification, restoration, and protection of specific habitat areas or types (Action Items H3, H5, and H10)

- Enhancement of planning initiatives for exotic species, sea level rise, and oil spills (Action Items H6, H7, and H8)
- Enhancement of scope and compliance with regulatory programs for wetlands and priority species (Action Items H4 and H9)

D. Toxic substances

Certain areas of the Delaware Estuary contain elevated concentrations of toxic substances in bottom sediments, the water column, and tissues of organisms. Table 3.5 shows a list of toxic substances of concern in the estuary. A major goal of the toxics action plan developed by the DELEP is to reduce contaminant loads to the estuary so that adverse effects of these substances on living resources can be mitigated. Another goal is to increase the protection of human health through fish/shellfish consumption advisories.

Action Items T1 to T6 (Table 3.12) are concerned with control, mitigation, and prevention strategies on toxic substances in the Delaware Estuary. The success of these items will be contingent upon a coordinated effort on the part of government agencies, business and industry, and citizens of three states to identify and mollify toxic contamination on an estuarywide basis. It is incumbent upon all parties to increase public awareness of the dangers of this contamination and the necessity to reduce the concentrations of these substances in the system. This will require further studies to document the sources of the toxic substances, to establish a long-term monitoring program for assessing their impacts, and to determine the most useful mitigation alternatives. Effective remedial action will require in part the enforcement of ongoing government regulations to protect the environment. Control strategies may require additional legislation, permit modification, and/or further research.

Figure 3.8 is a toxics management flow chart illustrating the decision-making process of the estuary program from problem identification to problem mitigation. Key elements of this chart are problem observation, problem identification, source characterization, and solution development. The long-term strategy is designed to yield estuarywide control and protection when dealing with toxic substances.

E. Education and involvement

Since 1989, the DELEP has strongly supported education and public involvement programs on the estuary. The Association of New Jersey Environmental Commissions and the Pennsylvania Environmental Council have conducted many public education activities for the program. An array of education topics has been addressed, such as wetlands values and functions, living resources of the estuary, pollution problems, and watershed development issues. Educational formats have included workshops, presentations, newsletters, fact sheets, public access guides, and media coverage. Hands-on educational activities (e.g., floating classrooms, outdoor classrooms, and guided walks) have been implemented. A minigrant program and action plan demonstration projects have been instituted to increase public awareness of the estuary and the remediation efforts needed to improve environmental quality. Demonstration projects may involve the dual objective of developing best management practices for reducing nonpoint source pollution as well as water quality management plans.

The success of the DELEP as well as all other national estuary programs is dependent on public involvement. An active Citizens Advisory Committee (CAC) and public outreach program serve as vehicles for increasing public awareness and participation. CAC members live in communities within the watershed and represent business, industry, municipal government, civic organizations, educational institutions, environmental groups, and

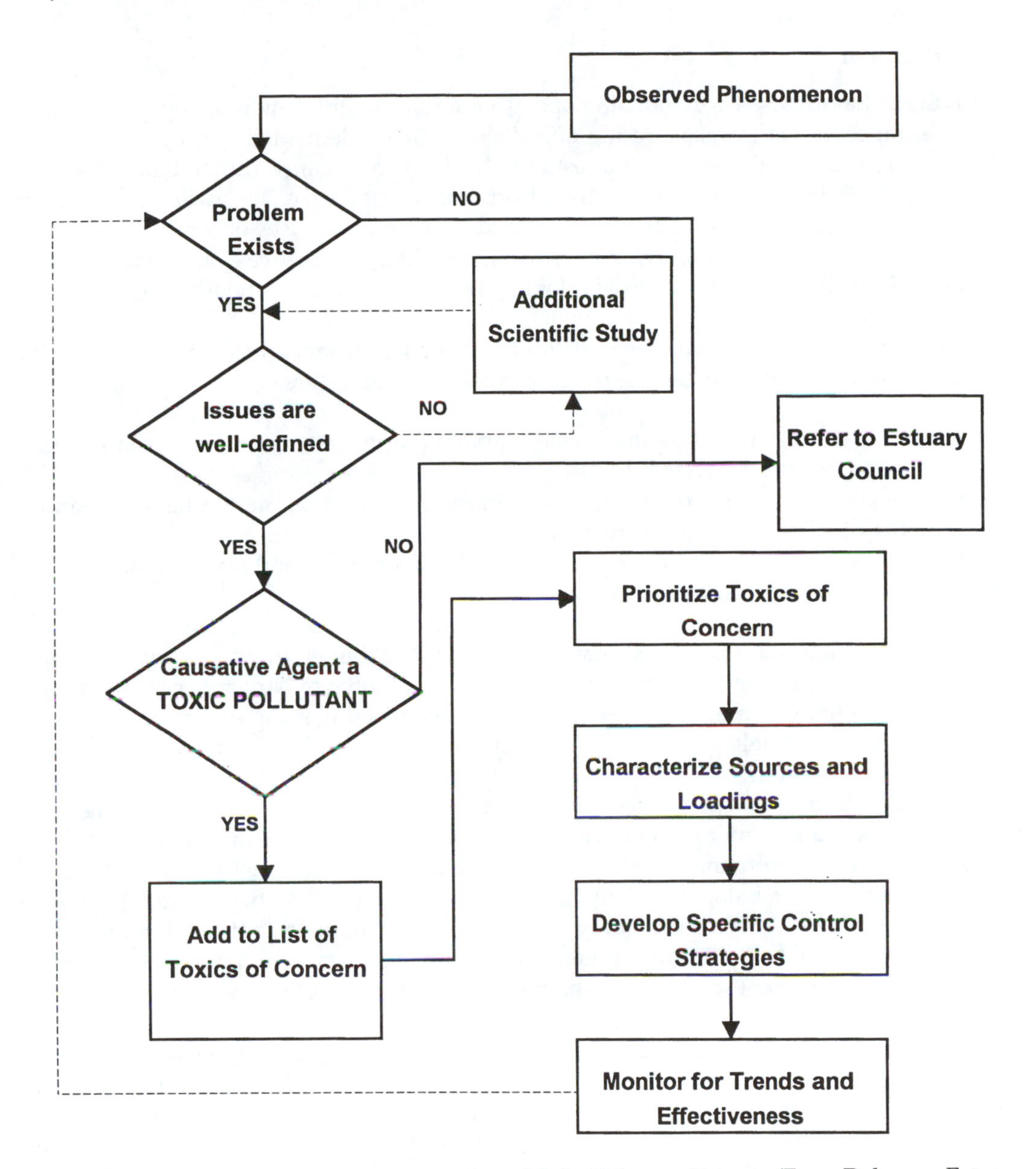

Figure 3.8 Toxic pollutant management strategy for the Delaware Estuary. (From Delaware Estuary Program, U.S. Environmental Protection Agency, New York).

other entities. They act as a direct link between the estuary program and the general public, informing their communities of estuarine issues and progress in addressing environmental problems. This open communication is an effective means of disseminating information that will foster greater protection of habitats and living resources of the estuary.

Action Items E1 to E23 (Table 3.12) address the proposed public education and involvement concerns of the program. They are designed to foster greater public understanding of the Delaware Estuary and its problems and to increase public participation in decisions and programs affecting the estuary. Recommendations of the CCMP cannot be successfully implemented without greater public involvement.

F. Monitoring

The DELEP has devised a regional cooperative environmental monitoring plan with a tripartite mission: (1) to measure the effectiveness of implemented action plans; (2) to evaluate the ecological health of the estuary; and (3) to enhance understanding of the ecosystem.[29,80,81] The plan is a cooperative effort of three states (i.e., Delaware, New Jersey, and Pennsylvania), the federal government, and industry. It focuses on several key target areas, specifically water quality, toxic substances, living resources, and habitat/land cover/land use. Action Items M1 to M6 (Table 3.12) have been recommended to implement the monitoring plan.

Monitoring objectives have been formulated for the aforementioned key areas. For water quality and toxic substances, the monitoring objectives are

- To measure status and trends in water quality parameters and toxic substances that relate to overall ecosystem health
- To evaluate areal extent and trends in parameters that define the habitat requirements of important aquatic resources
- To measure status and trends in water quality parameters and toxic substances that relate to impacts on public health

Monitoring should provide estimates of pollutant loadings to the estuary. Currently, water quality sampling entails biweekly sampling from March to November at 18 stations. This center-of-channel sampling program consists of collecting samples for water column chemistry and microbiology. The program may be augmented in the future by including winter monthly sampling.

Toxic substances are monitored, in conjunction with water quality sampling, at a subset of 10 stations. Water column samples are analyzed for trace metals (copper, lead, and zinc) and volatile organics. Finfish collected at five stations are also analyzed for trace metals as well as for halogenated hydrocarbons (chlorinated pesticides and PCBs). In addition, the Mussel Watch Program of NOAA generates data on toxic contamination in shellfish from the estuary. Future monitoring for toxic substances may be expanded to include bottom sediment sampling in shallow nearshore areas of the estuary, sampling of benthic organisms, and water column toxicity testing.

The monitoring of living resources is integral to the assessment of the total ecological health of the estuary. Aquatic populations not only respond to altered water quality but also habitat destruction; therefore, the condition of living resources provides valuable information for measuring the ecological integrity of the system. However, only limited routine monitoring of living resources has been conducted to date in the estuary, although efforts are under way to expand this program.

The objectives of the living resources monitoring plan are twofold:

- To estimate the relative abundance and trends of populations of living resources (i.e., harvestable fish and invertebrates, amphibians, reptiles, mammals, and birds)
- To estimate the overall health of the ecosystem in terms of diversity and production

To this end, several biotic monitoring programs will be continued. Included here are trawl sampling for demersal fish by the states of Delaware and New Jersey and beach seine sampling by the state of New Jersey. Ongoing sampling surveys for oysters (*Crassostrea virginica*), blue crabs (*Callinectes sapidus*), and horseshoe crabs (*Limulus polyphemus*) are expected to yield important information on the health of these populations. A new benthic monitoring program consisting of stratified random sampling of up to 68 stations at 5-year intervals is being proposed to characterize benthic assemblages. Sampling programs for amphibians, reptiles, mammals, and birds, although continuing, are being evaluated to determine if they should be expanded or if the populations need to be monitored more consistently.

Future changes in the Delaware Estuary ecosystem will be closely tied to alterations of habitat/land cover/land use in nearby watershed areas. Baseline land cover monitoring is invaluable for both habitat evaluation and land-use determinations. The DELEP recommends the application of large area mapping as the major monitoring tool to determine land cover. The use of aerial photography and satellite imagery, supplemented by ground-truth sampling, is deemed to be a critical element of the monitoring program.

The DELEP has proposed a basic survey of land cover. However, to properly assess land cover trends, remapping should be conducted along the estuary every 5 to 10 years. The state of New Jersey has undertaken a GIS mapping effort that will culminate in a comprehensive cover/habitat evaluation of the New Jersey portion of the estuary. It is hoped that a similar program can be implemented by the states of Delaware and Pennsylvania.

The objectives of habitat/land-cover/land-use monitoring of the Delaware Estuary are

- To delineate land cover/land use as a baseline using appropriate classification schemes developed on a cooperative basis for specific purposes, notably the areal extent and distribution of plant communities and critical habitat for priority species, growth and development, and population and economic trends
- To document changes in land use and land cover and to analyze trends in critical habitat for priority species, growth and development, and population and economic conditions
- To document the extent of fragmentation and connectivity of habitat, species composition, and substrate characteristics
- To update delineation on a frequency of 5-year intervals

G. *Regional Information Management Service*

The DELEP Data Management Committee developed the Regional Information Management Service (RIMS) to facilitate the use of existing environmental data sets on the estuary. The overall goal is to improve environmental data management for the program. Action Items R1 and R2 (Table 3.12) are devoted to a pilot-scale and expanded form of RIMS, respectively.

RIMS was implemented in 1995. It enables scientists, estuary managers, and other investigators to rapidly access and use numerous databases on the system. The key features of RIMS are: (1) an electronic index describing datasets and data providers; (2) a knowledgeable data manager who will maintain the data index and respond to questions; and (3) an electronic bulletin-board for data requests and other messages as well as data transmission. Internet connectivity and toll-free modem lines enhance access to data across the estuary. RIMS is constantly being improved. The long-term success of data management on the Delaware Estuary will be strongly dependent on development of RIMS.

IX. *Conclusions*

The CCMP contains recommendations for restoring and protecting the Delaware Estuary. It is the product of a collaborative effort. The document has been strongly shaped by the input of regional entities (e.g., Delaware River Basin Commission and Delaware Valley Regional Planning Commission) and nongovernment stakeholders (e.g., civic, conservation, and environmental organizations, industries, small businesses, the commercial and recreational fishing community, developers, boaters, and the public at large), with assistance from local, state, and federal government agencies. Successful implementation of the CCMP and the ability to effectively deal with environmental challenges in the future will require a concerted effort from all parties — a partnership approach — to efficiently guide and manage the natural resources of the estuary.

Although the CCMP has focused on many environmental issues, there are significant data gaps that must be filled to improve scientific understanding of the system. For example, information is generally lacking on certain estuarine processes and the cumulative environmental effects of pollutant loadings on these processes. Such shortcomings must be remedied to formulate additional corrective actions that will benefit the system.

Several other data gaps have been identified in the CCMP.[29] In regard to water use, insufficient information exists on the impacts of surface water withdrawal and discharge in watershed areas on the hydrological cycle and estuarine resources. Consequently, it is necessary to obtain more data on the incremental and cumulative impacts of water use on the estuary.

A significant land-use issue that must be resolved is how to protect natural resources without denying land owners economic use of their land. It is imperative to develop a strategy that will foster protection of estuarine resources while also maintaining economic viability of the region. Currently, land rendered undevelopable has little or no economic value to the owner. In the past, changing land uses associated with the conversion of open areas to developed lands in the upper watershed and elsewhere contributed to greater nonpoint source pollution to the estuary and the degradation of habitats and living resources.

Additional data are also needed on the acute and chronic effects of contaminated sediments on benthic communities and other organisms in the estuary. It is unclear whether organic or trace metal toxicity is limiting algal and bacterial growth in the riverine portion of the estuary. Uncertainties about biological responses to various pollutants limit the effectiveness of future management controls on the system. Additional research and monitoring should target these areas. The development of an estuarine community bioassessment protocol would be particularly valuable for biological monitoring. More emphasis must be placed on evaluating biological resources from an ecosystem health perspective.

The Delaware Estuary will continue to face threats from degraded water quality, habitat loss and alteration, and human development. As new problems emerge, it will be incumbent upon all interested parties to reevaluate and upgrade management of the estuarine system. This will require a unified effort from the general public, all levels of government, and various stakeholder organizations committed to the stewardship of the estuary. Only through unselfish cooperation and commitment can the integrity and health of this ecosystem be preserved for future generations.

References

1. Pennock, J. R. and Sharp, J., Phytoplankton production in the Delaware Estuary: temporal and spatial variability, *Mar. Ecol. Prog. Ser.*, 34, 143, 1986.
2. Sutton, C. C., O'Herron, J. C., III, and Zappalorti, R. T., *The Scientific Characterization of the Delaware Estuary*, Tech. Rept., Delaware Estuary Program, (DRBC Project No., 321; HA File No. 93.21), U.S. Environmental Protection Agency, New York, 1996.
3. Versar, Inc., *An Assessment of Key Biological Resources in the Delaware Estuary*, Tech. Rept., Delaware Estuary Program, U.S. Environmental Protection Agency, New York, 1991.
4. Sharp, J. H., Cifuentes, L. A., Coffin, R. B., Pennock, J. R., and Wong, K. C., The influence of river variability on the circulation, chemistry, and microbiology of the Delaware Estuary, *Estuaries*, 9, 261, 1986.

5. Marino, G. R., DiLorenzo, J. L., Litwack, H. S., Najarian, T. O., and Thatcher, M. L., *General Water Quality Assessment and Trend Analysis of the Delaware Estuary, Part 1. General Status and Trend Analysis*, Tech. Rept., Delaware Estuary Program, U.S. Environmental Protection Agency, Philadelphia, PA, 1991.
6. Sharp, J. H., Introduction to science chapters, in *The Delaware Estuary: Research as Background for Estuarine Management and Development*, Sharp, J. H., Ed., Tech. Rept., University of Delaware Sea Grant Program, Newark, DE, 1984, 1.
7. Smullen, J. T., Sharp, J. H., Garvine, R. W., and Haskin, H. H., River flow and salinity, in *The Delaware Estuary: Research as Background for Estuarine Management and Development*, Sharp, J. H., Ed., Tech. Rept., University of Delaware Sea Grant Program, Newark, DE, 1984.
8. Sharp, J. H., Culberson, C. H., and Church, T. M., The chemistry of the Delaware Estuary: general considerations, *Limnol. Oceanogr.*, 27, 1015, 1982.
9. Hires, R. I., Mellor, G. L., Oey, L. Y., and Garvine, R. W., Circulation of the estuary, in *The Delaware Estuary: Research as Background for Estuarine and Development*, Sharp, J. H., Ed., Tech. Rept, University of Delaware Sea Grant College Program, Newark, DE, 1984.
10. Pape, E. H. and Garvine, R. W., The subtidal circulation in Delaware Bay and adjacent shelf waters, *J. Geophys. Res.*, 87, 7955, 1982.
11. Greeley-Polhemus Group, Inc., *Delaware Estuary Program Land Use Management Inventory and Assessment*, Tech. Rept., Greeley-Polhemus Group, Inc., West Chester, PA, 1990.
12. DiLorenzo, J. L., Huang, P., Thatcher, M. L., and Najarian, T. O., *Effects of Historic Dredging Activities and Water Diversions on the Tidal Regime and Salinity Distribution of the Delaware Estuary*, Final Report for the Delaware Estuary Program, Najarian Associates, Inc., Eatontown, NJ, 1993.
13. Albert, R. C. and Pollison, D. P., The importance of water in the estuary: where do we stand today?, in *The State of the Estuary: Summary Report of the October 19, 1989 Workshop*, Tech. Rept., Delaware Estuary Program Scientific Technical Advisory Committee, University of Delaware, Lewes, DE, 1989.
14. Albert, R. C., The historical context of water quality management for the Delaware Estuary, *Estuaries*, 11, 99, 1988.
15. DRBC, *Delaware River and Bay Water Quality Assessment, 1988–1989 305(b) Report*, Tech. Rept., Delaware River Basin Commission, West Trenton, NJ, 1991.
16. DRBC, *Delaware River and Bay Water Quality Assessment: 1992–1993 305(b) Report*, Tech. Rept., Delaware River Basin Commission, West Trenton, NJ, 1994.
17. Sharp, J. H., Cifuentes, L. A., Coffin, R. B., Lebo, M. E., and Pennock, J. R., *Eutrophication: Are Excessive Nutrient Inputs a Problem for the Delaware Estuary?*, Tech. Rept., University of Delaware Sea Grant College Program, Newark, DE, 1994.
18. Lebo, M. E. and Sharp, J. H., Phosphorus distributions along the Delaware: an urbanized coastal plain estuary, *Estuaries*, 16, 291, 1993.
19. Sanders, J. G. and Reidel, G. F., *Factors Limiting Primary Production in the Urban Delaware River*, Final Report, Report No. 92-35, The Academy of Natural Sciences, Benedict, MD, 1992.
20. Academy of Natural Sciences of Philadelphia, *Status and Trends of Toxic Pollutants in the Delaware Estuary*, Report No. 91-14, Delaware Estuary Program, Academy of Natural Sciences of Philadelphia, Philadelphia, PA, 1991.
21. DRBC, *Sediment Contaminants of the Delaware River Estuary*, Tech. Rept., Estuary Toxics Management Program, Delaware River Basin Commission, West Trenton, NJ, 1993.
22. Costa, H. J. and Sauer, T. C., *Distributions of Chemical Contaminants and Acute Toxicity in Delaware Estuary Sediments*, Final Report, DELEP #94-08, Arthur D. Little, Inc., Cambridge, MA, 1994.
23. Frithsen, J. B., Strebel, D. E., and Schawitsch, T., *Estimates of Contaminant Inputs to the Delaware Estuary*, Tech. Rept., Versar Inc., Columbia, MD, 1995.
24. Riedel, G. F. and Sanders, J. G., *Trace Element Speciation and Behavior in the Tidal Delaware River*, Report No. 93-1, Benedict Estuarine Research Laboratory, Benedict, MD, 1993.
25. Rainbow, P. S., The significance of trace metal concentrations in marine invertebrates, in *Ecotoxicology of Metals in Invertebrates*, Dallinger, R. and Rainbow, P. S., Eds., Lewis Publishers, Boca Raton, FL, 1993, 3.

26. Viarengo, A., Heavy metals in marine invertebrates: mechanisms of regulation and toxicity at the cellular level, *Rev. Aquat. Sci.*, 1, 295, 1989.
27. Long, E. R. and Morgan, L. G., *The Potential for Biological Effects of Sediment-Sorbed Contaminants Tested in the National Status and Trends Program*, NOAA Tech. Mem. NOS OMA 52, National Oceanic and Atmospheric Administration, Rockville, MD, 1991.
28. Gottholm, B. W., Harmon, M. R., Turgeon, D. D., and Frew, S., *Assessment of Chemical Contaminants in the Chesapeake and Delaware Bays*, Tech. Rept., National Status and Trends Program, National Oceanic and Atmospheric Administration, Silver Spring, MD, 1994.
29. DELEP, *The Delaware Estuary: Discover Its Secrets*, Comprehensive Conservation and Management Plan for the Delaware Estuary, Delaware Estuary Program, U.S. Environmental Protection Agency, New York, 1996.
30. Church, T. M., Tramontano, J. M., Scudlark, J. R., and Murray, S. L., Trace metals in the waters of the Delaware estuary, in *Ecology and Restoration of the Delaware River Basin*, Majumdar, S. K., Miller, E. W., and Sage, L. E., Eds., Pennsylvania Academy of Science, Easton, PA, 1988, 93.
31. Rice, C. L., *Concentrations of Organochlorines and Trace Elements in Fish and Blue Crabs from the Delaware River, Easton to Deepwater*, Pennsylvania Field Office Special Project 93-5, U.S. Fish and Wildlife Service, State College, PA, 1992.
32. Guzzella, L. and De Paolis, A., Polycyclic aromatic hydrocarbons in sediments of the Adriatic Sea, *Mar. Pollut. Bull.*, 28, 159, 1994.
33. Sieger, T. L. and Tanacredi, J. T., Contribution of polynuclear aromatic hydrocarbons to Jamaica Bay ecosystem attributable to municipal wastewater effluents, in *Proceedings of the Second Conference on Scientific Research in the National Parks*, San Francisco, CA, Environmental Concerns in Urban Impacted Parks, 3, 1, 1979.
34. Hauge, P., *Polychlorinated Biphenyls (PCBs), Chlordane, and DDTs in Selected Fish and Shellfish from New Jersey Waters, 1988–1991: Results from New Jersey's Toxics in Biota Monitoring Program*, Tech. Rept., Office of Science and Research, New Jersey Department of Environmental Protection and Energy, Trenton, NJ, 1993.
35. Green, R. W. and Miller, R. W., *Summary and Assessment of Polychlorinated Biphenyls and Selected Pesticides in Striped Bass from the Delaware Estuary*, Tech. Rept., Delaware Department of Natural Resources and Environmental Control, Dover, DE, 1994.
36. USEPA, *Environmental Monitoring and Assessment Program: (EMAP) Virginian Province Estuaries (1990–1992), Delaware Bay and River Database*, Tech. Rept., Office of Research and Development, U.S. Environmental Protection Agency, Narragansett, RI, 1995.
37. DRBC, *Ambient Toxicity Study of the Delaware River Estuary, Phase 1, Estuary Toxics Management Program*, Delaware River Basin Commission, West Trenton, NJ, 1991.
38. DRBC, *Fish Health and Contamination Study, Delaware Estuary Use Attainability Project Element 10*, Delaware River Basin Commission, West Trenton, NJ, 1988.
39. Steidl, R. J., Griffin, C. R., and Niles, L. J., Differential reproductive success of ospreys in New Jersey, *J. Wildl. Manage.*, 55, 226, 1991.
40. Steidl, R. J., Griffin, C. R., Niles, L. J., and Clark, K. E., Reproductive success and eggshell thinning of a reestablished peregrine falcon population, *J. Wildl. Manage.*, 55, 294, 1991.
41. Niles, L. J., Clark, K., and Ely, D., Breeding status of bald eagle in New Jersey, *Records NJ Birds*, 17, 1, 1991.
42. Sneddon, L. A., Metzler, K. J., and Anderson, M., A classification and description of natural community alliances and community elements of the Delaware Estuary, in *Living Resources of the Delaware Estuary*, Dove, L. E. and Nyman, R. M., Eds., Tech. Rept., The Delaware Estuary Program, U.S. Environmental Protection Agency, New York, 1995.
43. Tiner, R. W., Jr., *Wetlands of New Jersey*, Tech. Rept., U.S. Fish and Wildlife Service, National Wetlands Inventory, Newton Corner, MA, 1985.
44. Tiner, R. W., Jr., *Wetlands of Delaware*, Cooperative Publication, National Wetlands Inventory Project, U.S. Fish and Wildlife Service Region 5, Newton Corner, Massachusetts, and Division of Environmental Control, Delaware Department of Natural Resources and Environmental Control, Dover, DE, 1985.

45. Tiner, R. W., Jr., *Pennsylvania's Wetlands: Current Status and Recent Trends*, Tech. Rept., U.S. Fish and Wildife Service, National Wetlands Inventory, Newton Corner, MA, 1990.
46. Ford, S. E. and Haskin, H. H., History and epizootiology of *Haplosporidium nelsoni* (MSX), an oyster pathogen, in Delaware Bay, 1957–1980, *J. Invert. Pathol.*, 40, 118, 1982.
47. Ford, S. E. and Haskin, H. H., Management strategies for MSX, *Haplosporidium nelsoni*, disease in eastern oysters, *Am. Fish. Soc. Spec. Publ.*, 18, 249, 1988.
48. Marshall, H. G., *Assessment of Phytoplankton Species in the Delaware River Estuary.* Technical Report to the Delaware River Basin Commission by the U.S. Environmental Protection Agency, Philadelphia, PA, 1992.
49. Marshall, H. G. Phytoplankton, in *Living Resources of the Delaware Estuary,* Dove, L. E. and Nyman, R. M., Eds., Tech. Rept., The Delaware Estuary Program, U.S. Environmental Protection Agency, New York, 1995, 25.
50. Hargreaves, B. R. and Kraeuter, J. N., The state of living resources in the Delaware Estuary, in *The State of the Delaware Estuary,* Sharp, J. H., Ed., Individual Papers from the October 19, 1989 Workshop, U.S. Environmental Protection Agency, New York, 1989.
51. Watling, L., Bottom, D., Pembroke, A., and Maurer, D., Seasonal variations in Delaware Bay phytoplankton community structure, *Mar. Biol.*, 52, 207, 1979.
52. Pennock, J. R., Food webs, in *The Delaware Estuary: Rediscovering a Forgotten Resource*, Bryant, T. L. and Pennock, J. R., Eds., Tech. Rept., University of Delaware Sea Grant College Program, Newark, DE, 1988, 55.
53. Herman, S. S., Zooplankton, in *The Delaware Estuary: Rediscovering a Forgotten Resource,* Bryant, T. L. and Pennock, J. R., Eds., Tech. Rept., University of Delaware Sea Grant College Program, Newark, DE, 1988, 60.
54. Stearns, D. E., Copepods, in *Living Resources of the Delaware Estuary,* Dove, L. E. and Nyman, R. M., Eds., Tech. Rept., The Delaware Estuary Program, U.S. Environmental Protection Agency, New York, 1995, 33.
55. Herman, S. S., Hargreaves, B. R., Lutz, R. A., Fritz, L. W., and Epifanio, C. E., Zooplankton and parabenthos, in *The Delaware Estuary: Research as Background for Estuarine Management and Development*, Tech. Rept., Delaware River and Bay Authority, University of Sea Grant College Program, Newark, DE, 1983.
56. Ford, S. E., Haskin, H. H., and Kraeuter, J. N., Eastern oyster, in *Living Resources of the Delaware Estuary,* Dove, L. E. and Nyman, R. M., Eds., Tech. Rept., The Delaware Estuary Program, U.S. Environmental Protection Agency, New York, 1995, 105.
57. Kinner, P., Maurer, D., and Leathem, W., Benthic invertebrates in Delaware Bay: animal-sediment associations of the dominant species, *Int. Revue Ges. Hydrobiol.*, 59, 685, 1974.
58. Steimle, F., Soft (mud/sand) bottom polyhaline communities, in *Living Resources of the Delaware Estuary,* Dove, L. E. and Nyman, R. M., Eds., Tech. Rept., The Delaware Estuary Program, U.S. Environmental Protection Agency, New York, 1995, 119.
59. Maurer, D., Watling, L., Kinner, P., Leathem, W., and Wethe, C., Benthic invertebrate assemblages of Delaware Bay, *Mar. Biol.*, 45, 65, 1978.
60. Foster, K. L., Steimle, F. W., Muir, W. C., Kropp, R. K., and Conlin, B., Mitigation potential of habitat replacement: concrete artificial reef in Delaware Bay — preliminary results, *Bull. Mar. Sci.*, 55, 783, 1994.
61. Steimle, F., Hard bottom polyhaline community, in *Living Resources of the Delaware Estuary,* Dove, L. E. and Nyman, R. M., Eds., Tech. Rept., The Delaware Estuary Program, U.S. Environmental Protection Agency, New York, 1995, 113.
62. Kreamer, G. R., Saltmarsh invertebrate community, in *Living Resources of the Delaware Estuary,* Dove, L. E. and Nyman, R. M., Eds., Tech. Rept., The Delaware Estuary Program, U.S. Environmental Protection Agency, New York, 1995, 81.
63. Paul, J. F., Gentile, J. H., Schimmel, S. C., Scott, K. J., and Campbell, D. E., *Assessment of Estuarine Conditions in the Virginian Province Using 1990–93 EMAP Data*, Tech. Rept., Environmental Research Laboratory, Narragansett, RI, 1994.
64. Botton, M. L., Horseshoe crab, in *Living Resources of the Delaware Estuary,* Dove, L. E. and Nyman, R. M., Eds., Tech. Rept., The Delaware Estuary Program, U.S. Environmental Protection Agency, New York, 1995, 51.

65. Botton, M. L. and Ropes, J. W., An indirect method for estimating longevity of the horseshoe crab (*Limulus polyphemus*) based on epifaunal slipper shells (*Crepidula fornicata*), *J. Shellfish Res.*, 7, 407, 1988.
66. Shuster, C. N., Jr. and Botton, M. L., A contribution to the population biology of horseshoe crabs, *Limulus polyphemus*, in Delaware Bay, *Estuaries*, 8, 363, 1985.
67. Botton, M. L., Loveland, R. E., and Jacobsen, T. R., Beach erosion and geochemical factors: influence on spawning success of horseshoe crabs (*Limulus polyphemus*) in Delaware Bay, *Mar. Biol.*, 99, 325, 1988.
68. Epifanio, C. E., Atlantic blue crab, in *Living Resources of the Delaware Estuary*, Dove, L. E. and Nyman, R. M., Eds., Tech. Rept., The Delaware Estuary Program, U.S. Environmental Protection Agency, New York, 1995, 43.
69. Laughlin, R. A., Feeding habits of the blue crab, *Callinectes sapidus* Rathbun, in the Apalachicola Estuary, Florida, *Bull. Mar. Sci.*, 32, 807, 1982.
70. O'Herron, J. C., II, Lloyd, T., and Laidig, K., *A Survey of Fish in the Delaware Estuary from the Area of the Chesapeake and Delaware Canal to Trenton*, Tech. Rept., U.S. Environmental Protection Agency, New York, 1994.
71. McBride, R. S., Marine forage fish, in *Living resources of the Delaware Estuary*, Dove, L. E. and Nyman, R. M., Eds., Tech. Rept., The Delaware Estuary Program, U.S. Environmental Protection Agency, New York, 1995, 211.
72. Price, K. S. and Beck, R. A., Finfish, in *The Delaware Estuary: Research as Background for Estuarine Management and Development*, Bryant, T. L. and Price, J. R., Eds., Tech. Rept., University of Delaware Sea Grant College Program, Newark, DE, 1988.
73. Daiber, F. C., Fisheries resources of the Delaware Estuary, in *The Delaware Estuary*, Majumdar, S. K., Miller, E. W., and Sage, L. E., Eds., Pennsylvania Academy of Science, Easton, PA, 1988, 169.
74. Frithsen, J. B., Killam, K., and Young, M., *An Assessment of Key Biological Resources in the Delaware River Estuary*, Tech. Rept., The Delaware Estuary Program, U.S. Environmental Protection Agency, New York, 1991.
75. Clark, K. E., *1988 Delaware Bay Shorebird Project*, Final Report, Endangered and Nongame Species Program, Division of Fish, Game, and Wildlife, New Jersey Department of Environmental Protection, Trenton, NJ, 1988.
76. Kibbe, D. P., Willet, in *Living Resources of the Delaware Estuary*, Dove, L. E. and Nyman, R. M., Eds., Tech. Rept., The Delaware Estuary Program, U.S. Environmental Protection Agency, New York, 1995, 465.
77. Jenkins, D. and Gelvin-Innvaer, L. A., Colonial wading birds, in *Living Resources of the Delaware Estuary*, Dove, L. E. and Nyman, R. M., Eds., Tech. Rept., The Delaware Estuary Program, U.S. Environmental Protection Agency, New York, 1995, 335.
78. Erwin, R. M. and Korschgen, C. E., *Coastal Waterbird Colonies: Cape Elizabeth, Maine to Virginia, 1977: An Atlas Showing Colony Locations and Species Composition*, U.S. Fish and Wildl. Serv., Off. Biol. Serv., FWS/OBS-79/08, Washington, D.C., 1979.
79. Kerlinger, P. and Widjeskog, L., Rails, in *Living Resources of the Delaware Estuary*, Dove, L. E. and Nyman, R. M., Eds., Tech. Rept., The Delaware Estuary Program, U.S. Environmental Protection Agency, New York, 1995, 425.
80. Niles, L. J., and Sutton, C., Migratory raptors, in *Living Resources of the Delaware Estuary*, Dove, L. E. and Nyman, R. M., Eds., Tech. Rept., The Delaware Estuary Program, U.S. Environmental Protection Agency, New York, 1995, 433.
81. Sharp, J. H. and Walsh, M., *A Cooperative Monitoring Plan for the Delaware Estuary*, DELEP Rept. #95-02, U.S. Environmental Protection Agency, New York, 1995.
82. Tetra Tech, Inc., *Delaware Estuary Regional Monitoring Plan, Vols. 1 and 2*, Tech. Rept., U.S. Environmental Protection Agency, New York, 1994.

chapter four

Case study 3: Galveston Bay National Estuary Program

I. Introduction

Galveston Bay ranks among the most important estuaries in the U.S. in terms of ecologic and economic value. It is a system critically important to the regional and state economies, supporting billions of dollars of commercial and recreational activities each year (e.g., oil and gas production, shipping, electric power generation, commercial and recreational fishing, and tourism). However, human development and overuse are now threatening the health of the Galveston Bay system. For example, wastewater discharges from more than 1400 municipal and industrial facilities, inputs of toxic substances from nonpoint sources, alteration of freshwater inflow, degradation and loss of wetlands habitat, overfishing, and the decline of some biotic populations have been detrimental to the estuary.

The Galveston Bay National Estuary Program (GBNEP) was established in 1990 to address environmental problems associated with human development, pollution, and overuse of the bay's resources. Many of these problems are complex and interrelated, involving watershed areas, tributaries, and the bay itself (Table 4.1). To develop a Comprehensive Conservation and Management Plan (CCMP) for the bay, the GBNEP employed an ecosystem-based approach, relying on both scientific and environmental management information to assess historical, current, and emerging problems and to formulate action plans to effectively remedy or control them. Between 1990 and 1995, intense investigations by the GBNEP identified specific problems existing in the bay and its watershed, compiled data on the status and trends of its resources, and generated a detailed list of management actions to improve the system. Of the hundreds of problems affecting the bay, the GBNEP selected 17 as most compelling.[1] These 17 priority problems, deemed to be most important by consensus of the GBNEP Management Conference, served as the focal point for developing CCMP goals, objectives, and actions that guide the future stewardship of the estuary.

Table 4.1 Comparison of Priority Environmental Problems in Galveston Bay to Those of Other National Estuary Program Sites

Project Location	Surface Area (sq. miles)	Drainage Basin Area (sq. miles)	Watershed Population	Priority Environmental Problems
Albemarle/ Pamlico Sound, NC	2,900	30,880	1,898,000	Wetlands, nutrients, fish disease, land use and population, freshwater flows, habitat loss, fisheries productivity, submerged aquatic vegetation
Barataria-Terrebonne, LA	2,141	5,460	695,000	Hydrological modification, eutrophication, pathogen contamination of shellfish, changes in biological resources, habitat loss and modification, toxics
Buzzards Bay, MA	228	432	236,000	Pathogens, nitrogen loading, shoreline development, habitat loss, toxic contamination
Casco Bay, ME	152	979	251,000	Toxic pollutants, nutrients, pathogens, habitat loss
Delaware Bay, DE/NJ/PA	768	475	6,000,000	Habitat loss, nonpoint source pollution, lack of public access, estuarine education, compliance
Delaware Inland Bays, DE	32	255	50,000	Habitat loss, eutrophication, land use, point/nonpoint pollutants
Galveston Bay, TX	**600**	**25,256**	**6,000,000**	**Habitat loss, urban runoff, toxic and bacterial contamination, inflow and circulatory modifications, subsidence and erosion**
Indian River Lagoon, FL	353	2,284	630,000	Nutrients, circulation, loss of wetlands, increased toxics, increased pathogens and suspended sediments
Long Island Sound, CT/MA/NY/RI	1,281	16,000	8,400,000	Eutrophication, hypoxia, toxicants, pathogens, floatable debris, impacts to living resources
Massachusetts Bay, MA	2,000	2,900	4,000,000	Toxics in water, sediments, fish and shellfish, pathogens, habitat loss and modification, sea level rise
Narragansett Bay, RI/MA	146	1,677	1,800,000	Pollutants, pathogens, living resources management, habitat protection, combined sewer overflow abatement
New York/ New Jersey Harbor, NY/NJ	298	8,467	17,000,000	Urban runoff, contaminated sediments, shoreline development, pathogens.

Table 4.1 (continued) Comparison of Priority Environmental Problems in Galveston Bay to Those of Other National Estuary Program Sites

Project Location	Surface Area (sq. miles)	Drainage Basin Area (sq. miles)	Watershed Population	Priority Environmental Problems
Puget Sound, WA	931	16,000	3,000,000	Pollutants, loss of aquatic habitats, eutrophication, dredging
San Francisco Bay/ Sacramento-San Joaquin Delta, CA	554	61,313	9,900,000	Decline of biological resources, altered freshwater flows, pollutants, dredging, land use
Santa Monica Bay, CA	266	414	9,000,000	Contaminants in fish and sediments, marine habitat, swimmable waters, municipal effluent, urban runoff
Sarasota Bay, FL	40	500	425,000	Nutrients, habitat loss, declines in living resources, population growth
Tampa Bay, FL	398	2,300	2,100,000	Habitat loss and modification, altered freshwater inflow, natural flushing

Source: The National Estuary Program, U.S. Environmental Protection Agency, Washington, D.C.

II. Physical description

A. The bay

Galveston Bay is a shallow, lagoon-type estuary located along the southeast coast of Texas. A barrier island system (i.e., Bolivar Peninsula and Galveston Island) separates the bay from the Gulf of Mexico (Figure 4.1). However, three inlets (i.e., Bolivar Roads, San Luis Pass, and Rollover Pass) provide conduits for tidal exchange between the gulf and bay. Primary tidal exchange occurs at Bolivar Roads (>80%) and San Luis Pass (slightly less than 20%). Rollover Pass, a man-made cut through Bolivar Peninsula, accounts for only minor tidal exchange (<1%).

With a surface area of ~1.55×10^5 ha, Galveston Bay forms an irregular tidal basin consisting of four major sub-bays (i.e., East Bay, West Bay, Trinity Bay, and Galveston Bay) and two relatively isolated secondary bays (i.e., Bastrop Bay and Christmas Bay). The present configuration of the estuary began to develop ~4500 years ago as sea level stabilized after a protracted period of rise. Bolivar Peninsula gradually formed via longshore drifting of sediment. Offshore bars migrated landward to create Galveston Island and Follets Island. These processes resulted in the enclosure of much of the bay and the formation of East, West, Christmas, and Bastrop Bays. As valleys of the San Jacinto and Trinity Rivers were drowned, they formed the inland portions of the estuary. Sediment infilling has subsequently modified them.

The average water depth is ~1 to 2 m in East Bay and West Bay, ~2 to 3 m in Trinity Bay, and generally <4 m (excluding channels) in Galveston Bay. Bolivar Roads, which is scoured by tidal currents, has a maximum depth of ~15 m; it marks the deepest part of the estuarine system. However, several deep man-made channels have been dredged in the bay bottom for navigation purposes, such as the 13-m Houston Ship Channel in Galveston Bay, which extends northward across the bay to the mouth of the San Jacinto

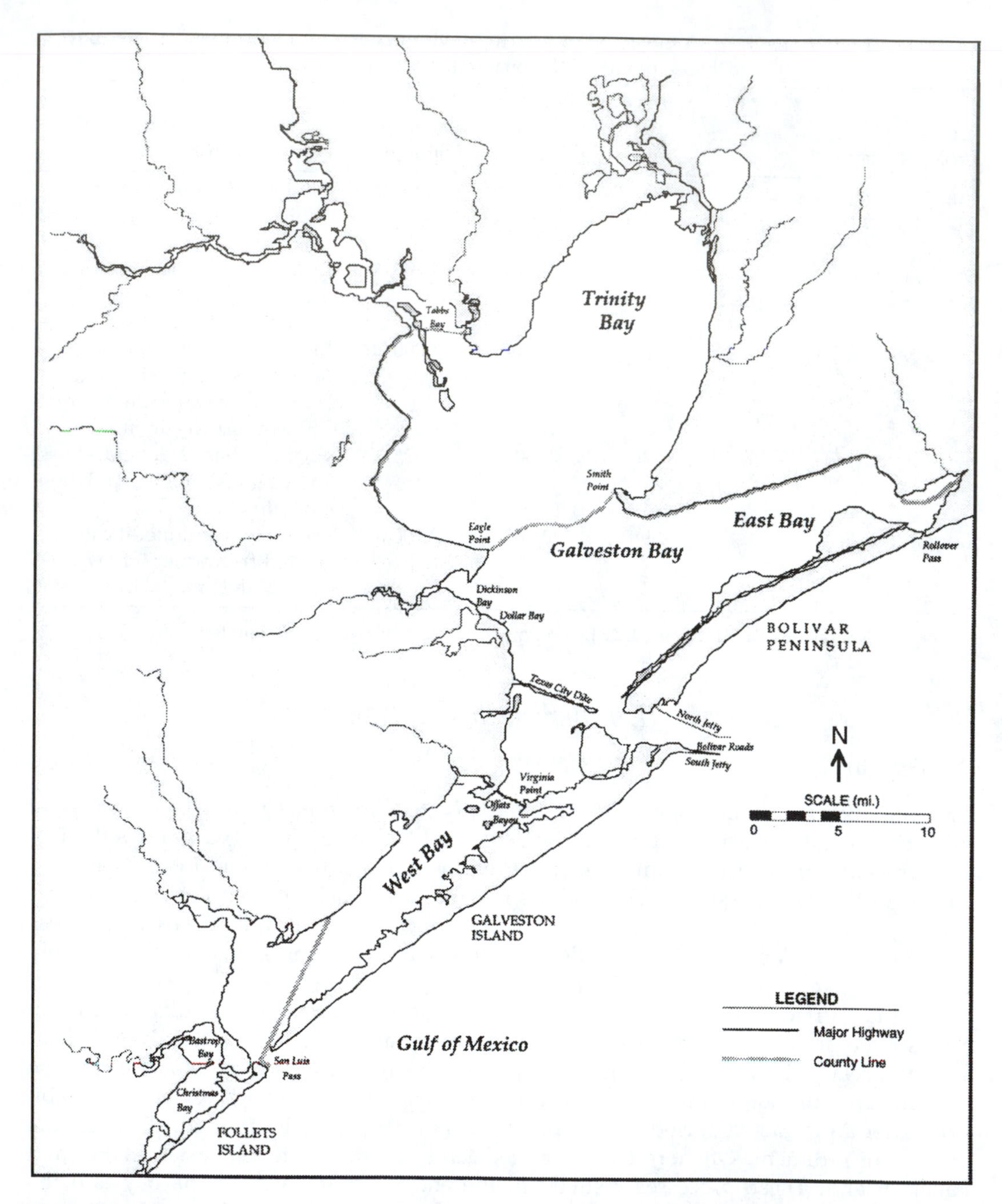

Figure 4.1 Map of Galveston Bay and surrounding watershed areas. (Modified from GBNEP, *The Galveston Bay Plan: The Comprehensive Conservation, and Management Plan for the Galveston Bay Ecosystem*, Galveston Bay National Estuary Program Publication GBNEP-49, Webster, TX, 1994.)

River and up the Buffalo Bayou. The dredged Gulf Intracoastal Waterway tracks parallel to the coast. The estuary is also regularly dredged in several other areas.

Because Galveston Bay is a shallow and largely enclosed system, winds dominate many physical processes and can significantly influence tidal currents, water levels, and bottom sediment dynamics. Summer and winter wind patterns differ dramatically. Prevailing summer winds are out of the south, whereas winter winds are mainly out of the north and northwest. During the winter months, brief and intense "northers" produce winds with speeds greater than 60 km/hr, which can drastically alter circulation in the estuary. According to Ward,[2] winds affect the circulation in Galveston Bay via the generation of internal wind-driven circulation, the development of windwaves, and large-scale shifts in water levels (denivellation). Winds may completely dominate short-term circulation patterns, especially in restricted embayments and secondary bays, and are largely responsible for the intense vertical mixing characteristic of the estuary. As a result, stratification is essentially absent in shallower waters.

Tides and freshwater inflows also influence circulation patterns in the estuary. The tides, which are relatively weak, consist of both diurnal and semidiurnal components. However, winds and changes in atmospheric pressure often disrupt the astronomical tidal cycles in the system. The tidal range is maximum at the inlets; damping reduces the mean tidal range in the bay to only ~0.6 m. Dissipation of the tide occurs rapidly upestuary.

The principal tributary systems include the Trinity and San Jacinto Rivers, which flow into Trinity Bay. East Bay receives runoff from Oyster Bayou and other drainage from Chambers County. West Bay receives discharges from Chocolate, Halls, and Mustang Bayous and several small streams.

The Trinity and San Jacinto Basins deliver 54 and 28% of the freshwater inflow to Galveston Bay, respectively. The largest volume of freshwater inflow from these two basins occurs during the spring freshet in April and May. The low-flow season takes place during the warmer months from July through October. The average freshwater inflow to the bay is $\sim 1.25 \times 10^{10}$ m^3/yr.[2]

Salinity in the bay ranges from 6 to 28‰ and averages ~17‰. It is influenced by an array of factors, most notably freshets, northers, density currents, the Houston Ship Channel, the Texas City Dike, and two large electric generating stations. The Houston Ship Channel acts as a conduit for the movement of higher salinity water up the bay.

The estuarine bottom consists of a wide range of sediment types (i.e., clay, silt, sand, gravel, and shell) derived from riverine influx, erosion of bay margins, the introduction of Gulf of Mexico sediments, and autochthonous biogenic material (i.e., skeletal remains of organisms). Mud, muddy sand, and sandy mud are overall the dominant sediment types.[3] Fine-grained sediments predominate in sections of East Bay, the center of Trinity Bay, the northwestern part of Galveston Bay, and other low-energy areas. Coarse-grained sediments are most abundant at the Trinity River delta front, lower Galveston Bay, and other high-energy areas near the tidal inlets. Aside from deltas and tidal inlets, other features that strongly influence local sediment patterns in the bay include dredged material islands, relict barrier islands, oyster reefs, and the Houston Ship Channel. Circulation patterns in the estuary largely control the resultant sediment distribution in which broad expanses of muds are interrupted by local areas of shell, oyster reefs, and sand.

Oyster reefs are conspicuous structures on the bay bottom, covering a total area of 1.08×10^4 ha (Figure 4.2).[4] They significantly affect circulation and sedimentation and provide important habitat for many organisms in the estuary. The reefs can be differentiated into natural and anthropogenic types. Natural reefs include barrier reefs (extending across the bay), patch reefs (isolated circular bodies), longshore reefs (oriented parallel to the shoreline), and perpendicular reefs (stretching from near the shoreline out into the bay). Anthropogenic reefs are those resulting from human activities (e.g., oyster leases, dredged material banks from channels, artificial reef construction, altered current flow as well as oil and gas development). Comparing 1971 and 1991 survey data, Powell et al.[4] showed that the areal extent of oyster reefs in Galveston Bay increased by 25.2% over the 2-decade period. Anthropogenic reefs now account for a major fraction of all present-day reefs in the system.

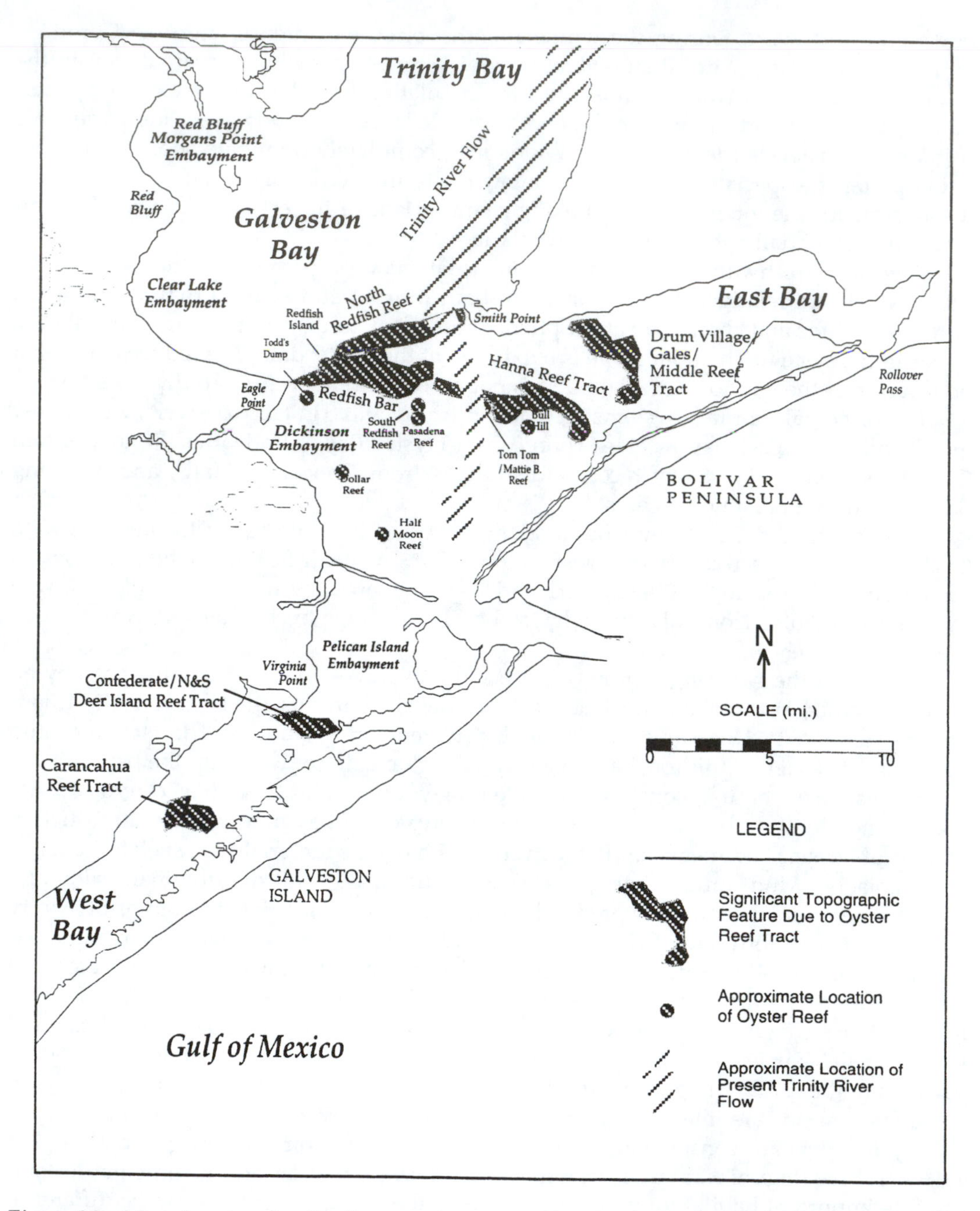

Figure 4.2 Map showing the distribution of major Galveston Bay oyster reefs. (From Powell, E.N., Song, J., and Ellis, M., *The Status of Oyster Reefs in Galveston Bay, Texas*, Galveston Bay National Estuary Program Publication GBNEP-37, Webster, TX, 1994.)

The estuarine shoreline is comprised of extensive marshes and bluffs. The marshes cover 61% of the natural estuarine shoreline, mainly on the East and West Bays, and clay bluffs occupy another 35%, primarily on Galveston and Trinity Bays. The remaining 4% consists of sand and shell beaches. Some areas are characterized by intertidal mudflats sparsely vegetated by micro- and macroalgae. Small patches of seagrass meadows cover ~1.81×10^5 ha of shallow subtidal sediments along the shoreline.

Shoreline erosion is an ongoing process in the Galveston Bay system. Between 1850 and 1982, erosion affected 78% of the estuarine shoreline. During this period, the average shoreline retreat amounted to 0.66 m/yr, which corresponds to more than 3240 ha of lost land. Sand beaches experienced the highest rate of retreat (1.84 m/yr).[5] Shoreline erosion is ascribed to both natural factors, such as sea level rise, meteorological forcing (hurricanes, tropical storms, and northern cold fronts), and wind-driven waves as well as anthropogenic effects, such as land subsidence due to groundwater withdrawal, upland reservoir construction, and altered land uses in the upper watershed (which have reduced the sediment load to the bay).[1]

B. The watershed

The Galveston Bay watershed encompasses an area of 8.56×10^6 ha, representing 12.6% of the entire surface area of Texas (Figure 4.3). Lake Houston and Lake Livingston are two major impoundments that tend to trap pollutants discharged from the upland areas. The watershed is subdivided into lower and upper portions. The lower watershed, defined as the 1.09×10^6 ha area draining to Galveston Bay below Lake Houston on the San Jacinto River and Lake Livingston on the Trinity River, has numerous nonpoint sources of pollution. Many riverborne pollutants entering the bay originate directly from the lower watershed, which is characterized by both rural drainages (i.e., Austin/Bastrop Bayous, Chocolate Bayou, and lower Trinity River) and urban drainages (i.e., Houston Ship Channel-Buffalo Bayou, Greens Bayou, Hunting Bayou, Sims Bayou, Brays Bayou, and White Oak Bayou). Because of the relatively high runoff potential of the lower watershed, nonpoint source pollution here remains a major concern for maintaining water quality in the bay.

In nonurban areas of the lower watershed, land cover consists of about equal portions of agriculture, open/pasture, and forest. The soils are primarily composed of fine-grained clays that have low permeability and high runoff potential. Runoff varies across the lower watershed, being principally dependent on land use/land cover and the frequency of rainfall. For example, runoff increases in areas of impervious cover, which precludes water infiltration, and it decreases where permeable soils (e.g., sandy loams) predominate. The volume of runoff is closely coupled to the amount of rainfall, which in the Trinity River (upper) watershed ranges from an average of 10.16 cm in the northwestern perimeter to 40.64 cm in the southern part of the basin. The rate of runoff in the Brays Bayou watershed is more than twice that in the lower Trinity River watershed. A disproportionately higher volume of runoff is contributed by urbanized land.

The upper watershed is subdivided into two components: the Lake Houston watershed and the Trinity River watershed/Lake Livingston. Lake Houston and Lake Livingston are two major impoundments located on the San Jacinto River and the Trinity River, respectively. The Lake Houston watershed includes 7.32×10^5 ha of land upstream of Lake Houston; land cover here is principally forest (73%) and open pasture. Some land in the lower drainage area is urbanized, most notably along Cypress Creek and Spring Creek in the southern part of the watershed.

The total drainage area of the Trinity River amounts to 6.73×10^6 ha. The river extends north past Dallas-Fort Worth, and it drains this metroplex. Lake Livingston to the south provides some attenuation of runoff and pollutant loads from the metroplex, acting as a filter and repository for many toxic substances originating upstream. Land cover in the Trinity River watershed varies from dry rangeland in the northwestern drainage area to agricultural land in many other locations. Wetlands and forested cover also lie along the Trinity River itself, mainly in the lower, wet portion of the watershed. Apart from Lake Livingston, several other reservoirs have been constructed in the upper watershed to meet water supply, flood control, and recreational needs. Among these basins are Benbrook Lake, Richland-Chambers Reservoir, Cedar Creek Reservoir, Lake Ray Hubbard, Lewisville Lake, and Lake Worth.

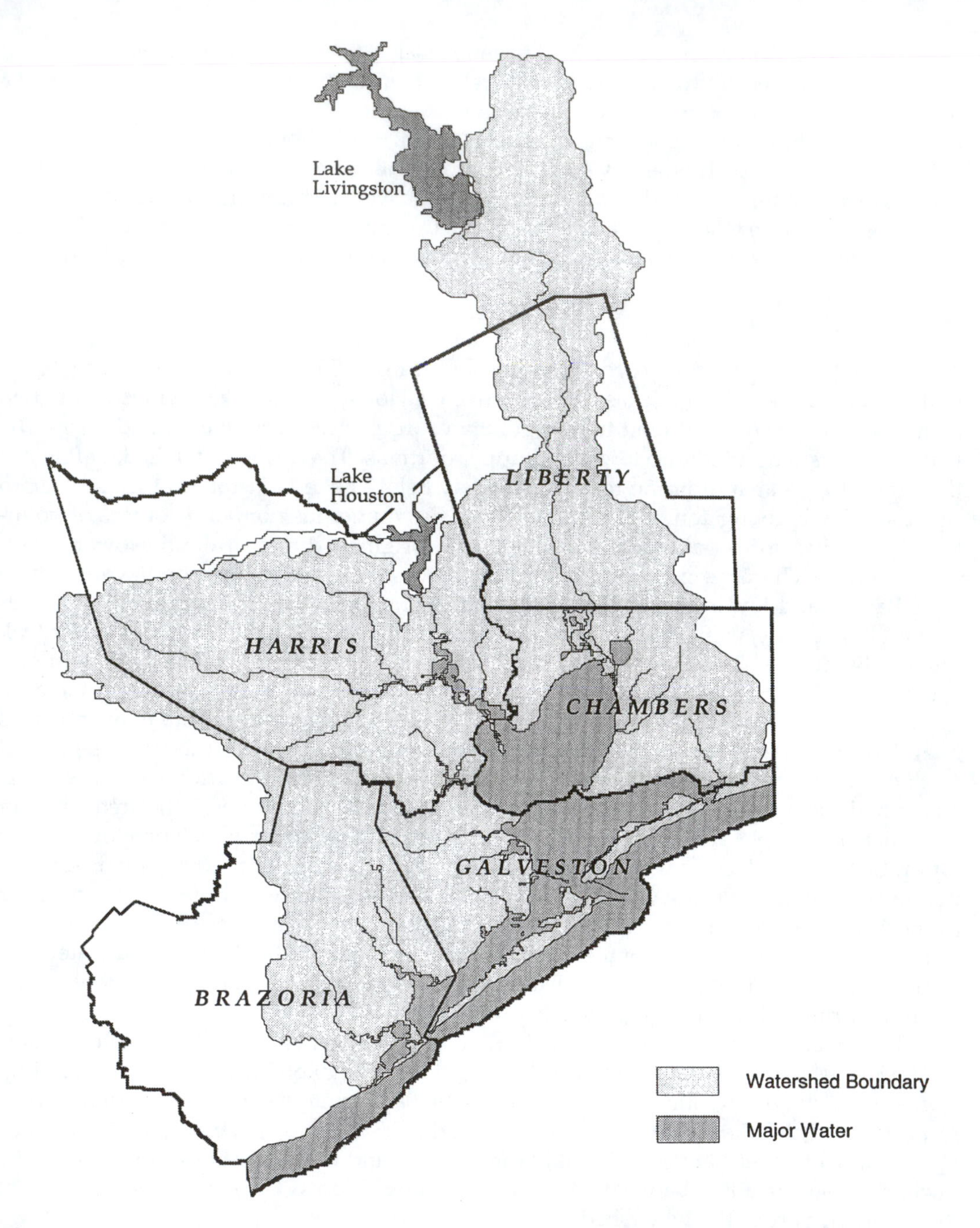

Figure 4.3 Map of the Galveston Bay watershed. (From GBNEP, *The Galveston Bay Plan: The Comprehensive Conservation, and Management Plan for the Galveston Bay Ecosystem*, Galveston Bay National Estuary Program Publication GBNEP-49, Webster, TX, 1994.)

III. Human uses and activities

A. Demography

Based on 1990 U.S. Bureau of the Census statistics,[6] 3.3 million people or 19% of the Texas population live in the five-county study area surrounding Galveston Bay (i.e., Chambers, Liberty, Harris, Brazoria, and Galveston Counties) (see Figure 4.3). The population is

Table 4.2 Land Use in the Galveston Bay Watershed

Category	Percent Cover
Open/pasture	23
Agriculture	22
Forest	18
Wetlands	15
High-density urban	10
Residential	9
Barren	1
Water	1

Source: Data from Newell, C. J., Rifai, H. S., and Bedient, P. B., *Characterization of Nonpoint Sources and Loadings to Galveston Bay*, Volume 1, Galveston Bay National Estuary Program Publication GBNEP-15, Webster, TX, 1992.

expected to reach ~4.4 million people by the year 2010.[1] Between 1980 and 1990, the population size and density both increased by ~16%, with Harris County experiencing the highest population change. This county accounts for ~85% of the total population in the region and nearly 20% of the state population. The trend in recent years has been toward greater urbanization, although larger cities (e.g., Galveston and Houston) have shown relatively low population growth. The population has expanded most rapidly in suburban communities.

B. Land uses

Much of the economic growth and development in the Galveston-Houston area during the past century has been attributed to the petroleum industry. Gas and oil production is extremely important to the regional economy. Ditton et al.[7] estimate that 30% of the total U.S. petroleum industry operates in areas adjacent to Galveston Bay, with the heaviest petroleum and industrial complexes located around the Houston Ship Channel in parts of Houston and Harris County as well as in the vicinity of Brazosport and Texas City.[8] Industrial growth has been particularly intense along the southwestern shore of the bay. The chemical and allied products industry has also contributed greatly to rapid urbanization of the Galveston-Houston area.

Although the Galveston Bay area has become increasingly urbanized over the years, considerable forested, agricultural, and open/pasture lands remain. For example, Chambers County is mainly characterized by agricultural land cover; however, several large national priority wetlands occur along the sector. Brazoria County likewise has two national priority wetlands (Hoskins Mound and Freshwater Lake). This county is also responsible for more than 40% of the natural gas production in the region. As in the case of Chambers County, both Brazoria and Liberty Counties contain significant amounts of agriculture.

Newell et al.,[9] examining major land-use categories, noted that open/pasture, agricultural, and forested lands dominate the Galveston Bay watershed, occupying 73% of the area. Urban and residential development cover 19%, and wetlands cover 15%. Table 4.2 provides a summary of land use in the watershed.

C. Bay uses

Galveston Bay is used for numerous commercial and recreational pursuits. Bay-related economic activities are closely coupled to commercial fishing, oil and gas production, transportation, shipping, and construction. Nearly 10% of the households in the counties surrounding Galveston Bay derive their income directly from activities involving the estuary.[10] Even more of the area households (34%) use the bay for recreational activities, such as fishing, crabbing, swimming, boating, sightseeing, and many other pursuits.[11]

Oil production has decreased appreciably since the 1970s but continues to be extremely important to the regional economy. The Texas Railroad Commission[12] reported a total of 550 producing gas wells and 1096 producing oil wells in four counties around the bay (i.e., Brazoria, Chambers, Harris, and Galveston Counties). Many producing wells are located in the bay itself. Nearly 40% of the oil refineries and 30% of the gas processing plants along the Texas coast are located in the bay region, and about 30% of the total petroleum industry is found here.[7] In 1992, total gas production was ~3.6×10^6 m^3 with an economic value of $193 million. In addition, the gas industry employed 4916 persons at this time. Total oil production, in turn, approached 12 million barrels worth $231.3 million. The petroleum industry also supported 5875 employees.

Growth of shipping has closely followed expansion of the petrochemical and related industries. For example, over the 30-year period from 1955 to 1985, the three major ports of the region (i.e., Houston, Galveston, and Texas City) grew substantially. Today, the Port of Houston ranks third nationally and sixth worldwide in terms of the total tonnage of materials transported. Shipping continues to be very heavy along the Houston Ship Channel and Gulf Intracoastal Waterway. The Houston Ship Channel, extending for ~80 km from the Port of Houston to the Gulf of Mexico, is a highly industrialized body of water. Petroleum and chemical products dominate the materials transported, comprising more than 50% of the cargo shipped along the channel.[10] Aside from the Houston Ship Channel and Gulf Intracoastal Waterway, other navigation channels (e.g., Anahuac Channel and Trinity River Channel) serve as minor shipping routes.

The commercial finfish and shellfish harvest in Galveston Bay comprises approximately one third of the state total. Mullet (*Mugil* spp.), black drum (*Pogonias cromis*), sheepshead (*Archosargus probatocephalus*), and southern flounder (*Paralichthys lethostigma*) constitute most of the commercial finfish catch (~75%), and white and brown shrimp (*Penaeus setiferus* and *P. aztecus*), blue crabs (*Callinectes sapidus*), and eastern oysters (*Crassostrea virginica*) constitute most of the shellfish catch (>99%).[13] In terms of total weight, shellfish account for more than 95% of the total commercial seafood harvest (Table 4.3). In terms of dollar value, the most important species typically are shrimp, oysters, and flounder, which usually comprise more than 90% of the total value of the landings. The mean annual inshore commercial harvest of finfish and shellfish reported for the bay by the GBNEP exceeded 5000 mt with an ex-vessel value greater than $12 million. The ex-vessel value reported for both the inshore and offshore catch, in turn, exceeded $60 million, which provided for more than 3450 full-time jobs.[1]

The recreational finfish catch is substantial, comprising ~86% of the total catch in the bay, with the commercial finfish catch constituting the remainder.[14,15] Most of the recreational catch (>75%) consists of Atlantic croaker (*Micropogonias undulatus*), sand seatrout (*Cynoscion arenarius*), and spotted seatrout (*C. nebulosus*).[16] About 2 million hours of sport fishing effort is exerted in the estuary each year, generating more than $350 million of revenue for the economy.[13] Approximately 50% of the sport fishing expenditures in Texas is ascribed to recreational fishing in Galveston Bay.

Recreational boating, swimming, and other water-contact activities also represent major uses of the estuary. Recreational boating statistics compiled by the Texas Water Commission[15] indicate that ~8000 boat slips and nearly 40 commercial marinas exist in the aforementioned four-county area surrounding the bay. The number of boat slips is significant, accounting for more than 60% of the total number of slips registered in commercial marinas on the Texas coast.

Table 4.3 Commercial Fish Landings from Galveston Bay for 1989

Species	Total Landings[a]
Fish	
Black drum	12.0
Southern flounder	8.0
Mullet	59.5
Sheepshead	8.9
Other species	33.3
Total fish	121.8
Shellfish	
Oyster	388.8
Crab	1184.7
Shrimp	2235.5
Other shellfish	7.4
Total shellfish	3816.4
Total fish and shellfish	3938.3

[a] Metric tons.

Source: Data from Green A., Osborn, M., Chai, P., Lin, J., Loeffler, C., Morgan, A., Rubec, P., Spanyers, S., Walton, A., Slack, R. D., Gawlik, D., Harpole, D., Thomas, J., Buskey, E., Schmidt, K., Zimmerman, R., Harper, D., Hinkley, D., Sager, T., and Walton, A., *Status and Trends of Selected Living Resources in the Galveston Bay System*, Galveston Bay National Estuary Program Publication-19, Webster, TX, 1992.

Tourism is a billion dollar industry for the local economy. In 1992, for example, the total expenditures for tourism in Harris, Galveston, Brazoria, Chambers, and Liberty Counties were nearly $4 billion. Harris County accounted for $3.5 billion, followed by Galveston ($306.7 million), Brazoria ($91.2 million), Liberty ($35.2 million), and Chambers ($10.4 million) Counties.[10] Ecotourism is now the most rapidly expanding sector of the tourist industry.[17]

IV. Anthropogenic effects

A. Pollution Impacts

Galveston Bay has experienced many of the same impacts associated with human activities that are observed at other National Estuary Program sites. The introduction of nutrients, fecal coliform bacteria, and toxic contaminants into the bay from point source discharges (i.e., municipal and industrial facilities) and nonpoint source loadings has degraded sediment and water quality in some areas of the bay, although the effects of point source discharges have declined substantially since the 1960s. Nitrogen and phosphorus concentrations have decreased throughout the bay concomitant with reduced wastewater discharges to the system. However, sewage bypasses and overflows remain a problem, and urban runoff contributes a significant fraction of the total nutrient, fecal coliform, and pesticide loadings to the bay. Toxic contaminants, e.g., polychlorinated biphenyls (PCBs) and polycyclic aromatic hydrocarbons (PAHs), pose a potential threat to living resources in the more heavily industrialized portions of the system, most notably in the upper bay (e.g., upper Houston Ship Channel). Problem pollution areas in the bay include points of surface runoff, shipping, and waste discharges as well as regions of intense human activity.[18] Marinas are also local sources of significant sediment and water quality problems.

Most physical alterations of Galveston Bay have occurred during the past 150 years. The principal modifications have been linked to the construction and maintenance of navigation channels, land subsidence, shell dredging, isolation of secondary bays and marshes, shoreline conversion, and construction of reservoirs (e.g., Lake Conroe, Lake Houston, and Lake Livingston). These modifications have had an important influence on the biological and chemical characteristics of the estuary.

1. *Pollution sources*

a. Point source discharges. There are four main categories of point source discharges into Galveston Bay: (1) permitted discharges (municipal and industrial); (2) unpermitted (illegal) discharges; (3) bypasses and overflows from municipal sewage treatment systems; and (4) produced water discharges.[1] As reviewed by Armstrong and Ward,[19] the total wastewater flow from permitted discharges to the Galveston Bay system in 1990 was 5.13 $\times 10^{12}$ l. The estimated pollutant loadings of major constituents were as follows:

- Biochemical and chemical oxygen demand loadings amounted to 4.5×10^6 kg/yr and 6.2×10^6 kg/yr, respectively (primarily into the Houston Ship Channel/San Jacinto River).
- Chlorinated hydrocarbon loadings equaled 1800 kg/yr (principally into the Houston Ship Channel and Cedar Bayou).
- Heavy metal loadings ranged from 0.19 kg/yr (silver) to 13,000 kg/yr (zinc) (largely into the Houston Ship Channel/San Jacinto River).
- Total phenol loadings were 2800 kg/yr (mainly into the Houston Ship Channel).
- Toluene loadings approached 37 kg/yr (into the Cedar Bayou and Chocolate Bayou).

Unpermitted (illegal) discharges are not considered to be a significant problem in the Galveston Bay system. Fay et al.,[20] surveying 9 shoreline segments of the bay, identified 117 unpermitted discharges. The Texas Natural Resource Conservation Commission later found that most of these site discharges were neither discharging illegally nor a likely important source of pollution in the estuary.

Municipal wastewater collection system overflows have been a chronic problem in the Galveston Bay watershed. The conveyance and overload of stormwater in wastewater collection systems invariably lead to the discharge of raw or partially treated sewage into the bay. The conditions are exacerbated during periods of intense rainfall, which culminate in the release of diluted sewage from overflow structures designed to provide system relief from excessive fluid volumes. Hence, these overflows promote fecal coliform bacteria inputs to the estuary.

Studies conducted since 1986 indicate a progressive improvement in stormwater overflow and bypass problems. In 1986, Winslow and Associates[21] registered a total BOD loading value of 3.1×10^6 kg/yr from stormwater overflows and bypasses to the Houston Ship Channel. They calculated that discharges from these sources contributed about 7% of the ammonia load, 7% of the total suspended solids load, and 11% of the annual BOD load to the channel.

Guillen et al.[22] documented 789 bypass incidents in 1991 and 578 incidents in 1992, primarily in the San Jacinto watershed. Partially treated effluent loadings associated with these incidents amounted to 1.71×10^9 l in 1991 and 8.98×10^8 l in 1992. The estimated BOD loadings decreased by 34.17 mt from 1991 to 1992.

Water quality in the estuary should improve further with completion of municipal wastewater collection system upgrades in the City of Houston. The upgrade program is multifaceted, incorporating the expansion of sewage treatment plants, the addition of wet weather facilities, the construction of new relief sewers, and other engineering features that provide greater structural integrity of the system. These improvements will ensure greater control of stormwater overflows.

Produced water or oil field brine is an unwanted by-product of petroleum production that typically contains toxic contaminants, such as PAHs, trace metals, and radioactive substances (e.g., radium-226). In addition to the toxic components, the produced water has a wide salinity range from slightly brackish to more than five times that of seawater. These discharges greatly alter ambient conditions in local areas of the estuary.

The Texas Railroad Commission issues permits for the discharge of produced water to the surface waters of Galveston Bay and other coastal systems of Texas. The total volume of produced water discharged to the bay varies considerably from year to year. In 1991, $\sim 5.75 \times 10^7$ l/d of produced brine were released to the bay and its tributaries, although only $\sim 2.20 \times 10^7$ l/d were considered to be active discharges.

Elevated hydrocarbon, metal, and salt concentrations of produced water often severely impact biotic communities — particularly benthos — in receiving waters. For instance, in an investigation of produced water discharges at Cow Bayou and Tabbs Bay, Roach et al.[23] observed high concentrations of brine salts, ammonia, and petroleum hydrocarbons in bottom sediments of outfall areas, rendering extensive areas unfit for benthic habitation. They reported substantial losses of invertebrates at both sites, with infaunal and epifaunal populations being depressed or nonexistent near the discharge points. Significant reductions in species abundance and diversity were discerned along a gradient for several hundred meters away from the discharge.

Toxicity tests of burrowing amphipods from two stations nearest the produced water discharge at Tabbs Bay also revealed only a 0 to 20% survival compared to ~90% survival for control specimens. Sublethal environmental stress associated with the discharge was evident in the stress protein compounds identified in grass shrimp (*Palaemonetes* spp.) exposed to contaminated resuspended and whole sediments. In addition, significantly higher concentrations of PAH metabolites were found in the bile of striped mullet (*Mugil cephalus*) collected near the discharge points compared to levels in striped mullet from control sites.

b. Nonpoint source pollution. Newell et al.[9] examined nonpoint source pollution loadings to the Galveston Bay system, which appear to have increased during the past several decades along with watershed development. Escalating urban land uses in the lower watershed has notably increased pollutant export to the estuary, promoting elevated fecal coliform bacteria levels that have culminated in the closure of about half of the bay waters to shellfish harvesting. Furthermore, dissolved oxygen levels have become deficient in some urban bayous and poorly flushed areas.

Surface runoff from the greater Houston area creates a potentially significant problem for the upper Houston Ship Channel. Highest nonpoint source loads from urban areas are coupled to human activities at construction sites, industrial centers, and high-density commercial zones. In rural areas, agriculture and silviculture operations foster high nonpoint source loads that may peak during certain periods (e.g., spring). Constituent loads of most concern are nutrients, oxygen-demanding compounds, oil and grease, fecal coliform bacteria, trace metals, organochlorine contaminants (i.e., pesticides), and sediments.

In assessing nonpoint source loadings, Newell et al.[9] focused on eight land-use types in the Galveston Bay watershed (Table 4.4). Subdividing the land area surrounding Galveston Bay into 21 watersheds and 100 subwatersheds, these investigators calculated the total nonpoint source loads to the bay for the aforementioned water quality constituents (Table 4.5). Contributions from high-density urban areas (i.e., Houston, Galveston,

Table 4.4 Nonpoint Source Loads to Galveston Bay by Land Use for a Year with Average Rainfall

	Land Use Type								
Constituent	High Density Urban	Residential	Open	Agricultural	Barren	Wetlands	Water	Forest	Total
				Runoff Volume					
1,000 ac-ft	766	371	567	593	21	187	164	345	3,014
(% of total)	(25)	(12)	(19)	(20)	(1)	(6)	(5)	(11)	(100)
				TSS					
Million kg	157	46	49	147	57	9	0	17	481
(% of total)	(33)	(10)	(10)	(31)	(12)	(2)	(0)	(3)	(100)
				Total Nitrogen					
1,000 kg	1,985	1,561	1,056	1,142	134	192	0	353	6,422
(% of total)	(31)	(24)	(16)	(18)	(2)	(3)	(0)	(5)	(100)
				Total Phosphorus					
1,000 kg	350	362	84	264	15	14	0	26	1,113
(% of total)	(31)	(32)	(8)	(24)	(1)	(1)	(0)	(2)	(100)
				BOD					
Million kg	8	7	4	3	0	1	0	3	26
(% of total)	(31)	(26)	(16)	(11)	(1)	(5)	(0)	(10)	(100)
				Oil and Grease					
Million kg	12	2	0	0	0	0	0	0	14
(% of total)	(87)	(13)	(0)	(0)	(0)	(0)	(0)	(0)	(100)
				Fecal Coliform					
$X10^{15}$ col	208	101	17	18	0	4	0	7	355
(% of total)	(59)	(28)	(5)	(5)	(0)	(1)	(0)	(2)	(100)
				Dissolved Cu					
kg	2,930	1,419	2,167	2,269	80	716	0	1,318	10,900
(% of total)	(27)	(13)	(20)	(21)	(1)	(7)	(0)	(12)	(100)
				Pesticides					
kg	378	183	70	73	3	0	0	43	749
(% of total)	(50)	(24)	(9)	(10)	(0)	(0)	(0)	(6)	(100)

Source: Newell, C. J., Rifai, H. S., and Bedient, P. B., *Characterization of Nonpoint Sources and Loadings to Galveston Bay*, Volume 1, Galveston Bay National Estuary Program Publication GBNEP-15, Webster, TX, 1992.

Texas City, and Baytown) accounted for the highest constituent loads per unit area. Particularly noteworthy in this respect were the urban contributions of oil and grease (87% of the total annual loading), fecal coliforms (59%), pesticides (50%), and total suspended solids (31%). Nitrogen and phosphorus loadings from high-density urban areas were also significant, comprising nearly one third of the total for each nutrient. Nonpoint source loads ascribable to residential and agricultural land use were of secondary importance.

Nonpoint source pollutant inputs to Galveston Bay vary greatly during the year, being dependent in large part on the frequency and duration of rainfall. Periodic rainstorms typically cause a pulsing of nonpoint source loads to the bay. However, large reservoirs (e.g., Lake Houston and Lake Livingston) can buffer the loading effects by filtering the discharge from the San Jacinto and Trinity Rivers.

Table 4.5 Nonpoint source Loads to Galveston Bay for an Average Year by Watershed

Watershed	Area (sq mi)	Runoff Volume (1,000 ac-ft)	Total Susp. Solids (Million kg)	Total Nitrogen (1,000 kg)	Total Phosphorus (1,000 kg)	Biochemical Oxygen Demand (Million kg)	Oil and Grease (Million kg)	Fecal Coliform ($\times 10^{15}$ Col)	Diss. Copper (kg)	Pesticides (kg)
Addicks Reservoir	134	82	22	195	36	0.7	0.4	9	312	20
Armand/Taylor Bayou	77	70	12	167	29	0.7	0.5	11	255	22
Austin/Bastrop Bayou	213	121	21	245	44	0.9	0.2	9	442	21
Barker Reservoir	122	71	32	181	31	0.6	0.2	6	271	14
Brays Bayou	127	147	29	406	75	1.7	1.7	34	561	63
Buffalo Bayou	105	116	22	337	65	1.4	1.3	27	445	51
Cedar Bayou	211	153	26	321	58	1.2	0.3	13	576	30
Chocolate Bayou	170	95	19	188	36	0.6	0.1	5	354	15
Clear Creek	182	138	22	301	51	1.2	0.7	16	503	34
Dickinson Bayou	101	60	8	130	21	0.5	0.2	6	223	13
East Bay	288	193	26	388	68	1.6	0.5	17	679	36
Greens Bayou	209	184	30	497	92	2.1	1.4	34	702	66
North Bay	25	25	4	65	11	0.3	0.2	5	94	9
San Jacinto River	68	65	8	126	22	0.5	0.2	7	202	14
Houston Ship Channel	166	198	34	498	90	2.0	1.9	39	713	74
Sims Bayou	93	91	16	235	41	1.0	0.8	17	346	33
South Bay	78	68	10	138	24	0.6	0.6	12	211	22
Trinity Bay	317	225	26	356	59	1.5	0.3	12	708	32
Trinity River	1,099	572	62	877	124	4.3	0.5	27	2,110	82
West Bay	344	212	30	405	68	1.6	0.9	21	706	44
White Oak Bayou	110	128	24	365	69	1.5	1.3	29	488	54
Total Project Area	4,238	3,010	481	6,420	1,110	26.3	14.2	355	10,900	749
Median	134	121	22	301	51	1.2	0.5	13	445	32
Maximum	1,099	572	62	877	124	4.3	1.9	39	2,110	82
Minimum	25	25	4	65	11	0.3	0.1	5	94	9

Source: Newell, C. J., Rifai, H. S., and Bedient, P. B., *Characterization of Nonpoint Sources and Loadings to Galveston Bay*, Volume 1, Galveston Bay National Estuary Program Publication GBNEP-15, Webster, TX, 1992.

Faulty septic tanks and marinas also contribute to water quality degradation in the estuary. Guillen et al.[22] chronicled 166 septic tank violations in the bay watershed during 1992. They estimated that ~1.2×10^7 l of partially treated effluent derived from malfunctioning septic systems entered the watershed at this time. Although not responsible for the major fraction of fecal coliform bacteria loading to the bay, faulty septic tanks can be of local importance in areas of restricted estuarine circulation.

Marinas are likewise locally important sources of fecal coliform bacteria as well as nutrients, trace metals, and other toxic contaminants. Guillen et al.,[24] for example, demonstrated that fecal coliform bacteria concentrations and trace metals (i.e., copper and lead) increased, whereas dissolved oxygen levels decreased, from open waters of the bay to marina sites. In some cases, state water quality standards (e.g., fecal coliforms) were exceeded near the marinas.

Because of its diffuse nature, nonpoint source pollution is difficult to control. Improved land management techniques and the implementation of conservation practices may be the most effective controls for minimizing nonpoint source pollutant inputs to Galveston Bay. To be successful in dealing with nonpoint source pollution, greater effort must be expended in educating the public regarding the impacts of watershed development and land-use activities. There is little understanding among watershed residents, for example, that development and urbanization of the watershed increase pollutant export by at least one order of magnitude over predevelopment levels.[25] It will require a concerted initiative by land-use planners, resource managers, government agencies, stakeholders, and inhabitants of the watershed to significantly reduce the impacts of nonpoint source pollution in the estuary.

c. *Baywide pollution assessment.*

Water column pollutant concentrations — Ward and Armstrong[18,26] have inventoried the concentrations of pollutants in the water column and bottom sediments of Galveston Bay by analyzing 26 separate data collection programs. The objective of this work was to produce a comprehensive long-term database on pollutants in the bay for future assessment of water and sediment quality. The focus was on large-scale distributions of pollutants throughout the bay.

A summary of the major water column pollutant concentrations in Galveston Bay, based on the work of Ward and Armstrong,[18,26] is as follows:

- Total phosphorus concentrations range from <0.2 to >2 mg/l.
- Nitrate concentrations range from 0.1 to >0.5 mg/l.
- Ammonia concentrations range from 0.2 to >2 mg/l.
- Fecal coliform concentrations range from <10 organisms per 100 ml to >100,000 organisms per 100 ml, with highest bacterial concentrations in the upper Houston Ship Channel.
- Oil and grease concentrations range from <2 to 20 mg/l, with highest concentrations at Texas City, the inlet to Galveston Bay, and the Houston Ship Channel.
- The relatively few measurements of water column concentrations of PAHs and organochlorine compounds (e.g., PCBs, DDT, and other pesticides) indicate highest concentrations in urbanized and industrialized areas of the estuary.
- Copper concentrations range from <20 to >100 μg/l.

Ward and Armstrong[18] also described water quality trends in the estuary over the past 3 decades based on key water quality parameters:

- Dissolved oxygen levels average near saturation over extensive areas of the estuary, but are low in poorly flushed tributaries, such as the upper Houston Ship Channel, where levels have declined by ~0.1 mg/l/yr during the past 2 decades.
- Nutrients have exhibited a gradually decreasing trend during the past 20 years (i.e., total nitrate ~0.01 mg/l/yr, total ammonia ~0.1 mg/l/yr, and total phosphorus ~0.05 mg/l/yr).
- Chlorophyll *a* concentrations, which range from <10 to >50 μg/l, have clearly fallen during the past several decades in response to lower nutrient and phytoplankton concentrations.
- The amount of total organic carbon has decreased by about one third during the past 25 years, in large part due to reduced phytoplankton biomass.
- The number of fecal coliform bacteria has generally dropped in the bay, except in some areas of West Bay and the western urbanized tributaries.
- Trace metal concentrations have diminished dramatically in the upper Houston Ship Channel, and generally throughout the system, over the past 2 decades.
- Biochemical oxygen demand loading, which historically has been lowest in the open bay and highest in the upper Houston Ship Channel, has shown a 20-fold reduction over the past 3 decades to levels $<5 \times 10^6$ kg/yr due largely to improved wastewater treatment.
- Total suspended solids with current concentrations of <10 to >60 mg/l have generally decreased over the past 2 decades.
- Water temperature has decreased at a nominal rate of 0.05°C/yr, and salinity has decreased at a rate of ~0.1 to 0.2‰ over the past few decades.

Figure 4.4 illustrates water quality trends in the heavily impacted upper Houston Ship Channel during the past 10- to 20-year period.

Sediment pollutant concentrations — As in the case of water column pollutants, bottom sediment pollutant concentrations tend to be highest in the Houston Ship Channel, lowest and more uniform in the open bay, and locally elevated in areas of wastewater discharges and high surface runoff. However, extensive areas of the bay bottom remain unsampled, and many of the measurements on specific pollutants are below analytical detection limits. Thus, it is difficult to establish data trends for large areas of the bay.

Sediment pollutant concentrations reported by Ward and Armstrong[18] are as follows:

- Total phosphorus levels range from <200 to >1000 mg/kg, being highest in the upper Houston Ship Channel.
- Peak concentrations of oil and grease, ranging from 1000 to >5000 mg/kg, also occur in sediments of the upper Houston Ship Channel, although significantly lower levels generally are found elsewhere in the system.
- Cedar Bayou, Clear Creek, Trinity River, and the upper Houston Ship Channel are sites of peak DDT sediment levels, which approach 500 μg/kg in the most impacted areas, but sediment concentrations typically range from <1 to >30 μg/kg.
- The mean sediment concentration of PCBs in the bay amounts to 76.8 μg/kg.
- Maximum concentrations of trace metals are usually found in sediments of the upper Houston Ship Channel, Trinity Bay, and the Texas City Dike. The mean sediment concentrations of specific trace metals are copper (10 to >50 μg/kg), chromium (20 to 100 μg/kg), lead (10 to 50 μg/kg), mercury (0.005 to 0.5 μg/kg), and zinc (50 to 250 μg/kg).

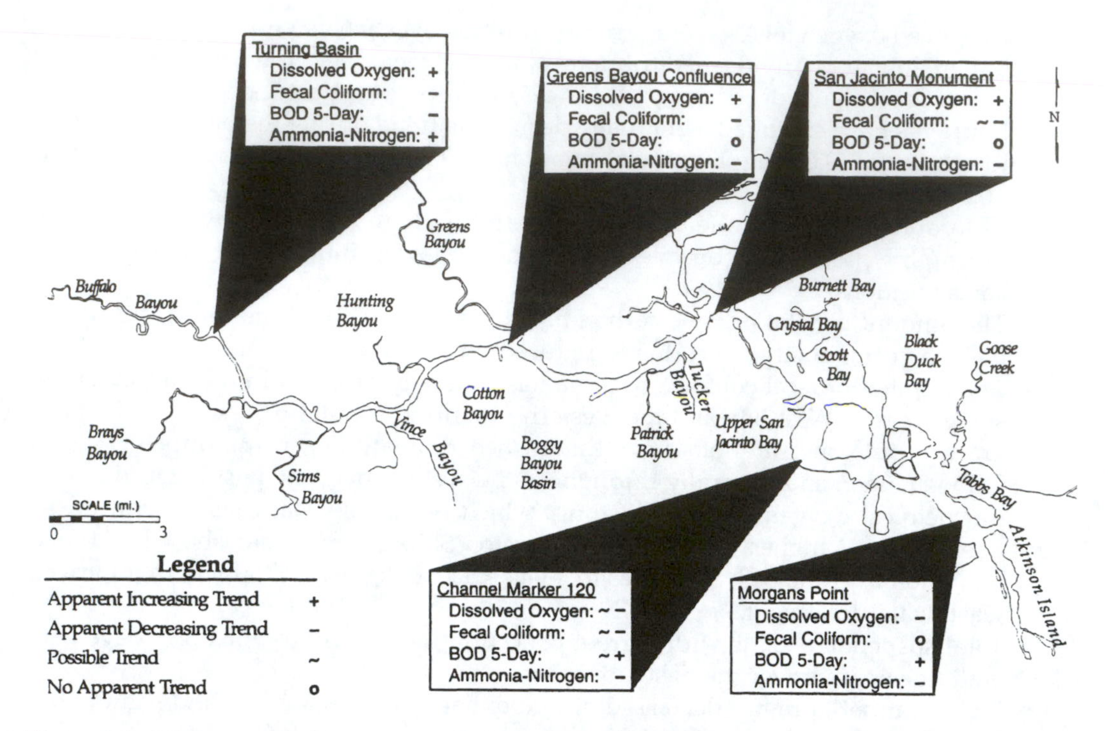

Figure 4.4 Water quality trends in the upper Houston Ship Channel. (From Crocker, P.A., Water quality trends for the Houston Ship Channel, in *Proceedings of the Second State of the Bay Symposium*, Jensen, R., Kiesling, R.W., and Shipley, F.S., Eds., Galveston Bay National Estuary Program Publication GBNEP-23, Webster, TX, 1993, 27.)

The bottom sediments of Galveston Bay are a repository and source of contaminants for overlying waters and organisms in the estuary. However, broad areas of the estuarine floor exhibit relatively low concentrations of contaminants, the most notable exceptions being areas impacted by various dredging, industrial, and petroleum activities. Sediments of the open bay contain lower concentrations of contaminants than those along the estuarine perimeter directly affected by runoff, inflow, and waste discharges.[1] The western urbanized tributaries have the most significant sediment and water quality problems. The Houston Ship Channel continues to exhibit some of the highest sediment contaminant concentrations in the system, although the trend is generally downward even at this historically impacted location.

Overall, sediment quality is improving in Galveston Bay. A trend of decreasing contaminant concentrations is evident baywide, but especially in the upper Houston Ship Channel. This is true for trace metals, excluding arsenic. Chlorinated hydrocarbons are below the detection limit (0.01 μg/g) over extensive areas of the bay bottom.[27] The exceptions include a few sites where detectable concentrations of PCBs have been recorded. Elevated and toxic levels of PAHs occur at sites adjacent to produced water separator platforms, but much lower concentrations are observed elsewhere in the bay.

Some sediment toxicity problems persist, however. For example, results of porewater/sea urchin tests show that all sites sampled in the upper Houston Ship Channel are highly degraded. In addition, all dredged material disposal sites display some degree of toxicity. However, only one third of the sampling sites affected by industrial or urban runoff exhibit toxicity.[27] These results have important food chain implications because some contaminants, notably chlorinated hydrocarbon compounds (e.g., PCBs and DDT) that concentrate in bottom sediments, bioaccumulate and biomagnify in estuarine biota. They ultimately can affect humans who consume contaminated seafood.

System analysis — The GBNEP compared the relative loadings from five of the bay's major pollution sources (i.e., municipal point sources, industrial point sources, the San Jacinto River, the Trinity River, and nonpoint source pollution from the lower watershed). Loading estimates for these pollutant sources differ considerably, reflecting in large part geographic differences in development, industrialization, and land use. The GBNEP did not consider other potential pollutant sources that are considered to be less significant, such as atmospheric deposition, groundwater, oil and chemical spills, boats, and internal bay sources.

In respect to municipal point sources, 559 permitted wastewater treatment plants discharge ~5.11×10^{14} l/yr of fluids to the system. Another 247 industrial facilities release ~1.48×10^{11} l/yr of "process water." This wastewater does not include cooling water inputs. Armstrong and Ward[19] and Goodman[28] provide loading estimates for both the municipal and industrial point sources.

As noted previously, the San Jacinto River drains ~8.28×10^4 ha of watershed above Lake Houston. The Trinity River, in turn, delivers flow from 6.73×10^6 ha of watershed, including Dallas and Fort Worth. Pollutant loadings from both of these watershed areas represent a mixture of point and nonpoint sources. Estimates were derived from Newell et al.[9] and Armstrong and Ward.[19]

Approximately 1.19×10^6 ha of watershed area immediately surrounding the bay are drained by a number of small streams, bayous, and stormwater outfalls. Runoff also enters the bay from general overland drainage. The loadings from the lower watershed represent nonpoint sources. Estimates of these loadings were likewise obtained from Newell et al.[9] and Armstrong and Ward.[19]

The GBNEP compared loading estimates for the aforementioned major pollutant sources in two geographic areas: the entire bay system and the upper Houston Ship Channel (upstream from the San Jacinto River mouth). Table 4.6 is a compilation of pollutant loading estimates to the entire bay system. These data indicate that the Trinity River and lower watershed are not only the largest sources of flow to Galveston Bay (46 and 25%, respectively) but also the primary sources of most pollutants to the system. For example, the lower watershed contributes more than 50% of the total suspended solids, BOD loading, oil and grease, total phosphorus, and PCBs to the bay, and the Trinity River contributes more than 50% of the total chromium, lead, and mercury. These two pollutant sources together account for more than 85% of the PCB loading and 75% of the total BOD loading. The Trinity River is the most significant source of all seven trace metals analyzed in this study, providing between 46 and 64% of the loading estimates. The San Jacinto River, comprising 24% of the total flow, yields 24% of the total copper, 22% of the total chromium, 17% of the total mercury and zinc, and 17% of the remaining constituents.

Municipal point sources constitute only 3% of the total flow but 37% of the total nitrogen and 30% of the total phosphorus loads. They are also responsible for 20% of the total chromium and 28% of the oil and grease inputs to the system. Although industrial point sources contribute 14% of the PCB load, they are a relatively insignificant source of all other constituents, ranging from 0 to 7% contribution.

Table 4.7 lists pollutant loadings to the upper Houston Ship Channel from point and nonpoint sources. Nonpoint source inputs dominate the total flow (73%) to the channel as well as the loadings of total suspended solids (96%), BOD (77%), oil and grease (59%), total lead (66%), total mercury (84%), and total zinc (55%). Municipal point sources contribute most of the total nitrogen (64%), total phosphorus (79%), total arsenic (54%), total cadmium (68%), total chromium (61%), and total copper (41%) to the urbanized tributary. Industrial discharges are responsible for relatively small loads of the pollutants, exceptions

Table 4.6 Baywide Analysis: Relative Contributions of Pollutants from Five Major Sources

Point Source Constituent	Municipal Point Sources	Industrial Point Sources	San Jacinto River	Trinity River	Lower Watershed Nonpoint Sources	Total
			Flow			
Million ac-ft/yr	0.4	0.1	2.8	5.5	3.0	11.8
(% of Total)	(3)	(1)	(24)	(46)	(25)	(100)
			Total Suspended Solids			
Million kg/yr	2.7	7.0	25.4	398.4	481.0	914.5
(% of Total)	(0)	(0)	(3)	(44)	(53)	(100)
			BOD			
Million kg/yr	1.4	3.3	3.5	13.0	26.3	47.5
(% of Total)	(3)	(7)	(7)	(27)	(55)	(100)
			Oil and Grease			
Million kg/yr	5.7	0.6	0.0	0.0	14.2	20.5
(% of Total)	(28)	(3)	(0)	(0)	(69)	(100)
			Total Nitrogen			
Million kg/yr	7.1	1.3	1.5	8.2	1.1	19.2
(% of Total)	(37)	(7)	(8)	(43)	(6)	(100)
			Total Phosphorus			
Million kg/yr	3.6	0.4	0.2	1.3	6.4	11.9
(% of Total)	(30)	(4)	(2)	(11)	(54)	(100)
			Total Arsenic			
1,000 kg/yr	16.3	3.8	3.9	50.7	29.0	103.7
(% of Total)	(16)	(4)	(4)	(49)	(28)	(100)
			Total Cadmium			
1,000 kg/yr	5.6	0.8	2.7	14.2	5.5	28.8
(% of Total)	(20)	(3)	(9)	(49)	(19)	(100)
			Total Chromium			
1,000 kg/yr	21.9	10.2	46.7	119.7	10.0	208.5
(% of Total)	(11)	(5)	(22)	(57)	(5)	(100)
			Total Copper			
1,000 kg/yr	18.8	9.1	65.4	129.1	49.5	271.9
(% of Total)	(7)	(3)	(24)	(47)	(18)	(100)
			Total Lead			
1,000 kg/yr	23.0	2.4	43.9	365.2	138.4	572.9
(% of Total)	(4)	(0)	(8)	(64)	(24)	(100)
			Total Mercury			
1,000 kg/yr	0.2	0.0	0.8	2.8	0.9	4.6
(% of Total)	(3)	(1)	(17)	(60)	(19)	(100)
			Total Zinc			
1,000 kg/yr	83.9	30.0	232.1	639.0	396.9	1381.9
(% of Total)	(6)	(2)	(17)	(46)	(29)	(100)

Table 4.6 (continued) Baywide Analysis: Relative Contributions of Pollutants from Five Major Sources

Point Source Constituent	Municipal Point Sources	Industrial Point Sources	San Jacinto River	Trinity River	Lower Watershed Nonpoint Sources	Total
			PCBs			
kg/yr	0.0	15.7	0.0	38.0	61.3	115.0
(% of Total)	(0)	(14)	(0)	(33)	(53)	(100)
			Pesticides			
kg/yr	588.0	—	170	575	749	2082
(% of Total)						(100)
			PAHs			
kg/yr	—	0.0	—	—	371.0	>371
(% of Total)						(100)

Source: Shipley, F.S. and Kiesling, R.W., Eds., *The State of the Bay: A Characterization of the Galveston Bay Ecosystem*, Publication GBNEP-44, the Galveston Bay National Estuary Program, Webster, TX, 1994.

being total chromium (29%) and total copper (20%). Municipal point sources, therefore, are the principal contributors of nutrients and certain metals (i.e., arsenic, cadmium, chromium, and copper) to the channel, and local nonpoint sources are the primary contributors of total suspended solids, BOD loadings, oil and grease, and the remaining metals (i.e., lead, mercury, and zinc). Thus, although municipal point sources and local nonpoint sources provide most of the metal loadings to the upper Houston Ship Channel, the Trinity River is responsible for much of the baywide metal inputs.

Water quality in the upper Houston Ship Channel has improved dramatically over the past 30 years due largely to upgrades in wastewater treatment leading to reduced pollutant inputs from municipal and industrial point source discharges. As a result, BOD loading to the channel decreased substantially from 7.6×10^6 kg/yr in 1968 to 3.1×10^6 kg/yr in 1990, representing more than a 50% reduction. Dissolved oxygen levels gradually increased during this period. The mean surface dissolved oxygen concentration now exceeds 5 mg/l. The percentage of water samples with dissolved oxygen levels of <2 mg/l has correspondingly decreased (Figure 4.5).

One of the most important changes in the Galveston Bay system in recent years has been the massive reduction in nitrogen loads, which commenced in 1973 (Figure 4.6).[29] Gradually increasing nitrogen concentrations in the Trinity River contributed greatly to a fourfold incease in total nitrogen loading to the bay between 1900 and 1973. However, construction of Lake Livingston in the late 1960s reduced overall nitrogen loadings by 50%. Industrial nitrogen loading also declined at this time, and municipal point source inputs leveled off. These events have decreased the probability of system eutrophication.

Despite the considerable improvement in water quality of the upper Houston Ship Channel and bay associated with upgrades in municipal and industrial wastewater treatment and reduced pollutant loadings, the overall conditions can be further improved. These waterbodies still receive substantial volumes of municipal and industrial wastewater discharges. As noted previously, for example, more than 5×10^{14} l of wastewater discharges (excluding cooling water input) enter the system each year. These discharges can be targeted in future pollution mitigation efforts.

In summary, the open waters of Galveston Bay generally have good water quality. Most pollution problems in the system occur in the western, urbanized tributaries that receive considerable contaminant inputs from both point and nonpoint sources. However,

Table 4.7 Upper Houston Ship Channel Analysis: Relative Contributions of Pollutants from Major Sources

Constituent	Municipal Point Sources	Industrial Point Sources	Lower Watershed Nonpoint Sources	Total
		Flow		
Million ac-ft/hr	0.3	0.1	1.0	1.4
(% of Total)	(22)	(5)	(73)	(100)
		Total Suspended Solids		
Million kg/yr	2.6	7.1	208.0	217.7
(% of Total)	(1)	(3)	(96)	(100)
		BOD		
Million kg/yr	1.0	2.2	10.9	14.4
(% of Total)	(7)	(16)	(77)	(100)
		Oil and Grease		
Million kg/yr	5.7	0.6	8.9	15.2
(% of Total)	(38)	(4)	(59)	(100)
		Total Nitrogen		
Million kg/yr	7.1	1.3	2.7	11.1
(% of Total)	(64)	(11)	(24)	(100)
		Total Phosphorus		
Million kg/yr	3.6	0.4	0.5	4.5
(% of Total)	(79)	(10)	(11)	(100)
		Total Arsenic		
1,000 kg/yr	16.3	3.8	10.3	30.4
(% of Total)	(54)	(13)	(34)	(100)
		Total Cadmium		
1,000 kg/yr	5.6	0.8	1.9	8.3
(% of Total)	(68)	(9)	(23)	(100)
		Total Chromium		
1,000 kg/yr	21.8	10.3	3.5	35.6
(% of Total)	(61)	(29)	(10)	(100)
		Total Copper		
1,000 kg/yr	18.8	9.2	17.4	45.4
(% of Total)	(41)	(20)	(38)	(100)
		Total Lead		
1,000 kg/yr	23.1	2.3	48.7	74.1
(% of Total)	(31)	(3)	(66)	(100)
		Total Mercury		
1,000 kg/yr	0.00	0.06	0.31	0.37
(% of Total)	(0)	(16)	(984)	(100)
		Total Zinc		
1,000 kg/yr	84.4	29.6	139.9	253.9
(% of Total)	(33)	(12)	(55)	(100)

Source: Shipley, F. S. and Kiesling, R. W., Eds., *The State of the Bay: A Characterization of the Galveston Bay Ecosystem*, Publication GBNEP-44, the Galveston Bay National Estuary Program, Webster, TX, 1994.

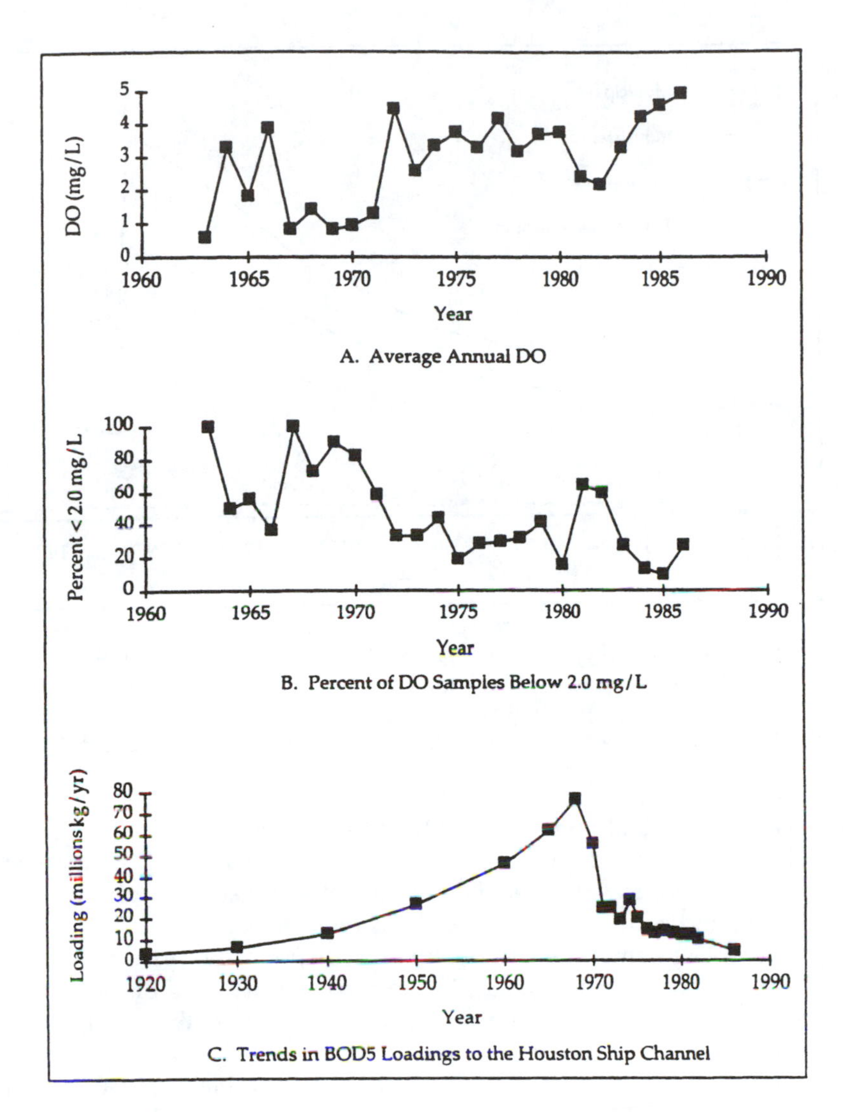

Figure 4.5 Improvements in dissolved oxygen in the upper Houston Ship Channel resulting from upgrades in wastewater treatment. Increased dissolved oxygen (A) and a reduction in the frequency of dissolved oxygen depletion below 2.0 mg/l (B) resulted from a greater than 95% reduction in BOD loading (C). (From Armstrong, N.E. and Ward, G.H., *Point Source Loading Characterization of Galveston Bay*, Galveston Bay National Estuary Program Publication GBNEP-36, Webster, TX, 1994.)

significant improvements in environmental quality have also been realized in the upper Houston Ship Channel, where urban tributary problems are most acute. BOD loadings have decreased by more than 95% in the channel since 1970, and dissolved oxygen levels have increased appreciably. Nevertheless, some degraded areas of this urban tributary continue to display dissolved oxygen levels of only 1 to 2 μg/l.

Although water and sediment quality have significantly improved in the system during the past 3 decades owing mainly to upgrades in wastewater treatment by point source dischargers, nonpoint source pollution remains an unresolved issue. Nonpoint sources in the lower watershed contribute more than 50% of the loadings of total suspended solids, BOD, oil and grease, total phosphorus, and PCBs to the bay. They also deliver most of the loadings of total suspended solids (96%), BOD (77%), oil and grease

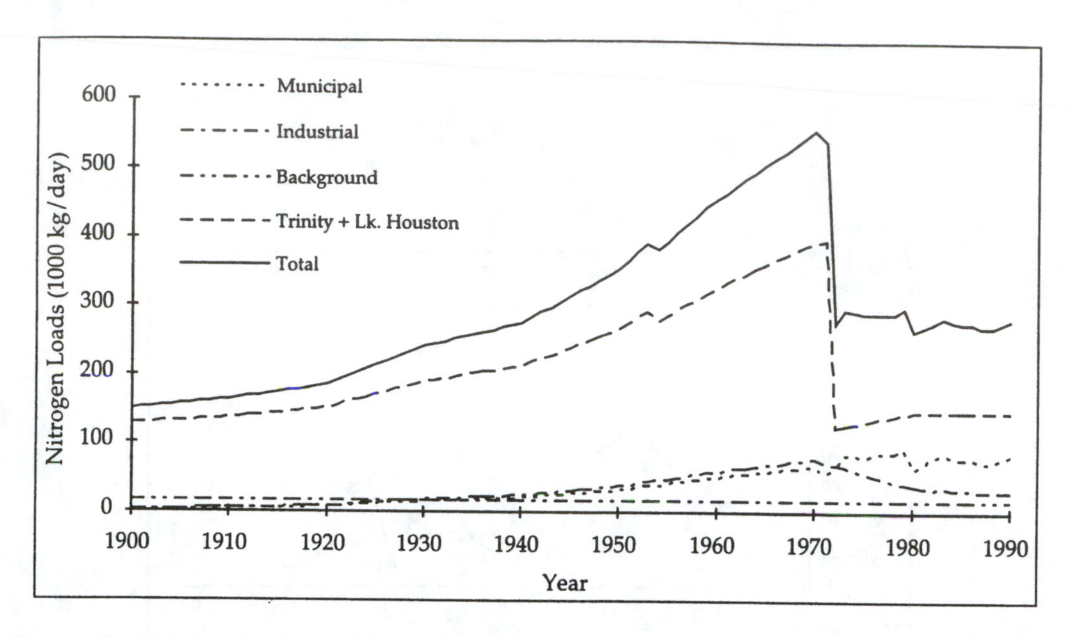

Figure 4.6 Nitrogen loads in the Galveston Bay system. Note acute reduction in nitrogen loads during the early 1970s. (From Jensen, P., Valentine, S., Garrett, M.T., Jr., and Ahmad, Z., Nitrogen loads to Galveston Bay, in *Proceedings of the Galveston Bay Characterization Workshop*, Shipley, F.S. and Kiesling, R.W., Eds., Galveston Bay National Estuary Program Publication GBNEP-6, Webster, TX, 1991, 99.)

(59%), total lead (66%), total mercury (84%), and total zinc (55%) to the upper Houston Ship Channel. Industrial point sources account for relatively minor pollutant loads. However, municipal point sources dominate the inputs of nitrogen, phosphorus, arsenic, cadmium, chromium, and copper to the upper Houston Ship Channel.

Among influent systems, the Trinity River transports more than 50% of the total chromium, lead, and mercury to the bay. It also carries nearly half of the total suspended solids and total nitrogen, arsenic, cadmium, copper, and zinc to the bay. In addition, ~575 kg/yr of pesticides enter the bay via the Trinity River.

The construction of Lake Livingston, which was completed in 1973, led to several significant changes in Galveston Bay water quality. Most notable, both nitrogen and sediment loads to the bay decreased by 40 to 50%, largely because the lake effectively traps these constituents from upstream and local nonpoint sources. The reduced nitrogen concentrations, in turn, were deemed responsible for a 50% reduction in phytoplankton biomass (as evidenced by declining chlorophyll *a*) since 1970.

In terms of toxic chemical substances, PAHs and PCBs are perhaps of greatest concern because they tend to bioaccumulate and biomagnify in estuarine organisms. Humans may be exposed to these compounds by consuming contaminated seafood. The primary source of PAHs and PCBs is urban runoff. However, bottom sediments in the estuary also act as a secondary source as well as a sink for the contaminants. The annual loading of PCBs to the bay is estimated at ~115 kg/yr, which amounts to 0.1 to 13% of the total contaminant concentration in bottom sediments.[1,19] Elevated levels of PCBs (exceeding state criteria for safe concentrations) have been documented in some waters of the Houston Ship Channel.

There are several other potential pollutant sources that can influence water and sediment quality in the Galveston Bay system, which, as yet, have not been adequately investigated. Included here are atmospheric deposition, groundwater influx, hazardous waste sites as well as oil and chemical spills. Future research efforts must focus on all of these potential pollution sources to further mitigate impacts on the estuary.

B. Physical impacts

Various human activities have substantially altered the physical structure of Galveston Bay. Most significant in this regard have been the construction and maintenance of navigation channels, shell dredging, water withdrawal (causing subsidence), shoreline modifications, and isolation of secondary bays and marshes. Major system modifications have been closely coupled to urbanization of the watershed, development of the petroleum and gas industries, and increased commercial and recreational use of the bay.

1. Navigation channels

Work on navigation channels commenced during the mid-1880s with construction of the north and south jetties at Bolivar Inlet. The formation of this jetty system by the U.S. Army Corps of Engineers resulted in successful maintenance of the inlet channel to a depth below 8 m. During the 1800s, the U.S. Army Corps of Engineers also commenced work on another major navigation project — dredging of a 3.9-m deep channel from Galveston to Houston. A third dredging project, undertaken by the Texas City Terminal Company, culminated in a 5.2-m deep navigation channel from Texas City to a point immediately north of Bolivar Inlet. Several other channel-dredging projects in the bay were also conducted prior to 1900.

The construction and maintenance of navigation channels in the estuary accelerated after 1900. The largest projects were associated with development of the Houston Ship Channel, Texas City Channel (together with the Texas City Dike), and Gulf Intracoastal Waterway. Although dredging of the Houston Ship Channel to a depth of ~8 m was completed in 1914, a series of later dredging projects deepened the channel in 1926 (~9.1 m), 1937 (~9.8 m), 1949 (~11.0 m), and 1963 (~12.2 m). The channel width also was increased to ~ 122 m by 1937. The U.S. Army Corps of Engineers is considering additional enlargement of the channel from the ~12.2 m × ~122 m configuration today to a future configuration of ~15.2 m × ~183 m.[30] However, simulations indicate that this expansion would significantly increase salinity in the estuary, thereby potentially impacting biotic communities.[1]

New dredging (as opposed to maintenance dredging) to create the Gulf Intracoastal Waterway and Texas City Channel was completed in the early 1900s. However, some routes of the Gulf Intracoastal Waterway have since changed. To reduce maintenance on the Texas City Channel, the 8.5-km Texas City Dike was completed in 1915. The dike had a significant effect on circulation in the West Bay reach.

New dredging for construction of the Houston Ship Channel has overshadowed that for the other two major channel systems. For example, $\sim 1.53 \times 10^8$ m^3 of sediment was dredged for construction of the Houston Ship Channel compared to only $\sim 3.82 \times 10^7$ m^3 for the Texas City Channel and 9.87×10^6 m^3 for the Gulf Intracoastal Waterway. Currently, most of the channels in the estuary (~80%) are deep draft (>~11 m), and they occupy ~2590 ha of bay bottom.[1]

Most navigation channels in Galveston Bay require frequent maintenance dredging. As a result, the mean volume of sediment removed annually from channels in the bay by maintenance dredging amounts to $\sim 6.53 \times 10^6$ m^3, with much of this sediment taken from the upper bay areas.[31] Over the past century, the total volume of bottom sediments extracted from the channels by maintenance dredging has exceeded 4.93×10^8 m^3 (more than twofold that associated with new dredging).

The dredged sediments have been dumped at designated disposal sites, mainly in open waters of the estuary. Dredged material islands (e.g., Atkinson Island) have been formed by this process. Approximately 1.01×10^4 ha of designated disposal area currently exists in the system. Alternatively, some of the dredged material has been used beneficially for the construction of bird islands and marsh restoration projects.

Aside from federal channel dredging projects, many dredge-and-fill operations in the Galveston Bay system are conducted by public agencies and private interests. Collectively, these nonfederal dredging efforts have removed $\sim 4.93 \times 10^7$ m^3 of sediment or one tenth that of the federal projects. Most of this dredging work has been conducted in areas of the San Jacinto River, Houston Ship Channel, and associated side bay areas. During the past 50 years, ~1135 ha of wetlands habitat (marsh) have been replaced by inland fill activities, and ~190 ha of wetlands have been destroyed by the construction of privately dredged channels.[31]

The dredging of navigation channels affects circulation in the estuary by increasing the bathymetry of the system. The larger cross section enables greater flow of water up and down the length of the bay. The breaching of reefs and bars (e.g., Karankawa Reef and Redfish Bar) likewise enhances flow. However, the formation of dredged material islands and shoals creates barriers or impediments to flow. Dredging of the Houston Ship Channel has provided a conduit for the upestuary flow of more saline waters. Salinity in the channel, for example, is ~2‰ greater than the surrounding shallow bay waters. Clearly, the construction of the Houston Ship Channel and the elimination of the Redfish Bar, which transected the middle bay, were major modifications responsible for the greater upestuary flow of higher salinity water. This alteration in circulation and salinity has likely had a significant effect on living resources of the system.

Historically, circulation in the lower bay was dramatically changed by construction of the Galveston jetties and the elimination of the natural bar at the Bolivar Roads Inlet during the late 1800s. The new inlet structure substantially increased the overall tidal flow in the bay, perhaps by 20%.[2] Construction of the Texas City Dike in 1915 and enlargement of the Texas City Channel in 1916 further modified circulation in the lower estuary, resulting in greatly reduced circulation in West Bay and substantially altered current flow in the lower bay.

2. *Shell dredging*

As mentioned previously, shell dredging prior to 1969 removed large quantities of oyster shell from the bottom of Galveston Bay. This activity effectively eliminated $\sim 1.67 \times 10^8$ m^3 of shell from the system, a volume comparable to that of the sediments removed during new work dredging of navigation channels. The overall effect of shell removal has been to increase the bay volume, thereby providing greater circulation in the system.

3. *Subsidence*

As the population inhabiting the Galveston Bay watershed and surrounding region has increased, demands for groundwater supplies have risen substantially. Excessive groundwater withdrawal during the past century has caused local subsidence of more than 3 m in some locations and an average subsidence of ~0.3 m in the upper Galveston Bay area. Lands have subsided over an area ~130 km in diameter. In response, bay waters have gradually inundated local lands, converting fringing salt marsh to open water and mudflat habitats. Between 1950 and 1990, $\sim 1.31 \times 10^4$ ha of vegetated wetlands were lost primarily because of subsidence. However, the implementation of new groundwater withdrawal management practices terminated excessive pumping in near-bay areas resulting in the apparent cessation of subsidence in this area during the 1990s.

Table 4.8 Changes in the Volume of Galveston Bay Associated with Physical Modifications of the System

Activity	Bay Volume Change Since 1900	
	Meters^3	Percent
Subsidence	$+9.86 \times 10^8$	+36.0
Deepdraft channels	$+1.79 \times 10^8$	+6.0
Shell dredging	$+1.67 \times 10^8$	+6.0
Other dredging and shallow channels	$+6.78 \times 10^7$	+2.5
Isolation	-2.47×10^7	+1.0
Dredged material disposal and fill	-2.28×10^8	–8.5
Siltation	-3.82×10^8	–14.0
Net	$+7.65 \times 10^8$	+30.0

Note: Current total bay volume: $\sim 2.71 \times 10^9$ m^3.

Source: Data from Ward, G. H., *Dredge-and-Fill Activities in Galveston Bay*, Galveston Bay National Estuary Program Publication GBNEP-28, Webster, TX, 1993.

As in the case of dredging, subsidence has increased the total volume of the bay, albeit to a much greater extent. The increase in bay volume attributable to subsidence has been threefold greater than that ascribed to all past channel dredging activities combined. It has contributed substantially to the 30% increase ($>2.7 \times 10^9$ m^3) in overall bay volume during the past 100 years (Table 4.8).

Subsidence affects a broader geographic area of the bay than the local impacts associated with channel dredging. However, both have had similar environmental impacts on the estuary, altering sediment distribution and modifying circulation. Shoreline erosion has escalated. Together these factors have produced considerable physical change in the bay over time.

4. *Shoreline modification*

Natural shoreline habitats have been altered by development practices that convert a sloping, vegetated land surface to an abrupt rectangular land–water barrier. This is most evident where an array of bulkheads, docks, and revetments abut the shoreline. Ward[31] estimated that nearly 120 km (~10%) of natural Galveston Bay shoreline have been converted to man-made shoreline during the past 50 to 60 years. Most of the altered shoreline is due to bulkheading (~72 km or 6% of the shoreline), with dock and revetment construction being of secondary importance (~48 km or 4% of the shoreline). Shoreline development and conversion cause several impacts on the Galveston Bay system, including the increase of point and nonpoint source pollution, the loss of wetlands area, accelerated shore erosion, and reduced public access.[1]

5. *Isolation of secondary bays and marshes*

Approximately 3×10^3 ha of bay bottom and 6.4×10^3 ha of tidal marsh have been removed from the original bay area due to the formation of "isolations." For example, the entrance to Turtle Bay was closed off in 1931 to protect rice irrigation systems, thereby creating an isolated freshwater lake (i.e., Lake Anahuac). An impoundment utilized by the Cedar Bayou (Electric) Generating Station for cooling water purposes as well as the development of the Delhomme hunting area isolated an additional 1.5×10^3 ha of bay bottom. These physical modifications represent permanent conversions and loss of original bay habitat.

V. Habitats and living resources

A. Habitats

Galveston Bay is a complex ecosystem comprised of a variety of habitats that support a multitude of organisms and natural communities. Major habitats include open bay water, open bay bottom, oyster reefs, submerged aquatic vegetation, small embayments, intertidal mudflats, and wetlands (i.e., saltmarsh, brackish marsh, freshwater marsh, and forested swamps). The GBNEP has assessed these habitats and the organisms occupying them.

Rich planktonic and nektonic communities inhabit the open bay waters that cover an area of $\sim 1.43 \times 10^5$ ha. Benthic communities along the open bay bottom are composed of highly diverse epifaunal and infaunal populations. Some members of these communities are of considerable recreational and commercial importance. Because many benthic populations have a meroplanktonic life stage, their long-term viability is closely coupled to conditions in the water column.

The oyster reef habitat is considered separately from the open bay bottom habitat because of its extreme ecological, hydrological, and commercial importance. Approximately 1.08×10^4 ha of bay bottom are covered by reefs and unconsolidated shell sediments.[4] The reef community consists of highly productive clusters of oysters and many other species of organisms (e.g., barnacles, bivalves, crabs, gastropods, polychaete worms, shrimp, etc.) that proliferate on and within masses of unconsolidated shell and soft sediments. Finfish populations are also important components of the oyster reef community.

In some shallow subtidal waters (e.g., Christmas Bay), seagrass meadows support a strong detritus-based system and highly productive benthic and nektonic communities. However, submerged aquatic vegetation has dramatically declined overall in Galveston Bay.[32] It once covered much more extensive areas of the bay bottom, forming continuous plant beds around Trinity River Delta (widgeon grass, *Ruppia maritima* and tape grass, *Vallisneria americana*), the west shoreline from Seabrook to San Leon (widgeon grass, *R. maritima*), and the southern shoreline of West Bay (widgeon grass, *R. maritima* and shoalgrass, *Halodule wrightii*). Several species of seagrass still occur in Chrismas Bay (widgeon grass, *R. maritima*; shoalgrass, *H. wrightii*; clovergrass, *Halophila engelmannii*; and turtlegrass, *Thalassia testudinum*). Of the estimated 280 ha of seagrass meadows remaining in Galveston Bay, most (~140 ha) exist in Christmas Bay.

Among the most productive areas of the Galveston Bay system are marsh-rimmed embayments or soft-bottom "lakes" (e.g., Alligator, Oyster, Greens, Swan, Clear, Robinson, Cotton, and Lost Lakes), which serve as valuable nursery habitat for numerous organisms. Located near the terminus of the drainage bayous, these waterbodies are characterized by deep and unconsolidated mud bottoms, high turbidity, and variable salinity. They are directly connected to the main bay system, thereby providing many endemic and migratory species ready access to their rich food supply and protective cover. The high biomass of fringing marsh and high concentration of organic matter in drainage bayous appear to support detritus-based food webs. A number of recreationally and commercially important species, such as the brown shrimp (*Penaeus aztecus*), white shrimp (*P. setiferus*), blue crab (*Callinectes sapidus*), pinfish (*Lagodon rhomboides*), spotted seatrout (*Cynoscion nebulosus*), Atlantic croaker (*Micropogonias undulatus*), southern flounder (*Paralichthys lethostigma*), and Gulf menhaden (*Brevoortia patronus*), heavily utilize these unique habitats.

Intertidal mudflats also occur in the Galveston Bay system. These open areas lack the emergent grasses and other vegetation that typify peripheral saltmarshes and submerged seagrass meadows; consequently, the biotic communities tend to be markedly different. Microalgae, macroalgae, and phytoplankton constitute the only flora here. The soft bottom sediments are primarily inhabited by bioturbating infauna. Although benthic epifauna may likewise populate the substrate, they usually attain low numbers except in localized areas. During rising tide, fish and crabs invade water-covered areas in search of food. During falling tide, shorebirds and wading birds prey on benthic invertebrates and fish in barren sediments and tidal pools.

Galveston Bay wetlands are subdivided into five dominant plant communities: (1) submerged aquatic vegetation (seagrass meadows); (2) saltmarshes; (3) brackish marshes; (4) freshwater marshes, and (5) forested wetlands (swamps). Although seagrass meadows have a restricted distribution in the system, as discussed previously, the emergent marsh communities are extensive. Forested wetlands, in turn, are confined almost exclusively to the Trinity River valley. In total, the wetlands cover ~5.6×10^4 ha, with saltmarshes and brackish marshes comprising ~4.4×10^4 ha, freshwater marshes ~9.0×10^3 ha, forested swamps ~2.3×10^3 ha, freshwater scrub/shrub wetlands ~8.1×10^2 ha, seagrass meadows ~2.8×10^2 ha, and estuarine scrub/shrub wetlands ~2.2×10^2 ha.[33]

Intertidal emergent plant communities inhabit ~61% of the Galveston Bay shoreline. The saltmarsh communities are dominated by smooth cordgrass (*Spartina alterniflora*), saltgrass (*Distichlis spicata*), and glasswort (*Salicornia* spp.). The low marsh community commonly consists of monotypic stands of *S. alterniflora*. Salt meadow cordgrass (*S. patens*) and Gulf cordgrass (*S. spartinae*) flourish at higher elevations. They also predominate in the brackish marshes.

Squarestem spikesedge (*Eleocharis quadrangulata*) is typically found in freshwater marshes. Other abundant species growing here include coastal arrowhead (*Sagittaria falcata*) and giant cutgrass (*Zizaniopsis miliacea*). Bald cypress (*Taxodium distichum*) represents the prinicipal species of the forested wetlands community.[34]

Wetlands are critical to the Galveston Bay system because of their extremely high production, unique hydrological characteristics, and invaluable biological functions. For example, they mitigate shoreline erosion, protect upland habitats from storms and storm surges, serve as flood control agents, and buffer the bay from excessive organic loadings. They also support a multitude of organisms ranging from microbes to amphibians, reptiles, mammals, and birds. Many species that are of recreational and commercial importance in both Galveston Bay and the Gulf of Mexico rely on the wetlands for shelter or as nursery areas. Finally, nutrient and organic carbon export from the fringing wetlands to the open bay appear to be significant to the overall production of the system.

Human activities have caused major losses of wetlands habitat over the years. From the 1950s to 1989, for example, 17 to 19% of vegetated wetlands were lost due to multiple anthropogenic impacts. The greatest losses were ascribed to: (1) subsidence coupled to groundwater withdrawal; (2) drainage and conversion to upland habitat for agricultural, urban, and industrial needs; (3) dredging and filling; and (4) shoreline modifications and resultant isolations.[1,32]

Sea level rise associated with man-induced subsidence has been principally responsible for wetlands loss along the northeast part of West Bay and the north, west, and south margins of Galveston Bay. Much of the loss of estuarine and freshwater emergent vegetation, in turn, resulted from the draining of wetlands. The bulk of upland conversion of wetlands has occurred landward from West and Christmas Bays. Most of the wetland conversions for agriculture and livestock grazing have taken place in the areas of Hitchcock, Hoskins Mound, and Oyster Bayou. The urbanization of wetlands has been concentrated along the south and west side of the bay (e.g., Virginia Point) as well as near Sea Isle, League City, Texas City, and Galveston.

Ward[31] estimated that ~2.9×10^3 ha of marsh habitat has been destroyed by dredging, filling, and disposal activities during the past century. Private dredge-and-fill operations have accounted for nearly half of this loss. The creation of designated dredged material disposal sites has been almost equally detrimental to wetlands habitat.

Although the total area of wetlands lost to dredging and filling has been significant, it is less than that incurred from shoreline development and isolation projects. The loss of wetlands due to the isolation of marshland from the bay has been conspicuous on the west side of Trinity Bay and along the Trinity River delta. The north shore of East Bay likewise has been the site of isolation activity and concomitant habitat loss.

Several factors appear to have contributed to the demise of submerged aquatic vegetation in different parts of the estuary. In western Galveston Bay, the decline of seagrass meadows has likely developed from gradual subsidence and the occurrence of Hurricane Carla. Development, dredging, chemical spills, and wastewater discharges have also played a role in the destruction of these vascular plant beds in perimeter areas of the estuary.[32]

Management strategies have been formulated during the past decade that specifically deal with the problems of habitat loss and alteration in the Galveston Bay system. The CCMP drafted by the GBNEP incorporates a number of action plans designed to mollify human impacts on sensitive habitat areas. The implementation of these proposed measures is necessary to ensure the preservation and long-term health of the bay and its living resources.

B. *Living resources*

The productivity of Galveston Bay arises from complex processes that cycle nutrients between organisms, sediments, and the water column. Autotrophs assimilate the nutrients, producing carbon; this carbon serves as an energy source for heterotrophs at higher trophic levels. Energy flows among aquatic organisms in the estuary via two interlocking components of the food web: (1) the phytoplankton (grazing)-generated component and (2) the detritus-generated component. These two components form a network or trophic interaction of producer, consumer, and decomposer organisms. This section examines the major biotic groups in the estuary beginning at the base of the food chain and progressing to higher levels (i.e., phytoplankton, zooplankton, benthos, finfish, avifauna, amphibians, reptiles, and mammals). Emphasis is placed on recent trends in the abundance and distribution of the principal taxa of the estuary.

1. *Phytoplankton*

Phytoplankton are free-floating microscopic plants (unicellular, filamentous, or chain-forming species of algae) that generate a large fraction of the total primary production in Galveston Bay. Although most of the phytoplankton consist of unicellular forms, some green and blue-green algae are filamentous (i.e., produce thread-like cell systems). Colonial diatoms provide examples. In addition, a number of diatoms and dinoflagellates form chains of loosely associated cells.

As reported by the Texas Department of Water Resources,[35] more than 130 phytoplankton species inhabit upper Galveston and Trinity Bays. Diatoms constitute most of the taxa (n = 54), followed by green algae (45 taxa), and blue-green algae (14 taxa). Examining salinity preferences, green and blue-green algae tend to dominate low-salinity waters, and diatoms dominate high-salinity waters.[35,36] During the cold months of the year, diatoms and green algae predominate and are replaced by dinoflagellates and other forms during the warmer months.[37]

The estimated productivity of phytoplankton in Galveston Bay (350 g dry wt/m^2/yr) is about two to seven times less than that of the other major primary producer communities in the system (i.e., benthic microflora, submerged aquatic vegetation, salt-brackish marsh,

Table 4.9 Primary Productivity in the Galveston Bay System

Flora	Average Estimated Primary Productivity (g dry/m²/yr)	Areal Coverage (km²)	Estimated Annual Production (metric tons)
Phytoplankton	350	1,425	498,750
Benthic microflora	500	1,425	712,500
Submerged vegetation	2,600	1	2,600
Fresh water marsh	820	40	32,800
Salt-brackish marsh	1,100	370	407,000
Woodlands/swamps	700	500	350,000

Source: Sheridan, P. F., Slack, R. D., Ray, S. M., McKinney, L. W., Klima, E. F., and Calnan, T. R., Biological Components of Galveston Bay, in *Galveston Bay: Issues, Resources, Status, and Management*, NOAA, Estuary-of-the-Month Seminar Series No. 13, National Oceanic and Atmospheric Administration, Washington, D.C., 1989, 23.

freshwater marsh, and woodlands/swamps) (Table 4.9).[36] However, because of the large areal coverage of phytoplankton in the estuary, the annual production of the community exceeds that of all other floral groups except the benthic microflora. Phytoplankton primary production in the bay varies considerably both on a spatial and temporal basis. Compared to most other estuaries, Galveston Bay exhibits slightly higher annual phytoplankton production.[38]

Long-term trends of phytoplankton abundance are evident in the estuary. For example, mean chlorophyll *a* concentrations, an indicator of phytoplankton biomass, appear to have increased from the late 1950s to the early 1970s. However, they decreased substantially (~50%) baywide between 1970 and 1991. Mean chlorophyll *a* dropped from 30 to 35 mg/m³ to 15 mg/m³ in Galveston and Trinity Bays between 1985 to 1991.[1] This decline has been attributed to reduced nutrient levels in the system associated with diminishing point source loadings.[18]

Both nuisance and toxic "red tide" blooms have been observed at various times in Galveston Bay. Nuisance phytoplankton blooms develop only occasionally in the system, primarily in tributaries with restricted circulation and high nutrient levels (e.g., Clear Lake and Dickinson Bayou). They typically trigger episodes of oxygen depletion. Red tide blooms, which are less common events, are usually caused by the dinoflagellate *Gonyaulax monilata*. Toxins produced by this species have resulted in kills of demersal fish and benthic invertebrates, as evidenced by the massive mortalities incurred during the red tide bloom of 1984.[39]

2. *Zooplankton*

Zooplankton are volumetrically abundant animals, typically several microns to 2 cm in size, that drift passively in currents due to their diminutive size and limited capabilities of locomotion. These lower-trophic-level consumers constitute the principal herbivorous component of the system, serving an essential link in the food web by converting plant to animal matter. They basically gather food via filter or raptoral feeding. By exerting grazing pressure, zooplankton can effectively regulate the standing crop of phytoplankton populations.

Zooplankton are classified by size or length of planktonic life. Based on size, zooplankton are differentiated into nanozooplankton (2 to 20 μm), microzooplankton (20 to 200 μm), mesoplankton (200 to 500 μm), and macrozooplankton (>500 μm). Based on the duration of planktonic life, zooplankton are grouped into three categories: (1) holoplankton; (2) meroplankton; and (3) tychoplankton. Holoplankton spend their entire life in the plankton, in contrast to meroplankton, which remain planktonic for only a limited period of time. Tychoplankton are primarily benthic organisms temporarily translocated into the water column by currents, behavioral activity (e.g., diurnal vertical migration), or other mechanisms.

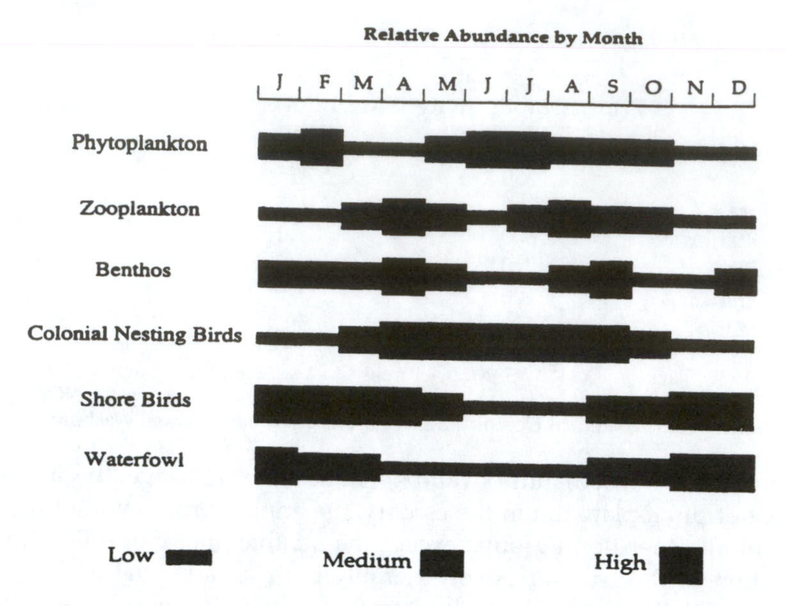

Figure 4.7 Relative abundance of various food web components of the Galveston Bay estuary. (From Sheridan, P.F., Slack, R.D., Ray, S.M., McKinney, L.W., Klima, E.F., and Calnan, T.R., Biological Components of Galveston Bay, in *Galveston Bay: Issues, Resources, Status, and Management*, NOAA, Estuary-of-the-Month Seminar Series No. 13., National Oceanic and Atmospheric Administration, Washington, DC, 1989, 23.)

Copepods as well as barnacle and polychaete larvae dominate the zooplankton community in Trinity Bay. Here, the density of zooplankton usually ranges from 1200 to 16,000 individuals/m^3.[38] The copepod *Acartia tonsa* is the most abundant species in Galveston Bay.[40] Peak abundance of zooplankton closely follows that of phytoplankton. Both groups experience large seasonal variations in absolute abundance (Figure 4.7).

3. *Benthic organisms*

The benthos includes those organisms, both plant and animal, living on, in, or near the sea bed. Attached macrophytes (e.g., seagrasses and salt marsh grasses) not only provide substantial amounts of primary production but also serve as habitat formers, supporting large numbers of benthic invertebrates and fish. Benthic algae are present as unicellular motile species on tidal flat surfaces, as multicellular forms attached to soft or hard surfaces, and as passively drifting populations unaffixed to any substrate. The microflora live on substrates as individual cells or filamentous colonies, and they adhere to the surfaces of other plants, animals, sediments, and man-made structures. They are often conspicuous in intertidal areas where prolific growth of minute, single-celled forms can discolor the sediments. Benthic diatoms are highly motile, migrating vertically within the sediments in response to variable light intensities. Mats of green or blue-green algae may also develop on tidal flats and saltmarsh surfaces.

Based on their life habits and adaptation, benthic fauna have been broadly classified into populations residing on the seafloor or on a firm substrate (epifauna), populations living in the sediment (infauna) as well as interstitial, boring, swimming, and commensal-mutualistic forms. Studies of bottom-dwelling invertebrates have progressed on two basic levels: (1) the species level, whereby data are gathered on the life history and population dynamics of the fauna, and (2) the community level, whereby an assemblage of organisms is assessed in terms of multiple variables, such as species composition, abundance, diversity, and productivity. Although many investigations of the benthos have concentrated on the benthic infauna, studies of the benthic epifauna and demersal finfish populations should also be considered in comprehensive assessments of benthic communities.

Benthic fauna have also been broadly classified according to their mode of obtaining food. Five categories are differentiated, specifically deposit feeders, suspension feeders, parasites, herbivores, and carnivores-scavengers.[41,42] Many researchers have focused on deposit and suspension feeders, such as polychaete worms, bivalves (e.g., clams, mussels, and oysters), and small crustaceans, which are usually abundant in estuarine benthic communities.

Benthic fauna are commonly grouped according to their size into microfauna, meiofauna, and macrofauna. The microfauna consist of those bottom dwelling animals that pass through sieves of 0.04- to 0.1-mm mesh. The meiofauna comprise smaller forms captured on sieves of 0.04- to 0.1-mm mesh. The macrofauna include larger metazoans retained by sieves of 0.5 to 2-mm mesh.

White et al.,[3] Roach et al.,[23] Carr,[27] Parker,[43] McBee,[44] and Ray et al.[45] investigated benthic communities in Galveston Bay between 1950 and 1992. During the 1950s, Parker[43] found that mollusks, polychaete worms, and crustaceans were major constituents of benthic macroinvertebrate assemblages in the estuary. During the 1970s, White et al.[3] likewise documented high abundances of bivalves, polychaetes, and crustaceans in the benthos, with one or two species usually dominating the community. Polychaetes attained highest abundance in muddy sediments, and crustaceans reached highest abundance in sandy sediments. Diversity peaked near tidal inlets (i.e., Bolivar Roads and Rollover Pass) in waters characterized by higher and more stable salinities. Species richness in Galveston and West Bays was double that in Trinity and East Bays. Benthic diversity was much greater in open bay waters than in tributaries and embayments, such as the Houston Ship Channel, San Jacinto River, and Clear Lake. Overall, Galveston Bay had low to moderate benthic diversity during this period.

Green et al.[38] concluded that the abundance of benthic populations in open bay waters decreases from the lower Galveston Bay to the Trinity Bay-upper bay region. Peak abundance occurs from February to May; abundance decreases markedly in October and November. According to Ray et al.[45] salinity appears to be the most important factor controlling the distribution of the benthic organisms, with the type of substrate being of secondary significance. Hence, species diversity increases along a seaward salinity gradient. Opportunistic species commonly dominate the benthic infauna, having adapted to the highly variable physical and chemical conditions of the estuary.

The benthic community is extremely valuable as an indicator of environmental or human-induced stress in the estuary. Carr[27] assessed anthropogenic impacts on the benthic community, concentrating on sites affected by urban and/or industrial runoff. He described altered benthic assemblages at two of six sampling stations (i.e., Burnett Bay and Black Duck Bay), both of which are located in the upper Houston Ship Channel. Degradation of the benthos at these locations was attributed to toxic contamination from industrial and urban sources. Of 10 reference stations in the open bay not associated with urban and/or industrial runoff, only one (in East Bay) exhibited an altered benthic assemblage. However, degradation of the benthos at this site was not contaminant induced.

As alluded to previously, Roach et al.[23] delineated severe impacts of produced water discharges on benthic organisms at a tributary outfall and a shoreline/open bay outfall. Due to high levels of sediment contamination (e.g., petroleum hydrocarbons), benthic organisms were absent at the tributary outfall and severely depressed at the shoreline/open bay outfall. The benthic community was most significantly altered within 50 m of the outfall sites.

Marsh habitats also support rich benthic communities. Green et al.[38] found that more than 90% of the infauna collected at six marshes in the system consisted of polychaete worms and small crustaceans. Mollusks accounted for <3% of the infaunal abundance at most sites. Benthic organisms were generally more numerous in vegetated marsh areas than in bare marsh or adjacent subtidal habitats.

Similar to conditions of the open bay, salinity appears to have a great influence on benthic assemblages in the marsh habitats. For example, the lower bay (high salinity) marshes have lower benthic abundance and species richness values than the mid-bay (mid-salinity) marshes but higher corresponding values than the upper bay (low salinity) marshes. Marsh infauna are most abundant in late winter and early spring and least abundant in the fall.

4. *Shellfish*

a. Oysters. Galveston Bay supports two species of oysters: the eastern oyster (*Crassostrea virginica*) and the Gulf oyster (*Ostrea equestris*). They inhabit areas where the long-term salinity ranges from ~10 to 30‰.[46] The commercially important eastern oyster prefers brackish water conditions, and occurs in large numbers in shallow embayments. The Gulf oyster requires higher salinity waters and thus occupies the open bay and areas closer to the inlets. Despite their distinct salinity preferences, both species must tolerate considerable short-term salinity changes in the estuary.

The success of oyster populations depends on a number of physical and biological factors, most notably a suitable habitat, competition, predation, parasites, and diseases. Oysters are most successful where existing reefs provide suitable substrate (shell) for attachment. Powell et al.[4] chronicled a large increase in oyster reef area (~100%) in Galveston Bay after 1976. Since that time, however, there has also been a substantial reduction in phytoplankton abundance that may have resulted in an inadequate food supply for the maintenance of a large market-size oyster population.

Several organisms compete with oysters for space and food, especially algae, barnacles, mussels, and worms. In addition, predators (e.g., oyster drills, crabs, mud crabs, and some finfish) can greatly reduce population numbers. Although these biological factors are important, they may be overshadowed by the impacts associated with parasites and diseases. Infection by the protozoan parasite *Perkinsus marinsus*, commonly known as "dermo," has been particularly problematic. For example, annual dermo mortality of market oysters in the estuary ranges from ~10 to 50%, with the most severe impacts (>50% infection rate) observed in West Bay.

Quast et al.[46] reviewed trends in the oyster fishery. They recounted a decrease in abundance of small and market oysters during the period from 1956 to 1977, a sharp increase between 1980 and 1982, and another decrease from 1983 through 1987. The oyster harvest fluctuated from <0.5 mt during some years in the 1950s to more than 1800 mt in 1966. The harvest exceeded 1300 mt in 1971 and 1976 but dropped precipitously to near 0 in 1979. The highest commercial catch was registered in 1983 (>3100 mt). However, the harvest gradually declined to ~450 mt during the early 1990s.[38]

b. Blue crab. One of the most intensely harvested shellfish species in Galveston Bay is the blue crab (*Callinectes sapidus*). Landings for this crustacean increased dramatically from ~90 mt in 1960 to >1300 mt in 1990. Although smaller crabs increased in abundance through the 1980s, larger individuals decreased. The chronic recent decline of larger and older crabs in recent years may be a consequence of overfishing of males in the upper estuary.[47] Alternatively, it may be the result of a decrease in food supply related to the loss of wetlands area and concomitant reduction in organic detritus input to the estuary.[38]

c. Shrimp. White shrimp (*Penaeus setiferus*) and brown shrimp (*P. aztecus*) account for a major fraction of the total shellfish harvest in Galveston Bay. During the 1980s, annual white shrimp (*P. setiferus*) and brown shrimp (*P. aztecus*) landings commonly exceeded 1300 and 900 mt, respectively. The landings of both of these species dropped sharply during the late 1980s, falling to lows of ~900 mt (*P. setiferus*) and <450 mt (*P. aztecus*) in 1990.[38,48] Commercial overfishing appears to have impacted the harvest at this time. However, landings rebounded during the early 1990s.[47]

5. *Finfish*

The Texas Parks and Wildlife Department collected more than 150 species of fish in a 2-year trawl survey of Galveston Bay.[48] Two of these species — the Atlantic croaker (*Micropogonias undulatus*) (51%) and bay anchovy (*Anchoa mitchilli*) — comprised nearly three fourths of all fish collected. The Atlantic croaker (*M. undulatus*) is also notable because it constituted more than one third of the total fish biomass.[36,38]

Approximately 75% of the total commercial finfish harvest is currently derived from four species (i.e, black drum, *Pogonias cromis*; striped mullet, *Mugil cephalus*; sheepshead, *Archosargus probatocephalus*; and southern flounder, *Paralichthys lethostigma*). In addition, two other species (i.e., red drum, *Sciaenops ocellatus* and spotted seatrout, *Cynoscion nebulosus*) yield most of the recreational finfish catch. These latter two species have experienced wide fluctuations in abundance in recent years. Some population changes have been ascribed in part to the implementation of harvest restrictions. In general, however, commercial finfish and shellfish are stable in the estuary.[1]

The Atlantic croaker (*Micropogonias undulatus*) and striped bass (*Morone saxatilis*) have exhibited significant population trends. Although the Atlantic croaker (*M. undulatus*) has shown an increase in adult numbers, the striped bass (*M. saxatilis*) has displayed a long-term decline. The cause of these trends is not completely understood. However, reductions in food supply, the loss and alteration of habitat, government regulations, and restocking programs may have played a role.

Several major fish kills have been reported in Galveston Bay. In a review of 220 fish kills in the bay over a 20-year period, Palafox and Wolford[49] determined that 46 (20.9%) were linked to point source pollution and 43 (19.5%) to nonpoint source pollution. The causes of the remaining 131 fishkills remain unresolved.

6. *Birds*

Arnold[50] identified 139 avifaunal species in open bay and wetland habitats of Galveston Bay. Texas Colonial Waterbird Surveys conducted from 1973 to 1990 uncovered no significant trend in the total number of colonial waterbirds. However, some important trends were apparent for individual species utilizing the estuary (Table 4.10). During the survey period, the number of pairs of colonial nesting waterbirds ranged from 39,000 pairs in 1978 and 1982 to 71,200 pairs in 1985.

The relative abundance of species in the colonial waterbird community has changed considerably during the past few decades. For instance, the total abundance and number of colonies of some open water birds (e.g., Caspian terns, *Sterna caspia* and Forster's terns, *S. fosteri*) have increased.[38] The number of colonies of some species of waders (e.g., black skimmers, *Rynchops niger*; great egrets, *Casmerodius albus*; snowy egrets, *Egretta thula*; and tricolored herons, *E. tricolor*) in marsh habitats has also increased. However, an overall decline in abundance of waders has occurred because of a large decrease in the number of individuals per colony. Among freshwater marsh feeders and generalists comprising the inland group of waterbirds (e.g., cattle egrets, *Bubulcus ibis*; great blue herons, *Ardea herodias*; and little blue herons, *Egretta caerulea*), no significant change has occurred in the number of individuals per colony.

Table 4.10 Estuarine-Dependent Bird Species Utilizing Galveston Bay: Results of Species Trend Studies

Species	Observations
	Apparent Decline
Tricolored heron	Fewer total birds nesting in more colonies; no trend in the Christmas Bird Count
Snowy egret	Decrease in numbers and numbers per nesting colony; Christmas Bird Count showed increase
Black skimmer	Decrease in numbers and numbers per nesting colony; no trend in the Christmas Bird Count
Roseate spoonbill	Decrease in numbers and numbers per nesting colony; Christmas Bird Count showed increase
Mottled duck	Mid-winter Waterfowl Transects showed decrease; Christmas Bird Count showed increase
Northern pintail	Mid-winter Waterfowl Transects showed decrease; Christmas Bird Count showed increase
Blue-winged teal	Mid-winter Waterfowl Transects showed decrease; Christmas Bird Count showed increase
	No Apparent Trend
Great egret	Constant or increasing total number nesting in more colonies with fewer birds
American avocet	No change on Bolivar Flats Survey; increase in Christmas Bird Count
Dunlin	No change on Bolivar Flats Survey; increase in Christmas Bird Count
American coot	No discernible trend
Green-winged teal	No discernible trend
Northern shoveler	No apparent trend; variability suggests local movements
American widgeon	No apparent trend; variability suggests local movements
Canvasback	No trend based on Christmas Bird Count
Fulvous whistling duck	No trend based on Christmas Bird Count
All blackbirds	No trend or insufficient data
	Apparent Increase
Olivaceous cormorant	Increase in total numbers and colonies; Christmas Bird Count showed variable increase
Forster's tern	Increase in colonies and Christmas Bird Count
Black-bellied plover	Apparent increase based on Bolivar Flats Survey and Christmas Bird Count
Willet	Apparent increase based on Bolivar Flats Survey and Christmas Bird Count
Sanderling	Apparent increase based on Bolivar Flats Survey and Christmas Bird Count
Western sandpiper	Apparent increase based on Bolivar Flats Survey and Christmas Bird Count
Mallard	No change or possible increase in wintering population
Gadwall	No change of possible increase in wintering population
All scaup	Christmas Bird Count showed increase
Ruddy duck	Christmas Bird Count showed increase
Bufflehead	Christmas Bird Count showed increase
Wood duck	Christmas Bird Count showed increase
Ring-necked duck	Christmas Bird Count showed increase
Black-crowned night heron	Increase in colonies and Christmas Bird Count

Source: Green, A., Osborn, M., Chai, P., Lin, J., Loeffler, C., Morgan, A., Rubec, P., Spanyers, S., Walton, A., Slack, R. D., Gawlik, D., Harpole, D., Thomas, J., Buskey, E., Schmidt, K., Zimmerman, R., Harper, D., Hinkley, D., Sager, T., and Walton A., *Status and Trends of Selected Living Resources in the Galveston Bay System*, Galveston Bay National Estuary Program Publication GBNEP-19, Webster, TX, 1992.

Censusing of waterfowl populations in the Galveston Bay system has revealed an annual average of 11,500 birds.[36] Migratory species are most abundant (i.e., ring-necked duck, *Aythya collaris*; lesser scaup, *A. affinis*; ruddy duck, *Oxyura jamaicensis*; red-breasted merganser, *Mergus serrator*; and green-winged teal, *Anas crecca*). Some species have declined in numbers. Included here are the blue-winged teal (*A. discors*) and northern pintail (*A. acuta*). However, their decreasing abundance is apparent throughout the country, and, therefore, local factors responsible for the diminishing numbers are more difficult to ascertain. Although some duck populations have decreased in abundance in the system, geese have increased due in large part to their successful adaptation to a range of habitats.

In regard to shorebirds, the most common species are the American avocet (*Recurvirostra americana*), black-bellied plover (*Pluvialis squatarola*), short-billed dowitcher (*Limnodromus griseus*), long-billed dowitcher (*L. scolopaceus*), dunlin (*Calidris alpina*), sanderling (*C. alba*), western sandpiper (*C. mauri*), and willet (*Catoptrophorus semipalmatus*). Surveys conducted by the U.S. Fish and Wildlife Service suggest a possible increase in shorebird abundance in the Galveston Bay area. However, data have not been analyzed sufficiently to unequivocally establish population trends.

7. Amphibians, reptiles, and mammals

A wide diversity of amphibians, reptiles, and mammals resides in wetland and upland habitats of the Galveston Bay system. In addition, various forms inhabit bay waters. More than 90 species of amphibians and reptiles have been recorded in land areas adjacent to the estuary. In addition, numerous mammals live in these areas (e.g., beaver, *Castor canadensis*; gray squirrel, *Sciurus carolinenses*; mink, *Mustela vison*; muskrat, *Ondatra zibethicus*; raccoon, *Procyon lotor*; and river otter, *Lutra canadensis*). Among important species utilizing bay waters are four species of threatened or endangered turtles (i.e., green sea turtle, *Chelonia mydas*; Kemp's Ridley sea turtle, *Lepidochelys kempi*; leatherback sea turtle, *Dermochelys coriacea*; and loggerhead sea turtle, *Caretta caretta*). Bottlenose dolphins (*Tursiops truncatus*) are inhabitants or frequent visitors of the estuary. Recent population counts indicate that ~100 individuals are permanent residents of the bay, and many more dolphins are frequent visitors. Most have been observed near the inlets, particularly Bolivar Roads. They are important predators of finfish in the estuary; for example, year-round residents consume an estimated 2.8×10^5 kg of fish each year.[1] The amount of additional fish consumed by the hundreds of transient dolphins visiting the bay annually has not been determined.

VI. Pathogens and contaminants

Jensen[51] reviewed the pathogens that have been identified in estuarine waters and seafood products of Galveston Bay. Brooks et al.[52] investigated the toxic chemical contaminants in aquatic organisms and birds inhabiting the estuary. Pathogens and toxic contaminants in bay waters and organisms of the estuary can pose some serious health risks to humans through contact and noncontact recreational activities as well as through consumption of contaminated seafood products. As a result, public health agencies closely monitor and regulate water quality conditions in the system.

A. Pathogens

The Texas Department of Health Shellfish Sanitation Program and the Texas Natural Resource Conservation Commission currently use fecal coliform bacteria as an indicator of human pathogens to assess recreational and shellfish growing waters of the estuary.

Acceptable levels are specified in the *Texas Surface Water Quality Standards*.[53] Runoff from upland areas, especially from urban sites, is the main source of fecal coliform bacteria to Galveston Bay. Point source discharges and septic systems can be locally important sources of fecal coliforms but are not of baywide significance.

Aside from fecal coliform bacteria, there are some bacterial pathogens that are unrelated to human wastes (e.g., *Vibrio parahaemolyticus* and *V. vulnificus*). However, these naturally occurring pathogens can be equally devastating to human health, occasionally causing severe illness and sometimes death. There does not appear to be a relationship between fecal coliform bacteria concentrations and *Vibrio* counts in the bay. Exposure to the various pathogens in Galveston Bay may lead to the development of serious maladies, such as cholera, dysentery, gastroenteritis, and infectious hepatitis.

The long-term database on fecal coliform bacteria concentrations in the estuary indicates that problem areas exceeding the contact recreation (e.g., swimming and water skiing) standards specified in the *Texas Surface Water Quality Standards*[53] occur in urbanized tributaries, particularly along the western shoreline. Urbanized bayous such as Baystrop Bayou, Buffalo Bayou Tidal, Cedar Bayou Tidal, Dickinson Bayou Tidal, and Sims Bayou provide examples. Over the long term, the open waters of the bay have been shown to meet the Texas water quality criterion for contact recreation of 200 colony forming units per 100 ml.

Noncontact recreation is also considered (e.g., power boating and sailing). The noncontact criterion allows 10 times greater fecal coliform bacterial concentrations (i.e, 2000 colony forming units per 100 ml) than the contact designation. The areas of the estuary found to violate the noncontact criterion are all urbanized tributaries with limited circulation.[51]

Due to the potential health risk from pathogenic bacteria, about half of the bay (~5.2 $\times 10^7$ ha) is permanently or provisionally closed to shellfish harvesting. The area of the bay subject to shellfish closure has generally remained the same for nearly half a century.[1] The principal concern involves contaminated oysters. *Vibrio* infections associated with Galveston Bay have periodically appeared in the adjacent counties. For example, over the period from May 1981 to September 1991, 68 *Vibrio* infections were reported there, with 12 being specifically linked to Galveston Bay. The most recent outbreak in June 1998 was caused by human consumption of raw bay oysters infected with the bacterium *V. parahaemolyticus*, which resulted in several hundred cases of gastrointestinal disorders in the region, notably diarrhea and abdominal cramps accompanied by nausea, vomiting, fever, and headache. In response to the outbreak, health agencies closed the entire bay to shellfish harvesting. Such outbreaks are testimony to the seriousness of pathogen contamination of seafood products in the estuary.

B. Toxic contaminants

Brooks et al.[52] analyzed toxic contaminants (i.e., PAHs, halogenated hydrocarbons, and heavy metals) in five species of organisms (i.e., the eastern oyster, *Crassostrea virginica*; blue crab, *Callinectes sapidus*; black drum, *Pogonias cromis*; southern flounder, *Paralichthys lethostigma*; and spotted seatrout, *Cynoscion nebulosus*) from four locations (i.e., Hanna Reef, Morgans Point, Eagle Point, and Carancahua Reef) in Galveston Bay. Highest contaminant levels of PAHs were found in oysters, especially those in urbanized areas of the upper bay. For example, contaminant levels in oysters from the upper bay and near the City of Galveston were up to 100 times higher than in oysters from nonurban portions of the system. Total PAHs in tissues of oysters from the estuary ranged from nondetectable to 1253 ng/g. The National Status and Trends Program of the National Oceanic and Atmospheric Administration reported that PAH concentrations in oysters

Table 4.11 Chlorinated Hydrocarbons in Oyster, Crab, and Fish Tissue from Four Sites in Galveston Bay

Compound	Findings
BHCs	Higher in oysters than crabs or fish tissue; highest in fish livers. Highest at Morgans Point (8.85 ng/g); similar at Eagle Point, Hanna Reef, and Carancahua Reef. Levels for oysters within range reported by National Status and Trends Program.
Chlordane	Similar in fish, crabs, and oysters; higher in fish livers. Highest at Morgans Point (58 ng/g) and decreased down-bay. Levels for oysters within range reported by National Status and Trends Program.
DDT	Similar in fish, crabs, and oysters; higher in fish livers. Highest at Morgans Point (average 87 ng/g) and decreased down-bay. Levels for oysters within range reported by National Status and Trends Program (which found DDT in Galveston Bay oysters among the highest 25 percent of concentrations Gulf-wide).
Dieldrin	Similar in fish and oysters; lower in crabs; highest in fish livers. No down-bay decreasing trend (range 2-11 ng/g). Levels for oysters within range reported by National Status and Trends Program.
PCBs	Similar in oysters and fish; slightly higher in crabs; much higher in fish livers. Decrease from Morgans Point down-bay (range 176-612 ng/g). Levels for oysters within range reported by National Status and Trends Program.

Source: Brooks, J.M., Wade, T.L., Kennicutt, M.C., II, Wiesenburg, D.A., Wilkinson, D., McDonald, T.J., and McDonald, S.J., *Toxic Contaminant Characterization of Aquatic Organisms in Galveston Bay: a Pilot Study*, Galveston Bay National Estuary Program Publication GBNEP-20, Webster, TX, 1992.

from Galveston Bay rank among the highest 25% for sites throughout the Gulf of Mexico region.

Table 4.11 shows chlorinated hydrocarbon contamination levels (including absolute concentrations) in oysters, crabs, and finfish from the estuary. PCBs and DDT are the most common contaminants. With regard to PCBs, DDT, and chlordane, the highest concentrations occur in biota at Morgans Point, with levels decreasing down-bay. Dieldrin does not exhibit such a down-bay gradient in concentrations.

Of particular concern is dioxin (polychlorinated dibenzo-*p*-dioxins) contamination of bay organisms. The Texas Department of Health issued a seafood consumption advisory in 1990 for blue crabs (*Callinectes sapidus*) and catfish in the upper bay and upper Houston Ship Channel because of elevated dioxin levels. At the San Jacinto Monument, dioxin concentrations in blue crabs (*C. sapidus*) and blue catfish (*Ictalurus furcatus*) in the late 1980s amounted to 54.8 and 3.2 pg/g, respectively. At Morgans Point located farther downstream, dioxin concentrations in sea catfish (*Arius felis* and *Bagre marinus*) and eastern oysters (*Crassostrea virginica*) were 14.8 and 6.2 pg/g, respectively. A later study conducted in 1990 found dioxin levels as high as 3.97 pg/g in blue crabs (*Callinectes sapidus*), eastern oysters (*Crassostrea virginica*), and red drum (*Sciaenops ocellatus*) in the upper bay.

Another seafood consumption advisory was issued by the Texas Department of Health in late 1993 for all fish from Clear Creek. In this case, three toxic chemical compounds — all industrial solvents (i.e., dichloroethane, trichloroethane, and carbon disulfide) — were documented in fish near a USEPA superfund site. These contaminants have been linked to cancer of the liver and kidney (i.e., dichloroethane and trichloroethene) as well as to nervous disorders (i.e., carbon disulfide).

The concentrations of trace metals recorded by Brooks et al.[52] in fish, crabs, and oysters were as follows: arsenic, 0.45 to 0.89 μg/g; cadmium, 0.168 to 0.310 μg/g; chromium, 0.022 to 0.211 μg/g; copper, 9.888 to 33.819 μg/g; lead, 0.04 to 0.06 μg/g; mercury, 0.079 to 0.123 μg/g; nickel, 0.04 to 0.20 μg/g; selenium, 0.88 to 2.27 μg/g; silver, 0.09 to 0.14 μg/g; and zinc, 68.7 to 115.2 μg/g. Crocker et al.[54] reported trace metal concentrations in fish and crabs from nine stations in the Houston Ship Channel. These concentrations were as follows: antimony, 3.3 to 4.2 mg/kg; arsenic, 0.25 to 4.2 mg/kg; chromium, 0.48 to 12 mg/kg; copper, 0.48 to 10 mg/kg; cyanide, <0.51 to 1.9 mg/kg; selenium, 0.61 to 14 mg/kg; silver, 0.48 to 1.5 mg/kg; and zinc, 4.8 to 51 mg/kg. The concentrations of antimony, arsenic, and selenium exceeded the USEPA fish tissue criteria. However, trace metals in the biotic samples from Galveston Bay rarely differed substantially from those of other Gulf of Mexico estuaries.

Aside from consumption advisories issued by health agencies that are designed to protect seafood consumers, a risk assessment study of seafood organisms has been conducted by Texas A&M University for the GBNEP. The purpose of this study was to determine the potential risk to seafood consumers. Eastern oysters (*Crassostrea virginica*), blue crabs (*Callinectes sapidus*), black drum (*Pogonias cromis*), southern flounder (*Paralicthys lethostigma*), and spotted seatrout (*Cynoscion nebulosus*) collected at the four locations in the estuary noted previously (see Brooks et al.[52]) were analyzed for contaminant concentrations. Results of this study reveal that most of the cancer risk to Galveston Bay seafood consumers is associated with PAHs and PCBs, with PCBs usually responsible for the larger portion of the overall risk. Of all organisms screened, eastern oysters (*Crassostrea virginica*) are the most contaminated, and blue crabs (*Callinectes sapidus*) are the least contaminated. The greatest risk is associated with consumption of seafood derived from the upper bay, where effects from the upper Houston Ship Channel and the Trinity River are apparent. The risk decreases for seafood obtained from down-bay locations.

VII. Management plan

Galveston Bay is an extremely valuable estuary — economically, ecologically, and aesthetically — that has experienced significant modifications for more than a century from a rapidly expanding human population. A major goal of the GBNEP has been to formulate an effective management plan to restore and maintain the estuarine system. To this end, the GBNEP produced the Galveston Bay Plan,[55] which is a Comprehensive Conservation and Management Plan designed to identify the most pressing problems of the estuary and to propose actions for resolving them. Clearly, degraded water quality, habitat loss and alteration, resource exploitation, and human-use activities threaten the ecological integrity of the system. It is essential to address these issues to assure that the resources of the estuary remain healthy and productive for future generations.

The GBNEP generated an Environmental Impact Matrix that illustrates the effect of natural and anthropogenic perturbations on resources and processes of the Galveston Bay ecosystem (Figure 4.8). This matrix demonstrates that the bay is a complex system affected by a myriad of factors. It has enabled the GBNEP to identify the key cause-and-effect relationships that require management attention to improve the system. Employing the matrix and various other information sources, the GBNEP developed a priority problems list for the estuary dealing with 17 different issues. Ranked in order from most to least important, these compelling problems are as follows:

1. Vital Galveston Bay habitats like wetlands have been lost or reduced in value by a range of human activities, threatening the bay's future sustained productivity.
2. Contaminated runoff from nonpoint sources degrades the water and sediments of bay tributaries and some nearshore areas.
3. Raw or partially treated sewage and industrial waste enters Galveston Bay due to design and operational problems, especially during rainfall runoff.
4. Future demands for freshwater and alterations to circulation may seriously affect productivity and overall ecosystem health.

Valued Ecosystem Components

Sources of Perturbation	Water Quality	Circulation	Sediment	Phytoplankton	Zooplankton	Oysters	Shellfish	Other Benthos	Finfish	Birds	Marine Mammals	Sea Turtles	Human Health	Wetlands	Submerged Plants	Shoreline	Aesthetic Appeal
Northers		**		?	?	*			**	*							
Hurricanes		**	*	?	?	*	*	**		*			?	?	***	***	
Inflow Modification	***	***	*	?	?	****	***	***	**			?		***	**		
Subsidence/Sea Level		**				*	**		*	*				****	***	****	
Shoreline Development	**	*	*	*			**		**	**				****	**	****	***
Dredging	***	****	****	?		**	*	**	**	***	?	?	?	***	**	***	**
Shipping	**		*								?			**		**	
Point Sources	****		****	***	**	***	**	**	**	**	?	?	****	*	**		**
Non-Point Sources	****		****	***	?	***	**	**	**	**	?	?	***	**	**		**
Commercial Fishing	?		?			**	****	?	***		?	?			**		
Recreational Fishing						*	*		***					?	*		
Boating/Marinas	***		***	?	?			**	*					*	*	*	?
Petroleum Activity	***		***	?	?	*	**	**	*	*	?	?	*	**	*		?
Oil/Chemical Spills	***		***	?	?	**	?	?	?	**	?	?	**	***	?		***
Marine Debris									?	*	*	**					***

NOTE: * = Slight influence; ** = Moderate influence; *** = Significant influence; **** = Major influence; ? = Unknown relationship; (shaded) = Possible management priority

Figure 4.8 Environmental Impact Matrix for Galveston Bay. (From GBNEP, *The Galveston Bay Plan: The Comprehensive Conservation, and Management Plan for the Galveston Bay Ecosystem*, Galveston Bay National Estuary Program Publication GBNEP-49, Webster, TX, 1994.)

5. Certain toxic substances have contaminated water and sediment and may have a negative effect on aquatic life in contaminated areas.
6. Certain species of marine organisms and birds have shown a declining population trend.
7. Shoreline management practices frequently do not address negative environmental consequences to the bay or the need for environmentally compatible public access to bay resources.
8. Bay habitats and living resources are impacted by spills of toxic and hazardous material during storage, handling, and transport.
9. Seafood from some areas in Galveston Bay may pose a public health risk to subsistence or recreational catch seafood consumers as a result of the potential presence of toxic chemicals.
10. Illegal connections to storm sewers introduce untreated wastes directly into bay tributaries.
11. Dissolved oxygen is reduced in certain tributaries and side bays, harming marine life.
12. About half of the bay is permanently or provisionally closed to shellfish harvesting because of high fecal coliform bacterial levels that may indicate risk to shellfish consumers.

13. Water and sediments are degraded in and around marinas from boat sewage and introduction of dockside wastes from nonpoint sources.
14. Some bay shorelines are subject to high rates of erosion and loss of stabilizing vegetation due to past subsidence/sea level rise and current human impacts.
15. Illegal dumping as well as waterborne and shoreline debris degrade water quality and aesthetics of Galveston Bay.
16. Some tributaries and nearshore areas of Galveston Bay are not safe for contact recreational activities such as swimming, wade-fishing, and sail-boarding due to risk of bacterial infection.
17. Some exotic/opportunistic species threaten desirable native species, habitats, and ecological relationships.[55]

The GBNEP established goals to address the aforementioned list of priority problems (Table 4.12). These goals form the basis for the management actions outlined in the *Galveston Bay Plan*.[55] They are subdivided into three categories: (1) water/sediment quality improvement; (2) habitat/living resources conservation; and (3) balance human uses. Within each management category, goals are listed in order of their priority. Priority levels are established to identify the relative importance of the goals in comprehensive planning to solve the 17 key problems facing the bay system.

Population growth, watershed development, and industrial expansion have all contributed significantly to the array of environmental problems that have plagued Galveston Bay for decades. Between 1850 and 1990, the population in the five-county area surrounding the bay increased by 3.3 million. Activities of this burgeoning population placed considerable stress on living resources and habitats of the system. Water quality was compromised as the bay became overburdened by wastes, especially in the upper Houston Ship Channel. By the 1960s, the channel was one of the most polluted waterways in the country, nearly devoid of oxygen and aquatic life in its upper 25 km. It was impacted by wastes from municipal facilities, petroleum and related industries, and shipping operations.

Water quality in the Galveston Bay system improved substantially after passage of the nation's first Water Quality Act in 1965. Municipal and industrial cleanup efforts, costing millions of dollars, accelerated through the 1970s to comply with stringent regulatory controls. As a consequence, many environmental problems in the upper Houston Ship Channel and inner bay were greatly mitigated or eliminated entirely.

Although point source pollution is no longer the principal cause of concern in the Galveston Bay system, a number of potentially serious problems still remain. Particularly noteworthy are: (1) pervasive habitat loss that threatens aquatic and wildlife populations; (2) increasing demands and overuse of the bay's resources; and (3) degraded water and sediment quality primarily associated with nonpoint sources of pollution. Not to be forgotten, however, are wastes that continue to enter the system from municipal and industrial sources. For example, the bay currently receives ~60% of all wastewaters discharged in Texas. Petroleum and related industries as well as shipping operations also are responsible for the inputs of some highly toxic chemical contaminants, which have accumulated in bottom sediments in some parts of the upper Houston Ship Channel and other urbanized areas.

Specific actions have been proposed by the GBNEP to address the priority problems noted previously that currently threaten the future health and productivity of the estuary. For example, the *Galveston Bay Plan*[55] proposes 9 actions for fish and wildlife habitat protection and 10 actions for species population protection. These actions are recounted subsequently together with those that target human uses of the bay (i.e., freshwater inflow, circulation, and shoreline development) as well as water and sediment quality issues (i.e., impacts coupled to point and nonpoint sources of pollution).

Table 4.12 Ranked Problems and Goals of the Galveston Bay Estuary Program

Priority Level	Water/Sediment Quality Improvement	Habitat/Living Resource Conservation	Balanced Human Uses
Very High	Reduce urban NPS pollutant loads. Reduce toxicity and contaminant concentrations in water and sediments. Eliminate wet weather sewage bypasses/overflows.	Increase the quantity and improve the quality of wetlands for fish and wildlife. Eliminate or mitigate the conversion of wetlands to other uses caused by human activities.	Ensure beneficial freshwater inflows necessary for a salinity, nutrient, and sediment loading regime adequate to maintain productivity of economically important and ecologically characteristic species in Galveston Bay.
High	Eliminate pollution problems from poorly operated wastewater treatment plants. Restore and/or compensate for environmental damage (injury) resulting from discharges of oil or the release of hazardous substances. Eliminate illegal connections to storm sewers, which result in introduction of untreated wastes directly to bay tributaries. Increase dissolved oxygen in problem areas.	Acquire existing wetland habitats and provide economic incentives for conversation. Reverse the declining population trend for affected species of marine organisms and birds, and maintain the populations of other economic and ecologically important species.	Reduce potential health risk resulting from consumption of seafood contamination with toxic substances. Reduce negative environmental consequences to the bay (i.e., human-induced erosion) from shoreline development.
Moderate	Reduce agricultural NPS pollutant loads. Reduce industrial NPS pollutant loads. Reduce marina water quality degradation associated with sewage. Reduce marina/dockside NPS loads.	Selectively moderate erosional impacts to the bay and associated shorelines. Increase productivity of oyster reefs in West Bay. Restore deteriorated colonial bird nesting islands to usefulness and create new islands for birds where nesting habitat is inadequate.	Reduce oyster reef harvest closures. Ensure that alterations to circulation do not negatively affect productivity and overall ecosystem health.
Low	Reduce construction NPS pollutant loads. Reduce the impact from spills on the natural environment. Eliminate illegal dumping. Eliminate waterborne debris.	Eradicate or reduce the populations of exotic/opportunistic species which threaten desirable native species, habitats, and ecological relationships. Prevent the introduction of additional exotic species.	Reduce risk of waterborne illness resulting from contact recreation. Increase environmentally compatible public access to bay resources.

Source: GBNEP, *The Galveston Bay Plan: The Comprehensive Conservation and Management Plan for the Galveston Bay Ecosystem*, Galveston Bay National Estuary Program Publication GBNEP-49, Webster, TX, 1994.

A. Galveston Bay actions

The following actions derived from the *Galveston Bay Plan*[55] have been proposed by the GPNEP to address priority environmental problems in the estuary. Funding to implement these actions will originate from federal, state, and private grants. Revenue from nonprofit foundations will also be used to support implementation of the actions.

1. Habitat protection

- HP-1 (high priority): Restore, create, and protect wetlands.
- HP-2 (high priority): Promote beneficial uses of dredged material to restore and create wetlands.
- HP-3 (high priority): Inventory degraded wetlands and fund remedial measures.
- HP-4 (high priority): Implement a coordinated system-wide wetland regulatory strategy.
- HP-5 (high priority): Acquire and protect quality wetlands.
- HP-6 (high priority): Develop economic and tax incentive programs to protect wetlands.
- HP-7 (high priority): Facilitate bird nesting on existing islands and beaches.
- HP-8 (high priority): Build nesting islands using dredged materials.
- HP-9 (low priority): Reduce erosional impacts on wetlands and habitats.

2. Species population protection

- SP-1 (medium priority): Implement a baywide effort to strengthen species management.
- SP-2 (medium priority): Return oyster shell to designated locations within the bay.
- SP-3 (medium priority): Promote the development of oyster reefs using alternate materials.
- SP-4 (medium priority): Set aside a portion of reef habitat as scientific research areas or preserves.
- SP-5 (medium priority): Encourage continued development of gear to reduce commercial by-catch.
- SP-6 (medium priority): Conduct educational programs about catch and release.
- SP-7 (medium priority): Investigate potential measures to reduce impingement and entrainment.
- SP-8 (medium priority): Develop management plans for endangered or threatened species.
- SP-9 (low priority): Improve enforcement of prohibitions against introduction of exotic species.
- SP-10 (low priority): Identify and implement techniques for the control of problem exotic species.

3. Public health protection

- PH-1 (medium priority): Develop a seafood consumption safety program.
- PH-2 (low priority): Enhance the Texas Department of Health Shellfish Sanitation Program.
- PH-3 (low priority): Develop a contact recreation advisory program.

4. *Freshwater inflow and bay circulation*

- FW-1 (high priority): Complete current studies to determine freshwater inflow needs for the bay.
- FW-2 (high priority): Expand streamflow, sediment loading, and rainfall monitoring.
- FW-3 (high priority): Establish management strategies for meeting freshwater inflow needs.
- FW-4 (high priority): Establish inflow regulations to protect the ecological needs of the estuary.
- FW-5 (high priority): Explore means of providing sediment to the estuary.
- FW-6 (high priority): Reduce water consumption.
- FW-7 (medium priority): Evaluate the effects of channels and structures on bay circulation, habitats, and species.

5. *Spills/dumping*

- SD-1 (medium priority): Promote planning to facilitate natural resource damage assessments.
- SD-2 (medium priority): Identify simplified procedures for damage assessment for small oil spills.
- SD-3 (medium priority): Facilitate effective restoration of Galveston Bay's natural resources damaged by spills.
- SD-4 (medium priority): Facilitate spill cleanup by advance shoreline characterization.
- SD-5 (low priority): Improve trash management near the shoreline.
- SD-6 (low priority): Remove trash from stormwater discharges.
- SD-7 (low priority): Publicize environmental harm caused by illegal dumping.

6. *Shoreline management*

- SM-1 (medium priority): Establish a planning program for shoreline development.
- SM-2 (medium priority): Identify appropriate residential shoreline development guidelines.
- SM-3 (medium priority): Identify appropriate commercial and industrial shoreline development guidelines.
- SM-4 (medium priority): Minimize negative effects of structures and dredging on publicly owned lands.
- SM-5 (medium priority): Improve access to publicly owned shorelines.

7. *Water and sediment quality*

- WSQ-1 (high priority): Reduce contaminant concentrations to meet standards and criteria.
- WSQ-2 (high priority): Determine sources of ambient toxicity in water and sediment.
- WSQ-3 (high priority): Establish and adopt sediment quality criteria.
- WSQ-4 (high priority): Perform Total Maximum Daily Load (TMDL) studies for toxics.
- WSQ-5 (high priority): Support the Clean Texas 2000 Pollution Prevention Program.
- WSQ-6 (medium priority): Reduce nutrient and BOD loadings to problem areas.
- WSQ-7 (medium priority): Perform TMDL loading studies for oxygen demand and nutrients.

8. Nonpoint sources of pollution

- NPS-1 (high priority): Implement stormwater programs for local municipalities.
- NPS-2 (high priority): Perform pilot projects to develop nonpoint source best management practices.
- NPS-3 (high priority): Identify and correct priority watershed pollutant problems.
- NPS-4 (high priority): Establish residential load reduction programs.
- NPS-5 (high priority): Correct malfunctioning shoreline septic tanks.
- NPS-6 (high priority): Implement a nonpoint source reduction plan program for new development.
- NPS-7 (high priority): Establish roadway planning to minimize nonpoint source effects.
- NPS-8 (high priority): Implement a NPDES stormwater program for area industries.
- NPS-9 (high priority): Prevent degradation of bay waters by known industrial groundwater plumes.
- NPS-10 (high priority): Develop inventory of agricultural nonpoint sources.
- NPS-11 (high priority): Coordinate/implement existing agricultural nonpoint source control programs.
- NPS-12 (high priority): Adopt regional construction standards for nonpoint source reduction.
- NPS-13 (high priority): Implement toxics and nutrient control practices at construction sites.
- NPS-14 (medium priority): Require sewage pumpout, storage, and provisions for treatment.
- NPS-15 (low priority): Require marine sanitary chemicals that can be treated in publicly owned treatment works.
- NPS-16 (low priority): Implement washdown controls and containment measures.

9. Point sources of pollution

- PS-1 (high priority): Determine the location and extent of bypass and overflow problems.
- PS-2 (high priority): Eliminate or reduce bypass and overflow problems.
- PS-3 (high priority): Regionalize small wastewater treatment systems.
- PS-4 (high priority): Improve compliance monitoring and enforcement for small dischargers.
- PS-5 (medium priority): Implement a dry-weather illegal connection program.
- PS-6 (medium priority): Issue NPDES Coastal General Permits or eliminate harm from oil field discharge.

10. Research

- RSC-1: Establish a research coordination board.
- RSC-2: Identify research needs from an ecosystem perspective.
- RSC-3: Continue the state-of-the-bay process.
- RSC-4: Increase funding for Galveston Bay research.

11. Public participation and education

- PPE-1: Establish citizen involvement as an integral part of the Galveston Bay Program.
- PPE-2: Continue and expand the state-of-the-bay symposia.

- PPE-3: Develop and implement a long-range adult education and outreach program.
- PPE-4: Develop specific curricula for use in Galveston Bay watershed school districts.
- PPE-5: Continue to develop effective volunteer opportunities for citizens.
- PPE-6: Maintain a citizen pollution reporting system.
- PPE-7: Develop and implement a strategy for informing, educating, and providing support for local government involvement.
- PPE-8: Provide assistance for user groups affected by implementation of the Galveston Bay Plan.

References

1. Shipley, F. S. and Kiesling, R. W., Eds., *The State of the Bay: A Characterization of the Galveston Bay Ecosystem*, Galveston Bay National Estuary Program Publication GBNEP-44, Webster, TX, 1994.
2. Ward, G. H., *Galveston Bay Hydrography and Transport Model Validation*, Tech. Rept., National Oceanic and Atmospheric Administration, Strategic Assessment Branch, Rockville, MD, 1991.
3. White, W. A., Calnan, T. R., Morton, R. A., Kimble, R. S., Littleton, T. G., McGowen, J. H., Nance, H. S., and Schmedes, K. E., *Submerged Lands of Texas, Galveston–Houston Area: Sediments, Geochemistry, Benthic Macroinvertebrates, and Associated Wetlands*, Tech. Rept., Bureau of Economic Geology, University of Texas, Austin, TX, 1985.
4. Powell, E. N., Song, J., and Ellis, M., *The Status of Oyster Reefs in Galveston Bay, Texas*, Galveston Bay National Estuary Program Publication GBNEP-37, Webster, TX, 1994.
5. Paine, J. G. and Morton, R. A., Historical shoreline changes in the Galveston Bay system, in *Proceedings of the Galveston Bay Characterization Workshop*, Shipley, F. S. and Kiesling, R. W., Eds., Galveston Bay National Estuary Program Publication GBNEP-6, Webster, TX, 1991, 165.
6. U.S. Bureau of the Census, *Census of Population: 1990*, U.S. Department of Commerce, U.S. Government Printing Office, Washington, D.C., 1991.
7. Ditton, R. B., Loomis, D. K., Fesenmaier, D. R., Osborn, M. O., Hollin, D., and Kolb, J. W., Galveston Bay and the surrounding area: human uses, production, and economic values, in *Galveston Bay: Issues, Resources, Status, and Management*, NOAA Estuary-of-the-Month Seminar Series No. 13, National Oceanic and Atmospheric Administration, Washington, D.C., 1989, 53.
8. Houston-Galveston Area Council, *Galveston Bay Area Socioeconomic Planning Data*, Houston-Galveston Area Council, Houston, TX, 1993.
9. Newell, C. J., Rifai, H. S., and Bedient, P. B., *Characterization of Nonpoint Sources and Loadings to Galveston Bay*, Volume 1, Galveston Bay National Estuary Program Publication GBNEP-15, Webster, TX, 1992.
10. Allison, R. C., Durand, R., Hill, R., Coppenbarger, K., Hameed, N., and Willey, B., *A Socioeconomic Characterization of the Galveston Bay System*, Tech. Rept., Galveston Bay National Estuary Program, Webster, TX, 1991.
11. Whittington, D., Amaral, D., and Cassidy, G., Research on economic valuation studies in Galveston Bay, in *Proceedings of the Second State of the Bay Symposium*, Jensen, R., Kiesling, R. W. and Shipley, F. S., Eds., Galveston Bay National Estuary Program Publication GBNEP-23, Webster, TX, 1993, 361.
12. Texas Railroad Commission, Personal Communication to the Galveston Bay National Estuary Program, Webster, TX, 1993.
13. Osburn, H. R., Quast, W. D., and Hamilton, C. L., *Trends in Texas Commercial Fishery Landings, 1977–1986*, Management Data Series No. 131, Texas Parks and Wildlife Department, Austin, TX, 1987.
14. Texas Department of Water Resources, *Plan Summary for the Trinity Basin Water Quality Management Plan*, Texas Department of Water Resources Report LP-149, Austin, TX, 1981.
15. Texas Water Commission, *Governor's Supplemental Nomination of Galveston Bay as an Estuary of National Significance*, Tech. Rept., Texas Water Commission, Austin, TX, 1988.
16. Osburn, H. R and Ferguson, M. O., *Trends in Finfish Landings by Sport-Boat Fishermen in Texas Marine Waters, May 1974–May 1985*, Management Data Series No. 90, Tech. Rept., Texas Parks and Wildlife Department, Austin, TX, 1986.

17. Eubanks, T. L., Ecotourism in Galveston Bay: an economic opportunity, in *Proceedings of the Second State of the Bay Symposium*, Jensen, R., Kiesling, R. W., and Shipley, F. S., Eds., Galveston Bay National Estuary Program Publication GBNEP-23, Webster, TX, 1993, 367.
18. Ward, G. H. and Armstrong, N. E., *Ambient Water and Sediment Quality of Galveston Bay: Present Status and Historical Trends*, Galveston Bay National Estuary Program Publication GBNEP-22, Webster, TX, 1992.
19. Armstrong, N. E. and Ward, G. H., *Point Source Loading Characterization of Galveston Bay*, Galveston Bay National Estuary Program Publication GBNEP-36, Webster, TX, 1994.
20. Fay, R. R., Sweet, S., and Wilson, R. J., *Shoreline Survey for Unpermitted Discharges to Galveston Bay*, Galveston Bay National Estuary Program Publication GBNEP-12, Webster, TX, 1991.
21. Winslow and Associates, Inc., *Houston Ship Channel Urban Runoff Nonpoint Source Study: Relative Significance of Waste Loads Entering the Houston Ship Channel*, Texas Water Commission Report, Austin, TX, 1986.
22. Guillen, G., Phillips, D., Harper, J., and Larson, J., *Partially Treated and Untreated Effluent Loadings on Galveston Bay*, Galveston Bay National Estuary Program Publication GBNEP-41, Webster, TX, 1994.
23. Roach, R. W., Carr, R. S., and Howard, C. L., An assessment of produced water impacts at two sites in the Galveston Bay system, in *Proceedings of the Second State of the Bay Symposium*, Jensen, R., Kiesling, R. W., and Shipley, F. S., Eds., Galveston Bay National Estuary Program Publication GBNEP-23, Webster, TX, 1993, 135.
24. Guillen, G., Smith, S., Broach, L., and Ruckman, M., The impacts of marinas on the water quality of Galveston Bay, in *Proceedings of the Second State of the Bay Symposium*, Jensen, R., Kiesling, R. W., and Shipley, F. S., Eds., Galveston Bay National Estuary Program Publication GBNEP-23, Webster, TX, 1993, 33.
25. Schueler, T. R., *Controlling Urban Runoff: A Practical Manual for Planning and Designing Urban BMPs*, Washington Metropolitan Water Resources Planning Board Report, Metropolitan Washington Council of Governments, Department of Environmental Programs, Washington, D.C., 1987.
26. Ward, G. H. and Armstrong, N. E., *Galveston Bay Data Base Inventory*, Galveston Bay National Estuary Program Publication GPNEP-40, Webster, TX, 1994.
27. Carr, R. S., *Sediment Quality Assessment Survey of the Galveston Bay System*, Galveston Bay National Estuary Program Publication GBNEP-30, Webster, TX, 1993.
28. Goodman, T. M., *Estimates of Toxic Material Loading to the Galveston Bay System*, M.S. Thesis, University of Texas, Austin, TX, 1989.
29. Jensen, P., Valentine, S., Garrett, M. T., Jr., and Ahmed, Z., Nitrogen loads to Galveston Bay, in *Proceedings of the Galveston Bay Characterization Workshop*, Shipley, F. S. and Kiesling, R. W., Eds., Galveston Bay National Estuary Program Pubication GBNEP- 6, Webster, TX, 1991, 99.
30. Martin, W. D., 3-D hydrodynamic model of Galveston Bay, in *Proceedings of the Second State of the Bay Symposium*, Jensen, R., Kiesling, R. W., and Shipley, F. S., Eds., Galveston Bay National Estuary Program Publication GBNEP-23, Webster, TX, 1993, 327.
31. Ward, G. H., *Dredge-and-Fill Activities in Galveston Bay*, Galveston Bay National Estuary Program Publication GBNEP-28, Webster, TX, 1993.
32. Pulich, W. M., White, W. A., Castiglione, M., and Zimmerman, R. J., Status of submerged vegetation in the Galveston Bay System, in *Proceedings of the Galveston Bay Characterization Workshop*, Shipley, F. S. and Kiesling, R. W., Eds., Galveston Bay National Estuary Program Pubication GBNEP-6, Webster, TX, 1991, 127.
33. White, W. A., Tremblay, T. A., Wermund, E. G., Jr., and Handley, L. R., *Trends and Status of Wetland and Aquatic Habitats in the Galveston Bay System, Texas*, Galveston Bay National Estuary Program Publication GBNEP-31, Webster, TX, 1993.
34. White, W. A. and Paine, J. G., *Wetland Plant Communities, Galveston Bay System*, Galveston Bay National Estuary Program Publication GBNEP-16, Webster, TX, 1992.
35. Texas Department of Water Resources, *Trinity-San Jacinto Estuary: A Study of the Influence of Freshwater Inflows*, Texas Department Water Resources Report LP-113, Texas Department of Water Resources, Austin, TX, 1981.

36. Sheridan, P. F., Slack, R. D., Ray, S. M., McKinney, L. W., Klima, E. F., and Calnan, T. R., Biological Components of Galveston Bay, in *Galveston Bay: Issues, Resources, Status, and Management*, NOAA Estuary-of-the-Month Seminar Series No. 13, National Oceanic and Atmospheric Administration, Washington, D.C., 1989, 23.
37. Zotter, J., *Species Composition and Seasonal Occurrence of Nanoplankton in the Galveston Bay Estuary,* M.S. Thesis, Texas A&M University, College Station, TX, 1979.
38. Green, A., Osborn, M., Chai, P., Lin, J., Loeffler, C., Morgan, A., Rubec, P., Spanyers, S., Walton, A., Slack, R. D., Gawlik, D., Harpole, D., Thomas, J., Buskey, E., Schmidt, K., Zimmerman, R., Harper, D., Hinkley, D., Sager, T., and Walton, A., *Status and Trends of Selected Living Resources in the Galveston Bay System*, Galveston Bay National Estuary Program Publication GBNEP-19, Webster, TX, 1992.
39. Harper, D. E. and Guillen, G., Occurrence of a dinoflagellate bloom associated with an influx of low salinity water at Galveston, Texas and coincident mortalities of demersal fish and benthic invertebrates, *Contr. Mar. Sci.*, 31, 147, 1989.
40. Lee, W. Y., Arnold, C. R., and Kalke, R. D., *Synthesis of Data on Acartia tonsa in Texas Bay Systems: Correlation Between its Abundance and Selected Environmental Factors*, Tech. Rept., Texas Water Development Board, Austin, TX, 1986.
41. Levinton, J. S., *Marine Ecology,* Prentice-Hall, Englewood Cliffs, NJ, 1982.
42. Pinet, P. R., *Invitation to Oceanography,* Jones and Bartlett, Boston, 1998.
43. Parker, R. H., Ecology and distributional patterns of marine macroinvertebrates, Northern Gulf of Mexico, in *Recent Sediments, Northwest Gulf of Mexico*, Sherpard, F. P., Phleger, F. B., and van Andel, T. H., Eds., American Association of Petroleum Geologists, Tulsa, OK, 1960, 203.
44. McBee, J. T., *Species Composition, Distribution, and Abundance of Macrobenthic Organisms in the Intake and Discharge Areas Before and After the Construction and Operation of the Cedar Bayou Electric Power Station*, Ph.D. Thesis, Texas A&M University, College Station, TX, 1975.
45. Ray, G. L., Clarke, D. G., and Bass, R. J., Characterization of benthic assemblages of Galveston Bay: 1990–1992, in *Proceedings of the Second State of the Bay Symposium*, Jensen, R., Kiesling R. W., and Shipley, F. S., Eds., Galveston Bay National Estuary Program Publication GBNEP-23, Webster, TX, 1993.
46. Quast, W. D., Johns, M. A., Pitts, D. E., Matlock, G. C., and Clark, J. E., *Texas Oyster Fishery Management Plan*, Tech. Rept., Texas Parks and Wildlife Department, Austin, TX, 1988.
47. Walton, A. H and Green, A. W., *Probable Causes of Trends in Selected Living Resources in the Galveston Bay System*, Galveston Bay National Estuary Program Publication GBNEP-33, Webster, TX, 1993.
48. Green, A., Walton, A., Osborn, M., Chai, P., Lin, J., Loeffler, C., Morgan, A., Rubec, P., Spanyers, S., Slack, R. D., Gawlik, D., Harpoole, D., Thomas, J., Buskey, E., Schmidt, K., Zimmerman, R., and Harper, D., Status, Trends, and Probable Causes for Changes in Living Resources in Galveston Bay System, in *Proceedings of the Second State of the Bay Symposium,* Jensen, R., Kiesling, R. W., and Shipley, F. S., Eds., Galveston Bay National Estuary Program Publication GBNEP-23, Webster, TX, 1993, 175.
49. Palafox, S. D. and Wolford, E. D., *Non-fishing/Human Induced Mortality of Fisheries*, Galveston Bay National Estuary Program Publication GBNEP-29, Webster, TX, 1993.
50. Arnold, K. A., *Checklist of Birds of Texas*, Texas Ornithological Society, Austin, TX, 1984.
51. Jensen, P., *Characterization of Selected Public Health Issues in Galveston Bay,* Galveston Bay National Estuary Program Publication GBNEP-21, Webster, TX, 1992.
52. Brooks, J. M., Wade, T. L., Kennicutt, M. C., II, Wiesenburg, D. A., Wilkinson, D., McDonald, T. J., and McDonald, S. J., *Toxic Contaminant Characterization of Aquatic Organisms in Galveston Bay: A Pilot Study,* Galveston Bay National Estuary Program Publication GBNEP-20, Webster, TX, 1992.
53. Texas Water Commission, *Texas Surface Water Quality Standards*, Sections 307.2-307.10, Permanent Rule Changes, Austin, TX, 1991.

54. Crocker, P. A., Guillen, G. J., Seiler, R. D., Petrocelli, E., Redmond, M., Lane, W., Hollister, T. A., Neleigh, D. W., and Morrison, G., *Water Quality Ambient Toxicity, and Biological Investigations in the Houston Ship Channel and Tidal San Jacinto River,* U.S. Environmental Protection Agency Region 6 Report, U.S. Environmental Protection Agency, Dallas, TX, 1991.
55. GBNEP, *The Galveston Bay Plan: The Comprehensive Conservation, and Management Plan for the Galveston Bay Ecosystem*, Galveston Bay National Estuary Program Publication GBNEP-49, Webster, TX, 1994.

chapter five

Case study 4: San Francisco Estuary Project

I. Introduction

San Francisco Bay and the Sacramento-San Joaquin Delta form the largest estuarine system on the Pacific Coast of the U.S. The bay itself covers an area of ~1125 km^2, and the delta covers an area of ~2985 km^2. The bay is subdivided into several distinct segments based on hydrographic and geographic characteristics: Suisun Bay, Carquinez Strait, San Pablo Bay, and San Francisco Bay (Figure 5.1, Table 5.1). The northern and southern parts of San Francisco Bay are informally referred to as the Central Bay and South Bay, respectively. Located at the confluence of the Sacramento and San Joaquin Rivers, the Sacramento-San Joaquin Delta appears as a complex triangular-shaped region of land and water consisting of interconnected embayments, sloughs, marshes, channels, and streams. It consists of three main segments: the northern delta (dominated by waters of the Sacramento River), southern delta (dominated by waters of the San Joaquin River), and eastern delta (dominated by waters of the Cosumnes and Mokelumne Rivers). The delta is a valuable agricultural area, although it once supported an enormous tidal freshwater marsh.

The estuary receives 47% of the total runoff of California from an extensive drainage basin (~153,000 km^2) that covers more than 40% of the state's land surface area. It supplies drinking water for 20 million people (approximately two thirds of the state population). In addition, it provides irrigation for ~1.82×10^6 ha of farmland.[1] Most of the freshwater inflow (~90%) derives from rivers and streams of the Central Valley and drainage from the Trinity River. Among the influent systems of the Central Valley, the Sacramento and San Joaquin Rivers are most important, accounting for 80 and 15% of the total discharge, respectively. The annual freshwater inflow from the Sacramento River amounts to ~2.38×10^{10} m^3, and that from the San Joaquin River, ~4.19×10^9 m^3.[2] The remaining 10% of the freshwater input originates from tributaries and local sources surrounding the bay.

The region bordering San Francisco Bay is the fourth largest metropolitan center in the U.S. During the past century, the estuary has been strongly influenced by urbanization of watershed areas. Nearly 30% of the land in nine counties surrounding the bay and 10% of the land in three delta counties are now urbanized. Several major cities (e.g., Oakland, Palo Alto, San Jose, and San Francisco) border the bay, and many residents utilize the system for diverse activities, such as recreational boating, commercial fishing, farming, and shipping. More than 7.5 million people reside in the 12 aforementioned counties. An additional 2 million people live in the Central Valley portion of the bay watershed. Approximately one third of the entire population of California inhabits land areas that

Table 5.1 Bathymetric Data for San Francisco Bay

Region	Surface Area[a,b]	Mean Depth[c]	Mean Volume[d]
Suisun Bay	93.2	4.3	323,000
Carquinez Strait	31.1	8.8	223,000
San Pablo Bay	271.8	2.7	605,000
Central Bay	266.6	10.7	2,307,000
South Bay	554.0	3.4	1,507,000
Total	1216.7	5.2	4,965,000

[a] At mean low water including saturated mudflats.

[b] Values in km^2.

[c] Values in m.

[d] Values in m^3.

Source: Cheng, R. T. and Gartner, J. W., Tides, Tidal and Residual Currents in San Francisco Bay, California — Results of Measurements, 1979–1980. Part 1: Description of Data, Water Resources Investigations Report 84-4339, U.S. Geological Survey, Menlo Park, CA, 1984.

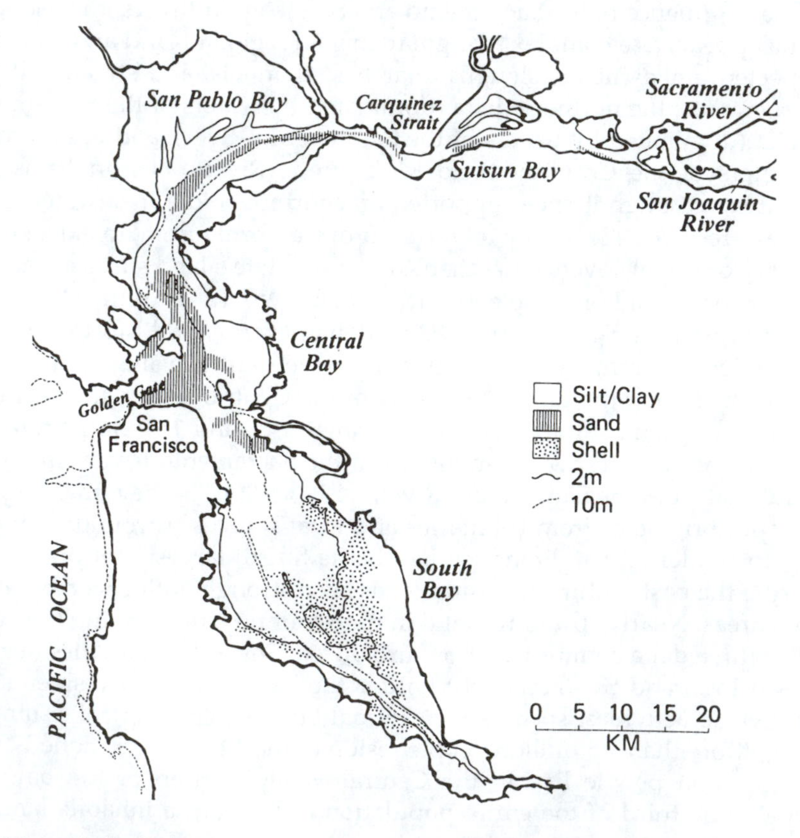

Figure 5.1 Map of the San Francisco Bay/Delta showing bottom sediment composition. (From Nichols, F.H. and Thompson, J.E., Persistence of an introduced mudflat community in south San Francisco Bay, California, *Mar. Ecol. Prog. Ser.*, 24, 83, 1985.)

drain into the delta and bay. The population in the bay watershed is projected to reach 12 million by the year 2005, which will undoubtedly result in additional conversion of productive agricultural, range, and forest habitat to urbanized lands.

With increased human development and intensified uses of resources, the estuary has experienced numerous anthropogenic impacts. For example, ~2020 km^2 of tidal wetlands have been lost or converted by human development. Freshwater diversions have dramatically altered the flow regime of the estuary and affected the water quality and habitat conditions for numerous organisms. Sewage treatment plants, industrial facilities, abandoned mines, farmlands, and urban streets contribute thousands of kilograms of pollutants to the delta and bay. Many chemical contaminants that reach the bay are carcinogenic, teratogenic, or mutagenic to large numbers of estuarine organisms. Dredging of waterways to maintain adequate depths for the navigation of commercial and recreational vessels damages benthic habitats. Dredging operations annually remove more than 6×10 m^3 of sediment from the estuary. These operations directly impact benthic communities, promote dispersal of sediment contaminants, and also modify circulation patterns and salinity distributions.

Some human-induced changes in San Francisco Bay and the surrounding watershed have been catastrophic, causing acute losses of valuable living resources. Water quality degradation and habitat losses have severely decreased the abundance of some commercially important species. For example, the American shad (*Alosa sapidissima*) and striped bass (*Morone saxatilis*) no longer support viable commercial fisheries. Shellfish are now only recreationally important. The introduction of exotic species has disrupted the natural ecological balance of the estuary. A number of these exotic forms have displaced resident taxa. State and federal government agencies have authorized special protection for 89 dwindling wildlife species. It is feared that more fish and wildlife populations will be threatened as development and human activities escalate in the estuary basin. Improved management of the watershed and estuary is necessary to reverse declining abundances of living resources in the system.

II. Goals

The San Francisco Estuary Project (SFEP) was established in 1987 to promote more effective management of the San Francisco Bay-Delta Estuary and to restore and maintain the water quality and natural resources of the system. The 5-year cooperative effort, jointly sponsored by the state of California and U.S. Environmental Protection Agency (USEPA), brought together more than 100 individuals from the public and private sectors, including representatives from government, industry, business, and academic institutions, to address the critical environmental issues of the estuary. To provide direction and purpose for this demanding effort, the SFEP Management Committee adopted four major goals in 1987:

- Develop a comprehensive understanding of environmental and public health values attributable to the bay and delta and how these values interact with social and economic factors
- Achieve effective, united, and ongoing management of the bay and delta
- Recommend priority corrective actions and compliance schedules addressing point and nonpoint sources of pollution
- Formulate a Comprehensive Conservation and Management Plan (CCMP) to restore and maintain the chemical, physical, and biological integrity of the bay and delta, including restoration and maintenance of water quality, as well as a balanced indigenous population of shellfish, fish, and wildlife, while supporting recreational activities in the bay and delta

Development of the CCMP was guided by the following 10 mission statements:

- Restore and protect a diverse, balanced, and healthy population of fish, invertebrates, wildlife, plants, and their habitats, focusing on indigenous species
- Assure that the beneficial uses of the bay and delta are protected
- Improve water quality, where possible, by eliminating and preventing pollution at its source while minimizing the discharge of pollutants from point and nonpoint sources and remediating existing pollution
- Oversee dredging and waterway modifications to minimize adverse environmental impacts
- Effectively manage and coordinate land and water use to achieve the goals of the estuary project
- Increase public knowledge about the estuary and use that knowledge to better manage the estuary
- Develop and expand nonregulatory programs, such as public–private partnerships and market incentives, in conjunction with regulatory programs, to achieve the goals of the project
- Preserve and restore wetlands to provide habitat for wildlife, to improve water quality, and to protect against flooding
- Ensure an adequate freshwater flow as one of the essential components to restore and maintain a clean, healthy, and diverse estuary

Five primary issues of environmental concern form the basis for the program areas of the CCMP. These include: (1) intensified land use; (2) decline of biological resources; (3) freshwater diversion and altered flow regime; (4) increased pollutants; and (5) dredging and waterway modification. Completed in 1993, the CCMP offers an effective strategy for improving environmental conditions and achieving and maintaining an ecologically diverse and productive estuarine system. Implementation of the recommended actions in the CCMP has been under way since its approval by the Administrator of the USEPA in December 1993.

III. Physical and chemical characteristics

A. Circulation

The volume and timing of freshwater inflow differ considerably in the northern and southern reaches of San Francisco Bay. In the northern reach (i.e., Suisun, San Pablo, and Central Bays as well as the 18-km-long Carquinez Strait) waters are well mixed during periods of low freshwater discharge from the delta and partially mixed at times of greater delta outflow. Tidal currents and winds play an important role in mixing the water column. San Francisco Bay is a shallow system, and strong winds common in the region can completely dominate short-term circulation patterns. The bay is characterized by semidiurnal tides. An average of 1.6×10^9 m^3 of water flows in and out of the bay during every tidal cycle.[1]

The position of the null zone (i.e., the most landward zone of gravitational circulation marking the location where landward- and seaward-flowing bottom currents converge)[1,3,4] in the northern reach depends on delta outflow and the strength of tidal currents. When flow from the delta is low ($< \sim 140$ m^3/s), the null zone is found in the deeper waters of the Sacramento River. At moderate outflow rates (~ 280 m^3/s), it extends to the upstream end of Suisun Bay, and at high flow rates (> 565 m^3/s) it retreats to San Pablo Bay. Tidal currents, in turn, shift the position of the null zone ~3 to 9 km upestuary and downestuary twice per day and also translocate entrained suspended materials.

Water residence time in the northern reach likewise is contingent upon freshwater outflow from the delta. For example, water passes from the delta to the ocean in only 5 days at high outflow rates (~9900 m^3/s). However, the transit time increases to 60 to 90 days at lower outflow rates (~400 m^3/s).[5]

In contrast to the northern reach, South Bay is a tidally oscillating lagoon typified by lower freshwater inflow and a longer residence time.[6] South Bay receives negligible natural freshwater inputs (1×10^{11} m^3/yr) relative to that of the entire system (2.1×10^{16} m^3/yr). Wastewater discharges in the southern reach (5×10^{11} m^3/yr) exceed natural freshwater inflow and may completely dominate during protracted dry periods.[7] As reported by Conomos,[8] the volume of treated water from agricultural, industrial, and residential sources that enters the estuary during summer is much greater than that of natural freshwater inflow. Although wastewaters have a greater potential impact on this area of the estuary than elsewhere, thorough mixing of the effluent with ambient bay waters by wind action mollifies the risk.

The volume of delta outflow influences both stratification and residence time in South Bay.[9] Higher delta outflow rates (>~1000 m^3/s) promote water column stratification throughout most of South Bay. At somewhat lower rates, stratification may only be evident in the northern part of the bay. Water in South Bay takes more than 90 days to move northward into Central Bay during periods of low delta outflow. Under high outflow conditions, however, the residence time in South Bay is reduced to 14 to 21 days.[5] The residence time can significantly influence water quality in the estuary by modulating the distribution and fate of contaminants in the system. In summary, the strength, duration, and direction of estuarine flows as well as the occurrence of water column stratification hinge on several major factors, notably freshwater diversions, river inflow, tides, wind, and bottom topography that interact to produce the observed circulation patterns in the estuary.

B. Salinity

The San Francisco Bay/Delta Estuary is relatively shallow, with 70% of the system being 6 m deep or less. Suisun, San Pablo, and South Bays all have a mean depth of <5 m (Table 5.1). Central Bay is deeper, averaging ~10.5 m. Deepest areas occur along narrow (natural and dredged) channels that incise the bays to depths of 15 to 20 m, reaching maxima of 27 m and 110 m at Carquinez Strait and the Golden Gate Bridge, respectively.[10] To maintain navigable waterways for large commercial and military oceangoing vessels, main channels must be dredged to at least 10 to 15 m. Dredging is also conducted along approach channels to marinas so that recreational vessels have free access to berths.

Aside from assuring increased commercial and recreational usage of waterways in the system, dredging improves circulation in many areas, leading to more even distribution of salinity, temperature, dissolved oxygen, and other physical–chemical conditions. Deeper channels (e.g., Carquinez Strait) serve as conduits for strong upestuary bottom flows of saltwater. It is in these deeper waters where the stratified, two-layered flow appears to be strongest. However, salinity is also affected by other important factors, such as seasonal delta outflow, freshwater storage and diversions, and wind. Consequently, salinity varies greatly from the tributaries flowing into the delta along the eastern perimeter to the Golden Gate at the western margin. Salinity of the Sacramento River and San Joaquin River averages <0.1 and 0.4‰, respectively. Proceeding westward, the mean salinity increases from ~7‰ in Suisun Bay to ~30‰ in the Central Bay (at the Presidio).[11] Salinity concentrations in South Bay are more consistent year round, approaching those of the nearshore ocean. Only a small amount of freshwater enters South Bay from the surrounding watershed.[2]

Delta outflow peaks during the winter and spring months, lowering salinities throughout the northern reach and, periodically, in extensive portions of South Bay. Dryer conditions during the summer and fall reduce freshwater inputs to the estuary, culminating in saline water intrusion from the bay into the delta. However, salinity intrusion into the delta at this time is limited by releases of freshwater from upstream reservoirs. These releases ensure that the salinity of water diverted from the delta is acceptable for agricultural, industrial, and municipal uses.[12] During these dry seasons, salinity in South Bay may occasionally exceed that of the nearshore ocean due to accelerated rates of evaporation.

Freshwater diversions were responsible for the consistent increase of mean monthly salinity observed in some parts of Central Bay between 1922 and 1986.[2,13] Despite the increased diversion of freshwater over this long interval, the average annual delta outflow did not decline due largely to increased precipitation in the Central Valley watershed.[11,13] Water imports from outside the watershed, increased runoff related to changes in land use, and the redistribution of groundwater are other factors that may have contributed to consistently high delta outflows.[2]

C. Temperature

Water temperature ranges from ~14 to 24°C in the delta and 10 to 20°C in the bay.[8] Waters in the northern reach of the estuary are warmer than those in the ocean during the summer and cooler during the winter. Water temperatures in South Bay are intermediate between river and ocean temperatures.[14] Shallow areas of the bay are highly responsive to ambient air temperatures. Water released from upstream storage areas, river discharges, and wetlands vegetation strongly influence water temperature in the delta. The volume of freshwater inflow and seawater intrusion primarily modulates temperatures of bay waters.

D. Dissolved oxygen

San Francisco Bay is generally a well-oxygenated system. Summer oxygen depletion was frequently observed 20 to 30 years ago in the extreme southern perimeter of South Bay. High water temperatures as well as poor tidal mixing and flushing were principally responsible for the diminished oxygen levels in this area. However, oxygen depletion has not been common in recent years.[15]

E. Nutrients

Bay waters receive nutrients (nitrate, ammonium, phosphate, and silicate) from several sources, including the nearshore ocean, river discharges, runoff, sewage treatment plants, atmospheric deposition, and the estuarine basin. River inflow, runoff, and atmospheric deposition provide most of the nutrient input to the northern reach, whereas sewage treatment plants supply most of the nutrients (e.g., 80 to 97% of the nitrates and phosphates) to the southern reach. The highest nutrient concentrations in the system occur in South Bay during the summer–fall period when river inflow and runoff are low and dilution and mixing with Central Bay waters decline.[16] Hence, municipal wastewaters assume greater significance at this time in regard to nutrient influx.

An important effect of the sewage treatment plants is to reduce the annual variation of nutrient concentrations in South Bay. This is particularly true for nitrogen because phosphate and silicate concentrations are somewhat higher here in the summer months.[8] In contrast to the relatively uniform temporal distribution of nutrients in the southern reach of the estuary, seasonal variations of nutrients are conspicuous in the northern reach, where maximum nutrient levels occur in winter and minimum nutrient levels take place in summer.

Nutrient inputs to the estuary are vital to primary production in the system. A large fraction of the total organic carbon in the San Francisco Bay/Delta Estuary derives from autochthonous sources, notably production of phytoplankton, benthic microalgae, seagrasses, macroalgae, and photosynthetic bacteria. More than 90% of the organic carbon in South Bay originates from these autocthonous sources. In marked contrast, allochthonous organic carbon predominates in the northern bays; the San Joaquin and Sacramento Rivers deliver as much as 75% of the organic matter found here.[17] Thus, it is clear that the surrounding watershed significantly affects the organic carbon concentrations in some areas of the northern reach.

F. Pollutants

As in most urbanized estuaries, San Francisco Bay is impacted to varying degrees by an array of chemical contaminants from numerous sources such as municipal and industrial wastewater discharges, urban and nonurban runoff, agriculture wastes, oil spills and leakages, dredged materials, and atmospheric deposition. Watershed modifications (e.g., deforestation and construction, landscape partitioning and paving, marsh infilling and bulkheading, and dredge material disposal) not only destroy natural habitats but also facilitate nonpoint source runoff of many chemical contaminants to the estuary. The gradual transformation of natural cover to impervious surfaces by development and urbanization has notably increased stormwater pollutant loads. Both urban and nonurban runoff release considerable quantities of pollutants to the estuary.

Pollutants in the estuary may be subdivided into four main categories: (1) biological pollutants (e.g., sewage and pathogens); (2) inorganic chemicals (e.g., nutrients and trace metals); (3) organic chemicals (e.g., halogenated hydrocarbons and polycyclic aromatic hydrocarbons); and (4) suspended solids (e.g., sediments and organic particulates). The estuary receives an estimated 9.07×10^6 to 7.25×10^7 mt of pollutants each year that attain highest concentrations in industrial waterways, harbors, harbor entrances, and marinas.[2] Despite improvements in waste disposal and treatment facilities, chemical contamination persists in bottom sediments and organisms. Because many pollutants identified in San Francisco Bay are carcinogenic, teratogenic, or mutagenic, there is concern that humans who regularly consume contaminated seafood from some areas of the bay may be at increased risk of developing serious health problems.

G. Sediments

Bottom sediments provide important habitat for benthic invertebrates and demersal finfish populations. They also serve as repositories for pollutants and other materials. Therefore, the overall health of the system is closely coupled to conditions in the benthic environment.

The estuarine floor consists of clay, silt, sand, and shell (Figure 5.1). In the delta area, the relative composition of bottom sediments varies considerably, with clay and silt predominating in sloughs and quiet embayments and sand concentrating in channels. Suisun and San Pablo Bays are mainly characterized by muddy sediments, although the channel between Carquinez Strait and San Pablo Strait is largely sand. Bottom sediments change markedly in Central Bay. Here, much of the bottom is comprised of coarse sand, which in some areas forms extensive standing sand waves. Along the eastern shoreline of Central Bay, silts and clays are most abundant. Proceeding to South Bay, silts and clays cover more than 80% of the bottom, primarily in the central and western sectors.

Heavy concentrations of shell fragments occur on the bottom along the east side of South Bay.[2,18]

The Sacramento and San Joaquin Rivers transport large amounts of sediment to the estuary, much of which is deposited in the delta area. An estimated 3.0 to 3.2×10^6 mt/yr of sediment are deposited in the estuary each year.[1] Farm fields in the Central Valley as well as urban land development in watershed areas promote sediment input to the estuary.

IV. Anthropogenic impacts

A. Development and land use

The San Francisco Bay/Delta Estuary is more heavily impacted by human activity than most estuarine systems largely because of the intense agriculture activity and the long history of urban and industrial development in the region. With the watershed population approaching 10 million people, urban expansion has converted thousands of hectares of forests, rangeland, and wetlands to towns and cities. Land reclamation has substantially reduced the area of freshwater and tidal wetlands. Specifically, diking-and-filling of wetlands bordering the bay have destroyed habitat for fish and waterfowl, thereby exacerbating the impact of low freshwater inflow. Only ~125 km^2 of undiked marsh remains from an original tidal marsh area of ~2200 km^2.[8] The diversion of freshwater for agricultural, municipal, and industrial uses has had dramatic ecological consequences in the estuary as noted previously. Increased runoff from urban and agricultural lands and wastewaters from municipal and industrial outfalls have transported considerable quantities of chemical contaminants to the estuary.

Many land-use changes have occurred concurrently with increased population growth in the 12-county estuary area. From 1900 to 1950, the rise of the industrial and manufacturing sectors fueled much of the urban and economic growth of the region. At this time, extensive wetlands habitat was filled for industrial and urban uses. A series of public and private water development projects was initiated for agricultural, municipal, and industrial needs as well as for electric power generation, flood control, and other purposes. Dams constructed on major tributary systems significantly reduced freshwater inflow to the estuary and allowed the diversion of large volumes of water for municipal and agricultural uses statewide. However, they precluded some economically important finfish populations (e.g., chinook salmon, *Oncorhynchus tshawytscha*; steelhead trout, *O. mykiss irideus*; and striped bass, *Morone saxatilis*) from reaching valuable spawning and rearing grounds. They also increased mortality of various finfish populations by altering migration routes. The entrainment of eggs and young in delta diversions added to the escalating mortality figures.

Development in the watershed accelerated dramatically after World War II. To meet the many needs of the rapidly growing bay area population, a boom in housing and roadway construction, public works projects, commercial and agricultural development, and shipping operations led to the degradation and loss of important habitat area, increased point and nonpoint source pollution, and overuse of estuarine resources.[19,20] Development was particularly conspicuous along major transportation corridors (e.g., between the bay/delta and Central Valley).[2] Current population and land-use projections indicate a high probability of continued habitat and aquatic impacts in the system. Adverse effects can only be averted by careful and effective long-term planning and improved management of freshwater supplies and pollutant sources.

The estuary region is vital to the economic health of the entire state. Nearly 4 million jobs are directly linked to agriculture, construction, services, retail trades, and other sectors.[21]

The estuary itself is critical to commercial and recreational fishing, shipping, marina operations, agriculture, and tourism, all of which inject billions of dollars into the local and regional economies. Ports in the estuary alone generate billions in gross sales transactions each year.

Tourism in the bay area accounts for $3 to 5 billion annually. Visitors are drawn to the area by scenic waterways, recreational fishing opportunities, the Golden Gate Bridge, wine country vineyards, and many other features. The city of San Francisco is one of the most popular urban tourist sites in the U.S., with 10 to 15 million people visiting the metropolitan center each year.[22]

The commercial fishing harvest, comprised of more than 80 species of fish, crustaceans, and mollusks, generally exceeds $20 million annually. Today, the Pacific herring (*Clupea harengus pallasi*) and chinook (or king) salmon (*Oncorhynchus tshawytscha*) are the two most important species landed in the fishery. The value of the herring catch alone has been $10 to 15 million during some years. Although the dungeness crab was once a valuable component of the fishery, it has not been harvested commercially in the bay since 1990. The Pacific herring continues to be the only commercial species taken in significant numbers within the estuary.

Closely coupled to fishing and boating interests in the bay are commercial marina operations. More than 200 marinas in the bay/delta area provide nearly 34,000 berths for boaters.[2] Marinas in the estuary generate more than $50 million in annual revenues.[23]

The success of agriculture throughout much of California depends greatly on water diverted from the estuary and its tributaries by the federal Central Valley Project and the State Water Project. The San Joaquin Valley receives much of the diverted water for farm use. In the delta, total crop and livestock production approaches half a billion dollars annually.[24] Beyond the delta, water diverted from the system supports billions of dollars worth of crop production each year.

Of even greater significance is the diversion of the estuary freshwater supply for drinking water to approximately 20 million state residents. An additional fraction of diverted water is used by industry for process and cooling operations at manufacturing facilities and other locations. Because pollutants in the diverted water may affect drinking water quality, emphasis has been placed on improved control of contaminant inputs to the water at its source (i.e., in the watershed). Greater source water protection will allay some of the concerns of the supplies and uses of the diverted water.

B. Pollution

San Francisco Bay, a highly urbanized system, receives thousands of kilograms of pollutants each day from a multitude of sources, including more than 50 municipal waste treatment plants, more than 65 industrial facilities, runoff from urban areas and farm fields, abandoned mines, dredging and dredged material disposal operations, and atmospheric fallout. Despite the large loads of pollutants entering the estuary every day, these substances generally occur in relatively low concentrations in water and bottom sediments. However, pollutant concentrations in bottom sediments throughout much of the estuary are slightly greater than those recorded at coastal reference sites. Due to bioaccumulation and biomagnification effects, pollutants may reach high concentrations in some estuarine animal tissues. This is most evident at highly impacted areas of the estuary (e.g., harbors, harbor entrances, marinas, and industrial waterways). Because many of these pollutants are carcinogenic, mutagenic, and teratogenic, there is concern not only for the protection of aquatic life but also for humans who consume contaminated seafood products from the estuary.

1. Type of pollutants

Pollution occurs in estuaries when the concentration of a waste substance exceeds the level at which damaging effects are manifested in the system.[25] Environmental quality may be significantly compromised due to adverse effects on physical, chemical, or biological properties. Although some pollutants derive from natural sources, the most acute and insidious impacts have usually been coupled to anthropogenic activities.

The Technical Advisory Committee of the SFEP identified four kinds of pollutants in the estuary. As noted previously, these are biological pollutants, organic chemicals, inorganic chemicals, and suspended solids. Each category is associated with serious environmental problems. For example, pollution impacts chronicled in the estuary during the 20th century have included numerous episodes of depleted dissolved oxygen concentrations linked to nutrient enrichment and excessive plant growth, degraded water quality due to high coliform bacteria counts, elevated levels of synthetic organic pesticide compounds in bottom sediments and organisms, bay area refinery loadings (i.e., oil and grease, chromium and zinc, BOD, and suspended solids), input of petroleum hydrocarbons from oil spills and leakages at terminals, and the influx of heavy metals (e.g., arsenic, lead, mercury, selenium, and zinc) from industrial facilities, farm fields, mines, and other sources.

Of all pollutants affecting the estuary, toxic chemicals are of greatest concern to the estuary program. The Pollutants Subcommittee of the SFEP has selected a list of toxic chemicals considered to pose the greatest threat to the system (Table 5.2). The subcommittee selected these pollutants based on five criteria: (1) the potential to cause toxicity or to affect beneficial uses of the estuary; (2) the extent of the database for each pollutant within the estuary; (3) whether the pollutant is found at high levels throughout the estuary; (4) whether the pollutant is found at high levels locally, and (5) whether the pollutant exerts or has the potential to exert detrimental effects on the estuary's biological resources. Management plans have been formulated in the CCMP to address toxic chemical issues.

a. Biological pollutants. Untreated municipal sewage, recreational boat discharges, and runoff from urban centers as well as farm fields and feedlots deliver enteric pathogens (i.e., bacteria, viruses, and parasites) to estuarine waters. These microbes cause serious illnesses such as cholera, dysentery, hepatitis, salmonella, and typhoid. Humans who swim in contaminated waters or consume contaminated seafood products are at risk of developing these maladies. Municipalities monitor fecal coliform bacteria levels in the estuary. When coliform levels exceed water quality standards, bathing beaches are closed and the harvest of shellfish beds is prohibited. All municipal wastewater treatment plants have tested for coliform bacteria since 1978. These microbes have been used as indicators that other harmful agents may be present in the system.

b. Organic chemicals. Among the most toxic pollutants in San Francisco Bay are an array of organic chemicals (e.g., pesticides, plastics, fertilizers, solvents, and pharmaceuticals). Particularly notable are lipophilic, high-molecular-weight halogenated hydrocarbons that tend to biomagnificate through the estuarine food chain and pose a potential health threat to humans. PCBs and certain pesticides (e.g., DDT, chlordane, dieldrin, and toxaphene) — synthetic organochlorine compounds — provide examples. These recalcitrant compounds persist for decades in the estuarine environment. Many halogenated hydrocarbons are broad-spectrum poisons that adversely affect entire biotic communities.

Based on findings of the SFEP in 1996, the levels of some synthetic organic compounds appear to be significantly elevated. For example, in 1993, 1994, and 1995, PCB concentrations

Table 5.2 Pollutants of Concern in the San Francisco Bay/Delta Estuary (Boldface Indicates Pollutants of Particular Concern)

Trace Elements	
Cadmium	Antimony
Copper	Arsenic
Mercury	Chromium
Nickel	Cobalt
Selenium	Lead
Silver	Zinc
Tin (Tributyl)	
Organochlorines and Other Pesticides	
Chlordane and its metabolites	Heptachlor and its epoxide
DDT and its metabolites	Hexachlorobenzene (HCB)
Polychlorinated biphenyls	Hexachlorobutadiene
Toxaphene	Hexachlorocyclohexane (HCH)
Aldrin	Methoxychlor
Chlorbenside	Polychlorinated terphenyls
Dacthal	2,4,6-trichlorophenol
Dieldrin	Malathion
Dioxins	Parathion
Endosulfan	
Endrin	
Petroleum Hydrocarbons	
(i) Monocyclic Aromatic Hydrocarbons (MAHs)	
Benzene	
Ethylbenzene	
Toluene	
Xylene	
(ii) Cycloalkanes	
(iii) Polynuclear Aromatic Hydrocarbons (PAHs)	
Acenaphthene	**2,6-Dimethylnaphthalene**
Acenaphthylene	**Fluoranthene**
Anthracene	**Fluorene**
Benz(b)fluoranthene	**1-Methylnaphthalene**
Benz(k)fluoranthene	**2-Methylnaphthalene**
Benz(g,h,i)perylene	**1-Methylphenanthrene**
Benzo(a)pyrene	**2-(4-morpholinyl)benzthiazole**
Benzo(e)pyrene	**Naphthalene**
Benzo(a)anthracene	**Phenanthrene**
Benzthiazole	**Pyrene**
Chrysene	**2,3,5-Trimethylphenanthrene**
Dibenzo(a,h)anthracene	**Indeno(1,2,3-c,d)pyrene**

Source: Monroe, M. W. and Kelly, J., State of the Estuary, Tech. Rept., San Francisco Estuary Project, Oakland, CA, 1992.

at nearly all 24 bay regional monitoring program sampling stations exceeded the USEPA criteria for human health. Laboratory analysis of edible marine and estuarine finfish collected at 13 stations in 1994 indicated that all muscle tissue samples exceeded the screening value (3 ng/g) for human consumption of PCBs. The highest concentrations were observed in samples from industrialized areas. Many of the finfish tissue samples also exceeded screening values for DDTs, chlordane, dieldrin, and dioxin. For instance, high levels of dioxin/furans (>0.15 pg/g) were found in 16 of 19 samples analyzed.[1] Highest contaminant levels were documented in fish (white croaker, *Genyonemus lineatus* and surfperch, *Archoplites interruptus*) with elevated lipid content in their tissues.

Sublethal effects of these persistent contaminants are particularly pronounced in biota at the top of the food web (i.e., fish, seals, and waterfowl). For example, effects of PCB and DDE uptake by the black-crowned night heron (*Nycticorax nycticorax*) include reduced eggshell thickness and embryo size. PCBs appear to be responsible for developmental malfunctions and decreased breeding success in the double-crested cormorant (*Phalacrocorax auritus*). They have been correlated with reduced egg and spleen mass in this species. Decreased reproductive success of starry flounder (*Platichthys stellatus*) inhabiting the eastern portion of the Central Bay has also been correlated with PCB contamination. The levels of PCBs in harbor seals (*Phoca vitulina*) may be high enough to cause immunosuppression and reduced reproduction.

PAHs are another important group of organic chemicals that pose a potential threat to aquatic communities in the estuary. These organic pollutants, ranging from lower-molecular-weight to higher-molecular-weight compounds (e.g., naphthalene to coronene), have been linked to an array of biochemical, behavioral, physiological, and pathological (sublethal) impacts in estuarine organisms. When exposed to high concentrations of PAHs, many estuarine organisms develop lesions, tumors, and other abnormalities.[26]

The highest concentrations of PAHs in the San Francisco Bay/Delta are found in the highly urbanized and industrialized areas of the system. These compounds originate from municipal and industrial effluents, urban and agricultural runoff, atmospheric deposition, fossil fuel combustion, chemical and microbial activity in sediments, and thermal conversion of chemically complex geological deposits.[1,2] Between 1993 and 1995, the concentrations of PAHs often exceeded water quality criteria at monitoring stations. Dissolved PAHs ranged from 0.07 to 17.3 ppt (parts per trillion) in 1994 and 0.39 to 8.24 ppt in 1995. Dissolved and suspended PAHs in bay water ranged from 2.8 to 504.9 ppt in 1995. At this time, PAHs in bottom sediments ranged from 16 to 3722 ppb.[1]

The National Status and Trends Program of NOAA ranked San Francisco Bay among the highest coastal systems in the U.S. in terms of total PAH concentrations in mussels (Table 5.3).[27] In some areas of San Francisco Bay, PAH concentrations in mussels reach levels comparable to those in highly contaminated systems (e.g., Los Angeles and San Diego Harbors).[28] Because PAHs accumulate in bottom sediments of the estuary, benthic organisms are continuously exposed to relatively high levels of the contaminants, especially in urban areas receiving large pollutant loads.

c. Inorganic chemicals. Nutrient elements (nitrogen and phosphorus) and trace metals (arsenic, cadmium, chromium, copper, lead, mercury, nickel, selenium, silver, and zinc) are the most important inorganic contaminants in the estuary. Prior to 1960, large nutrient loads were delivered by inadequately treated municipal wastes, leading to algal blooms and low dissolved oxygen concentrations in many parts of the bay and delta. With the upgrade of sewage treatment plant operations that commenced in the 1960s, nutrient inputs declined and water quality improved. However, runoff from farmlands has continued to introduce substantial quantities of nitrogen and phosphorus into the water column.

The largest fraction of trace metals entering the estuary originates from nonurban runoff. Secondary amounts derive from farmlands, abandoned mines, sewage treatment plant wastewaters, urban runoff, metal finishing and circuit board manufacturing industries, and oil refineries. In 1995 and 1996, the concentrations of chromium, copper, lead, mercury, and nickel in the water column occasionally exceeded water quality objectives. Throughout most of the estuary, the levels of arsenic, copper, chromium, mercury, and

Table 5.3 National Status and Trends Sites Ranking Among the Highest 20 in 1986, 1987, and 1988 and Overall for Total PAHs in Mussels (*Mytilus edulis, Mytilus californianus*) and Oysters (*Crassotrea virginica, Ostrea sandivicensis*)

Code	Location	State	Species	Ranking				Significant	
				1986	1987	1988	Overall	Difference?	Trend
SAWB	St. Andrew Bay	FL	cv	1	17t	<	3	no	no
EBFR	Elliott Bay	WA	me	2	13t	1	1	yes	no
HRUB	Hudson/Raritan Estuary	NY	me	3	2	<	4	yes	d
CBRP	Coos Bay	OR	me	4	<	<	9	yes	no
BHDB	Boston Harbor	MA	me	5	8	6	6	no	no
LITN	Long Island Sound	NY	me	6	10	15t	8	no	no
SGSG	Pt. St. George	OR	mc	7	<	<	16	yes	d
BHDI	Boston Harbor	MA	me	8t	4	3	5	yes	no
CBCH	Coos Bay	OR	mc	8t	<	<	13t	yes	no
BHHB	Boston Harbor	MA	me	10t	<	<	15	yes	no
PLLH	Pt. Loma	CA	mc	10t	—	<	<	yes	no
SIWP	Sinclair Inlet	WA	me	10t	<	<	19	yes	no
SCFP	Santa Cruz Island	CA	mc	13t	<	<	<	yes	no
SSSS	San Simeon Point	CA	mc	13t	<	<	<	yes	d
SCBR	S. Catalina Island	CA	mc	15	<	<	<	no	no
BPBP	Barber's Point	HI	os	16	<	<	<	no	no
HRJB	Hudson/Raritan Estuary	NY	me	17	3	4	7	no	no
NYSH	New York Bight	NJ	me	18	<	<	<	yes	no
NYLB	New York Bight	NJ	me	19t	15	<	<	yes	d
CFBI	Cape Fear	NC	cv	19t	<	<	<	no	no
CHSF	Charleston Harbor	SC	cv	19t	<	<	<	no	no
SDHI	San Diego Bay	CA	me	<	1	<	10	yes	no
MSBB	Mississippi Sound	MS	cv	<	5t	<	18	no	no
HHKL	Honolulu Harbor	HI	os	<	5t	<	<	yes	no
SFEM	**San Francisco Bay**	**CA**	**me**	**—**	**7**	**13**	**11t**	**—**	**—**
BBMB	Barataria Bay	LA	cv	<	9	<	<	no	no
CBSR	Choctawhatchee Bay	FL	cv	<	11	<	<	yes	no
BHBI	Boston Harbor	MA	me	<	12	5	13t	no	no
NYSR	New York Bight	NJ	me	<	13t	<	<	no	no
CBTP	Commencement Bay	WA	me	<	16	9t	<	no	no
BBPC	Biscayne Bay	FL	cv	—	17t	—	<	no	no
BBAR	Buzzards Bay	MA	me	<	19	17	<	yes	i
RSJC	Roanoke Sound	VA	cv	<	20	<	<	no	no
PCMP	Panama City	FL	cv	—	—	2	2	no	no
NMML	North Miami	FL	cv	—	—	7t	11t	no	no
APDB	Apalachicola Bay	FL	cv	<	<	7t	<	no	no
CBCI	Chincoteague Bay	VA	cv	<	<	9t	<	no	no
LICR	Long Island Sound	CT	me	<	<	9t	<	no	no
BBSM	Bellingham Bay	WA	me	<	<	9t	<	yes	i
DBAP	Delaware Bay	DE	cv	<	<	14	<	no	no
IRSR	Indian River	FL	cv	—	—	15t	17	—	—
UISB	Unakwit Inlet	AK	me	<	<	18	<	yes	i
LINH	Long Island Sound	CT	me	<	<	19	<	yes	i
SSBI	South Puget Sound	WA	me	<	<	20	<	yes	no
AIAC	Absecon Inlet	NJ	me	—	—	<	20	—	—

Note: cv = *Crassotrea virginica*; me = *Mytilus edulis*; mc = *Mytilus californianus*; os = *Ostrea sandivicensis*; t = two or more concentrations were equal; d = decreasing trends; i = increasing trends. Note high ranking for San Francisco Bay relative to other systems.

Source: NOAA, A summary of Data on Tissue Contamination from the First Three Years (1986–1988) of the Mussel Watch Project, NOAA Tech. Mem. NOS OMA 49, National Oceanic and Atmospheric Administration, Rockville, MD, 1989.

nickel in bottom sediments exceeded sediment guideline concentrations. Trace metals have remained highest in Suisun and South Bays.[1]

d. Suspended solids. Particulate organic and inorganic matter transported to the estuary from farmlands, wetlands, and other sources can adversely affect the system in several ways. The influx of particulate matter reduces sunlight penetration in the estuary, resulting in lower primary production rates. When this material settles to the estuarine floor in large concentrations, it can alter the behavior of demersal finfish and benthic invertebrates, hinder their feeding processes, and, in extreme cases, smother the organisms. Particle-bound contaminants accumulating in the sediments pose a potential threat to benthic communities. Various physical, chemical, and biological processes (e.g., current action, dissolution, and bioturbation) re-release some of the contaminants to the water column, thereby reducing their availability to organisms inhabiting bottom sediments.

2. *Pollutant loads*

Today, the San Francisco Bay/Delta Estuary receives large pollutant loads from multiple sources. For example, more than 50 publicly owned wastewater treatment plants and 65 industrial facilities discharge effluent to the estuary. The total wastewater flow of the municipal wastewater treatment plants ($>8 \times 10^9$ l/d) is more than 10 times that of the industrial facilities ($>2 \times 10^8$ l/d). Among the industrial discharges, petroleum refineries release the greatest volume of effluent. Despite improved wastewater quality during the past few decades, significant amounts of some pollutants continue to be discharged from the municipal and industrial facilities to the estuary.

Aside from the aforementioned point source discharges, rivers draining farmlands of the Central Valley carry substantial loads of certain pollutants (e.g., selenium).[28] Accidental spills, particularly petroleum products, account for significant, albeit intermittent pollutant loads. Oil spills release an average of more than 1.2×10^5 l/yr of petroleum products into the estuary. Several other sources may also contribute large pollutant loads to the estuary, but their inputs have been generally poorly characterized. Included here are atmospheric deposition, dredging and dredged material disposal, marine vessel discharges, and leakages from waste disposal sites.

Because of decreasing pollutant loads from point sources, the SFEP has focused greater attention on (uncontrolled) nonpoint source inputs, especially those associated with urban and nonurban runoff. Pollutants in urban runoff originate from impervious surfaces such as buildings, roadways, sidewalks, and macadam parking lots. Much of these pollutants enter the estuary via stormwater discharges. Those in nonurban runoff derive mainly from farmlands, pasture, forests, and natural range. Pesticides, solvents, and trace metals are pollutants of concern here. The frequency and amount of precipitation in watershed areas strongly influence the quantity and quality of both urban and nonurban runoff.

The SFEP reported that 4500 to 36,100 mt of (at least) 65 pollutants entered the estuary in 1991. According to the SFEP (p. 41),[1] "… Over the past eight years, most contaminant concentrations have remained constant, with some seasonal and annual fluctuations. Several long-term trends have emerged, however. Arsenic in the sediments at the confluence of the estuary's main rivers appears to be on the rise, for example. PCB concentrations in Central Bay water and sediments appear to be decreasing. Diazinon, a common orchard and garden pesticide, is turning up throughout the estuarine ecosystem at concentrations lethal to sensitive organisms. Mercury from abandoned mines and selenium from agricultural drainage continues to be a problem upstream. The largest biological effects resulting from estuary pollution are observed in the North Bay region at the Napa River, Suisun Bay, and the confluence of the Sacramento and San Joaquin Rivers (where trace metals

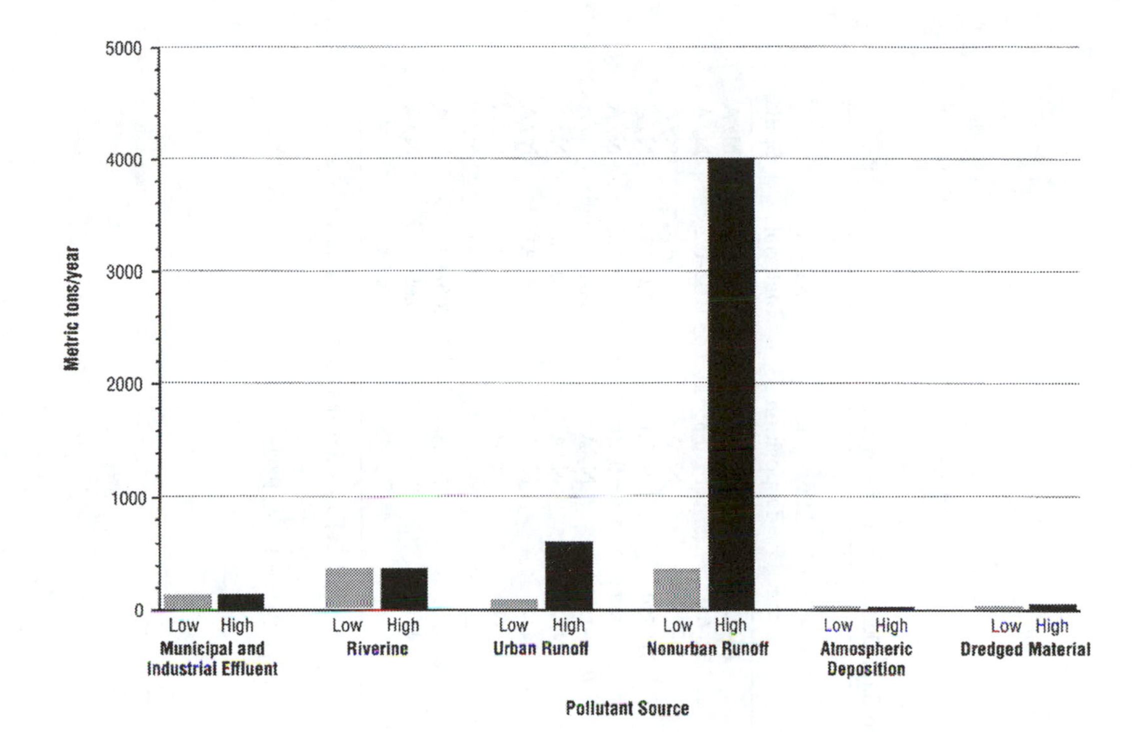

Figure 5.2 Combined loadings of selected pollutants (i.e., arsenic, cadmium, chromium, copper, lead, and zinc) by source to the San Francisco Bay/Delta Estuary. (From Monroe, M.W. and Kelly, J., State of the Estuary, Tech. Rept., San Francisco Estuary Project, Oakland, CA, 1992.)

from mine runoff may have accumulated in sediments and where pulses of pesticides from farm drainage converge). The incidence of biological effects is generally lower in Central Bay — flushed daily by strong tidal action — and moderate in South Bay (an enclosed, shallow area where pollutants may concentrate)."

Table 5.4 shows pollutant loadings to the estuary from primary sources based on both repeated measurements (i.e., for municipal and industrial effluent) and predictive models (i.e., for the other sources). Not all pollutants of concern to the SFEP are included in the table. For example, the loadings of organic pesticides are not addressed because no data exist on the quantity of the contaminants applied in the Central Valley. In addition, more loadings data have been determined for some sources (e.g., municipal and industrial effluent, urban runoff, and nonurban runoff) than others (e.g., Sacramento River, atmospheric deposition, and spills).

Nonurban runoff represents the largest quantified source of most trace metals to the estuary (Figure 5.2). This is true for arsenic (10 to 120 mt/year), cadmium (0.52 to 6.0 mt/yr), chromium (130 to 1500 mt/yr), copper (51 to 580 mt/yr), lead (31 to 360 mt/yr), mercury (0.15 to 1.7 mt/yr), and zinc (130 to 1450 mt/yr). Municipal and industrial effluent contains substantial amounts of cadmium (1.8 to 4.0 mt/yr), mercury (0.2 to 0.7 mt/yr), and silver (2.7 to 7.2 mt/yr). Urban runoff is responsible for most of the quantified loads of PCBs (0.006 to 0.40 mt/yr), PAHs (0.50 to 5.0 mt/yr), and total hydrocarbons (1100 to 11,000 mt/yr). Pesticides enter the estuary largely from agricultural runoff and urban landscapes. Some of these chemicals (e.g., diazinon, chlorpyrifos, malathion, and methidathion) have been applied in high concentrations in the Central Valley and now are detected in bay waters where they (especially diazinon and chlorpyrifos) pose a serious threat to aquatic organisms.

Table 5.4 Summary of Pollutant Loadings to the San Francisco Bay/Delta Estuary from Major Sources[a]

Pollutant	Municipal and Industrial Effluent	San Joaquin River	Sacramento River	Urban Runoff	Total Nonurban Runoff	Atmospheric Deposition	Dredged Material	Spills
Arsenic	1.5–5.5	12	N/A	1.0–9.0	***10–120***	N/A	N/A	N/A
Cadmium	***1.8–4.0***	N/A	N/A	0.3–3.0	***0.52–6.0***	0.14–0.35	0.02–0.2	N/A
Chromium	12–13	66	N/A	3.0–15	***130–1,500***	N/A	N/A	N/A
Copper	19–30	80	N/A	7.0–59	***51–580***	1.9–3.1	1.0–10	N/A
Lead	11–16	51–55	N/A	***30–250***	***31–360***	6.0–21	1.0–10	N/A
Mercury	***0.2–0.7***	N/A	N/A	0.026–0.15	***0.15–1.7***	N/A	0.01–0.1	N/A
Nickel	19–27	***51***	N/A	N/A	N/A	N/A	2.0–20	N/A
Selenium	2.1	***4.2***	1.1	N/A	N/A	N/A	N/A	N/A
Silver	***2.7–7.2***	N/A	N/A	N/A	N/A	N/A	N/A	N/A
Zinc	77–80	164–175	N/A	34–268	***130–1,450***	N/A	3.0–30	N/A
PCBs	N/A	N/A	N/A	***0.006–0.40***	N/A	N/A	0.00067–0.0067	N/A
PAHs	N/A	N/A	N/A	***0.50–5.0***	N/A	***0.8–4.8***	0.05–0.47	N/A
Total Hydrocarbons	N/A	N/A	N/A	***1,100–11,000***	N/A	2.1–45	N/A	94

Note: Values in boldface indicate the largest quantified source (or sources where ranks are relatively ambiguous) of each pollutant. N/A = Data not available.

[a] Values in metric tons/year.

Source: Monroe, M. W. and Kelly, J., State of the Estuary, Tech. Rept., San Francisco Estuary Project, Oakland, CA, 1992.

The use of pesticides in the Central Valley has reached extreme levels. More than 20,000 mt of pesticides (about 500 different varieties representing 10% of the total pesticides used each year in U.S. agriculture) are applied annually to farmlands in the Central Valley.[29] The widespread use of herbicides, pesticides, and fertilizers in Central Valley farmlands results in the runoff of considerable quantities of chlorinated hydrocarbon contaminants into influent systems of San Francisco Bay, particularly the San Joaquin River. More than 20% of the total San Joaquin River flow consists of agricultural wastewater returned to the river in subsurface pipe drainage. This wastewater contains salts and contaminants leached from the soil. The Sacramento River also transports chlorinated hydrocarbon contaminants from farmlands into the bay. The Colusa Basin Drain, an artificial channel that discharges to the Sacramento River at Knights Landing, directs runoff and agricultural return-flows from more than 4×10^5 ha of farmland. DDT, chlordane, dacthal, nonachlor, and PCBs have been documented in fish, invertebrates, and sediments from the Colusa Basin Drain, and particle-bound fractions of these organochlorine compounds eventually accumulate in bottom sediments of the estuary. This contaminant pool is augmented by PCBs and DDT originating from point sources located on the shore of the the bay and delta as well as from atmospheric deposition.[30]

3. *Pollutant concentrations*

The SFEP has compiled data on priority pollutant concentrations in water, sediments, and biota of the estuary.[1,2] Pollutants that can significantly affect living resources (e.g., trace metals, PAHs, DDT, and PCBs) have been found at elevated levels in water, sediment, and biotic samples relative to background and reference sites. Highest concentrations have been recorded in heavily utilized areas, notably harbors, marinas, and industrial waterways. The accumulation of toxic chemicals in bottom sediments is a major concern because some of the contaminants are slowly leaching out of the sediments into the water column (e.g., cadmium, copper, zinc, and PCBs), exposing numerous organisms to potentially hazardous conditions. The effects of the contaminants on these organisms may range from subtle physiological changes to death. Some pollutants occur in very high concentrations — among the highest worldwide — in sediments and biota of the estuary.

a. Polluants in water. Table 5.5 summarizes available data on pollutant concentrations in waters of the estuary. The most comprehensive database is on trace metals. Far less data are available on organic compounds. Some pollutants (i.e., copper, lead, mercury, and nickel) have exceeded state water quality objectives that establish permissible levels of pollutants in California waters. Pollutants in estuarine waters are important because they can accumulate in bottom sediments, where they contaminate valuable habitat and pose a danger to organisms that assimilate them.

b. Pollutants in sediments. Table 5.6 records existing data on pollutant concentrations in sediments downstream of the delta. The concentrations of a number of pollutants are slightly elevated in bottom sediments throughout the system when compared to background levels or reference sites. Several pollutants have relatively uniform concentrations estuarywide.[31] However, at heavily contaminated locations such as harbors, marinas, and industrial waterways, some sediment pollutant concentrations are dangerously high. Specific areas of the estuary exhibiting elevated pollutant concentrations in bottom sediments include the delta, South Bay, areas off the Richmond/Berkeley shore, and effluent outfall sites.[1,2]

Table 5.5 Concentrations of Selected Pollutants in Waters of the San Francisco Bay/Delta Estuary (Values in ppb)

Pollutant	Range of Total Concentrations[a]	State Water Quality Objective Downstream of Carquinez Strait	State Water Quality Objective Upstream of San Pablo Bay	Any Samples Exceeding State Water Quality Objectives?
Arsenic	0.5–4.5[b]	36 (4D) 69 (1H)	190 (4D) 360 (1H)	No
Cadmium	0.005–0.159	9.3 (4D) 43 (1H)	1.1 (4D) 3.9 (1H)	No
Chromium	0.540–3.600	—	—	—
Copper	0.9–7.2	—	6.5 (4D) 9.2 (1H)	Yes
Lead	0.15–3.54	5.6 (4D) 140 (1H)	3.2 (4D) 82 (1H)	Yes
Mercury	0.001–0.032	0.025 (4D) 2.1 (1H)	0.025 (4D) 2.4 (1H)	Yes
Nickel	1.22–11.28	7.1 (1D) 140 (Inst.)	56 (1D) 100 (Inst.)	Yes
Selenium	0.013–4.700[2]	—	—	—
Silver	0.003–0.100	2.3 (Inst.)	1.2 (Inst.)	No
Tributyltin	0.004–0.570	—	0.04 (1D) 0.06 (Inst.)	Yes
Zinc	1.4–17.4	58 (1D) 170 (Inst.)	38 (1D) 170 (Inst.)	No
PAH	—	15 (1D)	—	—
DDT	—	—	—	—
PCB	0.0004–0.0066	—	—	—

Note: Dashes indicate that either reliable data or water quality objectives do not exist. Inst. — instantaneous value; 1H — one-hour average; 1D — one-day average; 4D — four-day average.

[a] All concentrations are from Flegal et al., 1991, except chromium, selenium, tributyltin, and PCB, which are from Davis et al., 1991.

[b] Concentrations of unfiltered samples.

Source: Monroe, M. W. and Kelly, J., State of the Estuary, Tech. Rept., San Francisco Estuary Project, Oakland, CA, 1992.

c. *Pollutants in biota.* Since the 1970s, there have been several major studies of pollutant levels in biota of the estuary. These studies have focused on edible aquatic organisms (finfish and shellfish) and wildlife populations at the top of the food chain (waterfowl, waterbirds, and seals). Table 5.7 contains the results of these studies.

Mussels (*Mytilus edulis, M. californianus,* and *M. senhousia*) are particularly valuable, serving as sentinel organisms in pollution surveys. They are hardy, sessile, and easily sampled. In addition, they have a limited capacity for metabolizing contaminants and, hence, accumulate them. As a consequence, their pollutant concentrations typically reflect current loadings to the system. For certain contaminants, the tissue levels are elevated. Silver enrichment, for example, was ascertained in *M. californianus* and *M. senhousia* samples from Central Bay and lower South Bay during the 1980s. Silver concentrations in *M. californianus* transplanted to Central and South Bays at this time ranked among the highest 15% of 358 determinations by the California State Mussel Watch Program.[32] In South Bay, silver concentrations in bottom sediments are as high as 5 μg/g dry weight and in benthic organisms at least one order of magnitude above the baseline concentrations of estuarine organisms.[33,34]

Table 5.6 Concentrations of Selected Pollutants in San Francisco Bay Sediments (Values in ppm)[a]

Pollutant	Mean	Range
Arsenic	—	13–66[b]
Cadmium	1.06	0.02–17.3
Chromium	89	8–769
Copper	51	1–1,500
Lead	56	1–10,000
Mercury	0.5	<0.01–6.80
Nickel	—	84–189[b]
Selenium	—	0.001–.035[b]
Silver	1.13	<0.01–16
Tributyltin	—	0.003–0.09[b]
Zinc	≈100[c]	<100–1,255[b]
PAH	4.1	0.02–80.9
DDT and metabolites[a]	0.1	0.00025–1.96
PCB	0.115	0.006–0.824

[a] Does not include data on extremely contaminated sediment in the Lauritzen Canal. The overall mean including the additional samples from the Lauritzen Canal is 7.5 ppm dry weight. — indicate data are not available.

[b] From SWRCB, 1990.

[c] From Long et al., 1988; Phillips, 1987.

Source: Monroe, M. W. and Kelly, J., State of the Estuary, Tech. Rept., San Francisco Estuary Project, Oakland, CA, 1992.

As evident from Table 5.7, trace metals are significantly elevated in animal tissues, commonly exceeding alert levels. This is especially true for arsenic, cadmium, chromium, copper, lead, mercury, and selenium in shellfish. These data indicate that the pollutant levels exceed state or international safety levels established to protect human health.

In regard to organochlorine contaminants, the principal concern has been with biotic uptake of PCBs and DDT. Analysis of historical datasets on these contaminants reveals several general trends in the estuary. Surveys conducted during the 1960s showed a well-developed, 100-fold gradient of pesticide concentrations in bay water, decreasing from the San Joaquin and Sacramento Rivers through the northern bays to the Golden Gate Bridge. Highest PCB levels occurred in fish and shellfish in South Bay, and lowest levels occurred in fish and shellfish in San Pablo Bay. There is evidence of increasing PCB concentrations in fish liver samples collected after 1976 and decreasing PCB concentrations in bivalve samples collected after 1981. Compared to PCB levels in fish liver samples from other estuarine and coastal marine systems, those reported from South Bay samples during the mid-1980s by NOAA's National Status and Trends Program are relatively high.[35]

DDT contamination in fish and shellfish declined sharply throughout the bay between 1969 and 1977 and then leveled off. Mean total DDT concentrations in native mussels (*Mytilus edulis*) and transplanted mussels (*M. californianus*) in the bay between 1979 and 1986 were >100 ng/g dry weight. At some locations, DDT concentrations in the mussels exceeded 1000 ng/g dry weight.[36-38] The highest levels of DDT were found in samples from the Sante Fe Channel in Richmond Harbor, which lies in close proximity to the former site of a pesticide formulation and packaging plant on the banks of the Lauritzen Canal. Pereira et al.,[30] examining chlorinated hydrocarbon contamination in striped bass (livers) collected in 1992, reported total DDT concentrations as high as 396 ng/g wet weight and total chlordane as high as 42 ng/g dry weight.

Table 5.7 Concentrations of Selected Pollutants in San Francisco Bay/Delta Estuary Biota[a]

Pollutant	Mussel	Clam	Fish	Bird	Seal	Concentrations Exceeding Alert Levels[b]
Arsenic	1.16–2.16 (1,9)	—	0.13–1.20 (2)	—	—	Yes. Levels in some bay shellfish exceed MIS.
Cadmium	0.11–4.91 (3)	—	0.03–0.48 (2)	4.17 (5)	<.06–.33 (13)	Yes. Levels in some bay shellfish exceed MIS.
Chromium	0.014–2.114 (3)	0.15–3.92 (4)	0.02–0.1 (2) 1.8 (striped bass) (7)	—	—	Yes. Levels in some bay shellfish exceed MIS.
Copper	0.314–4.385 (3)	10–100 (6)	1.3–30 (2)	7.14–13.86 (5)	3.0–8.7 (13)	Yes. Levels in some bay shellfish exceed MIS. Levels in some Suisun Bay and delta fish exceed MIS.
Lead	0.03–74 (3)	—	0.02–0.2 (2)	64–102 (5)	0.13–1.22 (13)	Yes. Levels in some bay shellfish exceed MIS.
Mercury	0.01–0.46 (3)	—	0.13–0.94 (2)	0.16–0.6 (2)	0.40–3.65 (13)	Yes. Levels in some bay shellfish and delta fish exceed MIS.
Nickel	0.5–2.4 (1,11)	—	0.8 (2)	0.1 (8)	0.11–4.10 (13)	No alert levels established for tissue.
Selenium	0.19–0.66 (1)	0.3–1.30 (9)	0.28–22.0 (10)	24–58 (10)	2.07–6.49 (13)	Yes. Levels in some bay shellfish exceed MIS. Levels in some bay fish exceed MARL. Levels in some bay ducks exceed MARL.
Silver	0.02–22.5 (3)	0.14–28.57 (6)	0.13–0.94 (2)	0.33–3.70 (8)	—	No alert levels established for tissue.
Tributyltin	0.120–2.960 (1)	—	—	—	—	No alert levels established for tissue.
Zinc	11.0–45.8 (1)	—	16.0–43.0 (2)	21.6 (8)	—	No alert levels established for tissue.
PAH	0.025–13 (3)	—	0.017–14 (3)	—	—	No.
DDT and metabolites	<.002–3.21 (3)	—	0.020–5.18 (2)	—	5–34 (13)	Yes. Levels in some delta fish exceed FDA action level.
PCB	0.009–0.657 (3)	—	0.05–6.99 (2,12)	—	0.05–330 (13)	Yes. Levels in some bay and delta fish exceed FDA action level.

Note: Concentrations are shown for wet weight; data originally given for dry weight have been converted by dividing by seven. For seals, trace element data represent concentrations in dry whole blood; data for DDT and PCB represent concentrations in blood plasma lipids.

[a] Values in ppm wet weight.

[b] The alert levels referred to in this table are the maximum tissue residue levels that are protective of human health. They include: (1) the median international standard (MIS), which is a general guideline of what other nations consider to be elevated contaminant levels in fish and shellfish tissue; (2) the U.S. Food and Drug Administration (FDA) action levels, which represent maximum allowable concentrations for some toxic substances in human foods; and (3) the State Department of Health Service's maximum allowable residue levels (MARL), established to ensure that a consumer of specified fish or wildlife species does not exceed the permissible intake level for particular contaminants.

From data in: (1) State Mussel Watch Program in SWRCB, 1990; (2) State Toxic Substances Monitoring Program in SWRCB, 1990; (3) Long et al., 1988; (4) Hayes and Phillips, 1986 in SWRCB, 1990; (5) Ohlendorf, 1985 in SWRCB, 1990; (6) Luoma et al., 1985 in SWRCB, 1990; (7) Saiki and Palawski, 1990 in SWRCB, 1990; (8) Ohlendorf et al., 1986 in SWRCB, 1990; (9) Girvin et al., 1975 in SWRCB, 1990; (10) DFG, 1991; (11) Risebrough et al., 1978 in SWRCB, 1990; (12) NOAA, 1987; (13) Kopec et al., 1991.

Source: Monroe, M. W. and Kelly, J., State of the Estuary, Tech. Rept., San Francisco Estuary Project, Oakland, CA, 1992.

Elevated PCB levels have also persisted in finfish and shellfish populations in the bay despite being banned from use for more than 2 decades. In the 1970s, for example, Risebrough et al.[39] disclosed high concentrations of PCBs (~500 to 1500 ng/g dry weight) in native mussels from South Bay. Later surveys of native and transplanted mussels (*Mytilus edulis* and *M. californianus*) uncovered similar PCB contaminant levels (~150 to 1000 μg/g dry weight).[40] PCB concentrations in mussels ranged from ~50 to 1800 ng/g dry weight during the 1979 to 1986 period.[36-38]

During the mid-1980s, NOAA[40] documented high PCB levels in fish (livers) from San Pablo Bay, Southampton Shoal, and Hunter's Point, totaling 1191, 3734, and 6990 μg/g dry weight, respectively. As noted previously, more recent findings of PCBs in fish tissues by the SFEP in 1994 continued to show persistent high levels of contamination.[1] Chlorinated hydrocarbons tend to accumulate in the hepatobiliary tract and other lipid rich tissues of striped bass and other finfish species, where they may generate hepatoxic effects that can be lethal. Another effect of PCB and DDT contamination of fish, in general, is the potential inhibition of the ATPase system responsible for the regulation of ion balance. At sufficiently high levels, these chlorinated organic compounds may also hinder fish adaptation to salinity changes.

The accumulation of chlorinated hydrocarbons in edible finfish of the estuary is a major human health concern. Pollutant levels of PCBs and DDTs in some finfish and shellfish from the estuary exceed the U.S. Food and Drug Administration action levels, which represent maximum allowable concentrations for toxic substances in human foods (see Table 5.7). It is clear that these pollutants have resisted degradation and continue to bioaccumulate in living resources of the estuary.

4. *Pollutant trends*

Table 5.8 describes the toxic effects of the 10 trace metals, PAH compounds, DDT, and PCBs that have been the focus of the SFEP. It is apparent from these data that most of the pollutants are carcinogenic, mutagenic, or teratogenic to estuarine organisms, depending on their concentration and chemical form. Those pollutants deemed to be of greatest concern in the estuary include cadmium, copper, mercury, selenium, silver, and PCBs.

The trends of each of the aforementioned pollutants in sediments and biota are summarized in Table 5.9. Many of the pollutants attain highest concentrations in samples from Central and South Bays. Based on available data, pollutant concentrations are minimum in the central portions of the estuarine basins and maximum along perimeter areas at or in close proximity to harbors, harbor entrances, marinas, and industrial waterways.

There have been significant changes in the concentrations of pollutants in the estuary during the past 60 years. Due to accelerated effluent discharges from municipal and industrial facilities and runoff from mines in the watershed, pollutant concentrations increased substantially in bay waters and sediments between 1940 and 1975. By the 1960s and early 1970s, organisms in the estuary were exposed to high levels of trace metals, PAHs, and organochlorine contaminants (PCBs and DDTs), and food web impacts were conspicuous. However, an array of federal and state regulations enacted in the 1970s, together with the advent of advanced waste treatment, the termination of certain types of industrial activity, reduction of pollutant loads from oil refineries, the cessation of mining, and improved land-use management practices, resulted in a dramatic decrease in the quantity of pollutants entering the system after 1975.

In spite of the diminished contaminant inputs to the estuary from point sources of pollution after 1975, some serious problems have continued into the 1980s and 1990s. For instance, high levels of PAH compounds have persisted in some bottom sediments and biota. Pervasive problems coupled to selenium and tributyltin contamination have not been resolved. DDT and PCBs are still bioaccumulating in shellfish, finfish, waterfowl,

Table 5.8 Effects of Selected Pollutants that Occur in the San Francisco Bay/Delta Estuary

Pollutant	Effects	Comments
Arsenic	Carcinogenic/mutagenic. Toxicity dependent on chemical form. Acutely toxic to most marine organisms. (1,2)	Effect on estuary biota unknown. Probably a pollutant of less concern. (9)
Cadmium	Carcinogenic/mutagenic/teratogenic. Highly toxic in aquatic environments. Bioaccumulates up to 250,000 times concentration in water. Of exceptional toxicity to mammals, including humans. (1,3,4)	A pollutant of greatest concern. Ubiquitous in bay. Levels in biota warrant health concern and further investigation. (1,9)
Chromium	Carcinogenic/mutagenic/teratogenic. Strongly accumulates in sediments and biota. Detrimental effects in biota at levels in water of 10 ppb. Accumulates highly in sediments. (1,5,6)	Poorly characterized in estuary. Large industrial source in Suisun Bay area. Concentrations in Bodega and Tomales Bay sediments also high. Elevated levels cause for concern and further investigation. (1,3)
Copper	Chronically toxic to marine organisms at concentrations in water of .01–10.0 ppm. Acutely toxic at concentrations in water greater than 0.1 ppm. Bioaccumulates in shellfish up to 30,000 times concentration in water. Highly bioavailable in the estuary. (1,3,4,5)	A pollutant of greatest concern. Elevated levels in water, sediment, and biota cause for further investigation. (3,9)
Lead	Carcinogenic/teratogenic. Chronically toxic to marine organisms at concentrations in water of 0.1 ppm. Bioaccumulates readily. Highly toxic to mammals. (1,3,4)	Given moderate toxicity and relatively even distribution, a problem only at specific sites. (3)
Mercury	Teratogenic. Most toxic of all trace elements. Effects occur at low parts per billion level. Wide range of acute and chronic toxicities to aquatic biota. Chronic toxicity to marine organisms occurs at concentrations in water of 1 ppb. Bioaccumulates in some aquatic biota at levels 100,000 times that in water. (1,3,4)	Possibly a pollutant of greatest concern. Given effect and high concentrations in biota, further investigation warranted. (3,9)
Nickel	Carcinogenic/mutagenic. Chronically toxic in water at levels greater than 0.1 ppm. Acutely toxic at concentrations above 1.0 ppm. (1,3)	Poorly characterized in estuary. Enrichment in sediments and biota is localized. (3)
Selenium	Teratogenic. Toxicity depends greatly on chemical form. Toxic effects occur at concentrations of 10 ppb in freshwater, 1 ppm dry mass in sediments, and 0.3 ppm wet weight in shellfish. (1,3,4)	A pollutant of greatest concern. Effects on biota, especially those higher in food web, and levels in water and biota warrant further investigation. (3,9)
Silver	One of the most hazardous trace elements, ranking second after mercury. Retards growth of sea urchin larvae at levels in water of 0.36 ppb. Kills American oysters at levels in water of 6 ppb. Kills clam embryos at levels in water of 13 ppb. Bioaccumulates at levels up to 3,000 times its concentration in water. (1,4)	A pollutant of greatest concern. High toxicity and levels in Central and South bay sediment and shellfish warrant further investigation. (3,9)
Tributyltin	Mutagenic/teratogenic. Toxicity highly dependent on chemical form. Toxic to aquatic biota at the parts per trillion range. Bioaccumulates in some biota to levels thousands of times greater than in water. (1,3,5)	Levels at marinas and harbors are sufficiently high to cause toxic effects in sensitive biota. (1)

Table 5.8 (continued) Effects of Selected Pollutants that Occur in the San Francisco Bay/Delta Estuary

Pollutant	Effects	Comments
Zinc	Moderately toxic. Chronically toxic to marine organisms at concentrations in water of about 0.05 ppm. Acute toxicity to marine and freshwater animals occurs at concentrations in water above 0.1 ppm. Bioaccumulates in shellfish to levels 100,000 times that of water. (1,3,4)	Toxicity and concentrations in sediment and biota indicate minor concern. (3)
PAH	Carcinogenic/mutagenic/teratogenic. Toxicity varies among chemicals. May bioaccumulate. (1,3)	Poorly characterized in estuary; sampling has occurred only since 1983. Effects on biota possible, but not well defined. (1,7)
DDT	Carcinogenic/teratogenic. Highly toxic and extremely persistent. Effects occur in many species of biota, and over a large range of concentrations. Causes reproductive impairment in fish and birds. Bioaccumulates at levels up to one million times that in water. (1,4,10)	Although contamination levels seem to have dropped in biota since the early 1980s, this chemical continues to enter the estuary from Central Valley soils. Localized contamination continues, especially at Lauritzen Canal. Overall impact on estuary biota is probably low. (3,7)
PCB	Carcinogenic. More persistent than DDT. Effects occur at extremely low concentrations. Bioaccumulates at levels up to one million times that in water. May affect reproduction in birds and mammals. (1,4)	Elevated levels in sediments and tissue are cause for concern. Increasing levels in black-crowned night heron linked to decreasing embryo weights and thin eggshells. (1,8)

Sources: (1) SWRCB, 1990; (2) PSWQA, 1988; (3) Phillips, 1987; (4) Callahan et al., 1979 in CBE, 1987; (5) CBE, 1987; (6) Eisler, 1986a in Phillips, 1987; (7) Long et al., 1988; (8) Davis et al., 1991; (9) Luoma and Phillips, 1988; (10) SCCWRP, 1988.

Source: Monroe, M. W. and Kelly, J., State of the Estuary, Tech. Rept., San Francisco Estuary Project, Oakland, CA, 1992.

and seals even though they have been banned from use for more than 20 years. They also may pose a threat to humans consuming certain contaminated seafood products from the estuary. In general, the concentration of most pollutants in bottom sediments and animals of the estuary do not appear to have decreased appreciably with the lower pollutant loads in recent decades.

Bioassay tests have indicated that effluent discharges from municipal and industrial facilities as well as runoff from some urban and rural areas are at times toxic to test organisms, such as sea urchins, sand dollars, mussels, oysters, and algae. These tests have also revealed that ambient bay waters may be moderately toxic themselves, although the toxicity may periodically worsen. In regard to sediment toxicity, samples collected from several sites in the bay have likewise elicited toxic responses in bioassays, including developmental abnormalities and high mortality of amphipods, mussels, and oysters.[2]

Future investigations must address several unanswered questions. For example, many pollutants have not been adequately surveyed in water, sediment, and biotic media of the estuary. This is necessary to establish if the concentration of these pollutants is actually increasing or decreasing. Aside from determining temporal trends, geographic trends in the concentrations of many pollutants must also be better characterized.

Table 5.9 Pollutant Trends in Sediments and Biota of the San Francisco Bay/Delta Estuary

Pollutant	Trends in Sediments	Trends in Biota
Arsenic	Few sites highly contaminated.	Data unavailable to determine geographic or temporal trend. (2)
Cadmium	Ubiquitous in the bay; patchy distribution. Possible increasing concentration from north to south. Highest concentrations in South Bay. Slight decrease in mean sediment concentrations since mid-1970s. (3)	Concentrations in mussels fairly uniform among various basins of San Francisco Bay. Highest concentrations in South Bay. Possible general pattern of slightly decreasing concentrations in mussels during the 1980s. Wide variation in concentrations in biota from year to year. (3)
Chromium	Spread throughout system. Concentrations higher in basins than on periphery. Highest levels in San Pablo Bay. No temporal trend apparent. (3)	Concentrations in mussels highest in Central and South Bays. There are no baywide temporal trends apparent among mussels. (3)
Copper	Spread throughout system. Concentrations higher on periphery than in basins. Data unavailable to determine temporal trend. (3)	Appears to be in similar concentrations in bivalves throughout San Francisco Bay; very patchy distribution. Mean concentrations similar in basins and peripheral areas, but highest levels occur in peripheral areas. No temporal trends in concentrations in biota are apparent. (3)
Lead	Spread throughout system at low concentrations. Concentrations highest on peripheral areas. No temporal trend apparent. (3)	Concentrations in mussels highest in peripheral areas. Concentrations in mussels highest in Central Bay and South Bay. Data unavailable to determine temporal trend. (3)
Mercury	Patchy distribution. Concentrations higher in peripheral areas. Highest mean concentrations on South Bay periphery. No temporal trend apparent. (1,3)	Concentrations fairly uniform in biota throughout San Francisco Bay. Highest levels in biota of South Bay. No significant temporal trend of increasing or decreasing concentrations. (3)
Nickel	Increasing concentrations from north to south. Highest concentrations in South Bay. No temporal trend apparent. (1)	Concentrations elevated in mussels from Carquinez Strait area and in clams from South Bay. In general, levels in biota poorly characterized. Data unavailable to determine temporal trend. (2)
Selenium	Few data available. Concentrations 3–4 times that in shales. Highest concentration in San Pablo Bay. Data unavailable to determine temporal trend. (2)	Concentrations in shellfish highest in northern and southern reaches of San Francisco Bay. Concentrations in ducks in South Bay and Suisun Bay are comparable to ducks from Kesterson National Wildlife Refuge that had reproductive problems. Recent increase in concentrations in North Bay scaup and sturgeon. (1,4)
Silver	Increasing concentrations from delta to South Bay. Highest concentration in Central and South Bays. No temporal trend apparent. (1,3)	Concentrations in shellfish increase along gradient from Delta to South Bay. No significant temporal trend of increasing or decreasing concentrations in biota. (1,3)
Tributyltin	Concentrations highest at marinas and harbors. No temporal trend apparent. (1)	Concentrations of TBT are highest in marinas and harbors throughout the estuary; however, data are unavailable to determine geographic and temporal trends in concentrations in biota. (2)

Table 5.9 (continued) Pollutant Trends in Sediments and Biota of the San Francisco Bay/Delta Estuary

Pollutant	Trends in Sediments	Trends in Biota
Zinc	Concentrations generally moderate and, with few exceptions, fairly uniform. Highest concentrations at sites in Central and South Bays. No temporal trend apparent. (1,2)	Concentrations in biota are moderately elevated. Highest concentrations occur in biota inhabiting peripheral areas of Central and South Bays. High concentrations in Sacramento River water above the estuary cause mortality in young salmon. Data unavailable to determine temporal trend. (1,2)
PAH	Concentrations higher in peripheral areas. Data unavailable to determine temporal trend. (3)	Concentrations in mussels highest in South Bay. Concentrations in fish highest in East Bay and lowest in San Pablo Bay. There is no apparent temporal trend in concentrations in biota. (3)
DDT	Concentrations higher in peripheral areas, with few exceptions. Data unavailable to determine long-term temporal trend. (3)	Concentrations in clams historically highest in Suisun Bay and Delta biota—lowest in San Pablo Bay. Concentrations in fish relatively similar at various sites, but somewhat lower in San Pablo Bay than in delta. Concentrations in oysters, clams, and mussels have declined steadily since the early 1980s. Possible decline in concentrations in striped bass. (3)
PCB	Widespread in system. Concentrations higher in peripheral areas. Concentrations lowest in San Pablo Bay. Data unavailable to determine temporal trend. (3)	Concentrations in clams and bottomfish highest in eastern Central Bay and in South Bay. Concentrations in San Pablo Bay typically low. There was an apparent peak in PCB levels in mussels in 1981, then a decline to current levels. Data are insufficient to determine trends in other biota. (3)

Sources: (1) SWRCB, 1990; (2) Phillips, 1987; (3) Long et al., 1988; (4) DFG, 1991.

Source: Monroe, M. W. and Kelly, J., State of the Estuary, Tech. Rept., San Francisco Estuary Project, Oakland, CA, 1992.

Based on current farming practices in the Central Valley, agricultural pollutants (primarily pesticides) should remain a major environmental issue in the San Francisco Bay/Delta Estuary. Urban runoff will take on greater importance as the lands surrounding the estuary become increasingly urbanized. With escalating population growth projected for the Central Valley and bay watersheds during the next several decades, anthropogenic impacts in the estuary will likely accelerate. Nonurban runoff, atmospheric deposition, accidental spills, vessel wastes, floatable debris, exploitation of marine resources, and various direct human activities in the estuary are also expected to increase. Effective pollution prevention programs must be established to mitigate future pollution impacts on the system and to minimize losses of valuable habitat. This will require careful planning by local, state, and federal administrators, scientists, and the general public to improve the way the estuary's land and water are managed.

C. Freshwater diversions and altered flows

1. Freshwater inflow

The estuary's freshwater supply is critical for several reasons. First, it supplies irrigation for nearly 2 million hectares of farmland and drinking water for more than 20 million people. Second, freshwater inflow greatly influences the physical and chemical characteristics of the

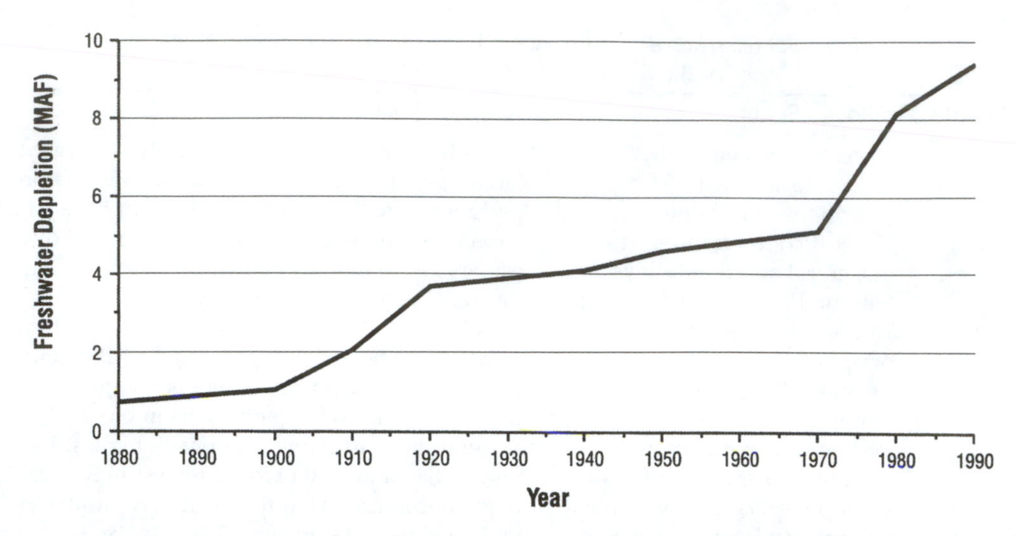

Figure 5.3 Long-term freshwater depletions upstream of the Sacramento-San Joaquin Delta. (From Monroe, M.W. and Kelly, J., State of the Estuary, Tech. Rept., San Francisco Estuary Project, Oakland, CA, 1992.)

estuary, including estuarine circulation patterns, water quality, and salinity. The interaction of freshwater inflow and tidal exchange modulates the salinity regime of the bay. Third, the abundance, distribution, and reproductive success of many estuarine organisms are contingent upon the amount and seasonal variation of freshwater flow through the bay. Since the 1920s, the total annual freshwater inflow to the estuary has ranged from $<7.4 \times 10^9$ m^3 to 7.4×10^{10} m^3 and averaged $\sim 2.96 \times 10^{10}$ m^3.

Most of the freshwater supply (~90%) for the estuary originates as precipitation in the Central Valley watershed. Of all the influent systems, the Sacramento River delivers ~80% of the freshwater flow to the estuary. However, the amount of freshwater entering the system not only depends on precipitation but also on the quantity of freshwater diverted within the delta and upstream of the delta in the estuary's watershed to meet agricultural, municipal, and industrial needs of the state. The annual total volume of freshwater diverted from the estuary's supply currently amounts to $>1.73 \times 10^{10}$ m^3. The largest volume of freshwater diverted to date occurred in 1987 and 1990 (>50% of the inflow). In 1993, 1994, 1995, and 1996, the freshwater supply diverted from the delta amounted to 5.67×10^9 m^3, 4.93×10^9 m^3, 6.17×10^9 m^3, and 6.41×10^9 m^3, respectively. Upstream diversions, in turn, reduced delta inflow by an average of $\sim 1.11 \times 10^{10}$ m^3 during these years, which represented about one third of the total annual inflow.[41]

Freshwater diversions upstream of the delta gradually increased through the 20th century (Figure 5.3). Along with in-delta uses and delta exports, upstream diversions have substantially reduced the annual delta outflow (Figure 5.4). During wet years diversions have reduced delta outflow by 10 to 30%, and during dry years they have reduced the outflow by 50 to 70%. The rate of diversions has been increasing during the past 2 decades to meet agricultural, industrial, and domestic demands. In extreme years (e.g., 1987 to 1988), the amount of water exported from the delta exceeds that flowing into the bay.

The total volume of freshwater that reaches the estuary varies considerably with the amount of precipitation in watershed areas. Seasonal variations of freshwater inflow are also conspicuous. For example, the rate of freshwater inflow into the bay ranges from a low of ~100 to 300 m^3/s in the summer and fall when dry conditions predominate to a high of ~1200 m^3/s in the winter when wet conditions prevail. As a consequence, the

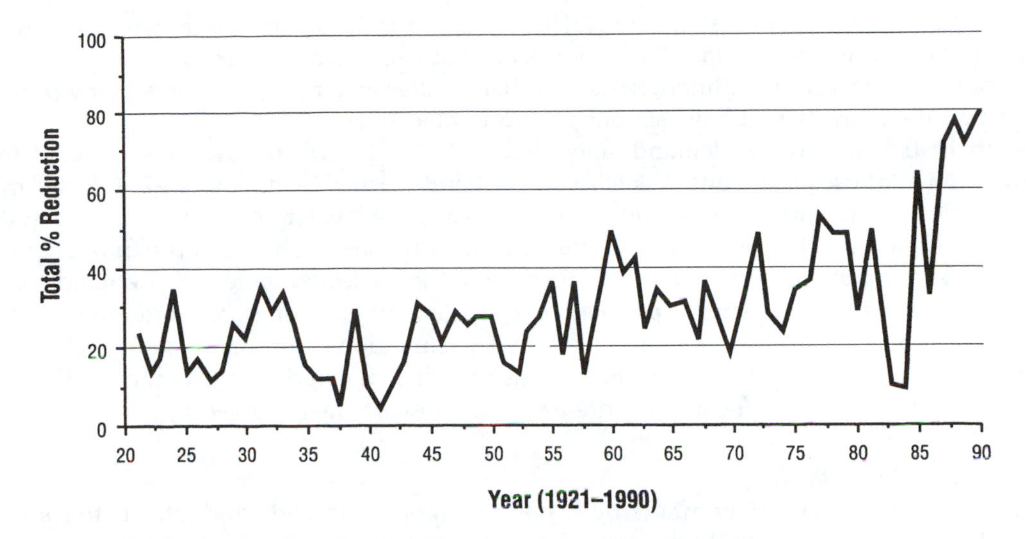

Figure 5.4 Long-term reduction in delta outflow caused by upstream diversions, in-delta uses, and delta exports, 1921 to 1990. (From Monroe, M.W. and Kelly, J., State of the Estuary, Tech. Rept., San Francisco Estuary Project, Oakland, CA, 1992.)

residence time of water in northern San Francisco Bay ranges from ~1 day during periods of peak winter freshwater inflow to ~2 months during times of restricted summer inflow.[42]

Protracted drought conditions, such as during the 1917 to 1920, 1923 to 1924, 1928 to 1934, 1976 to 1977, and 1987 to 1992 periods, greatly reduce riverine inputs.[43] Because of the pressing need for reliable water supplies during drought conditions, state and federal agencies have constructed canals, dams, and reservoirs to increase storage capacity. There are currently more than 100 reservoirs in the Central Valley watershed with a collective storage capacity of ~3.33×10^{10} m^3. Reservoir releases generally provide the principal source of freshwater inflow into the bay during the summer months.

The overall effect of the Central Valley reservoirs is to reduce the seasonal variation of freshwater flow to the estuary. Storage and flow-release operations of the reservoirs are such that the volume of water flowing downstream increases from late fall through spring. The reservoirs effectively capture snowmelt runoff in the spring from the Sierras and release it for agricultural use in the summer. They generally supply most of the freshwater flow to the bay during the summer months. In the winter, they operate primarily for flood control.[44]

Private entities, local municipalities, and state and federal governments operate the diversion and water storage systems. The two largest diversion sources, the federal Central Valley Project and the State Water Project, remove nearly 1.23×10^{10} m^3/yr from the estuary's freshwater supply, with ~85% used by agriculture and the remainder used by municipal, industrial, and other entities.[2] Much of the diversion of freshwater from the Sacramento and San Joaquin Rivers and their tributaries is coupled to irrigation needs of Central Valley farmlands during the dry season (May through October). Some of the water is exported in aquaducts for agricultural and municipal consumption in southern California. Because 90% of the freshwater entering the bay passes through the delta and Carquinez Strait and only 10% derives from the local bay watershed, the volume of water diverted for human uses has a significant effect on estuarine processes and biological communities.[45]

The federal Central Valley Project, the largest water development project in the world, is a multipurpose endeavor designed to regulate river flows, control floods, reclaim arid and semiarid lands, facilitate electric generation, improve navigation, and enhance fish and wildlife in the region. The State Water Project similarly is a multipurpose project that supplies water needs, provides flood control and recreational opportunities, and generates hydroelectric power. In addition, there are other smaller water projects operated by public and private interests that have similar goals and objectives.

In 1990, urban water demand amounted to 8.39×10^9 m^3, whereas the demand for environmental uses (wild and scenic rivers, wetlands, fisheries, etc.) was ~2.96×10^{10} m^3. Both are expected to increase substantially during the next two decades as the state population approaches 40 million. The bay/delta watershed satisfies approximately two thirds of the statewide water demand. Because of the projected increasing demand associated with a rapidly expanding population, management programs are promoting greater municipal and industrial water recycling and reclamation efforts and improved water conservation planning. However, the success of water management programs will continue to rely heavily on the aforementioned water development projects.

2. *Biotic impacts*

Freshwater diversions, river inflow, tides, bottom topography, and wind interact to control circulation and salinity patterns in the estuary. Changes in diversions and altered flow regimes have a great influence on the physical and chemical properties of the system. Freshwater flux from the delta causes an array of physical, chemical, and biological responses downestuary, which may be manifested by changes in salinity stratification, net flow direction of estuarine waters, pollutant concentrations, and biological production.

The abundance and production of many estuarine organisms have been correlated with the volume of freshwater input, which also largely regulates the salinity gradient downestuary. For example, high delta discharge reduces salinity, and it promotes the development of a stratified water column. Peak phytoplankton production and biomass have historically occurred during the spring months in South Bay when high delta outflow coincides with periods of low tidal current velocity resulting in well-developed water column (salinity) stratification. Under these conditions, phytoplankton populations are isolated in the upper sunlit waters of the estuary away from filter feeding bivalves and other bottom-dwelling grazers, and they attain high growth rates.[46-48]

Periods of extended drought, extremely low river inflow, and diminished water column stratification significantly lower primary production and suppress the pelagic food web of the system. Peterson et al.[13] demonstrated that escalating water diversions have accounted for rising salinity in Central Bay and concomitant changes in phytoplankton and other organisms residing there. Severely restricted freshwater inflow to the estuary in the summer of 1977 caused the phytoplankton biomass in the upper estuary to decline by 80% from normal levels, and the abundance of zooplankton also dropped markedly. Other notable effects included the absence of summer phytoplankton blooms in northern San Francisco Bay, low striped bass recruitment, and a general depression of the pelagic food web.[16]

In the northern reaches of the estuary, the location of the entrapment zone — which is strongly influenced by freshwater inflows — modulates phytoplankton production and abundance. When the entrapment zone is positioned at the upstream end of San Pablo and Suisun Bays, phytoplankton production and abundance peak in both embayments. Maximum phytoplankton production in Suisun Bay occurs when the delta outflow ranges from ~140 to 225 m^3/s.[49] Reduction in phytoplankton abundance and the number of desirable diatom species in the delta and Suisun Bay during the 1970 to 1990 period have been attributed to water diversions, natural perturbations (e.g., drought conditions), and introduced undesirable aquatic species.

The abundance and survival of many organisms in the estuary have been coupled to the location of the 2 psu (practical salinity units) near-bed isohaline, commonly referred to as "X2". The tidally averaged distance of the 2 psu isohaline from the Golden Gate Bridge has been used as an index of flow. In the past, periods of drought and low inflow have displaced the isohaline upestuary into the delta, whereas periods of heavy precipitation and high inflow have shifted the isohaline downestuary into the Carquinez Strait. The 2 psu isohaline usually occurs in the vicinity of the entrapment zone where nutrients and biota accumulate. As noted by Armor and Herrgesell,[50] the abundance of most of the estuarine species they investigated increased as a function of reduced salinity associated with greater freshwater inflow.

The significance of the 2 psu isohaline is reflected in new state water quality standards. Under the 1994 Bay-Delta Accord and the resulting 1995 State Water Quality Plan, the 2 psu isohaline was established as a key water quality standard to protect the estuarine environment. The standard restricts the upestuary displacement of the isohaline. To ensure that the standard is maintained, freshwater inflow must be regulated so that the isohaline remains within a range of positions near the Carquinez Strait. This requires careful monitoring of the isohaline position.

Freshwater diversions and outflow variability are major factors affecting abundance, distribution, reproductive success, and survival of finfish populations. These effects are most conspicuous in the delta and upstream areas. Freshwater diversions involve elaborate pumping operations that remove nutrients, phytoplankton, zooplankton, and finfish from the estuary. Agricultural diversions are largely unscreened, and thus millions of fish eggs and larvae as well as numerous juvenile and adult forms pass through the pumps. Although the diversion pumps of the State Water Project and federal Central Valley Project are screened, they do not prevent the entrainment of fish eggs and larvae. As a result, losses to entrainment for some species can be substantial.

The Department of Fish and Game in California has reported heavy mortality of American shad, salmon, striped bass, and several other species due to diversion pumping. For example, nearly 20 million striped bass were lost in 1986 due to diversion pumping operations. Annual losses of salmon approached 200,000 individuals between 1976 and 1986. With annual diversions from the delta by the State Water Project expected to increase by at least 1.11×10^9 m^3, entrainment will continue to be a significant source of mortality for many fish species.

Apart from increased mortality of eggs and young due to diversion pump entrainment, several other impacts on finfish populations have been ascribed to water resources development. These include altered water temperature, habitat loss and degradation, displacement of species into new areas, and the alteration of migration patterns of spawning adults and outmigrating young.[51] One or more of these factors have contributed to reduced abundance of chinook salmon (*Oncorhynchus tshawytscha*), striped bass (*Morone saxatilis*), American shad (*Alosa sapidissima*), delta smelt (*Hypomesus transpacificus*), and several other species. When comparing recent population abundance figures to historic levels, the following changes are evident: (1) the salmon population has declined by 70% since the early 1900s; (2) the adult striped bass has decreased by 83.3% since the early 1960s; (3) the American shad has dropped substantially since the 1940s; and (4) the delta smelt population has diminished markedly since the 1980s. Reductions in some wildlife species have also been attributed by many investigators to water development projects.

Reduced flow upstream and in the delta has clearly decreased the abundance and reproductive success of many fish species in the Sacramento-San Joaquin estuary. Flow variability can adversely affect recruitment in the following ways:

1. Low flows during incubation following high flows during spawning often result in dewatering of salmonid redds, causing mortality of eggs, embryos, and alevins of salmon. Many other fishes spawn around submerged objects, and their adhesive

eggs would then be subject to the same sort of mortality during years with sharp differences in outflow across a short time span.

2. Low flows expose a higher proportion of fish populations to possible entrainment by water diversions. A higher proportion of water is taken in years of low inflow, and greater numbers of fish are thus entrained.
3. Smaller river volumes increase the density of young fish in the river channels, thereby permitting more efficient foraging by predators.
4. Moderately high flows increase the diversity of habitats available, especially increasing the availability of shallow habitats where young fish enjoy greatly reduced predation pressure.
5. Moderately high spring/summer flows increase zooplankton abundance in the bay, resulting in more food available for larval striped bass and smelt.[52]

The impacts of altered freshwater inflow are more difficult to delineate in populations of fish in the bay than those in the delta. However, some trends in bay populations are beginning to emerge based on long-term studies conducted by the Department of Fish and Game of California. These studies have shown that certain species increase in abundance during wet years and high freshwater inflow (e.g., starry flounder, *Paralichthys stellatus*; delta smelt, *Hypomesus transpacificus*; longfin smelt, *Spirinchus thaleichthys*; staghorn sculpin, *Leptocottus armatus*; Pacific herring, *Clupea harengus pallasi*; and Pacific tomcod, *Microgadus proximus*), whereas others increase in abundance during dry years and low freshwater inflow (e.g., bay goby, *Lepidogobrius lepidus*; California halibut, *Paralichthys californicus*; and English sole, *Parophrys vetulus*). Many other species have exhibited no significant difference in abundance between wet (high flow) and dry (low flow) years. Although data continue to be compiled on the responses of finfish in the bay to altered flow regimes, many more investigations must be conducted to unequivocally establish the relationship between altered flow and changes in the ichthyofaunal structure of the estuary.

D. Waterway modification

1. Diking

European methods of agriculture were adapted in the Central Valley during the 1800s, resulting in the reclamation of vast areas of aquatic and wetlands habitat for cropland production. A part of this transformation entailed the diking of rivers and clearing of riparian vegetation in the lower valley that not only destroyed or degraded valuable habitat, but also directly impacted numerous species of organisms. Freshwater marsh habitat was severely damaged to the extent that its contribution of productivity to the downstream food web was essentially eliminated. Although reclamation of saltmarshes continued into the early 1970s, filling of freshwater marshlands in the delta was essentially complete by the early 1920s. Diked marsh has been greatest at the Suisun Bay area. Aside from being remnants of land reclamation, some diked marshes are the product of salt pond construction, flood control projects, and other development programs.[51]

Only 125 km^2 of undiked marsh still exist along the perimeter of the estuary, and, hence, there is considerable concern regarding the impact of past marshland elimination on the overall ecology of the bay. For example, tidal marshes are ecologically and economically important to the system because they serve as valuable nursery grounds for commercially and recreationally important fish. In addition, the loss of the marsh has removed potential spawning habitat for the Sacramento blackfish (*Orthodon microlepidotus*), Sacramento splittail (*Pogonichthys macrolepidotus*), prickly sculpin (*Cottus asper*), and possibly the longfin smelt (*Spirinchus thaleichthys*) and delta smelt (*Hypomesus transpacificus*), causing a dramatic reduction in abundance, especially of juveniles, in shallow areas. Waterbirds and wildlife populations also utilize the marsh habitat. About one half of the migratory birds on the Pacific Flyway overwinter on or near San Francisco Bay.[53] The loss of marsh as food and habitat can have a detrimental impact on waterfowl and other populations residing along the margins of the bay. Because tidal marshes serve as a source or sink of nutrients and organic matter, their losses may have a significant effect on system productivity. Today, $<3\%$ of the delta habitat remains in a state similar to that which existed 150 years ago, with $\sim 10\%$ of the delta consisting of aquatic habitat.[54]

2. *Dredging and dredged material disposal*

The San Francisco Delta/Bay Estuary is a system heavily used by commercial, recreational, and military vessels. However, because of the shallow expanse of most of the estuary ($\sim 70\%$ of the bay is <3 m deep) and the generally deep waters (~ 10 to 15 m or more) required by oceangoing vessels to effectively navigate, many areas (e.g., navigation channels, turning basins, marinas, etc.) must be dredged on a regular basis. The estuary receives more than 4.5×10^6 m^3 of sediment each year, mainly from the Sacramento and San Joaquin Rivers. This sediment tends to accumulate in waterways of the delta and bay, and it must be systematically excavated to maintain design depths.

The U.S. Army Corps of Engineers and the U.S. Navy conduct most dredging operations, although county flood control and reclamation districts, commercial marina operators as well as ports and refineries also remove significant volumes of sediment. All dredging projects require appropriate permits from the U.S. Army Corps of Engineers. Hopper, cutterhead, and clamshell dredges are the three types of dredging devices used in the estuary, with the hopper dredge being most commonly employed.

The U.S. Army Corps of Engineers dredged nearly 1.35×10^8 m^3 of sediment from the bay between the 1930s and mid-1980s. Together with the U.S. Navy, the U.S. Army Corps of Engineers dredged an annual mean sediment volume of more than 3.5×10^6 m^3 from both the delta and bay between 1975 and 1985. More than 1.2×10^7 m^3 of sediment were dredged from the estuary during 1986 and 1987.[2] In 1987, the U.S. Army Corps of Engineers was responsible for 14 major dredging projects in the estuary (Figure 5.5). Over the next 50 years, another 2.3×10^8 m^3 of sediment are expected to be dredged from the estuary.

During the past century, dredged material has been dumped at more than 30 sites in the delta, bay, ocean, and surrounding uplands. Figure 5.6 depicts the aquatic dredged material disposal sites in the estuary. The number of aquatic dredged material disposal sites in the bay has decreased through time from dozens between the 1930s and 1960s, to six in 1972, and only four in 1975. Three sites in the bay have received nearly all of the dredged material since 1975: San Pablo Bay (Site 4, Figure 5.6), Carquinez Strait (Site 8, Figure 5.6), and Alcatraz Island (Site 21, Figure 5.6). Of these sites, Alcatraz Island has been the most frequently used, receiving about two thirds of all dredged material during some years. Another aquatic disposal site in Suisun Bay (Site 10, Figure 5.6) has received limited amounts of material dredged from the Suisun Bay Channel. The total annual volume of dredged material dumped by the U.S. Army Corps of Engineers and U.S. Navy at the San Pablo Bay, Carquinez Strait, and Alcatraz Island sites has generally ranged from $\sim 1.5 \times 10^6$ m^3 to 8.5×10^6 m^3. Uplands that have received substantial amounts of dredged material over the years include the Mare Island, Petaluma River, San Leandro Marina, and Napa River sites. The annual cost of dredged material disposal operations is ~$24 to $46 million.

In 1989, the San Francisco Bay Regional Water Quality Control Board adopted amendments to the Regional Water Quality Control Plan, which revised the policy on dredging and dredged material disposal in the bay area. These amendments were designed to minimize water quality and biotic impacts of dredging and dredged material disposal

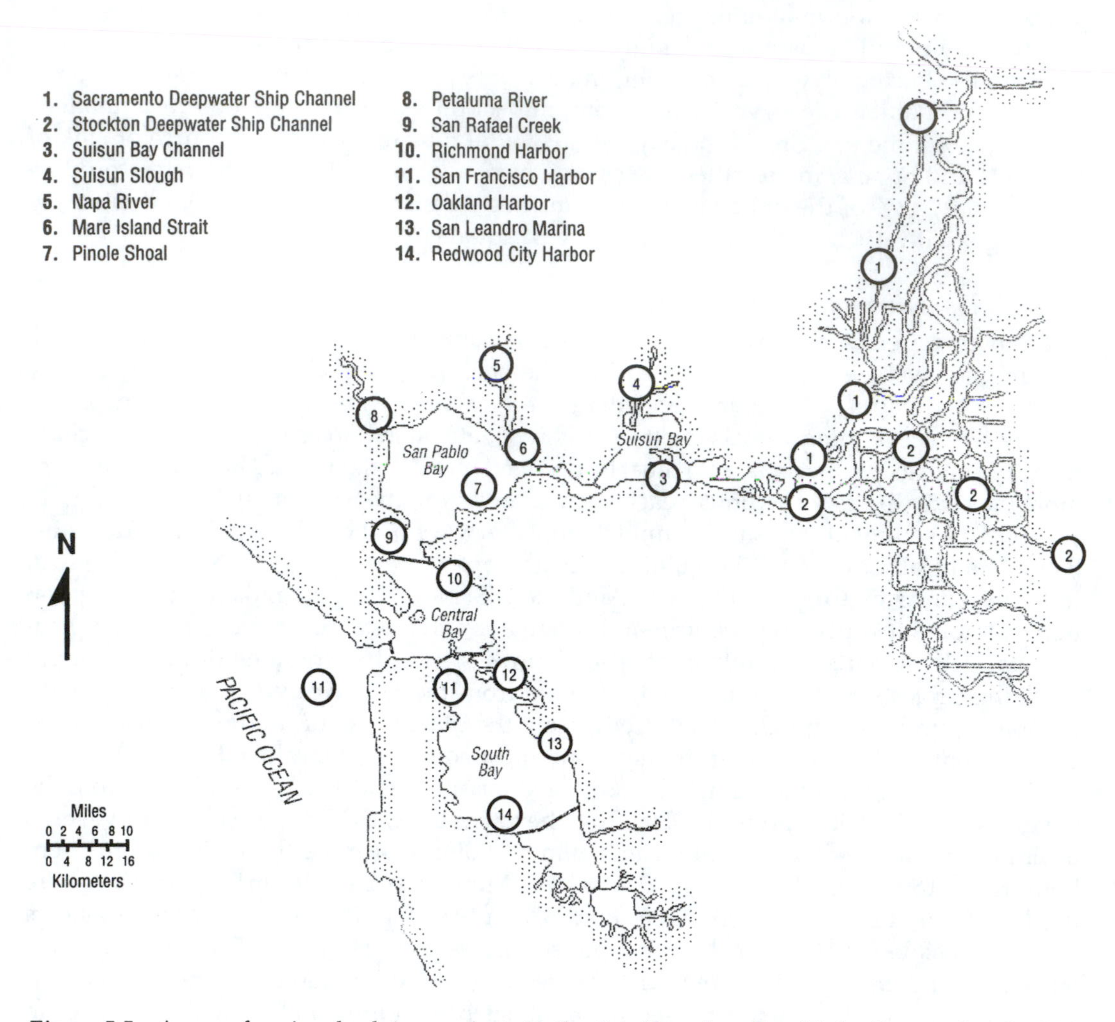

Figure 5.5 Areas of major dredging projects in the San Francisco Bay/Delta Estuary by the U.S. Army Corps of Engineers. (From Monroe, M.W. and Kelly, J., State of the Estuary, Tech. Rept., San Francisco Estuary Project, Oakland, CA, 1992.)

operations in the bay. The State Water Resources Control Board approved the amended policy of the Regional Board in 1990. Since that time, emphasis has been placed on land and ocean disposal of dredged material rather than mainly in-bay disposal. The strategy that has been accepted is one that minimizes environmental risks at any one disposal environment; it prescribes the use of a mix of bay, ocean, and upland sites rather than the current reliance on in-bay sites for disposal of ~90% of the dredged material.

The long-term management strategy proposes the following dredged material disposal mix: 20% at in-bay sites, 40% at ocean sites, and 40% at upland/wetland reuse sites. A deepwater ocean disposal site was approved in 1994. Another shift in policy stresses beneficial reuse of dredged material for such projects as landfill cover, construction fill, and wetland restoration. In recent years, 22 highly feasible potential reuse sites have been identified in the bay region by the principal long-term management strategy participants (i.e., U.S. Army Corps of Engineers, USEPA, Bay Conservation and Development Commission, San Francisco Bay Regional Water Quality Control Board, State Lands Commission, and

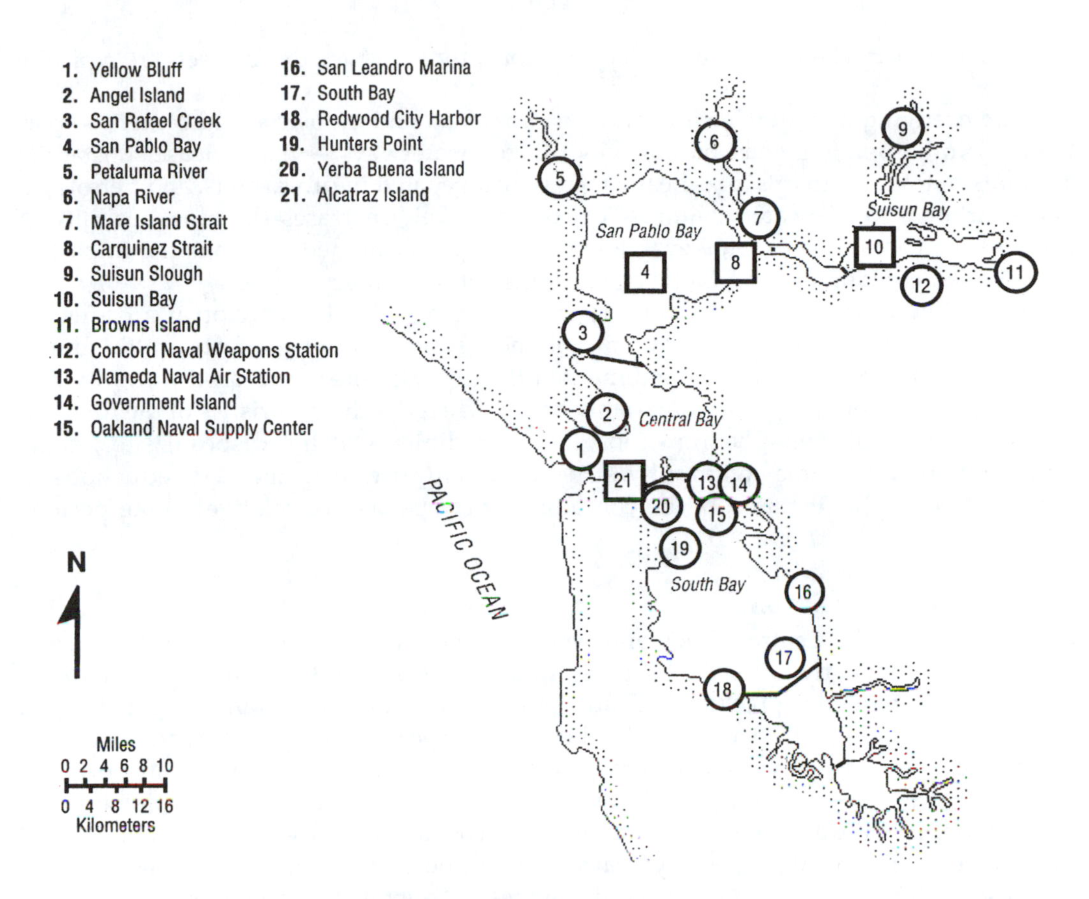

Figure 5.6 Subaqueous dredged material disposal sites in the San Francisco Bay/Delta Estuary. Squares indicate sites used after 1975. (From Monroe, M.W. and Kelly, J., State of the Estuary, Tech. Rept., San Francisco Estuary Project, Oakland, CA, 1992.)

California Environmental Protection Agency). Research by these participants indicates that of the estimated 2.3×10^8 m^3 of sediment expected to be dredged from the bay region during the next half century, 80 to 90% is clean enough for unconfined bay or ocean disposal, 10 to 20% needs alternative management, and <1% is deemed to be hazardous.

Potential impacts of dredging and dredged material disposal on water quality and living resources in the estuary are considerable. Dredging destroys the benthic habitat. Sediments are removed from the estuarine bottom, and some particle resuspension occurs due to the action of the dredge. Benthic organisms are excavated along with the dredged sediment, often resulting in their mass mortality. The entire benthic community is thereby disrupted until recolonization of the dredged site takes place. In some cases, the recolonization process may be protracted over one or more years.

Although the greatest immediate impacts on the benthic community and habitat are attributable to sediment removal from the estuarine floor, other adverse effects are associated with increased turbidity at the dredged site, redeposition of suspended sediment subsequent to dredging, and the release of toxic substances concomitant with the roiling of sediments during the dredging process. Plants appear to be particularly susceptible to increased turbidity and redeposition of suspended particulates. When turbidity and light attenuation are sufficiently high in the vicinity of the dredged site, primary production of phytoplankton and benthic flora can decrease abruptly. However, these effects are typically ephemeral.

The dredging of bottom sediments near urbanized and industrialized centers — particularly in harbors, marinas, and heavily utilized waterways — may release substantial amounts of toxic chemicals (e.g., heavy metals, halogenated hydrocarbons, and petroleum hydrocarbons) to the water column. This process likely increases the bioavailability of some of the contaminants. However, few studies have examined the potential impact of the released contaminants on biotic communities of the estuary.

Biotic impacts similar to those observed at dredging sites also develop at subaqueous dredged material disposal sites. Among the most serious adverse effects are the burial and smothering of organisms by sediment at the disposal sites. Increased turbidity and changes in sediment properties invariably occur. When the characteristics of the dredged material differ substantially from those of the sediments at the disposal sites, acute perturbations in benthic community structure often arise (e.g., marked reductions in species diversity). Alteration of the community may persist for relatively long periods of time.

3. Flood control projects

Development of the Central Valley and bay watershed areas accelerated rapidly during the late 1800s and early 1900s. At this time, tributary systems and surrounding lands were increasingly modified to preclude high stream flows and tides from inundating developed areas and damaging infrastructure. Flood control projects were devised to afford greater protection of homes, farmlands, commercial and industrial facilities, and transportation arteries. Today, flood control projects implemented in the Central Valley watershed are often multimillion dollar efforts funded through cost-sharing of local, state, and federal sources. The largest projects are typically federal flood control initiatives planned and constructed by the U.S. Army Corps of Engineers. Several dozen of these projects are either planned or under way in the bay area as well as in the Delta-Central Sierra, Sacramento, and San Joaquin Basins. In addition, there are numerous smaller, flood control projects on many influent systems. These projects usually employ several major flood control features, such as the construction of levees adjacent to channels to confine high flows within the channel, the lining of channels with concrete and the placing of rock revetment along banks to mitigate erosion, the deepening and straightening of channels to increase their capacity, the removal of riparian vegetation to facilitate high flows, and the use of dams and reservoirs to regulate discharges.[2] Of all these features, levees are most common in the delta. They play a crucial role in preventing thousands of hectares of seasonal formed wetlands from flooding and converting the delta to open water conditions. However, they drastically modify hydrological conditions.

Flood control features frequently impact natural stream channels and adjacent riparian vegetation. Habitat values and biodiversity of associated aquatic and terrestrial communities generally decline. Recognizing the unavoidable negative environmental effects of channelization, bank protection, and the removal of riparian vegetation, the general public is beginning to seek alternatives to traditional flood control methods. There is growing interest in restoring altered urban creeks, and pursuing more environmentally friendly practices to control flooding and erosion problems. An example is the use of wetland vegetation — rather than the application of engineering structures — to minimize erosion. Future flood control efforts in the delta and Central Valley watershed will likely include a mix of traditional engineering practices and a new generation of environmentally friendly initiatives.

4. *Sea level rise*

Increasing rates of sea level rise, attributable by many atmospheric and marine scientists to global warming, will gradually affect shallow water and shoreline habitats of the estuary. Mean sea level may rise by as much as 0.5 to 1.5 m during the 21st century. The most notable impacts anticipated from the rising sea level are

- Saltwater intrusion in tidal marshes, freshwater tributaries, and groundwater
- Submergence of tidal marshes
- Increased periodic flooding of previously protected low lying areas around the bay and in the delta
- Increased shoreline and beach erosion[2]

5. *Mining and siltation*

Hydraulic mining in the Sierras had a devastating effect on siltation of influent systems and the introduction of large volumes of sediment into the estuary during the past century. An estimated 1.15×10^9 m^3 of sediment have entered the Sacramento-San Joaquin Estuary due to hydraulic mining, and much has been deposited in San Pablo Bay (4.36×10^8 m^3), South and Central Bays (2.49×10^8 m^3), and Suisun Bay (1.55×10^8 m^3). The influence of hydraulic mining on sediment flux to the estuary gradually abated through the 20th century.

E. *Introduced species*

One of the most serious biological problems in San Francisco Bay is the proliferation of introduced species that threaten the ecological balance of the system. At least 234 invertebrate, fish, bird, and other non-native species have been introduced into the San Francisco Bay/Delta Estuary, and many of them have become remarkably successful, dominating many habitats in terms of number of species, number of individuals and biomass, and rate of invasion.[55] These introductions have been the result of either deliberate attempts to create a profitable fishery (e.g., crayfish, oysters, and striped bass) or unintentional actions, such as the entry of exotic species in cargo ship ballast water. Today most of the common macroinvertebrates along the inner shallows of the bay are introduced species.[56,57] Exotic species now comprise more than half of the fishes in the delta.[57] In addition, some zooplankton, algal, and bird populations are also introduced forms. Biological communities in the estuary currently consist of a large fraction of hardy, adaptable, temporally variable, and spatially patchy introduced populations.[58] Cohen and Carlton[55] have suggested that this estuary may be the most heavily invaded in the world due to a large number and variety of transport vectors, a depauperate native biota, and extensive natural and anthropogenic disturbances.

A few of the earliest introductions in the estuary were east coast invertebrates accidentally shipped with oysters (*Crassostrea virginica*) to the bay area during the 19th century. Recreationally and commercially important shellfish introduced at this time included the soft-shelled clam, *Mya arenaria*. The Japanese littleneck clam (*Tapes japonica*) introduced from other regions also established a viable population in the estuary. Less-desirable invertebrate populations that expanded rapidly in the bay were oyster drills (*Urosalpinx cinerea*) and shipworms (*Teredo navalis*). Damage to wooden structures, boats, bridges, docks, and piers by shipworms soon developed into a major crisis in the late 1800s and early 1900s.

New invertebrate species continue to arrive. The most notable is the Asian clam (*Potamocorbula amurensis*), which has rapidly altered the planktonic and benthic communities in various reaches of the estuary and has had far reaching impacts on fish and wildlife as well. First discovered in the estuary in 1986, this small bivalve has overgrazed and decimated the phytoplankton community in Suisun Bay and has outcompeted and displaced more favorable clam species (e.g., *Macoma balthica* and *Mya arenaria*). Its rapid feeding and reproductive rates could also have a profound effect on the rest of the food web. Two other invading invertebrate species of note include the Chinese mitten crab (*Eriocheir sinensis*), first collected in the bay in 1992, and the green crab (*Carcinus maenas*), first collected in 1989 or 1990.[1]

In regard to introduced fish, 30 species have successfully invaded the estuary, mainly in brackish and freshwater habitats (e.g., inland silversides, *Menidia beryllina*).[55] The striped bass was introduced in the estuary in 1879 from the east coast of the U.S., and it now is a prized sport fish. Some 20 of the 42 fish species inhabiting the sloughs of Suisun Bay marshes are introduced forms. As in the case of introduced invertebrates, these non-native fish populations may create instability in the estuary's biotic communities.

V. Habitat and living resources

A. Habitats

Bottom sediments, open waters, wetlands, and riparian zones represent critically important habitat in the San Francisco Bay/Delta system. Many organisms utilize these habitats for food, shelter, and reproduction. Included here are an estimated 300 species of birds and mammals, 150 species of fish, 35 species of reptiles and amphibians, and hundreds of species of invertebrates and plants.[59] In addition, 97 endangered, threatened, rare, and special status species, or candidates for such listings, are either directly or indirectly dependent on wetland and aquatic habitats for some part of their life cycle.[60] Tables 5.10 through 5.12 provide a list of these species types found in wetlands habitat of the estuary.

The SFEP has identified the following wetlands and related habitats of significance in the system:

- Intertidal mudflats and rocky shore (estuarine)
- Tidal salt and brackish marsh, and tidal channels or pond connections (estuarine)
- Seasonal and permanent marshes (palustrine)
- Tidal freshwater marshes (palustrine)
- Riparian woodland (palustrine)
- Salt ponds (lacustrine)
- Lakes and ponds (lacustrine/palustrine)
- Perennial and intermittent streams (riverine)
- Adjacent upland habitat (and transition area)
- Open water

Table 5.13 shows the areal distribution of these habitats.

Wetlands are vital to the health and long-term viability of the system. Apart from providing habitat for hundreds of aquatic and wildlife species, they serve a significant role in water quality maintenance, shoreline stabilization, flood control, floodpeak reduction, stormwater flow attenuation, and food chain support. Wetlands also act as an important filter of pollutants which derive from runoff, sewage treatment plant effluent, and other sources.

There are four major wetland systems in the San Francisco Bay/Delta: estuarine, riverine, lacustrine, and palustrine (Table 5.14). Palustrine and estuarine wetlands are most common, comprising >75 and 17% of the estuary's wetlands, respectively. Farmed wetlands form the most extensive wetlands habitat type (1562 km^2). In total, the wetlands cover 2546 km^2. This is greater than the areal coverage of the open water habitat (1197 km^2) but much less than that of the upland habitat (6420 km^2).

Wetlands have been severely impacted by decades of reclamation and conversion activities. Of the 2212 km^2 of tidal marsh occurring in the system during the 1850s, only 180 km^2 remained by the mid-1980s, representing a 92% loss of wetlands habitat. Most of the tidal wetlands were converted to agriculture and pasture lands, salt ponds, and diked wetlands. The losses were particularly acute in the delta, where more than 97% of the original tidal wetlands area was converted to farmland or other uses. However, 82% of the original tidal wetlands area in the bay was also lost or converted. In addition, there have been significant losses of riparian woodland, freshwater marsh, and other wetland types. Habitat areal changes in the lower estuary are recorded in Table 5.15.

The rate of wetlands loss and degradation has slowed dramatically since 1970. Nevertheless, conservation efforts have been accelerated to preserve the remaining habitat. For example, ~19% of the wetlands area of the estuary (490 km^2) is now protected in parks, preserves, and refuges. The cumulative effects of the loss and degradation of wetlands are: (1) the reduction of habitat for numerous aquatic and wildlife populations; (2) the decrease in biodiversity of the system; and (3) the reduction in functional values, such as flood control, water quality improvement, and food web energy availability.[59]

Although wetland losses have declined appreciably since 1970, they still continue primarily as a result of urban expansion. Development in watershed areas has contributed to ongoing encroachment fragmentation and habitat degradation. Particular threats include shoreline development, highway construction, and airport expansion. To counter these effects, restoration and enhancement programs have been initiated to expand the area of permanently protected wetlands. Local interests, private organizations, and government agencies have also pursued the acquisition of additional wetlands habitat for long-term preservation and protection. Since 1993, for example, 32 km^2 of wetlands habitat have been restored and enhanced, and more than 80 km^2 of wetlands habitat have been acquired for the public trust. As of 1996, ~275 km^2 of wetlands had been protected in the Central Valley and Suisun Marsh under perpetual conservation easements (private, state, and federal).[1]

The future trend is to place threatened wetlands in the domain of public ownership, although private entities will continue to play an important role in safeguarding these critically important habitats. It is expected that restoration of tidal action to former wetlands will be a high priority, especially along salt ponds and diked farmed baylands. The continuation of no net loss policies emphasizing the avoidance of destruction or degradation of wetlands will be stressed.

B. *Living resources*

1. *Habitats and biological communities*

The SFEP has grouped habitats into two broad categories for the purpose of describing the estuary's living resources: (1) wetlands and deepwater habitats and (2) uplands. These habitats sustain the entire estuarine ecosystem. Wetlands and deepwater areas encompass a range of habitats that support numerous species of organisms, including many valuable species of recreational and commercial importance. Table 5.16 provides a list of these habitats and the representative plant and animal populations occupying them.

Most land area in the estuary region consists of upland habitats. Included here are seven principal habitat types: grassland, coastal scrub, mixed chaparral, oak woodland, broad-leaved evergreen forest, agricultural land, and urban land. All are extremely important for

Table 5.10 Endangered, Threatened, and Rare Plants Found in the Wetlands of the San Francisco Bay/Delta Estuary

Species	Status[a]	Location[b]	Wetland Type
Listed Species			
Palmate-bracted bird's-beak (*Cordylanthus palmatus*)	FE, SE	SB	Seasonal wetlands, alkali sinks
Solano grass (*Tuctoria mucronata*)	FE, SE	D	Vernal pools
Soft-haired bird's beak (*Cordylanthus mollis* subsp. *mollis*)	SR, FC 1	NB	Tidal salt marshes
Mason's quillwort (*Lilaeopsis masonii*)	SR, FC 2	NB, SM, D	Tidal brackish marshes and pilings
Colusa grass (*Neostapfia colusana*)	SE, FC 2	D	Vernal pools
White sedge (*Carex albida*)	SE, FC 1	NB	Freshwater marshes
Pitkin marsh indian paintbrush (*Castilleja uliginosa*)	SF, FC 1	NB	Freshwater marshes
Fountain thistle (*Cirsium fontinale* var. *fontinale*)	SE, FC 2	SB	Wet seeps and streams
Delta button celery (*Eryngium racemosum*)	SE, FC 1	D	Riparian habitat
Burke's goldfields (*Lasthenia burkei*)	SE, FC 2	NB	Vernal pools
Pitkin marsh lily (*Lilium pitkinense*)	SE, FC 1	NB	Freshwater marshes
Sebastopol meadowfoam (*Limnanthes vinculans*)	SE, FC 2	NB	Vernal pools
Slender orcutt grass (*Orcuttia tenuis*)	SE, FC 1	D	Vernal pools
Sticky orcutt grass (*Orcuttia viscida*)	SE, FC 1	D	Vernal pools
Kenwood marsh checkerbloom (*Sidalcea oregana* subsp. *valida*)	SE, FC 2	NB	Freshwater marshes
Napa blue grass (*Poa napensis*)	SE, FC 2	NB	Moist meadows

Candidate Species			
Gairdner's yampah (*Perideridia gairdneri* subsp. *gairdneri*)	FC 2	SB, NB, SM	Wet meadows
Glabrous allocarya (*Plagiobothrys glaber*)	FC 2	SB, NB, SM	Vernal pools
Bearded allocarya (*Plagiobothrys hystriculus*)	FC 2	NB, SM, D	Vernal pools
Calistoga allocarya (*Plagiobothrys strictus*)	FC 2	NB	Vernal pools
Marin knotweed (*Polygonum marinense*)	FC 2	NB	Tidal salt marsh
California beaked-rush (*Rhynchospora californica*)	FC 2	NB, SM	Wet meadows
Showy indian clover (*Trifolium amoenum*)	FC 2	SB, NB, SM, D	Vernal pools
Swamp sandwort (*Arenaria paludicola*)	FC 2	SB	Freshwater marsh
Mt. Hamilton thistle (*Cirsium fontinale* var. *campylon*)	FC 2	SB	Seeps
Legenere (*Legenere limosa*)	FC 2	SB	Vernal pools
Suisun aster (*Aster chilensis* var. *lentus*)	FC 2	NB, SM, D	Riparian habitat
Suisun thistle (*Cirsium hydrophilum* var. *hydrophilum*)	FC 1	NB, SM	Brackish tidal marshes
North Coast bird's-beak (*Cordylanthus maritimus* subsp. *palustris*)	FC 2	NB, SM	Tidal salt marshes
Hispid bird's-beak (*Cordylanthus mollis* subsp. *hispidus*)	FC 2	NB, SM, D	Brackish and salt tidal marshes
San Francisco gumplant (*Grindelia maritima*)	FC 2	SB, NB	Tidal salt marshes
California hibiscus (*Hibiscus californicus*)	FC 2	NB, SM, D	Riparian habitat
Delta tule-pea (*Lathyrus jepsonii* subsp. *jepsonii*)	FC 2	SB, NB, SM, D	Riparian habitat
Valley spearscale (*Atriplex patula* subsp. *spicata*)	FC 2	NB, SM, D	Alkali sink
Hinds' walnut (*Juglans hindsii*)	FC 2	NB, SM	Riparian habitat
Contra Costa goldfields (*Lasthenia conjugens*)	FC 2	SB, NB, SM	Vernal pools

[a] FE — federally endangered; SE — state endangered, SR — state rare; FC 1 — federal category 1: taxa for which the Fish and Wildlife Service has sufficient biological information to support a proposal to list as endangered or threatened; FC 2 — federal category 2: taxa for which existing information indicated may warrant listing, but for which substantial biological information to support a proposed rule is lacking.

[b] SB — South San Francisco Bay; NB — North San Francisco Bay (including San Pablo Bay); SM — Suisun Marsh and Bay; D — Sacramento–San Joaquin Delta.

Source: Meiorin, E. C., Josselyn, M. N., Crawford, R., Calloway, J., Miller, K., Richardson, T., and Leidy, R. A., Status and Trends Report on Wetlands and Related Habitats in the San Francisco Estuary, Tech. Rept., San Francisco Estuary Project, U.S. Environmental Protection Agency, San Francisco, CA, 1991.

Table 5.11 Listed Endangered, Threatened, and Rare Animals Found in Wetlands of the San Francisco Bay/Delta Estuary

Species	Status[a]	Location[b]	Wetland Type
		Birds	
Aleutian Canada goose (*Branta canadensis leucopareia*)	FE, SE	SB, NB, SM, D	Seasonal and permanent marshes, palustrine farmed wetlands, reservoirs
American peregrine falcon (*Falco peregrinus anatum*)	FE, SE	SB, NB, SM, D	Foraging over all wetland types, except riparian
California brown pelican (*Pelecanus occidentalis californicus*)	FE, SE	SB, NB	Salt ponds, tidal lagoons, open bay
California clapper rail (*Rallus longirostris obsoletus*)	FE, SE	SB, NB, SM	Tidal salt marshes
California least tern (*Stema antillarum browni*)	FE, SE	SB, NB, SM	Salt ponds, tidal lagoons, open bay
California black rail (*Laterallus jamaicensis coturniculus*)	FC, ST	NB, SM	Tidal salt marshes
Swainson's hawk (*Buteo swainsoni*)	ST	D	Riparian woodland
Greater sandhill crane (*Grus canadensis tabida*)	ST	D	Farmed wetlands
		Mammals	
Salt marsh harvest mouse (*Reithrodontomys raviventris*)	FE, SE	SB, NB, SM	Tidal salt marshes, diked seasonal salt marshes and transitional habitat
		Reptiles	
San Francisco garter snake (*Thamnophis sirtalis tetrataenia*)	FE, SE	SB	Freshwater ponds, lakes, marshes, sloughs, and slow-moving streams
Giant garter snake (*Thamnophis couchi gigas*)	FC, ST	D	Freshwater marshes, slow-moving streams
		Fish	
Chinook salmon–winter run (*Oncorhynchus tshawytscha*)	FT, SE[c]	SB, NB, SM, D	Open water
		Invertebrates	
Valley elderberry longhorn beetle (*Desmocerus californicus dimorphus*)	FT	D	Riparian habitat
California freshwater shrimp (*Syncaris pacifica*)	FE, SE	NB	Freshwater streams with underwater vegetation and exposed roots of riparian vegetation
Delta green ground beetle (*Elaphus viridus*)	FT	D	Vernal pools

[a] FE — federally endangered; FT — federally threatened; SE — state endangered; ST — state threatened; SR — state rare; FC — federal candidate species for listing.

[b] SB — South San Francisco Bay; NB — North San Francisco Bay (including San Pablo Bay); SM — Suisun Marsh and Bay; D — Sacramento–San Joaquin Delta.

[c] Emergency listed, effective 8/4/89–11/28/90.

Source: Meiorin, E. C., Josselyn, M. N., Crawford, R., Calloway, J., Miller, K., Richardson, T., and Leidy, R. A., Status and Trends Report on Wetlands and Related Habitats in the San Francisco Estuary, Tech. Rept., San Francisco Estuary Project, U.S. Environmental Protection Agency, San Francisco, CA, 1991.

Table 5.12 Candidate Endangered Animals Found in Wetlands of the San Francisco Bay/Delta Estuary

Species	Status[a]	Location[b]	Wetland Type
		Birds	
Alameda song sparrow (*Melospiza melodia pusillula*)	2	SB	Tidal salt marshes
Suisun song sparrow (*Melospiza melodia maxillaris*)	2, SC	SM	Brackish marshes
San Pablo song sparrow (*Melospiza melodia samuelis*)	2	NB	Tidal salt marshes
Tricolored blackbird (*Agelaius tricolor*)	2	D	Freshwater emergent marshes
Western snowy plover (*Charadrius alexandrinus nivosus*)	2	SB	Salt ponds and their levees, intertidal mudflats
Saltmarsh common yellowthroat (*Geothlypis trichas sinuosa*)	2	SB, NB, SM	Tidal and diked salt and brackish marshes, freshwater marshes, riparian woodland
Long-billed curlew (*Numenius americanus*)	2	SM, D	Palustrine farmed, freshwater marshes
		Mammals	
Suisun ornate shrew (*Sorex ornatus sinuosus*)	1	NB, SM	Tidal salt and brackish marshes
Salt marsh vagrant shrew (*Sorex vagrans halicoetes*)	1	SB	Tidal salt marshes
		Reptiles	
California tiger salamander (*Ambystoma tigrinum californiense*)	2	D	Ponds, reservoirs, seasonal wetlands, slow-moving streams
Red-legged frog (*Rana aurora draytoni*)	2	D	Permanent marshes, slow-moving streams, ponds, and reservoirs
		Fish	
Delta smelt (*Hypomesus transpacificus*)	1	NB, SM, D	Dead-end sloughs
Sacramento splittail (*Pogonichthys macrolepidotus*)	2	NB, SM, D	Dead-end sloughs, tidal brackish marshes
Sacramento perch (*Archoplites interruptus*)	2	D	Dead-end sloughs
		Invertebrates	
San Francisco forktail damselfly (*Ischnura gemina*)	2	SB	Still-water emergent wetlands
Sacramento Valley tiger beetle (*Cicindela hirticollis abrupta*)	2 R	D	Sandy areas adjacent to creeks and streams
Ricksecker's water scavenger beetle (*Hydrochara rickseckeri*)	2	SB, NB	Creeks, ponds, vernal pools
Curved-foot hygrotus diving beetle (*Hygrotus curvipes*)	2	D	Alkali vernal pools and wetlands

[a] 1 — Category 1: taxa for which the Fish and Wildlife Service has sufficient biological information to support a proposal to list as endangered or threatened; 2 — Category 2: taxa for which existing information indicated may warrant listing, but for which substantial biological information to support a proposed rule is lacking; 2R — recommended for Category 2 status; SC — state candidate for listing.

[b] SB — South San Francisco Bay; NB — North San Francisco Bay (including San Pablo Bay); SM — Suisun Marsh and Bay; D — Sacramento–San Joaquin Delta.

Source: Meiorin, E. C., Josselyn, M. N., Crawford, R., Calloway, J., Miller, K., Richardson, T., and Leidy, R. A., Status and Trends Report on Wetlands and Related Habitats in the San Francisco Estuary, Tech. Rept., San Francisco Estuary Project, U.S. Environmental Protection Agency, San Francisco, CA, 1991.

Table 5.13 Approximate Area of Wetland, Open Water, and Upland Habitats within the San Francisco Estuary[a]

Habitats	Delta	Suisun Bay	San Francisco Bay
Wetland Habitat			
Estuarine intertidal mudflat/rocky shore	130	2,428	23,399
Estuarine intertidal salt/brackish/freshwater marsh	3,330	4,326	10,314
Palustrine seasonal wetlands	6,683	19,230	8,566
Palustrine farmed wetlands	141,890	3,266	11,074
Riparian woodland, riverine	3,964	164	940
Salt ponds	22	11	14,824
Nonwetland Habitat			
Perennial lakes and ponds	5,055	1,428	5,411
Open water	18,550	11,440	77,804
Upland	121,932	105,903	414,460

[a] Values in hectares

Source: Meiorin, E. C., Calloway, J., Josselyn, M. N., Crawford, R., Miller, K., Pratt, R., Richardson, T., and Leidy, R., *Status and Trends Report on Wetlands and Related Habitats in the San Francisco Estuary*, Tech. Rept., San Francisco Estuary Project, U.S. Environmental Protection Agency, San Francisco, CA, 1991.

Table 5.14 Classification of Bay/Delta Estuary Wetlands

System	Subsystem	Estuary Wetland Type
Estuarine	Intertidal	Mudflats
		Rocky shore
		Saltmarsh
		Brackish marsh
Riverine	Tidal	Permanent and intermittent streams
	Perennial	
	Intermittent	
Lacustrine	Limnetic	Salt ponds
	Littoral	Lakes and ponds
Palustrine	—	Freshwater marsh
		Riparian forest
		Seasonal wetlands
		Farmed
		Diked marsh
		Vernal pools
		Abandoned salt ponds

Source: Meiorin, E. C., Calloway, J., Josselyn, M. N., Crawford, R., Miller, K., Pratt, R., Richardson, T., and Leidy, R., *Status and Trends Report on Wetlands and Related Habitats in the San Francisco Estuary*, San Francisco Estuary Project, U.S. Environmental Protection Agency, San Francisco, CA, 1991.

living resources of the region, serving as critical nesting, foraging, resting, and reproductive habitat for many wildlife populations.

Habitat reclamation and conversion activities during the past 150 years have significantly affected biotic communities inhabiting wetlands, deepwaters, and uplands of the system. As noted previously, development has drastically reduced wetlands habitat area.

Table 5.15 Changes in areal extent of intertidal and subtidal habitats in the lower estuary from historical to modern times[a,b]

Habitat	Historical Period	Modern Period
Intertidal Bay Flat	198	116
Young Tidal Marsh	5	69
Mature Tidal Marsh	773	68
Muted Tidal Marsh	?	23
Lagoon	0.2	9
Shallow Bay	680	680
Deep Bay	387	334

[a] Data from San Francisco Estuary Project, Bay Area EcoAtlas, 1997.

[b] Data in km^2.

Open water habitat has also decreased by one third. More than one half of native upland habitat has likewise been lost, with urban land conversion being largely responsible for the diminished habitat area. A decline in living resources has accompanied these long-term habitat impacts.

a. Deepwater habitats. These habitats include permanent and often deep open water areas in the bay, delta, and tributary river channels. Some of the richest biological resources in the system occur in these habitats. Extensive communities of phytoplankton, zooplankton, benthos, fish, birds, and mammals have been described (see subsequent detailed discussion on aquatic biotic communities). However, these communities have experienced substantial change during the past century due to intensified human activities in the Central Valley and bay watersheds.

b. Wetlands. Wetlands habitat is characterized by undrained, waterlogged substrate and water-adapted plants.[61] The delta as well as Suisun, San Pablo, and South Bays, exhibit the most extensive wetlands habitat. In the delta, farmed wetlands predominate, although freshwater marsh, riparian forest, and other significant habitat also exist and support many aquatic and wildlife species. Diked saltmarshes and brackish marshes dominate in the Suisun Bay area, whereas farmed wetlands and intertidal mudflats dominate in San Pablo Bay. Intertidal mudflats, seasonal wetlands, and salt ponds are abundant in South Bay.

The large expanses of intertidal mudflats in San Pablo and South Bays support algal populations and diverse communities of benthic invertebrates, fish, and wildlife. Green algae, red algae, and diatoms dominate the plant communities. The benthic macroinvertebrate communities consist of large numbers of bivalves (clams and mussels), polychaete and annelid worms, and crustaceans (amphipods, crabs, and shrimp). Opportunistic species with high reproductive potential often dominate these communities. Shorebirds feed heavily on the benthic invertebrates during low tide, and fishes consume them during high tide.

As with intertidal mudflats, rocky shore habitats are physically stressed. However, they are more disturbed environments subjected to the high energies of breaking waves. All of the rocky shores in the system occur in San Pablo and Central Bays. Aside from the ribbed mussel (*Ischadium* [*Modiolus*] *demissum*), few invertebrate populations proliferate along these shores, and there is a limited number of wildlife species. Among the latter are several species of cormorants, brown pelicans, gulls, shorebirds, harbor seals, and sea lions. Green and red algae are the most abundant flora.

Table 5.16 Representative Plants and Animals of the Estuary's Wetland and Deep Water Habitats

Plants	Invertebrates	Fish	Birds	Mammals
		Open Water		
Diatoms	Opossum shrimp	Chinook salmon	Western grebe	Harbor seal
Dinoflagellates	Bay shrimp	Striped bass	Brown pelican	California sea lion
Blue-green algae	Asian clam	American shad	Scaup	
Green algae	(*Potamocorbula*)	Green sturgeon	Canvasback	
Eelgrass		Pacific herring	Surf scoter	
		Northern anchovy	Osprey	
		Intertidal Mudflat		
Sea lettuce	Clams	Sharks	Western sandpiper	Harbor seal
Green algae	Amphipods	Rays	Dunlin	
Red algae	Polychaete worms	Longfin smelt	Marbled godwit	
Diatoms	Bay mussel	Staghorn sculpin	Willet	
	Annelid worms	Starry flounder	American avocet	
		Rocky Shore		
Green algae	Ribbed mussel	—	Brown pelican	Harbor seal
Red algae			Cormorant	California sea lion
Sea lettuce			Black oystercatcher	
			Western gull	
		Salt Marsh		
Pickleweed	Ribbed mussel	Topsmelt	California clapper rail	Salt marsh harvest mouse
Cordgrass	Baltic clam	Arrow goby	California black rail	Salt marsh vagrant shrew
Saltgrass	Hornsnail	Yellowfin goby	Salt marsh song sparrow	Harbor seal
	Yellow shore crab	Staghorn sculpin	Black-necked stilt	
			Mallard	
		Brackish Marsh		
Tule	Eastern soft-shelled clam	Splittail	Sora rail	Salt marsh harvest mouse
Cattail	Asian clam (*Corbicula*)	Delta smelt	Snowy egret	Salt marsh vagrant shrew
Alkali bulrush	Amphipods	Stickleback	American coot	River otter
	Annelid worms			Muskrat

		Freshwater Marsh		
Cattail	Red swamp crayfish	Longfin smelt	Great blue heron	Muskrat
Reeds	Asian clam (*Corbicula*)	Largemouth bass	Song sparrow	Beaver
Tule		Splittail	American bittern	Mink
		Striped bass	Marsh wren	River otter
		Chinook salmon	Common yellowthroat	
		Diked Marsh		
Pickleweed	Blood worm	Mosquitofish	Northern shoveler	Salt marsh harvest mouse
Alkali heath	Red swamp crayfish	Carp	Canada geese	Black-tailed jackrabbit
Fat hen	Midge	Green sunfish	Cinnamon teal	Striped skunk
Alkali bulrush			Marsh hawk	Gray fox
		Vernal Pool		
Rushes	Midge	—	Mallard	Western harvest mouse
Sedges	Damselfly		Killdeer	California meadow mouse
Meadow foam	Mosquito		Common snipe	Striped skunk
	Fairy shrimp		Snowy egret	Coyote
		Farmed Wetland		
Corn	Leafhopper	—	Tundra swan	California vole
Hay	Thrip		White-fronted geese	California ground squirrel
Potato	Weevil		Long-billed curlew	Coyote
Sugarbeet	Moth		Western sandpiper	Red fox
	Amphipods			
		Abandoned Salt Pond		
Wigeongrass	Brine shrimp	Carp	Bufflehead	Red fox
Pickleweed	Water boatmen	Mosquitofish	Snowy plover	
	Brine fly	Sculpin	Caspian tern	
		Topsmelt	American avocet	

Table 5.16 (continued) Representative Plants and Animals of the Estuary's Wetland and Deep Water Habitats

Plants	Invertebrates	Fish	Birds	Mammals
		Riparian Woodland and Streams		
Cottonwood	Midge	Carp	Wood duck	Black-tailed deer
Sycamore	Mosquito	Bluegill	Belted kingfisher	Raccoon
Willow	Water boatmen	Squawfish	Green heron	Opossum
Elderberry		White catfish	Great egret	Ornate shrew
				Deer mouse
		Salt Pond		
Blue-green algae	Brine shrimp	Topsmelt	Eared grebe	—
Diatoms	Water boatmen	Rainwater killifish	Forster's tern	
Green algae	Brine fly	Yellowfin goby	Wilson's phalarope	
Wigeon grass		Staghorn sculpin	Black-necked stilt	
		Lakes and Ponds		
Duckweed	Opossum shrimp	Black crappie	Mallard	Beaver
Green algae	Red swamp crayfish	Bluegill	Ruddy duck	Muskrat
Pondweed	Amphipods	Largemouth bass	Pied-billed grebe	
Smartweed			Northern shoveler	
			American coot	

Source: Monroe, M. W. and Kelly, J., State of the Estuary, Tech. Rept., San Francisco Estuary Project, Oakland, CA, 1992.

Emergent vegetation dominates plant communities in tidal, brackish, and freshwater marshes. Commonly encountered plant groups here include grasses, sedges rushes, and succulents. Well-developed floral zonation typifies the tidal marshes. For example, there are three zones of plant growth in tidal saltmarshes. The low marsh is dominated by cordgrass (*Spartina foliosa*), which extends to mean high water. Pickleweed (*Salicornia virginica*) replaces cordgrass as the predominant form above mean high water, but several other plant species (e.g., saltgrass, *Distichlis spicata*; alkali heath, *Frankenia grandifolia*; and gumplant, *Grindelia humilis*) occur at higher elevations. In brackish marshes, bulrushes (*Scirpus olneyi* and *S. robustus*) predominate in the low marsh, bulrushes and cattails (*Typha* sp.) in the middle marsh, and Baltic rush (*Juncus balticus*) and saltgrass (*D. spicata*) in the high marsh. The tidal freshwater marshes are more diverse with species of cattails (*Typha* sp.), reeds (*Phragmites communis*), tule (*Scirpus californicus*), and willows (*Salix hindsiana*) growing in a mixed assemblage. Although dense stands of emergent vegetation can be found in many areas of the marsh habitats, the continuity of plant growth is often broken by open bare areas, pannes, and tidal channels.

Common invertebrate and fish populations recorded in the tidal saltmarshes are ribbed mussels (*Ischadium* [*Modiolus*] *demissum*), Baltic clams (*Macoma balthica*), shore crabs (*Hemigrapsus oregonensis*), gobies (*Clevelandis rios*), topsmelt (*Atherinops affinis*), and staghorn sculpin (*Leptococcus armatus*). In brackish marshes, amphipods, annelid worms, Asian clams (*Potamocorbula amurensis*), soft-shelled clams (*Mya arenaria*), delta smelt (*Hypomesus transpacificus*), splittails (*Pogonichthys macrolepidotus*), and threespine stick-lebacks (*Gasterosteus aculeatus*) are common. Freshwater marshes, in turn, are frequently inhabited by Asian clams (*P. amurensis*), red swamp crayfish (*Procambarus clarki*), chinook salmon (*Oncorhynchus tshawytscha*), striped bass (*Morone saxatilis*), longfin smelt (*Spirinchus thaleichthys*), and splittails (*P. macrolepidotus*). In addition to the invertebrate and fish populations occupying these habitats, an array of birds (e.g., rails, ducks, egrets, herons, etc.) and mammals (e.g., seals, otters, muskrat, beaver, etc.) proliferate here (Table 5.16).

Tidal marshes rank among the most severely impacted habitats in the San Francisco Bay/Delta system, with 80% of them having been filled or converted to other wetland uses. This alteration has adversely affected many organisms, especially those dependent on salt- or brackish marshes during a portion of their life cycle. Many birds that rely on these habitats (e.g., saltmarsh song sparrow, *Melospiza melodia pusillula*; saltmarsh yellowthroat, *Geothlypis trichas sinuosa*; California black rail, *Laterallus jamaicensis coturniculus*; and California clapper rail, *Rallus longirostris obsoletus*) are now listed as endangered or candidates for endangered status.[2]

Seasonal wetlands (i.e., diked marsh, vernal pools, farmed wetland, and abandoned salt ponds), which predominate in the delta and Suisun Marsh, are habitats subject to radical fluxes in environmental conditions. In these habitats, shallow depressions fill with rain water during wet seasons (winter and spring), and they dry out during dry seasons (summer and fall). Excluding farmed wetlands with common cultivated crops (e.g., corn, potatoes, sugar beets, tomatoes, etc.), seasonal wetlands are areas characterized by high diversity of plant species. For example, more than 200 species of plants have been identified in vernal pools alone. Some seasonal wetlands, such as diked marsh habitat, harbor many of the same plant species found in the aforementioned tidal marshes (e.g., alkali bulrush, cattails, pickleweed, and saltgrass). These habitats also support considerable numbers of invertebrates (e.g., worms; crayfish; and brine shrimp, *Artemis gracilis*). Insects (e.g., damselflies, leafhoppers, mosquitos, and weevils) attain high densities in farmed wetlands and vernal pools.

Some wetlands area has been converted to salt ponds for commercial purposes. These habitats are extreme environments, characterized by salinities ranging from brackish to ten times that of seawater. They are common features in former tidal marsh habitat areas of San Pablo and South Bays. Salinity is a dominant factor controlling the floral and faunal composition of the salt ponds. Plant communities are relatively simple, being comprised of diatoms, green and blue-green algae, and widgeon grass. Invertebrate and fish populations inhabiting salt ponds with low to intermediate salinities include small clams, brine shrimp (*Artemis gracilis*), topsmelt (*Atherinops affinis*), gobies (*Clevelandis ios*), and killifish (*Lucania parva*). These organisms provide forage for many shorebirds, waterbirds, and waterfowl.

Freshwater lakes, ponds, and reservoirs are scattered throughout the region. They support a range of submergent, emergent, and floating plant species (e.g., cattails, *Typha* sp.; duckweed, *Lemna minor*; and pondweed, *Potamogeton pectinatus*). Commonly observed invertebrates and fish are red swamp crayfish (*Procambarus clarki*), bluegills (*Lepomis macrochirus*), and black crappies (*Pomoxis nigromaculatus*). Significant numbers of waterfowl (e.g., mallards, *Anas platyrhynchos* and ruddy ducks, *Oxyura jamaicensis*) feed in these habitats.

Several hundred streams flow through watershed areas. Although many have surface water year round, most are intermittent with significant flows only during the wet season. The intermittent streams support several species of plants such as bulrush (*Scirpus olneyi* and *S. robustus*), black mustard (*Brassica nigra*), wild oat (*Avena fatua*), and wild radish (*Raphanus sativa*). Duckweed (*Lemna minor*), pondweed (*Potamogeton pectinatus*), and several other plant species grow along perennial streams. Midge, mosquitos, and other insects as well as submerged plants, provide forage for tens of fish species.[62] Among the common fish populations in the streams are bluegill (*Lepomis macrochirus*), carp (*Cyprinus carpio*), mosquitofish (*Gambusia affinis*), Sacramento perch (*Archoplites interruptus*), splittail (*Pogonichthys macrolepidotus*), and rainbow trout (*Salmo gairdneri*).

Along stream and river channels, corridors of broad-leaved deciduous trees and shrubs comprise riparian forest. This habitat is particularly valuable for wildlife populations (e.g., opossum, *Didelphis virginiana*; raccoon, *Procyon lotor*; shrew, *Sorex pacificus*; and deer, *Odocoileus hemionus*). The greatest diversity of bird species in the estuary region occurs in riparian habitat.

c. Upland habitats. As noted previously, there are seven major kinds of habitat in the uplands: grassland, coastal scrub, mixed chaparral, oak woodland, broad-leaved evergreen forest, agricultural land, and urban land. Proceeding from grasslands at the estuary's edge to hardwood forests inland, a wide range of vegetation (e.g., ferns, herbs, shrubs, cultivated crops, evergreen forests, and oak woodlands) supports a multitude of wildlife populations (e.g., rabbits, deer, skunks, foxes and opossums). Numerous bird populations (e.g., ducks, geese, swans, raptors, and sparrows) nest, feed, or reproduce in these habitats.

Despite their unequivocal ecological value, upland habitats have been greatly impacted by development. For instance, 7% of the original upland habitat area has been converted to farmland and more than 50% to urban land. Accompanying this habitat conversion has been a corresponding reduction in abundance and production of many biotic populations.

2. Bay/Delta biotic communities

The shallow and deepwater habitats of the San Francisco Bay/Delta are inhabited by several major biotic groups that interact to form a complex food web. Included here are phytoplankton, zooplankton, benthos, fish, mammals, and birds. These groups of organisms play a critically important ecological role in the estuarine ecosystem. In addition, some have great commercial and recreational value. The following discussion provides a description of the biotic communities in the estuary.

a. Phytoplankton. Of the several hundred species of phytoplankton inhabiting the San Francisco Bay/Delta estuary, most belong to three major algal groups: diatoms (Bacillariophyceae), dinoflagellates (Dinophycea), and cryptomonads (Cryptophyceae). The abundance, biomass, and production of phytoplankton in the estuary vary both spatially and temporally due to constant changes in physical–chemical conditions, particularly light, temperature, nutrients, salinity, turbidity, and mixing processes. Zooplankton grazing must also be considered.

Phytoplankton abundance typically peaks during the spring and summer months. Concentrations are higher in South Bay than Central Bay. Since the mid-1970s, phytoplankton levels have declined appreciably in the northern bays and delta. Reduced freshwater inflows, increased water transparency, and the proliferation of the Asian clam (*Potamocorbula amurensis*) have contributed considerably to this decrease.[2] In Suisun Bay and the western delta, phytoplankton abundance dropped nearly 10-fold within 2 years of the first appearance of the Asian clam in 1986.

Net phytoplankton production in the estuary ranges from ~95 to 150 $gC/m^2/yr$, with highest values recorded in South Bay. Peak biomass values have been registered in Suisun Bay.[56,63] For the bay as a whole, phytoplankton production is the dominant source of organic carbon, accounting for ~50% of the total organic carbon concentration.

Diatoms dominate spring phytoplankton blooms in the estuary.[64,65] For example, the most abundant phytoplankton populations during spring blooms in South, Central, San Pablo, and Suisun Bays include *Cyclotella* spp., *Skeletonema costatum*, and *Thalassiosira* spp. In the delta, *Melosira granulata* is the principal species comprising most of the blooms. Species of *Achnanthes*, *Amphora*, *Fragilaria*, *Chroomonas*, *Cryptomonas*, *Melosira*, and *Pyramimonas* predominate in different areas of the estuary during other seasons of the year.

b. Zooplankton. As in the case of phytoplankton, zooplankton have decreased in abundance in recent years, particularly in the northern reach of the estuary. Declining phytoplankton abundance, increased water transparency, low delta outflow, and introduced species may be largely responsible for the diminished zooplankton abundance. Some introduced zooplankton species such as the copepods *Psuedodiaptomous forbesi* and *Sinocalanus doerri* have increased markedly in abundance during the past decade, whereas the once dominant native copepod form, *Eurytemora affinis*, has decreased greatly (Figure 5.7). The mysid, *Neomysis mercedis*, has likewise decreased in abundance, especially in the northern bays. Because this mysid provides forage for many estuarine fishes, its declining abundance must be viewed with concern because of the potential adverse impacts on the entire food chain.

Copepods, cladocerans, protozoans, and rotifers are the primary groups of zooplankton in the estuary. Ambler et al.[66] have investigated the seasonal changes in abundance and distribution of copepods in the system. They reported that in the northern reach copepod populations are distributed according to their salinity tolerances. Some species preferences are as follows: *Sinocalanus doerri* in the Sacramento and San Joaquin Rivers, *Eurytemora affinis* in Suisun Bay, *Acartia* spp. in San Pablo Bay, and *Paracalanus parvus* in Central Bay. The magnitude of river inflow modulates the distribution of zooplankton such that *S. doerri* is found at the riverine boundary, *E. affinis* at the oligohaline mixing zone, *Acartia* spp. at polyhaline waters, and *P. parvus* at the seaward boundary. *Acartia hudsonica* (formerly *A. clausi*) concentrates in San Pablo Bay nearly year round, but *A. californiensis* appears there only from August to October. South Bay exhibits large densities

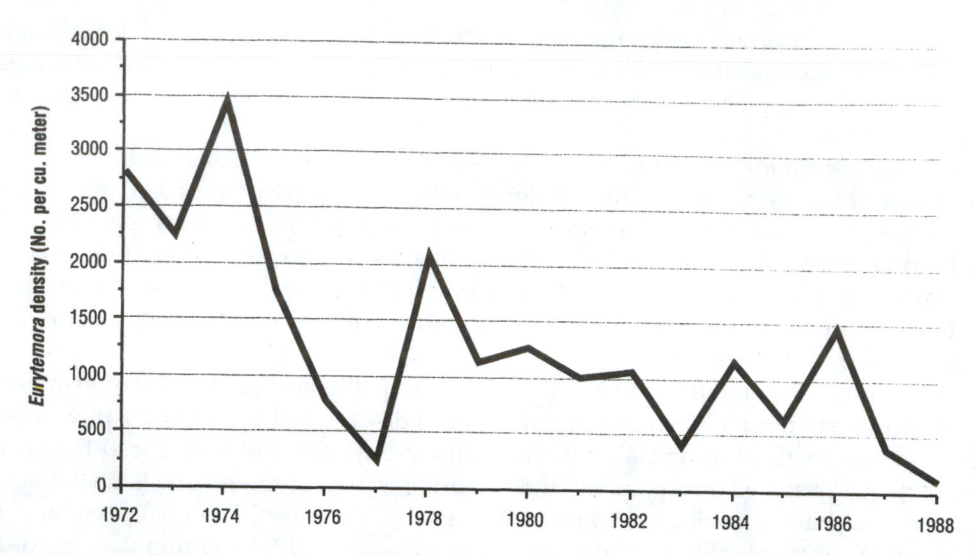

Figure 5.7 Densities of the copepod species *Eurytemora affinis* in Suisun Bay from 1972 to 1988. (From Monroe, M.W. and Kelly, J., State of the Estuary, Tech. Rept., San Francisco Estuary Project, Oakland, CA, 1992.)

of both species. *Acartia hudsonica* attains maximum densities during the wet season, and *A. californiensis* reaches maximum densities during the dry season. A seasonal succession of these two populations is evident in South Bay, with the warm-water form (*A. californiensis*) replacing the cold-water form (*A. hudsonica*) in the spring. *Oithona davisae* is most conspicuous in the fall.

Copepod populations introduced into the bay during the 1970s and 1980s have rapidly increased in abundance at the expense of native forms. For example, populations of *Limnoithona sinensis*, *Pseudodiaptomus forbesi*, and *Sinocalanus doerii* expanded markedly after being unintentionally introduced into the bay from China. As these populations grew, native species, such as *Eurytemora affinis*, experienced sharp declines.[67,68]

San Pablo Bay and South Bay not only differ in the succession of *Acartia* species, but also in overall microzooplankton structure. Rotifers (*Synchaeta* sp.) and tintinnids (*Tintinnopsis* sp. A, *Tintinnopsis* sp. B, and *Eutintinnus neriticus*), together with *Acartia* spp. nauplii, are the most frequently occurring and abundant microzooplankton in the bay. Although rotifers and tintinnids are distributed baywide, they occur in lower numbers in oligohaline areas (e.g., Suisun Bay). The dominant tintinnid, *Tintinnopsis* sp. B, reaches densities in excess of $10^5/m^3$. In South Bay, maximum counts of *Tintinnopsis* sp. B are more variable from year to year than those of *Tintinnopsis* sp. A, which peak at the time of phytoplankton blooms. Tintinnids persist nearly year-round in San Pablo Bay, attaining highest numbers in shoal waters.

Seasonal occurrences of meroplankton contribute to the temporal patchiness of the zooplankton community. In the winter–spring, meroplankton of barnacles, bivalves, gastropods, and polychaetes comprise a substantial fraction of the total zooplankton biomass in South Bay. During summer–fall, however, meroplankton of these organisms account for a large portion of the zooplankton biomass in the northern reach.

Herbold et al.[51] indicate that *Acartia* spp. and *Oithona davisae* remain the dominant copepods of South and Central Bays. Both *Acartia* spp. and *Eurytemora affinis* occur in San Pablo Bay. Species of *Brachionus*, *Keratella*, and *Synchaeta* are the dominant rotifers. Similar to protozoans, rotifers are relatively unimportant in terms of their contribution to the total zooplankton biomass in the bay. Larger zooplankton, including species of *Bosmina*, *Diaphanosoma*,

Cyclops, *Diaptomus*, *Eurytemora*, and *Pseudodiaptomus*, account for most of the zooplankton biomass. There has been a general long-term decrease in abundance of several major zooplankton groups (i.e., copepods, cladocerans, and rotifers) in the estuary and delta. Human activities appear to have played a significant role in the decline of native zooplankton populations.

c. Benthos.

Benthic flora — Bottom-dwelling flora and fauna of mudflats, tidal creeks, and open water areas support numerous recreational and commercial species in the estuary. The benthic microalgal flora of the estuary includes an assemblage of diatoms, blue-green algae, and flagellates. Numerous species occur in the system, including some typical estuarine forms (e.g., *Nitzchia acuminata* and *N. pusilla*).[69] Intertidal mudflats, which cover an area of ~260 km^2, are ideal sites for growth of these minute plants.

More than 160 species of benthic macroalgae have also been chronicled in the estuary. Green algae are the most common forms, including species of *Cladophora*, *Enteromorpha*, and *Ulva*. Red algae of the genus *Polysiphonia* are also common.[70]

Dense stands of vascular plants occur in saltmarshes bordering the bay. Cordgrass (*Spartina foliosa*), saltgrass (*Distichlis spicata*), pickleweed (*Salicornia virginica*), and baltic rush (*Juncus balticus*) are abundant species. The saltmarshes serve several major ecological roles. They produce a large amount of organic matter, much of which enters the detritus pool, and are sources or sinks for nutrients and other substances. Well-developed root systems anchor the sediments, thereby stabilizing the substrate and mitigating erosion. Finally, many benthic invertebrates, fish, and wildlife populations use saltmarsh habitats for feeding, nesting, and reproduction.

Benthic fauna — The benthic macrofaunal community varies significantly in the estuary, largely because of differences in salinity and substrate types. Species diversity of the community generally increases from the delta to the more saline waters downestuary. Herbold et al.[51] reported only 82 benthic species in the delta, with 5 species comprising ~90% of all individuals collected at most sites. The benthic community of Suisun Bay is also typified by relatively low species diversity. Taxa commonly found here include bivalves (*Corbicula fluminea*, *Macoma balthica*, *Mya arenaria*, and *Potamocorbula amurensis*), annelids (*Limnodrilus hoffmeisteri* and *Nereis succinea*), and amphipods (*Corophium spinicorne* and *C. stimpsoni*). *Potamocorbula amurensis*, the introduced Asian clam, reaches densities up to 30,000 m^{-2}. Species diversity increases in San Pablo Bay, with mollusks (*Gemma gemma*, *Ilyanassa obsoleta*, *Musculus senhousia*, and *Tapes japonica*), amphipods (*Ampelisca abdita*, *Corophium* spp., and *Grandidierella japonica*), and polychaetes (*Glycinde* sp., *Heteromastus filiformis*, and *Streblospio benedicti*) being abundant. These species also occur in the deeper waters of South Bay, along with the tube-dwelling polychaete, *Asychis elongata*. Numerical dominants in shallower areas of the bay are *Ampelisca abdita*, *Gemma gemma*, and *S. bendicti*. The benthic community of Central Bay is likewise highly diverse. Some of the species encountered in Central Bay derive from coastal ocean waters of California.

Abundance and biomass of benthic fauna in the northern reach of the estuary also differ appreciably from those in the southern reach. According to Nichols and Pamatmat,[56] macrofaunal biomass in the northern bays is low, ranging from 60 to 630 g wet wt/m^2. In the channel of the northern bays, the Asian clam (*Potamocorbula amurensis*) attains a biomass of 200 g wet wt/m^2.[71] In South Bay, benthic macrofaunal abundance and biomass values are much higher than in the northern bays. This is particularly true for deposit feeders, such as polychaetes (*Asychis elongata*), which exist in very dense patches in some areas.

Nearly all of the common benthic invertebrate species in the estuary are introduced forms, including shellfish species. In the last 150 years, more than 100 benthic invertebrates have been accidentally or intentionally introduced into the estuary. They have had a

Table 5.17 Biomass of Mollusca, Annelida, Arthropoda, and All Other Phyla in San Francisco Bay

Location	Biomass in Grams[a]	
	Winter	Summer
Lower South Bay	88	207
Upper South Bay	549	402
Central Bay	59	42
San Pablo Bay	40	64
Suisun Bay	16	6

[a] Data are averages of 4 to 13 stations in each region of the bay.

Source: Nichols, F. H. and Pamatmat, M. M., The Ecology of the Soft-Bottom Benthos of San Francisco Bay: A Community Profile, Biological Rept. 85 (7-23), U.S. Fish and Wildlife Service, Washington, D.C., 1988.

dramatic impact, shifting community composition and altering the ecological balance of the system.[72] Most notable in this regard is the Asian clam (*Potamocorbula amurensis*), which appears to have caused significant reductions in abundance of native populations.

Arthropods, annelids, and mollusks predominate in intertidal mudflats.[60] In this habitat, salinity and sediment stability exert strong control on the species composition and distribution of the benthic invertebrate community. In the delta, bivalves (*Corbicula fluminea, Gemma gemma,* and *Tapes japonica*), polychaete worms (*Boccardia berkelyorum* and *Neanthes limnicola*), and amphipods (*Corophium spinicorne* and *C. stimpsoni*) appear to be most abundant. Proceeding downestuary, bivalves (*Macoma balthica* and *Mya arenaria*), annelids (*Limnodrilus hoffmeisteri* and *Nereis succinea*), and amphipods (*C. spinicorne* and *C. stimpsoni*) predominate in Suisun Bay. *Ampelisca abdita, Streblospio benedicti,* and *G. gemma* occur in mudflat sediments of San Pablo Bay. The mudflat community in Central Bay is more diverse, with a greater number of marine species. In South Bay, the numerical dominants are *A. abidita, G. gemma,* and *S. benedicti*. Less abundant forms include *Illyanassa obsoleta, M. balthica,* and *M. arenaria*. The highest biomass of mudflat infauna exists in South Bay, where salinity, sediment, and food supply are more favorable for the development of high faunal biomasses (Table 5.17).

Among the epifauna which frequent intertidal mudflats are dungeness crabs (*Cancer magister*) and caridean shrimp (*Crangon franciscorum, C. nigricauda,* and *Palaemon macrodactylus*). Crayfish (*Pacificasticus leniusculus* and *Procambarus clarki*) occur along intertidal habitats of the delta region. Availability of prey greatly influences the abundance of epifauna on the mudflats.

A number of benthic invertebrates are relatively abundant in tidal salt and brackish marsh habitats. Bivalves (*Gemma gemma* and *Tapes japonica*), polychaetes (*Capitella capitata, Eteone californica,* and *Streblospio benedicti*) and amphipods (*Corophium spinicorne* and *Grandidierella japonica*) have been documented in these habitats. Species preferences are clearly evident. For example, the ribbed mussel (*Ischadium* [*Modiolus*] *demissum*) inhabits the Pacific cordgrass (*Spartina foliosa*) zone, and isopods (*Sphaeroma quoyana*), the higher mudflat-pickleweed marsh edge.[73] Gastropods (*Assiminea californica* and *Ovatella myosotis*) occupy pickleweed marshes. Together with the snail *Cerithidea californica, A. californica* and *O. myosotis* are also locally abundant in the higher tidal salt marsh habitat.[60]

The benthic macrofauna in San Francisco Bay, as in many estuaries, exhibit large within-year and between-year variations in abundance and distribution in response to seasonal changes of environmental conditions and aperiodic fluctuations from year to year.[69] Spatial and temporal changes in some environmental factors (e.g., freshwater inflow, temperature, salinity, bottom currents, etc.) can greatly influence periodicities of reproduction and recruitment that contribute substantially to the observed patterns. Anthropogenic impacts (e.g., dredging, freshwater diversion, and pollutant impacts) may also significantly affect the structure and dynamics of the benthic community.

d. Fish. The San Francisco Bay/Delta Estuary supports more than 130 species of fish. Of these species, the delta smelt (*Hypomesus transpacificus*) is the only completely estuarine form.[1] Some species occur in the estuary seasonally or during a particular life stage. Others reside there year-round but maintain part of their population outside of the estuary. Many of the most abundant species are introduced forms. Table 5.18 provides detailed information on nearly 30 representative fish species of the estuary.

Herbold et al.[51] examined the freshwater and marine species which inhabit the estuary. Important native freshwater species are the Sacramento blackfish (*Orthodon microlepidotus*), Sacramento squawfish (*Ptychocheilus grandis*), Sacramento sucker (*Catostomus occidentalis*), Sacramento splittail (*Pogonichthys macrolepidotus*), hardhead (*Mylopharodon conocephalus*), and thicktail chub (*Gila crassicauda*). Some introduced freshwater forms have become dominant members of the ichthyofauna in some areas of the estuary. Several species of ictalurids (*Ameirus*) and centrarchids (*Lepomis*, *Micropterus*, and *Pomoxis*) are examples.

Marine species can be divided into two groups based on when they occur in the estuary: (1) species present seasonally (e.g., Pacific herring, *Clupea harengus pallasi*; northern anchovy, *Engraulis mordax*; and striped bass, *Morone saxatilis*), and (2) resident species (e.g., bay gobies, *Lepidogobrius lepidus*; chameleon gobies, *Tridentiger trigonocephalus*; shiner perch, *Cymatogaster aggregata*; and staghorn sculpins, *Leptocottus armatus*). The northern anchovy and Pacific herring are the two most abundant fishes in the estuary. Among the resident marine species, the most abundant forms include bay gobies (*L. lepidus*), shiner perch (*C. aggregata*), and staghorn sculpins (*L. armatus*).

Another category of ichthyofauna — anadromous fish — is represented by several recreationally important species. The chinook salmon (*Oncorhynchus tshawytscha*), striped bass (*Morone saxatilis*), American shad (*Alosa sapidissima*), and starry flounder (*Platichthys stellatus*) are anadromous and recreationally important. Some valuable anadromous species are introduced forms (e.g., striped bass and American shad).

A number of anadromous species once supported significant commercial fisheries in the estuary during the 20th century. For example, striped bass (*Morone saxatilis*), chinook salmon (*Oncorhynchus tshawytscha*), American shad (*Alosa sapidissima*), and white sturgeon (*Acipenser transmontanus*) were all commercially important in the estuary during the first half of the 20th century, but their fisheries were terminated in 1917, 1935, 1956, and 1956, respectively. Today, only the the Pacific herring supports a major commercial fishery (particularly for roe). Several smaller commercial fishery operations exist in the bay–delta region, notably bait fisheries (e.g., shiner perch, *Cymatogaster aggregata*; mudsucker, *Gillichthys mirabilis*; bullhead, *Ictalurus* sp.; and threadfin shad, *Dorosoma petenense*) and a small "hook-and-line" fishery for California halibut (*Paralichthys californicus*), white croaker (*Genyonemus lineatus*), and striped bass (*M. saxatilis*). By comparison, the recreational fishery is much more active; the chinook salmon (*O. tshawytscha*), striped bass (*M. saxatilis*), American shad (*A. sapidisima*), starry flounder (*Platichthys stellatus*), staghorn sculpin (*Leptococcus armatus*), white sturgeon (*Acipenser transmontanus*), green sturgeon (*A. medirostris*), white catfish (*Ictalurus cactus*), leopard shark (*Triakis semifasciata*), brown smoothhound shark (*Mustelus henlei*), brown rockfish (*Sebastes auriculatus*), white croaker (*Genyonemus lineatus*), jacksmelt (*Atherinopsis californiensis*), California halibut (*Paralichthys californicus*), and several other species all support the estuary's sport fishery.

The abundance and distribution of fish populations in the estuary are strongly affected by freshwater inflow. For example, Armor and Herrgesell[50] demonstrated that 24 species of fish attained highest abundance in the estuary during wet years when the volume of

Table 5.18 Representative Fishes of the San Francisco Bay/Delta Estuary

Species	Species Origin	Species Type	Life History: Spawning Time	Life History: Spawning Location	Life History: Nursery Area	Center of Population	Importance of Species	Preferred Habitat	Use of Bay or Delta	Major Food Source: Adult	Major Food Source: Juvenile	Recent Population Trend
Pacific herring	N	M	Fall–winter	Bay	SSFB–SPB	Ocean	Commercial, forage	Pelagic	Bay–spawning, nursery	P	P	Variable
Longfin smelt	N	E	Winter, spring	Rivers	Delta, bay	SPB	Forage	Pelagic	Bay–nursery, residence	P	P	Variable, recently down
Pacific staghorn sculpin	N	E	Fall, winter	Bay, ocean	Bay	CSFB–SPB	Forage	Littoral/ Demersal	Bay–nursery, residence	F, B	B	Variable
Starry flounder	N	E	Winter	Ocean	SB–Delta	Ocean–bay	Commercial, recreation	Demersal	Bay–nursery, residence	B	B	Down
Speckled sanddab	N	M	All year	Ocean	Ocean–CSFB SPB SSFB	Ocean	Forage	Demersal	Bay–nursery, residence	B	B	Variable
English sole	N	M	Winter	Ocean	Ocean–bay	Ocean	Commercial	Demersal	Bay–nursery	B	B	Variable
White croaker	N	M	Summer–fall	Ocean	Ocean–CSFB	Ocean	Forage	Demersal	Bay–nursery, residence	B	B	Up
Yellowfin goby	I	E	Winter	Bay	SB–Delta	SPB–SB	Forage, commercial	Demersal	Bay–residence	B	B	Down
Plainfin midshipman	N	M	Spring–summer	Bay	SSFB–SPB	SSFB–SPB	Forage	Demersal	Bay–nursery, residence	B	B	Up
Bay goby	N	M	Summer–fall	Bay	SSFB–SPB	CSFB	Forage	Demersal	Bay–nursery, residence	B	B	Variable
Topsmelt	N	M	Summer	Bay	SSFB–CSFB	SSFB	Forage	Littoral/ Pelagic	Bay–residence	B	B	Variable
Jacksmelt	N	M	Spring–summer	Bay, ocean	SSFB–CSFB	Ocean	Recreation, forage	Pelagic	Bay–spawning, nursery	F	P	Variable

Northern anchovy	N	M	Spring–summer	Bay, ocean	Bay, ocean	Ocean	Commercial, forage	Pelagic	Bay–spawning, nursery	P	P	Variable
Pacific lamprey	N	A	Spring	Rivers	Rivers, upper delta	—	Parasite, forage	Pelagic	Upper delta–nursery, migration	F	B veg.	Down
White sturgeon	N	A	Spring	Rivers	Estuary	—	Recreation	Demersal	Delta–residence	P	P	Down
American shad	I	A	Spring	Rivers, upper delta	Rivers, delta	—	Recreation	Pelagic	Bay and delta–nursery, migration	P	P	Down
Threadfin shad	I	FW	Spring	Delta	Delta	—	Forage, bait, commercial	Pelagic	Delta–residence	P	P	Down
Steelhead trout	N	A	Spring	Rivers	Rivers	—	Recreation	Pelagic	Bay and delta–migration	P	P, B insects	Down
Chinook salmon	N	A	All months, greatest nos. in fall	Rivers	Rivers, upper delta	—	Recreation, commercial	Pelagic	Delta–nursery, migration; Bay–migration	F, P	P, insects	Down
Delta smelt	N	E	Spring	Delta	Delta, Suisun Bay	—	Forage	Pelagic	Delta–spawning, nursery, residence	P	P	Down
Splittail	N	E	Spring	Delta	Delta, Suisun Bay	—	Recreation, forage	Pelagic	Delta–spawning, nursery, residence	P	P	Down
Carp	I	FW	Spring	Delta	Delta	—	Recreation	Pelagic	Delta–spawning, nursery, residence	P	P	Down?
Sacramento sucker	N	FW	Spring	Rivers	Rivers	—	Educational	Demersal	Delta–nursery	B	B	?

Table 5.18 (continued) Representative Fishes of the San Francisco Bay/Delta Estuary

Species	Species Origin	Species Type	Life History: Spawning Time	Life History: Spawning Location	Life History: Nursery Area	Center of Population	Importance of Species	Preferred Habitat	Use of Bay or Delta	Major Food Source: Adult	Major Food Source: Juvenile	Recent Population Trend
White catfish	I	FW	Spring, summer	Delta	Delta	—	Recreation	Demersal	Delta–spawning, residence	B	B	Down
Striped bass	I	A, E	Spring	Rivers, delta	Delta, San Pablo Bay	—	Recreation	Pelagic/ Demersal	Delta–spawning, nursery; Bay–nursery	P, B, F	P	Down
Bluegill	I	FW	Spring	Delta	Delta	—	Recreation	Littoral	Delta–spawning, nursery	P, B	P, B	?
Black crappie	I	FW	Spring	Delta	Delta	—	Recreation	Pelagic	Delta–spawning, nursery, residence	P, F	P	?
Largemouth bass	I	FW	Spring	Delta	Delta	—	Recreation	Littoral	Delta–spawning, nursery, residence	F, B	P, B	?
Tule perch	N	E	Spring, live bearer	Delta, marsh sloughs	Delta, marsh sloughs	—	Educational	Littoral	Delta–nursery	B, P	B, P	Down?

Note: N = native; I = introduced; E = estuarine; M = marine; A = anadromous; FW = freshwater; SSFB = South San Francisco Bay; CSFB = Central San Francisco Bay; SPB = San Pablo Bay; SB = Suisun Bay; P = plankton; B = benthos; F = fish; Pelagic = open water; Littoral = shoreline; Demersal = bottom.

Source: Monroe, M. W. and Kelly, J., State of the Estuary, Tech. Rept., San Francisco Estuary Project, Oakland, CA, 1992.

freshwater inflow increased. In contrast, 12 other species of fish reached peak abundance during dry years when freshwater inflow declined. The coupling of freshwater diversions to delta outflow, therefore, indicates that human activities in the delta and Central Valley watershed can have a profound influence on fish dynamics in the estuary.

The abundance of a number of fish species in the estuary has varied substantially in recent years. The SFEP has focused its attention on four species of exceptional commercial, recreational, or ecological importance: chinook salmon, striped bass, American shad, and delta smelt. The following discussion deals with recent trends in these populations.

Chinook salmon — Salmon stocks have decreased appreciably in the estuary over the years. During spawning runs in the early 1900s, nearly 900,000 salmon entered Central Valley streams to spawn each year. This number dropped to ~400,000 fish during the early 1950s and ~275,000 during the mid-1990s. Winter-run chinook salmon showed the most dramatic decline; consequently, it has been listed as both a state and federal endangered species. In 1994, the number of winter-run salmon in the Central Valley dropped to an all-time low of 189, although the population rebounded to 1361 in 1995 and 900 in 1996. The decrease in abundance of natural salmon stocks in the system has been attributed to dam construction (which has blocked fish migration), reduced spawning habitat, inadequate stream spawning flows, intermittent poor water quality, acid mine drainage, unsuitable spawning gravel, high stream temperature, and losses of young fish to freshwater diversions.[2]

Hatchery production of chinook salmon has compensated for the long-term decline in natural stocks of the species. Five Central Valley hatcheries release ~30 million young fish each year into the estuary. Hatchery production has enabled commercial and recreational salmon fisheries to maintain seasonally high levels despite the reduced natural production. These fish account for about one third of the total ocean catch of the species.[74]

Striped bass — The striped bass has also experienced an appreciable decrease in abundance. The estimated number of striped bass currently in the estuary (~600,000 fish) is the lowest level recorded during the past century, having dropped from nearly 2 million individuals in the early 1970s. The decline of this species has been most acute since 1977. Possible causes for the diminishing striped bass abundance are reduced carrying capacity of the system, delta water diversions and associated fish entrainment losses, reduced delta outflows, low San Joaquin River inflow, water pollution, dredging and sediment disposal, wetland infilling, illegal take and poaching, diseases and parasites, annual die-off of adult bass, commercial bay shrimp fishery, exotic aquatic organisms, and greater migration of adult bass out to sea with El Niño events.[1,2] Of these factors, increased losses associated with freshwater diversions in the delta appear to be a primary factor in the observed decline of the species.

American shad — First introduced into tributaries of the estuary in 1871, the American shad rapidly increased in abundance to ~3 million individuals by the early 1900s. The number of American shad remained high until 1945 when the population began to decrease. The adult population today is ~1 to 2 million individuals, which supports a recreational fishery in the delta as well as the San Joaquin River, Sacramento River, and a few other influent systems.

Delta smelt — The abundance index of delta smelt decreased from more than 1500 in 1980 to less than 500 through the 1980s. Between 1993 and 1995 the abundance index fluctuated markedly from 1078.4 in 1993 to 101.2 in 1994 and 898.7 in 1995.[1] The delta smelt is a planktivore, and the declining abundance of this species may be linked to the concomitant decreases in plankton levels of the estuary during the past two decades. Other potential causes for the diminishing numbers of delta smelt include entrainment losses to water diversions, habitat modification, invasions of exotic species, pollution, changes in outflow, disease, competition, and predation.[1,2] The delta smelt is now listed as endangered under the Endangered Species Acts of the California Fish and Game Commission and the U.S. Fish and Wildlife Service.

Other species — Abundance of the longfin smelt (*Spirinchus thaleichthys*) has likewise dropped in the estuary, falling to minimum levels in 1992. However, abundance of this species increased again between 1993 and 1995. The Sacramento splittail has followed a similar abundance pattern, increasing from low levels during the late 1980s and early 1990s to much higher levels in the mid-1990s.

Two species that continue to maintain high abundances in the estuary are the Pacific herring and northern anchovy. The Pacific herring has rebounded from lower population levels during drought years in the late 1980s and early 1990s. The northern anchovy remains the most abundant fish in San Francisco Bay, migrating into the bay seasonally from a large coastal population.

Spatial trends — Herbold et al.[51] have summarized the fish assemblages inhabiting the delta and open bay waters. Assemblages of fish in the delta consist principally of striped bass (*Morone saxatilis*), catfish (*Ictalurus* sp.), black crappies (*Pomoxis nigromaculatus*), sunfish (*Lepomis cyanellus*), and carp (*Cyprinus carpio*). Those of Suisun Bay are dominated by delta smelt (*Hypomesus transpacificus*), longfin smelt (*Spirinchus thaleichthys*), striped bass (*M. saxatilis*), and bay gobies (*Lepidogobrius lepidus*). Species characteristic of San Pablo Bay include starry flounder (*Platichthys stellatus*), longfin smelt (*S. thaleichthys*), striped bass (*M. saxatilis*), and bay gobies (*L. lepidus*). More marine species (e.g., English sole, *Parophrys vetulus*) occur in Central Bay. South Bay fish assemblages are typified by the northern anchovy (*Engraulis mordax*), topsmelt (*Atherinops affinis*), white croaker (*Genyonemus lineatus*), brown smoothhound shark (*Mustelus henlei*), and bay gobies (*L. lepidus*).

Native fishes dominate parts of many watershed streams. Most of these species (~75%) are maintaining healthy population levels. Table 5.19 reveals their status.

An array of freshwater, euryhaline, marine, and anadromous fishes frequent intertidal mudflats. Some freshwater species recorded in this habitat include carp (*Cyprinus carpio*), goldfish (*Carassius auratus*), Sacramento blackfish (*Orthodon microlepidotus*), Sacramento squawfish (*Ptychocheilus grandis*), Sacramento sucker (*Catastomus occidentalis*), black crappie (*Pomoxis nigromaculatus*), and tule perch (*Hysterocarpous traski*). Euryhaline species found there are threespine stickleback (*Gasterosteus aculeatus*), Sacramento smelt (*Spirinchus thaleichthys*), pond smelt (*Hypomensus transpacificus*), staghorn sculpin (*Leptocottus armatus*), starry flounder (*Platichthys stellatus*), and bay gobies (*Lepidogobrius lepidus*). Examples of marine forms registered in the intertidal mudflats are the spring dogfish (*Squalus acanthias*), leopard shark (*Triakis semifasciata*), brown smoothhound shark (*Mustlus henlei*), Pacific herring (*Clupea harengus pallasi*), northern anchovy (*Engraulis mordax*), and diamond turbot (*Hypsopsetta guttulata*). Among anadromous species that occasionally occupy the mudflats, several recreationally important forms have been noted, namely chinook salmon (*Oncorhynchus tshawytscha*), steelhead trout (*O. mykiss irideus*), striped bass (*Morone saxatilis*), white sturgeon (*Acipenser transmontanus*), green sturgeon (*A. medirostris*), and American shad (*Alosa sapidissima*).

Freshwater and anadromous species documented in tidal salt and brackish marshes include juvenile striped bass (*M. saxatilis*), mosquitofish (*Gambusia affinis*), tule perch (*Hysterocarpous traski*), and splittail (*Pogonichthys macrolepidotus*). Marine and euryhaline species that can attain significant abundances in these habitats are staghorn sculpin (*Leptococcus armatus*), topsmelt (*Atherinops affinis*), arrow goby (*Clevelandis ios*), threespine stickleback (*Gasterosteus aculeatus*), mosquitofish (*Gambusia affinis*), rainwater killifish (*Lucania parva*), delta smelt (*Hypomesus transpacificus*), starry flounder (*Platichthys stellatus*), and shiner perch (*Cymatogaster aggregata*). Commercially and recreationally important species as well as non-game fishes utilize tidal marsh habitats. Surveys have shown that 16 of the more than 120 total fish species recorded in the bay probably occur in the tidal marshes.[75]

Table 5.19 Status of Native Fishes in Streams of the San Francisco Estuary

Species	Status[a]
Petromyzontidae	
Pacific lamprey, *Lampetra tridentata*	R?
River lamprey, *L. ayresi*	R?
Pacific brook lamprey, *L. pacifica*	0, R?
Acipenseridae	
Green sturgeon, *Acipenser medirostris*	0, S
Osmeridae	
Delta smelt, *Hypomesus transpacificus*	0, R
Salmonidae	
Coho salmon, *Oncorhynchus kisutch*	0, EX?
Chinook salmon, *O. tshawytscha*	R
Chum salmon, *O. keta*	0, S
Steelhead trout, *O. mykiss irideus*	R, D
Resident rainbow trout, *O. mykiss mykiss*	C
Cyprinidae	
Hardhead, *Mylopharadon conocephalus*	R
Splittail, *Pogonichthys macrolepidotus*	0, D
Thicktail chub, *Gila crassicauda*	0, EX
Hitch, *Lavinia exilcauda*	LC, D
California roach, *Lavinia symmetricus*	C
Speckled dace, *Rhinichthys osculus*	0, EX?
Sacramento blackfish, *Mylopharodon conocephalus*	LC
Sacramento squawfish, *Ptychocheilus grandis*	LC
Catostomidae	
Sacramento sucker, *Catostomus occidentalis*	C
Gasterosteidae	
Threespine stickleback, *Gasterosteus aculeatus*	C
Cottidae	
Prickly sculpin, *Cottus asper*	C
Riffle sculpin, *C. gulosus*	LC
Coastrange sculpin, *C. aleuticus*	0, R
Pacific staghorn sculpin, *Leptocottus armatus*	LC
Centrarchidae	
Sacramento perch, *Archoplites interruptus*	0, R
Embiotocidae	
Tule perch, *Hysterocarpus traskii*	LC
Gobiide	
Tidewater goby, *Eucyclogobius newberryi*	0, EX
Longjaw mudsucker, *Gillichthys mirabilis*	LC

[a] Status within the drainage is abbreviated as follows: EX = extinct; R = rare; D = depleted and declining, range and numbers substantially reduced; C = common throughout range; LC = locally common; S = stray; 0 = not collected during this study; ? = current status unknown.

Source: San Francisco Estuary Project, State of the Estuary: 1992–1997, Tech. Rept., San Francisco Estuary Project, Oakland, CA.

3. *Wildlife*

Development and associated human activities in the Central Valley and bay watersheds have resulted in the irretrievable loss and alteration of wildlife habitats, leading to the reduction in population abundance of most wildlife species and the extirpation of others.

The loss of tidal marshes has been particularly detrimental to wildlife. Aside from habitat impacts, several other factors have also influenced wildlife abundance and distribution, notably the introduction of non-native species, pollution, human disturbance, hunting, predation, competition, and disease. Five mammal species, three bird species, one reptile species, and seven insect species have been eliminated from the estuary basin, mainly because of habitat loss. State and federal governments have listed 22 wildlife species as threatened or endangered, and numerous others have experienced population declines (Table 5.20).[2] In addition, nearly 70 wildlife species are considered as candidates for federal listing or as state species of special concern.

Despite the downward trend in abundance of wildlife from historic levels, the estuary's wildlife resources remain considerable. About 380 species of wildlife currently inhabit the estuary basin; the number of bird species (255) far exceeds that of mammals (81), reptiles (30), and amphibians (14). Table 5.21 summarizes the status and trends of representative wildlife species in the system.

a. Birds. San Francisco Bay lies within the Pacific Flyway and, therefore, is an important stopover for migrating shorebirds. Waterfowl populations also heavily utilize the estuary basin. The California Department of Fish and Game and U.S. Fish and Wildlife Service conduct midwinter waterfowl surveys in the estuary. These surveys have revealed a total of at least 32 waterfowl species in the estuary. The abundance of waterfowl has declined from an average of more than 6×10^5 birds in 1991 to $\sim 3 \times 10^5$ birds in 1997.[1] Statewide and nationwide population declines of some species (e.g., scaup, *Aythya affinis* and *A. marila*, and northern pintail, *Asa acuta*) have occurred concomitantly with baywide declines.

The estuary represents a major staging and wintering area for migrating waterfowl populations. Swans, geese, mergansers, dabbling ducks, and diving ducks inhabit the bay and delta region during all or part of the year. Included among the dabbling ducks are mallards (*Anas platyrhynchos*), northern pintails (*A. acuta*), gadwalls (*A. strepera*), northern shovelers (*A. clypeata*), American widgeons (*A. americana*), green-winged teals (*A. crecca*), and cinnamon teals (*A. cyanoptera*). Diving ducks, in turn, include ruddy ducks (*Oxyura jamaicensis*), buffleheads (*Bucephala albeola*), Barrow's goldeneyes (*B. islandica*), common goldeneyes (*B. clangula*), lesser scaups (*Aythya affinis*), greater scaups (*A. marila*), ring-necked ducks (*A. collaris*), and redheads (*A. americana*).[76]

Inventories of shorebird populations have been conducted less regularly than those of waterfowl. More than 30 species of shorebirds occupy the estuary on a regular basis, notably avocets, plovers, oystercatchers, sandpipers, and stilts. Censusing between 1988 and 1993 revealed a moderate decrease of shorebird abundance. The number of shorebirds during spring surveys ranged from lows of 5.88×10^5 in 1991 to highs of 9.31×10^5 in 1989. Abundance of shorebirds during fall surveys ranged from 2.25×10^5 in 1989 to 8.38×10^5 in 1988 (Table 5.22). Shorebird activity is greatest in South Bay, where 60% of bird use in the system occurs. Here up to 1.0×10^6 shorebirds have been counted during spring surveys, and 3.75×10^5 have been counted during fall surveys.[76] Extensive mudflats and salt ponds along South Bay provide ideal habitats for many shorebirds.

In regard to seabirds, the five most abundant species recorded by a comprehensive censusing effort during 1989 and 1990 were (in decreasing order) California gulls (*Laurus californicus*), Forster's terns (*Sterna forsteri*), Caspian terns (*S. caspia*), western gulls (*L. occidentalis*), and double-crested cormorants (*Phalacrocorax auritus*). The creation of artificial habitats, such as salt ponds, have enabled some populations to become establish as nesting species. Examples are California gulls (*L. californicus*), Forster's terns (*S. forsteri*), Caspian terns (*S. caspia*), and California least terns (*S. antillarum*).

Raptors appear to be declining in abundance. The loss of habitat (e.g., seasonal wetlands, grasslands, oak woodlands, alfalfa fields, and riparian woodlands) has likely contributed to decreases in Cooper's hawk (*Accipiter cooperii*), Swainson's hawk (*Buteo swainsoni*), red-shouldered hawk (*B. lineatus*), short-eared owls (*Asio flammeus*), long-eared owls (*A. otus*), northern harriers (*Circus cyaneus*), and golden eagles (*Aquila chrysaetos*). Continued urban expansion threatens other raptors as well.

The loss of tidal wetlands has had a notable impact on other bird populations. Human development has been devastating to tidal wetlands habitat. Approximately 82% of the historic tidal and brackish wetlands habitat area has been converted to other wetland types and to nonwetland uses. Even more historic freshwater wetlands area (97%) has been converted to other uses. As a consequence, many bird species dependent on these habitats are now listed on endangered species lists or are candidates for endangered status. The endangered California clapper rail (*Rallus longirostris obsoletus*) is an example. This once prominent resident species declined from tens of thousands of individuals in tidal marshes in the early 1900s to fewer than 1500 individuals in the 1990s. This species utilizes cordgrass, pickleweed, and other marsh plants for nesting.[60] Losses of emergent vegetation in marsh habitats together with increased levels of toxic contaminants in marsh sediments and predation by the red fox (*Vulpes vulpes*) and other wildlife populations have all contributed to recent population declines. The California black rail (*Laterallus jamaicensis coturniculus*), listed by the state as a threatened species, is another bird that inhabits tidal salt and brackish marshes. Other bird species that use these habitats for nesting and feeding are the Virginia rail (*R. limicola*), sora (*Porzana carolina*), great blue heron (*Ardea herodias*), great egret (*Casmerodius albus*), snowy egret (*Egretta thula*), black-crowned night heron (*Nycticorax nycticorax*), northern harrier (*Circus cyaneus*), marsh wren (*Cistothorus palustris*), red-winged blackbird (*Agelaius phoeniceus*), savannah sparrow (*Passerculus sandwichensis*), American bittern (*Botaurus lentiginosus*), mallard (*A. platyhynchos*), cinnamon teal (*Anas cyanoptera*), and common yellowthroat (*Geothlypis trichas*).

Some of the same species that occupy tidal salt and brackish marshes also occur in tidal freshwater marsh habitat. For example, the Virgina rail (*Rallus limicola*), California black rail (*Laterallus jamaicensis coturniculus*), sora (*Porzana carolina*), and common yellowthroat (*Geothlypis trichas*) utilize tidal freshwater marshes. Some other bird species found here include the pied-billed grebe (*Podilymbus podiceps*), American bittern (*Botaurus lentiginosus*), marsh wren (*Cistothorus palustris*), yellow-headed blackbird (*Xanthocephalus xanthocephalus*), and tricolored blackbird (*Agelaius tricolor*).[60] These species depend on the marshes for breeding, migration, resting, or roosting. The conversion of marsh habitat represents a potentially serious threat to the long-term viability of these bird populations.

b. Amphibians, reptiles, and mammals. Widespread agricultural conversion of natural habitat, urban encroachment, and human disturbance have caused serious declines in mammal, amphibian, and reptile populations. In particular, much of the habitat required by amphibians and reptiles for foraging, resting, and breeding has been reduced or eliminated entirely. Thus, many amphibians and reptiles have declined considerably and are now listed among special status, extirpated, and extinct wildlife species of the study area. Included here are the California tiger salamander (*Ambystoma tigrinum californiense*), California red-legged frog (*R. aurora draytoni*), foothill yellow-legged frog (*Rana boylii*), western spadefoot (*Scaphiopus hammondii*), blunt-nosed leopard lizard (*Gambelia sila*), California horned lizard (*Phrynosoma coronatum frontale*), San Francisco garter snake (*Thamnophis sirtalis tetrataenia*), giant garter snake (*T. gigas*), and Alameda striped racer (*Masticophis laterialis euryxanthus*).

Table 5.20 Wildlife Species Listed as Threatened or Endangered, or Known or Believed to be Experiencing Population Declines

Species	Status	Habitat
		Insects
Lange's metalmark butterfly	FE	Dunes
Mission blue butterfly	FE	Grasslands, scrub
San Bruno elfin butterfly	FE	Grasslands, scrub
Bay checkerspot butterfly	FT	Grasslands, serpentine soils
Valley elderberry longhorn beetle	FT	Riparian
Delta green ground beetle	FT	Seasonal marshes, grasslands
		Amphibians and Reptiles
San Francisco garter snake[a]	FE, SE	Freshwater marshes, sagponds
Alameda striped racer[a]	ST, FC2	Coastal scrub, chaparral
Western spadefoot[a]	SSC	Grasslands, vernal pools
California tiger salamander[a]	FC2, SSC	Vernal pools
Red-legged frog[a]	FC2, SSC	Freshwater marshes, riparian
Giant garter snake[a]	ST, FC2	Lakes, freshwater marshes
		Birds
California brown pelican	FE, SE	Open water, salt ponds
Greater sandhill crane	ST	Brackish/freshwater/seasonal marshes, grasslands, agriculture
Aleutian Canada goose	FT, SE	Seasonal marshes, farmed wetlands, grasslands
Swainson's hawk[a]	ST	Riparian, grasslands
Bald eagle	FE, SE	Open water, brackish/freshwater marshes, riparian
Golden eagle[a]	SSC	Open country, cliffs
Burrowing owl[a]	SSC	Grasslands
Short-eared owl[a]	SSC	Seasonal wetlands, grasslands
Northern spotted owl	FT, SSC	Evergreen forest
Long-eared owl[a]	SSC	Riparian
Cooper's hawk[a]	SSC	Woodlands
Sharp-shinned hawk[a]	SSC	Woodlands
Northern harrier[a]	SSC	Seasonal wetlands, grasslands
American peregrine falcon	FE, SE	Salt/brackish/freshwater marshes, riparian, chaparral
Black rail[a]	ST, FC1	Salt marshes
California clapper rail[a]	FE, SE	Salt marshes
Western snowy plover[a]	FC2, SSC	Salt ponds
Long-billed curlew[a]	FC2, SSC	Grassland, farmed wetlands
California least tern[a]	FE, SE	Salt ponds, sandy bayshore
Saltmarsh yellowthroat[a]	FC2, SSC	Marshes, riparian
Alameda song sparrow[a]	FC2, SSC	Salt marshes
San Pablo song sparrow[a]	FC2, SSC	Salt marshes
Suisun song sparrow[a]	FC2, SSC	Salt marshes
Tri-colored blackbird[a]	FC2, SSC	Freshwater marshes
Bank swallow	ST	riparian
		Mammals
San Joaquin kit fox[a]	FE, ST	Grasslands
Salt marsh harvest mouse[a]	FE, SE	Salt marshes
Salt marsh wandering shrew[a]	FC1, SSC	Salt marshes
Suisun ornate shrew[a]	FC1, SSC	Salt marshes

Table 5.20 (continued) Wildlife Species Listed as Threatened or Endangered, or Known or Believed to be Experiencing Population Declines

Species	Status	Habitat
San Joaquin Valley woodrat[a]	FC2, SSC	Riparian
Riparian brush rabbit[a]	FC1, SSC	Riparian
American badger[a]	SSC	Grasslands
Townsend's big-eared bat[a]	SSC	Conifer-hardwood, structures

[a] Species with believed or known population decline; FE — federally endangered; SE — state endangered; ST — state threatened; SSC — state species of concern; FC — federal candidate for listing; Category 1 — taxa for which the Fish and Wildlife Service has sufficient biological information to support a proposal to list as endangered or threatened; Category 2 — taxa for which existing information may warrant listing, but for which substantial information to support a proposed rule is lacking.

Source: Harvey, T. E., Miller, K. J., Hothem, R. L., Rauzon, M. J., Page, G. W., and Keck, R. A., Status and Trends Report on Wildlife of the San Francisco Estuary, Tech. Rept., San Francisco Estuary Project, U.S. Environmental Protection Agency, San Francisco, CA, 1992.

Even more mammalian populations are listed among special status, extirpated, and extinct wildlife species of the estuarine system. Examples are the saltmarsh wandering shrew (*Sorex vagrans halicoetes*), Suisun ornate shrew (*Sorex ornatus sinuosus*), Pacific western big-eared bat (*Plecotus townsendii townsendii*), pallid bat (*Antrozous pallidus*), greater western mastiff bat (*Eumops perotis californicus*), riparian brush rabbit (*Sylvilagus bachmani riparius*), saltmarsh harvest mouse (*Reithrodontomys raviventris*), San Joaquin valley woodrat (*Neotoma fuscipes riparia*), San Pablo vole (*Microtus californicus sanpabloensis*), gray wolf (*Canis lupus fuscus*), San Joaquin kit fox (*Vulpes macrotis mutica*), grizzly bear (*Ursus arctos*), badger (*Taxidea taxus*), sea otter (*Enhydra lutris*), Roosevelt elk (*Cervus elaphus roosevelti*), and pronghorn (*Antilocapra americana*). The number of endangered and threatened mammalian species is expected to increase as the loss and alteration of wetland and upland habitats continue from urban encroachment, agricultural conversion, and increased pollution.

Some mammalian populations using the estuary are either stable or increasing in size. For example, the number of harbor seals (*Phoca vitulina*) in the estuary is generally stable, and the number of California sea lions (*Zelophus californianus*) in the system is increasing.

Tidal salt, brackish, and freshwater marshes continue to provide valuable habitat for amphibians, reptiles, and mammals. Tidal salt and brackish marshes are important habitat for an array of snakes, lizards, and turtles. The harbor seal is the largest mammal that utilizes tidal salt marshes of the bay. Other mammals using these habitats are the saltmarsh harvest mouse (*Reithrodontomys raviventris*), saltmarsh vagrant shrew (*Sorex vagrans halicoetes*), Suisun ornate shrew (*S. ornatus sinuosus*), river otter (*Lutra canadensis*) (brackish only), muskrat (*Ondatra zibethica*) (brackish only), mink (*Mustela vison*) (brackish only), beaver (*Castor canadensis*) (eastern part of Suisun Marsh only), California vole (*Microtus californicus*), Norway rat (*Rattus norvegicus*), and house mouse (*Mus musculus*).[60]

In the delta freshwater marshes, several amphibians, reptiles, and mammals are abundant. The bullfrog (*Rana catesbeiana*) is an important amphibian in this habitat. Among reptile populations, the western pond turtle (*Clemmys xanthocephalus*) and giant garter snake (*Thamnophis gigas*) frequent the area. Beaver (*Castor canadiensis*), muskrat (*Ondatra ziabethicus*), and mink (*Mustela vison*) are mammals of significance in the marshes.

Most wildlife species in the estuary are less abundant than in the past. Approximately 90 species and subspecies of wildlife are now designated by federal or state agencies as being in need of special attention.[77] It is incumbent upon the San Francisco Estuary Project to implement sound and effective management programs to ensure the long-term success of these wildlife populations.

Table 5.21 Status and Trends of Selected Wildlife Species of the San Francisco Bay/Delta Estuary

Species	Status[a]	Seasonal Occurrence[b]	Habitat Use[c]	Current Abundance	Recent Trends in Numbers or Use of Estuary	Comments
Birds						
Common loon	SSC, MC, SBS	R-W, M-Sp, F	SM, BM, LP	Common	1984–1989: 21–83 birds in Oakland 1975–1989: 10–40 birds in Marin	Recent declines on west coast. Human disturbance during breeding, oil spills, and lake acidification causes of decline throughout range.
Eared grebe	—	R-W, M-Sp, F	OW, FM, SP	Abundant	>40,000 in estuary	Abundant on salt ponds; sporadic nesting in South Bay at seasonal marsh/ponds.
Double-crested cormorant	SSC	R-Yr	OW, RS, SM, BM, FM, DM, SP, LP	Common	Breeding population in bay increasing over past 5 years; 1989–1990 = 1,185 nesting pairs	Nests on bridges and other man-made structures; large colonies on S.F.-Oakland and Richmond-San Rafael bridges.
Great blue heron	—	R-Yr	IM, RS, SM, BM, FM, DM, FH, others	Common	160 nesting pairs in many habitats	Adversely affected by human disturbance, loss of nesting trees, and possibly pesticides.
Black-crowned night heron	—	R-Yr	IM, SM, BM, FM, FH, FW, SP	Common	At least 1,500 pairs nesting; greatest number in North Bay	Abnormal embryos and crushed eggshells in some nests suggest contaminant-related reproductive problems.
Western gull	—	R-Yr	OW, IM, RS, SM, BM, SP	Abundant	1,690 breeding pairs in 1990	Population increasing; nesting on rocky shores and bridges.
California gull	SSC	R-Yr	OW, IM, RS, BM, FH, FW, SP, LP	Common	2,221 breeding pairs in 1990	Began nesting in bay in 1981. Today most abundant nesting seabird. Nests at salt ponds.
Caspian tern	SBS	R-Su	OW, SM, BM, FM, SP, LP	Common	1,409 nesting pairs in 1989–1990	Red fox predation in South Bay cause for nesting failure.
California least tern	FE, SE	R-Su	OW, SM, SP	Uncommon	Started nesting in bay in 1967. Recent average nesting population = 74 pairs	Red fox and other predators caused total nesting failure of Oakland Airport population in 1990.

Tundra swan	—	R-W	OW, DM, FW, LP	Common	About 12,000 birds wintered in delta in 1990.	Delta is most important wintering area in Pacific Flyway; population increasing.
Northern pintail	HA	R-SU, W	SM, BM, FM, DM, FW, SP, LP, AL	Common	97,000 birds wintered in estuary in 1990; 10% of 1977 population	In early 1980s, drought in northern prairie caused massive population drop; now rebounding.
Canvasback	HA	R-W, M-Sp, F	OW, SM, BM, FM, SP, LP, GR	Common	During 1980s, wintering population declined, then increased to 40,000	10–15% of U.S. wintering population occurs in estuary; affected by wetland loss in northern prairie.
Greater scaup	HA	R-W, M-Sp, F	BM, FM, SP, LP	Common	In 1990, 150,000 wintered in estuary, an increase from preceding decade	Most abundant diving duck in estuary; 50% occur in North Bay; statewide population increasing.
Swainson's hawk	FC, ST, SSC, SBC	R-Su	RW, GR, WS, AL	Uncommon	Statewide population is 550; 78% in Central Valley, 9% in delta	Conversion of grasslands to agricultural and urban; pesticides caused massive decline; population stable past 10 years.
American peregrine falcon	FE, SE	R-Su, W, M-Sp, F	SM, BM, FM, RW, CS, MC	Uncommon	During past 20 years, local population up by tenfold; 10–20 birds in estuary region	Estuary population increasing; however, no successful reproduction has occurred.
California clapper rail	FE, SE	R-Yr	SM, BM	Rare	During past decade, population declined from 1,500 to 300–500; 90% occur in South Bay	Red fox predation a major threat; high mercury concentrations in eggs.
Snowy plover	FC2, SSC, SBS, BMC	R-Yr	IM, SP, AL	Uncommon	Breeding population has declined from 1970 levels; now about 200–350 winter in estuary	Red fox predation and habitat loss major threats; soon may be listed as threatened.
Long-billed curlew	FC, SSC	R-W, M-Sp, F	IM, SM, BM, FM, DM, FH, FW, SP, LP, GR	Uncommon	Fall population as high as 2,300; fewer in winter	Uses seasonal wetlands, especially in Central Valley; population declining due to agricultural conversion in Great Basin.
Western sandpiper	—	R-W, M-Sp, F	IM, RS, SM, BM, FM, FH, FW, SP, LP, AL	Abundant	475,000–700,000 in bay during spring	Most abundant shorebird in estuary.

Table 5.21 (continued) Status and Trends of Selected Wildlife Species of the San Francisco Bay/Delta Estuary

Species	Status[a]	Seasonal Occurrence[b]	Habitat Use[c]	Current Abundance	Recent Trends in Numbers or Use of Estuary	Comments
Black-necked stilt	—	R-Su, W	IM, SM, BM, FM, DM, FH, FW, SP, LP	Common	8,000–12,000 in bay during fall; mostly in South Bay salt ponds	The increasing South Bay breeding population is threatened by introduced predators.
Alameda song sparrow	FC2, SSC	R-Yr	SM	Rare	Habitat stable, except at Coyote Creek	Conversion of South Bay salt marshes has greatly reduced available habitat.
Tri-colored blackbird	FC2, SSC	R-Yr	BM, FM, FH, FW, LP, GR, AL	Common	Steep population decline throughout range; several colonies in Fremont eliminated in past decade	Nests in freshwater marsh; Central Valley population declined nearly 90% during 1930–1980.
Mammals						
Salt marsh wandering shrew	FC1, SSC	R-Yr	SM, DM	Rare	Stable, except where marsh erosion and conversion occurring	One of the most endangered species in the estuary basin.
Salt marsh harvest mouse	FE, SE	R-Yr	SM, BM, DM	Rare	Great seasonal and annual population fluctuations	Only 760 acres of diked habitat available in South Bay; 6,000 acres of tidal marsh in North Bay; vulnerable to flooding/genetic isolation.
Harbor seal	—	R-Yr	OW, IM, RS, SM, BM	Common	Population stable at about 300–500; more than 350 used two South Bay haulout sites in 1990	Recent study detected PCB and DDT in tissue.
Red fox	FC2, ST, IN	R-Yr	SM, BM, FM, FN, GR, CS, MC, AL	Uncommon	Population increasing since arrival in bay area in early 1980s	Major predator on bayshore nesting birds, especially rails and terns.
Amphibians and Reptiles						
California tiger salamander	FC2, SSC	R-Yr	FM, FH, WS	Uncommon	Populations small, isolated, and declining	Loss of vernal pools is a major cause of population decline.
California red-legged frog	FC2, SSC	R-Yr	BM, FM, DM, FH, RW, LP, GR, BF	Uncommon	Populations small, isolated, and declining	May be extirpated from the delta; still occurs at a few locations in bay area.

Giant garter snake	FC2, ST	R-Yr	FH, LP, GR, WS	Uncommon	Populations small, isolated, and declining	Loss of sloughs and marshes in Central Valley and delta, possibly pesticides, cause for decline.
San Francisco garter snake	FE, SE	R-Yr	SM, LP	Rare	Populations small, isolated, but stable during past decade	Loss of major prey, the red-legged frog, due to habitat conversion and introduced bullfrog.
Western pond turtle	FC	R-Yr	FM, DM, FH, FW, RW, LP, GR	Uncommon	Population greatly reduced, but probably stable	Loss of riparian vegetation and natural shorelines in delta cause of decline.

[a] FE — federal endangered; FT — federal threatened; SE — state endangered; ST — state threatened; FC — candidate for federal threatened/endangered listing; SSC — California Dept. of Fish and Game species of special concern; SBS — U.S. Fish and Wildlife Service sensitive bird species; HA — harvested species; MC — federal management concern.

[b] R — resident; M — migrant; Sp — spring; Su — summer; F — fall; W — winter.

[c] Wetland Habitats: OW — open water; IM — intertidal mudflat; RS — rocky shore; SM — salt marsh; BM — brackish marsh; FM — fresh marsh; DM — diked marsh; FH — other freshwater habitat; FW — farmed wetlands; RW — riparian woodland; SP — salt pond; LP — lakes and ponds. Upland Habitats: GL — grassland; AL agricultural land; UR — urban land; CS — coastal scrub; BF — broad-leaved forest; MC — mixed chaparral; WS — oak woodland.

Source: Monroe, M. W. and Kelly, J., State of the Estuary, Tech. Rept., San Francisco Estuary Project, Oakland, CA, 1992.

Table 5.22 Spring and Fall Inventories of Shorebirds in San Francisco Bay

Year	Spring Survey	Fall Survey
1988	—	838,470
1989	931,561	225,427
1990	663,790	357,754
1991	588,964	342,504
1992	692,959	325,449
1993	627,093	—

Source: San Francisco Estuary Project, State of the Estuary: 1992–1997, Tech. Rept., San Francisco Estuary Project, Oakland, CA, 1998.

VI. Management plan

The San Francisco Estuary Project completed a Comprehensive Conservation and Management Plan (CCMP) in 1993 focusing on the bay/delta estuary's principal environmental problems. As noted in Section II, the CCMP addresses five management issues that pose a potential threat to the estuary: (1) intensified land use; (2) decline of biological resources; (3) freshwater diversion and altered flow regime; (4) increased pollutants; and (5) dredging and waterway modification. It also proposes action items designed to restore and maintain the chemical, physical, and biological integrity of the bay and delta. In so doing, the CCMP provides a blueprint for the long-term protection and improvement of the estuarine system.[77]

The implementation of management programs during the past 2 decades has resulted in the upgrade of a number of system components. For example, municipal and industrial pollutant loads have decreased appreciably, wetland losses have diminished, and the regulation of dredging activities has been more effective. Despite these improvements, other problems persist and may be getting worse. For instance, nonpoint source pollution continues to degrade water quality in the estuary. Urban expansion is depleting valuable upland wildlife habitat. In addition, the effects of freshwater diversions are systemic, adversely affecting habitat quality, primary production, recreational and commercial fish populations, and other living resources baywide.

The CCMP consists of 145 action items to protect fish, wildlife, wetlands, and watersheds of the estuarine system. These action items constitute a coordinated and comprehensive strategy to achieve the goals of the project. They are organized under nine different program areas: (1) aquatic resources; (2) wildlife; (3) wetlands; (4) water use; (5) pollution prevention and reduction; (6) dredging and waterway modification; (7) land use; (8) public involvement and education, and (9) research and monitoring. The subsequent list of actions derives from the CCMP.[77]

A. Aquatic resources

The goals of the aquatic resources program area are

- To stem and reverse the decline in the health and abundance of estuarine biota (indigenous and desirable non-indigenous), with an emphasis on natural production
- To restore healthy estuarine habitat conditions to the bay–delta, taking into consideration all beneficial uses of bay–delta resources

- To ensure the survival and recovery of listed and candidate threatened and endangered species as well as other species in decline
- To optimally manage the fish and wildlife resources of the estuary to achieve the purpose of the goals stated previously

The aquatic resources action items are as follows:

AR-1.1. Refine and coordinate existing monitoring programs to: (1) better evaluate ecosystem responses to immediate, phased, and long-term water quality and flow standards; (2) more fully characterize ecosystem processes and properties; and (3) enhance predictive capabilities of ecosystem models.

AR-2.1. Develop, implement, and enforce stringent regulations to control discharges of ship ballast water within the estuary or adjacent waters.

AR-2.2. Prohibit the intentional introduction of aquatic exotic species into the estuary and its watershed.

AR-2.3. Control problem exotic species already in the estuary.

AR-2.4. Develop programs to educate the public about the problems with exotic species and their incidental transport or introduction.

AR-2.5. Strengthen programs to reduce the poaching of species within the estuary.

AR-2.6. Review and modify, if necessary, harvest regulations for aquatic species of concern.

AR-2.7. Identify and control sources and sinks of contaminants that may affect fish populations or ecosystem health.

AR-2.8. Research and develop methods to reduce the incidental take of nontarget species in commercial activities.

AR-3.1. Prepare/update recovery plans for all listed species. This includes designation of critical habitat.

AR-3.2. Monitor status of all candidate species and list them if warranted.

AR-3.3. Initiate consultations with all federal agencies that propose or are continuing actions that may affect listed species.

AR-3.4. Review all nonfederal proposals and continuing actions that may result in take of listed species and take appropriate actions.

AR-3.5. Investigate the feasibility of developing a habitat conservation plan (or plans) for the bay and delta that promotes the recovery of the species and addresses incidental take associated with nonfederal actions.

AR-3.6. Adopt listed species recovery as a policy for all public agencies whose actions affect them.

AR-4.1. Adopt water quality and flow standards and operational requirements designed to halt and reverse the decline of indigenous and desirable nonindigenous estuarine biota and to contribute to the attainment of Objective AR-5. Implement these standards and requirements in at least three phases: (1) immediate, interim standards and requirements consistent with current legal requirements that would be in place with the delta in its existing configuration; (2) standards and requirements linked to South Delta Water Management facilities; and (3) standards and requirements, as may be necessary, linked to off-stream storage south of the delta to facilitate water banking and water-transfer activities, so long as the last two phases significantly reduce impacts on aquatic estuarine resources and meet all environmental requirements.

AR-4.2. Establish conditions on industrial facilities to control entrainment of eggs, larvae, and juvenile fish.

AR-4.3. Design and install gates or other facilities at channel openings known to be associated with the loss of fishes.

AR-4.4. Design, install, and effectively operate fish screens or other protective devices at diversions associated with fish mortality.

AR-4.5. Improve screen efficiencies at state and federal water project pumping and fish salvage facilities.

AR-4.6. Develop and implement a management plan to reduce predation in Clifton Court Forebay and near the John E. Skinner Delta Fish Protection Facility.

AR-4.7. Protect existing shaded riverine aquatic habitats to ensure no net loss of acreage, lineal coverage, and habitat value within the estuary. Activities within the "legal delta" should be conducted consistent with California's Delta Levees Flood Protection Act of 1988.

AR-4.8. Increase the quantity of shaded riverine aquatic habitat by 1000%.

AR-4.9. Promote the maintenance and development of tule islands, tidal wetlands, and offshore berms to protect against erosion and to provide detrital input and juvenile fish nursery habitat.

AR-4.10. Work with the dredging and flood control interests to reduce or eliminate practices that adversely affect fish habitat.

AR-4.11. Identify and protect remnant stream habitats containing indigenous and endemic fishes and other native aquatic species.

AR-4.12. Protect and maintain marshes, wetlands, shallow water areas, and tidal sloughs to protect fisheries values.

AR-5.1. Based on information developed in AR-1.1, identify alternative long-term water quality and flow standards, water management measures, operational changes, habitat improvements, and facilities as needed to manage the estuarine aquatic resources (including water) for optimum benefit.

AR-5.2. Develop an EIS/EIR to display the alternatives and trade-offs identified in AR-5.1 and to initiate the selection of a preferred alternative.

AR-5.3. Implement the alternatives from AR-5.2 (including the adoption long-term water quality and flow standards and operational requirements) that best optimizes conditions for aquatic resources, efficiently conserves scarce water resources, and restores an equitable balance to the estuarine ecosystem.

AR-6.1. Provide necessary instream flows and temperatures to benefit salmon and steelhead in the Central Valley to support the implementation of the state and federal mandates to double the natural production of anadromous fishes.

AR-6.2. Implement the Upper Sacramento River Management Plan.

AR-6.3. Develop and implement the San Joaquin River Management Plan to identify reservoir operational changes, habitat improvement measures, and other items to improve habitat and health of the aquatic ecosystem in the San Joaquin River watershed.

AR-6.4. Screen upstream diversions that individually or cumulatively result in significant mortality to fishes that utilize the estuary.

AR-6.5. Seek damages for all impacts to trust resources from spills and discharges affecting them and use the funds to improve the resource base.

B. *Wildlife*

The goals of the wildlife program area are

- To stem and reverse the decline of estuarine plants and animals and the habitats on which they depend

- To ensure the survival and recovery of listed and candidate threatened and endangered species as well as special status species
- To optimally manage and montior the wildlife resources of the estuary

The wildlife action items are as follows:

WL-1.1. Preserve, create, restore, and manage large, continuous expanses of tidal salt marsh and necessary adjacent uplands for the California clapper rail and the salt marsh harvest mouse.
WL-1.2. Complete the expansion of the San Francisco Bay National Wildlife Refuge and its satellite refuges and acquire the proposed Stone Lakes National Wildlife Refuge.
WL-1.3. Implement concerted efforts to acquire wetlands already degraded or destroyed and restore them so that wetlands in the estuary are increased by 50% by 2000.
WL-1.4. Restore tidal marshes in San Francisco Bay.
WL-1.5. Identify and convert or restore nonwetland areas to wetland or riparian-oriented wildlife habitat.
WL-2.1. Prepare a comprehensive management plan for the San Francisco Bay National Wildlife Refuge.
WL-2.2. Enhance the biodiversity within all publicly owned or managed wetlands and other wildlife habitats as appropriate.
WL-2.3. Complete and implement a wildlife habitat restoration and management plan for the estuary.
WL-3.1. Implement predator control programs in areas where introduced predators are a constraint to maintenance and restoration of native populations.
WL-4.1. Update, and, where necessary, prepare recovery plans for all listed wildlife species.
WL-4.2. Provide secure colony sites, allow for population recover, control predators, and protect adjacent foraging areas for the California least tern.
WL-4.3. Monitor status of all candidate species and list them if warranted.
WL-4.4. Continue hunting closures to protect the Aleutian Canada goose. Investigate the need for hunting closures for other waterfowl species as necessary.
WL-4.5. Implement a captive breeding program for the clapper rail.

C. Wetlands

Goals of the wetlands program area are

- To protect and manage existing wetlands
- To restore and enhance the ecological productivity and habitat values of wetlands
- To expedite a significant increase in the quantity and quality of wetlands
- To educate the public about the values of wetland resources

The wetlands actions items are as follows:

WT-1.1. Prepare regional wetlands management plan(s).
WT-1.2. Encourage geographically focused cooperative efforts to protect wetlands.
WT-2.1. Establish a comprehensive state wetlands program for the estuary which, in addition, includes a coordinated regulatory and policy framework.

WT-2.2. Increase enforcement efforts to curtail illegal wetland alteration and to ensure compliance with permit conditions.
WT-2.3. Develop and adopt uniform compensatory mitigation policies.
WT-2.4. Improve wetlands protection provided under the Clean Water Act.
WT-3.3. Encourage wetland protection by-laws.
WT-4.1. Identify and convert/restore nonwetland areas to wetland- or riparian-oriented wildlife habitat. Purchase nonwetland areas to create wetlands. This should be guided by and consistent with the Regional Wetlands Management Plan.

D. Water use

The goal of the water use program area is

- To develop and implement aggressive water management measures to increase freshwater availability to the estuary

The water use action items are as follows:

WU-1.1. Water reclamation and reuse feasibility studies should be completed by each publicly owned treatment works, municipality, and/or water district.
WU-1.2. Municipalities and counties should adopt water reclamation ordinances encouraging the use of reclaimed water, to the maximum extent practicable, while providing for the protection of public health and the environment.
WU-1.3. Local entities interested in implementing reclamation projects should develop and conduct public education programs.
WU-1.4. Ensure that state water quality standards and basin plans encourage water reclamation and reuse.
WU-1.5. If practical, use existing facilities and develop new facilities to deliver reclaimed and recycled water for beneficial reuse.
WU-1.6. Address and resolve, as appropriate, the impacts on water reclamation and water conservation caused by the discharge of brine from self-regenerating water softeners and other sources into the wastewater stream.
WU-2.1. Governmental, agricultural, public, and environmental interests should work together to develop a mechanism to ensure implementation of efficient agricultural water management practices.
WU-2.2. New methods of agricultural water conservation should be researched through pilot projects and implemented where feasible.
WU-2.3. Water conservation feasibility studies shall be completed and implemented by municipalities and/or water districts.
WU-2.4. Maximize conjuctive use of water through groundwater recharge.
WU-2.5. Study storage of surface water on delta islands.
WU-2.6. Evaluate and adopt, where appropriate, mechanisms to manage groundwater to protect the long-term integrity of groundwater basins.
WU-3.1. More fully utilize the existing and expand, where appropriate, the legal and regulatory framework to facilitate voluntary water-marketing agreements among agricultural, urban, and environmental interests.
WU-3.2. The state should continue to negotiate with the federal government to determine whether, and to what extent, it is appropriate for the federal government to transfer the ownership or operational control of the Central Valley Project to a nonfederal entity.

E. Pollution prevention and reduction

The goals of the pollution prevention and reduction program area are

- To promote mechanisms to prevent pollution at its source
- To control and reduce pollutants entering the estuary, where pollution prevention is not possible
- To clean up toxic pollution throughout the estuary
- To protect against toxic effects, including bioaccumulation and toxic sediment accumulation

The pollution prevention and reduction action items are as follows:

PO-1.5. Reinforce existing programs and develop new incentives where necessary to reduce selenium levels in agricultural drainage.
PO-1.6. Develop a comprehensive strategy to reduce pesticides coming into the estuary.
PO-2.1. Pursue a mass emissions strategy to reduce pollutant discharges into the estuary from point and nonpoint sources and to address the accumulation of pollutants in estuarine organisms and sediments.
PO-2.2. Adopt water quality objectives that effectively protect estuarine species and human health.
PO-2.3. Identify and control sources and sinks of selenium and mercury where they are accumulating in aquatic populations in the estuary.
PO-2.4. Improve the management and control of urban runoff from public and private sources.
PO-2.5. Develop control measures to reduce pollutant loadings from energy and transportation systems.
PO-2.6. Improve the management and control of agricultural sources of toxic substances.
PO-2.7. Reduce toxic loadings from mines.
PO-2.8. Establish a model environmental compliance program at federal facilities within the jurisdiction of the estuary project.
PO-3.1. Clean up contaminants presently affecting fish, wildlife, their habitats, and food supplies.
PO-3.2. Expedite the cleanup of toxic hot spots in estuarine sediments.

F. Dredging and waterway modification

The goals of the dredging and waterway modification area are

- To eliminate unnecessary dredging activities
- To maximize the use of dredged material as a resource
- To conduct dredging activities in an environmentally sound fashion
- To adopt a sediment management strategy for dredging and waterway modification
- To manage modification of waterways to avoid or offset the adverse impacts of dredging, flood control, channelization, and shoreline development and protection projects

The dredging and waterway modification action items are as follows:

DW-1.1. Conduct studies, research, and models of sediment dynamics.
DW-1.2. Conduct studies on sediment changes aimed to define accumulation and erosion processes in marsh and mudflat areas.
DW-1.3. Adopt policies to manage modification of estuarine sediment production, movement, and deposition.
DW-2.1. Conduct laboratory and field bioaccumulation investigations and studies on suspended sediment effects on sensitive life stages throughout the food chain.
DW-2.2. Develop and set sediment quality objectives.
DW-3.1. Develop a dredge project needs assessment and, as necessary, a prioritization plan, including structural and nonstructural methods to minimize volume requirements.
DW-3.2. Identify dredged material reuse and nonaquatic disposal opportunities and constraints.
DW-3.3. Develop regulatory land use procedures to promote reuse of dredged material, wetlands restoration and/or creation, and other beneficial uses.
DW-3.4. Identify the aquatic and terrestrial resources that are affected by dredging and disposal and are to be protected in the bay and delta.
DW-3.5. Designate dredged material reference sites for use in development of sediment testing protocols.
DW-3.6. Evaluate retention and removal needs for derelict structures in the bay and delta.
DW-3.7. Adopt regulatory and management policies for estuary dredging activities and develop dredging and disposal projects that are consistent with the state's existing policies in the San Francisco Bay Plan and in the San Francisco Bay and Central Valley Basin Plans.
DW-4.1. Identify dredged material disposal options, including cost estimates and alternative disposal methods. Conduct periodic review as necessary.
DW-4.2. Conduct modeling and field studies to determine the saltwater intrusion impacts caused by dredging projects.
DW-4.3. Revise Public Notice 87-1, "Interim Testing Procedures for Evaluating Dredged Material Suitability for Disposal in San Francisco Bay," and develop testing procedures and protocols for ocean and upland environments.
DW-5.1. Determine areas subject to flooding and erosion and identify causes.
DW-5.2. Implement waterway modification policies that protect shoreline areas from detrimental flooding and erosion while maintaining natural resource values.
DW-5.3. Establish a program to acquire diked historic baylands listed as buffer areas for coastal flooding and sea level rise.

G. Land use

The goals of the land use program area are

- To establish and implement land use and transportation patterns and practices that protect, enhance, and restore the estuary's open waters, adjacent wetlands, adjacent essential uplands habitat, and tributary waterways
- To coordinate and improve planning, regulatory, and development programs of local, regional, state, and federal agencies to improve the health of the estuary
- To adopt and utilize land-use policies that provide incentives for more active participation by the private sector in cooperative efforts that protect and improve the estuary

The land-use action items are as follows:

LU-1.1. Local general plans should incorporate watershed protection plans to protect wetlands and stream environments and reduce pollutants in runoff.
LU-1.2. Amend the California Environmental Quality Act Guidelines to add simple and concise criteria for assessing the cumulative environmental impacts on the estuary when adopting or reviewing general plans.
LU-1.3. Integrate protection of the estuary with other state land-use related initiatives.
LU-2.1. Regional agencies should assist in identifying and developing consistent policies that provide an integrated framework for local governments to protect the resources of the estuary.
LU-2.2. Adopt policies and plans to promote compact, contiguous development, in both the nine-county bay area and the three-county delta region.
LU-2.3. Compile and analyze data pertaining to future population and land-use change in the nine-county bay area and the three-county delta region to provide information for improved decision-making.
LU-3.1. Prepare and implement watershed management plans that include the following complementary elements: (1) wetlands protection; (2) stream environment protection; and (3) reduction of pollutants in runoff.
LU-3.2. Develop and implement guidelines for site planning and best management practices.
LU-4.1. Educate the public about how human actions impact the estuary.
LU-4.2. Provide training workshops for local government officials and other key stakeholders to improve land-use decision-making that affects the estuary.
LU-5.1. Create economic incentives that encourage local governments to implement measures to protect and enhance the estuary.
LU-5.2. Develop new funding mechanism to pay for plans, physical improvements, and program administration to protect the resources of the estuary.
LU-5.3. Investigate and create market-based incentives that promote active participation by the private sector in cooperative efforts to implement goals for protection and restoration of the estuary.
LU-5.4. Identify financial barriers to implementing the actions recommended in this land-use management program and propose alternative taxation and funding arrangements.
LU-5.5. Create a forum to improve communication and resolve disputes regarding land-use management among different interest groups that have a stake in the protection and enhancement of the estuary.

H. Public involvement and education

The goals of the public involvement and education program area are

- To build public understanding of the value of the estuary's natural resources and the need to restore, protect, and maintain a healthy estuary for future generations
- To increase public involvement in the ongoing stewardship of the estuary

The public involvement and education action items are as follows:

PI-1.1. Build awareness, interest, and support in the general public and decision-makers for the CCMP's goals and plans.

PI-1.2. Provide and encourage opportunities for direct citizen involvement in implementing the CCMP.

PI-1.3. Provide and encourage opportunities for direct citizen involvement in following the CCMP and making any necessary revisions to it.

PI-1.4. Serve as a public involvement and education resource for government agencies taking the lead in CCMP management actions.

PI-1.5. Ensure provisions for a central collection and distribution (clearinghouse) point for communication and coordination of all information concerning CCMP issues and the estuary.

PI-1.6. Develop and/or promote community designed model projects for public education and participation activities aimed at implementing the CCMP.

PI-1.7. Seek, encourage, and, where appropriate, actively support environmental projects and/or programs that are consistent with CCMP goals and objectives.

PI-2.1. Develop, promote, and support multicultural understanding of and involvement in estuary issues and the decision-making process for these issues.

PI-2.2. Work with education groups, interpretive centers, decision-makers, and the general public to build awareness, appreciation, knowledge, and understanding of the estuary's natural resources and the need to protect them. This would include how these natural resources contribute to and interact with social and economic values.

PI-2.3. Promote, support, and cooperate with existing public education and involvement programs concerned with protecting and restoring the estuary's biological resources.

PI-2.4. Develop or promote necessary public education tools, such as a general education speakers bureau, bay-delta "estuary watch" bulletin boards, slide shows, brochures, and other support materials on a variety of topics.

PI-2.5. Assist in the development of long-term educational programs designed to prevent pollution to the estuary's ecosystem and provide assistance to other programs as needed.

PI-2.6. Hold a state of the estuary conference at least every other year.

PI-3.1. Increase public opportunities to contribute directly to the protection and management of fish and wildlife populations and their habitats within the estuary.

PI-3.2. Using government agencies and citizens, promote the continued development of needed citizen monitoring programs to assist in the restoration and protection of the estuary.

PI-3.3. Provide opportunities for hands-on citizen involvement and estuary restoration activities.

PI-3.4. Assess the need and, if appropriate, develop and organize an estuary conservation corps.

PI-4.1. Develop, plan, and facilitate the transition from the estuary project to a community-based entity or entities that would help carry out public involvement and education goals and objectives of the estuary project and the CCMP in ways that do not duplicate the efforts of other organizations.

PI-4.2. Work to fund and support existing and new public involvement, education, research, and monitoring activities that seek to fulfill the goals of the CCMP.

PI-4.3. Ensure that a technical/scientific/academic entity has responsibility to promote scientific research on and monitoring of the estuary and provide advice and guidance related to those activities.

I. Research and monitoring

The goal of the research and monitoring actions is

- To improve the scientific basis for managing natural resources within the estuary through an effective monitoring and research program.

The research and monitoring action items are as follows:

RM-1.1. Establish and operate a San Francisco Estuarine Institute for research on and monitoring of land use, biological resources, flow regime, pollutants, and dredging and waterway modification.
RM-1.2. Provide a long-term administrative home and regular funding for the research enhancement program.
RM-2.1. Develop and implement the regional monitoring strategy, which will integrate and expand on existing efforts and eventually be part of a comprehensive regional monitoring program.

J. Conclusions

The recommended actions of the CCMP listed above represent only an initial step toward achieving the primary goal of the project: to restore and maintain the estuary's water quality and natural resources while maintaining its many beneficial uses. An equally significant part of "revitalizing" the estuary is to effectively implement these actions. This will require a cooperative effort on the part of local, state, and federal government agencies, stakeholders, educational institutions, and the general public.

Between 1993 and 1996, 59 of the CCMP's 145 action items (40%) to protect fish, wildlife, wetlands, and watersheds had moderate to full implementation.[1,78] Substantial progress has been made in most of the nine program areas. In regard to the 35 aquatic resources actions, the most notable progress relates to the adoption of new water quality and flow standards to mitigate impacts on estuarine organisms. Moderate progress has been accomplished on a number of wildlife actions, particularly the expansion of wildlife refuges. Some progress has also been realized on the protection of endangered species and the development of recovery plans for various wildlife populations (e.g., California clapper rails, *Rallus longirostris obsoletus*; salt marsh harvest mice, *Reithrodontomys ravinentris*; and western snowy plover, *Charadrius alexandrinus*).

In wetlands management, moderate progress has been achieved on the acquisition and restoration of wetland parcels. Some degree of progress was shown in the adoption of no-net loss policies and other regulatory controls. Water-use actions that have advanced forward include water reclamation and conservation initiatives and the formulation of policies that promote water reuse (e.g., Porter-Cologne Act). There has been moderate progress associated with the promotion of agricultural and urban water conservation programs. Of all actions coupled to pollution prevention and reduction, the formulation and implementation of municipal stormwater management programs and public education concerning pollution prevention have exhibited the greatest progress.

The Long-Term Management Strategy for Dredged Material Disposal, a public-private cooperative planning effort, has facilitated substantive progress on dredging and waterway modification actions. Considerable progress has been made in promoting beneficial reuse of dredged material. Plans have been developed for reuse of dredged material for wetland restoration, landfill cover, beach enhancement, road building, and other projects through federal regulatory procedures as well as regional and state policies.

In respect to the remaining actions, the most significant progress regarding land-use management has been accomplished on the formulation of plans, policies, and programs that protect the estuary's resources. Among public involvement and education actions, significant progress has occurred in the development of a strong public education and involvement program for teachers, students, and communities under Friends of the San Francisco Estuary. There has been moderate progress in increasing the opportunities of individuals in the public sector to contribute directly to fish and wildlife protection. Progress on research and monitoring actions has been accelerated by the creation of the San Francisco Institute in 1994, which led to the implementation of the Regional Monitoring Strategy that identified information needs for aquatic resources, wildlife, wetlands, water use, land use, pollution, and dredging. Finally, there has been notable progress in improving data accessibility to scientists, administrators, and decision-makers dealing with environmental problems in the estuary.[1]

In conclusion, an immense effort has been undertaken during the 1990s to protect the water quality, natural resources, and economic vitality of the San Francisco Bay/Delta Estuary. Despite the involvement of numerous scientists, resource managers, government agencies, industry, and the general public, many complex environmental problems persist. Some of the most compelling issues involve ongoing freshwater diversions and their impacts on fish populations and other estuarine biota, nonpoint source pollution, dredging and dredge material disposal, the introduction and proliferation of exotic plants and animals, and wildlife habitat destruction. They pose challenging tasks for all members of the estuary project. Finally, successful restoration and maintenance of the estuary depends greatly on communities inhabiting the Central Valley and bay watersheds, which must sustain a strong desire to conserve and protect the estuarine environment.

References

1. San Francisco Estuary Project, *State of the Estuary: 1992–1997*, Tech. Rept., San Francisco Estuary Project, Oakland, CA, 1998.
2. Monroe, M. W. and Kelly, J., *State of the Estuary*, Tech. Rept., San Francisco Estuary Project, Oakland, CA, 1992.
3. Arthur, J. F. and Ball, M. D., *The Significance of the Entrapment Zone Location to the Phytoplankton Standing Crop in the San Francisco-Delta Estuary*, Tech. Rept., U.S. Department of the Interior, Bureau of Reclamation, Sacramento, CA, 1980.
4. Cohen, A. N., *An Introduction to the Ecology of the San Francisco Estuary*, Tech. Rept., San Francisco Estuary Project, U.S. Environmental Protection Agency, San Francisco, CA, 1990.
5. Smith, L. W., *A Review of Circulation and Mixing Studies of San Francisco Bay*, California, U.S. Geological Survey Circular 1015, Menlo Park, CA, 1987.
6. Pereira, W. E., Hostettler, F. D., and Rapp, J. B., Bioaccumulation of hydrocarbons derived from terrestrial and anthropogenic sources in the Asian clam, *Potamocorbula amurensis*, in San Francisco Bay estuary, *Mar. Pollut. Bull.*, 24, 103, 1992.
7. Smith, G. J. and Flegal, A. R., Silver in San Francisco Bay estuarine waters, *Estuaries*, 16, 547, 1993.
8. Conomos, T. J., Properties and circulation of San Francisco Bay waters, in *San Francisco Bay: The Urbanized Estuary*, Conomos, T. J., Ed., Pacific Division, American Association for the Advancement of Science, San Francisco, CA, 1979, 47.
9. Imberger, J., Kirkland, W. B., Jr., and Fischer, H. B., *The Effect of Delta Outflow on the Density Stratification in San Francisco Bay*, Report HBF-77/02, Hugo B. Fischer, Inc., Association of Bay Area Governments, Berkeley, CA, 1977.
10. Kennish, M. J., *Pollution Impacts on Marine Biotic Communities*, CRC Press, Boca Raton, FL, 1998.
11. Fox, J. P., Mongan, T. R., and Miller, W. J., Trends in freshwater inflow to San Francisco Bay from the Sacramento-San Joaquin Delta, *Water Res. Bull.*, 26, 101, 1990.

12. SWRCB, *Water Quality Control Plan for Salinity*, Rept. 91-15WR, San Francisco Bay/Sacramento-San Joaquin Delta Estuary, State Water Resources Control Board, Sacramento, CA, 1991.
13. Peterson, D. H., Cayan, D. R., Festa, J. F., Nichols, F. H., Walters, R. A., Slack, J. V., Hager, S. E., and Schemel, L. E., *Climate Variability in an Estuary: Effects of Riverflow on San Francisco Bay*, Geophysical Monograph 55, American Geophysical Union, Washington, D.C., 1989.
14. NOAA, *Marine Environmental Assessment San Francisco Bay, 1985 Annual Summary*, Tech. Rept., National Environmental Satellite, Data, and Information Service, National Oceanographic and Atmospheric Administration, Washington, D.C., 1985.
15. Kockelman, W. J., Conomos, T. J., and Leviton, A. E., Eds., *San Francisco Bay: Use and Protection*, Pacific Division, American Association for the Advancement of Science, San Francisco, CA, 1982.
16. Nichols, F. H., Cloern, J. E., Luoma, S. N., and Peterson, D. H., The modification of an estuary, *Science*, 231, 567, 1986.
17. Jassby, A., Cloern, J. E., and Powell, T. M., Organic carbon sources and sinks in San Francisco Bay: freshwater flow induced variability, *Mar. Ecol. Prog. Ser.*, 93, 39, 1993.
18. Krone, R. B., Sedimentation in the San Francisco Bay system, in *San Francisco Bay: The Urbanized Estuary*, Conomos, T. J., Ed., Pacific Division, American Association for the Advancement of Science, San Francisco, CA, 1979, 31.
19. Davis, J. A., Gunther, A. J., Richardson, B. J., O'Connor, J. M., Spies, R. B., Wyatt, E., Larson, E., and Meiorin, E. C., *Status and Trends Report on Pollutants in the San Francisco Estuary*, Tech. Rept., San Francisco Estuary Project, U.S. Environmental Protection Agency, San Francisco, CA, 1991.
20. USFWS, Wetlands of the Central Valley, Status and Trends: 1939 to mid-1980s, Tech. Rept., U.S. Fish and Wildlife Service, Portland, OR, 1989.
21. ABAG, Projections 90: *Forecasts for the San Francisco Bay Area to the Year 2005*, Tech. Rept., Association of Bay Area Governments, Oakland, CA, 1989.
22. SFCVB, San Francisco Visitor Statistics, San Francisco Convention and Visitors Bureau, San Francisco, CA, 1990.
23. Ogden Beeman & Associates, *Long-term Management Strategy for Dredged Material in the San Francisco Bay Region: Benefits Related to Navigation Channel Maintenance*, Tech. Rept. Ser., Ogden Beeman & Associates, San Francisco, CA, 1990.
24. DWR, *Delta Gross Value, Crop and Livestock, Comparison with State, 1985*, Exhibit 340, California State Water Resources Control Board, 1987 Water Quality/Water Rights Proceeding on the San Francisco Bay/Sacramento-San Joaquin Delta, California, California Department of Water Resources, Sacramento, CA, 1987.
25. Bacci, E., *Ecotoxicology of Organic Contaminants*, Lewis Publishers, Ann Arbor, MI, 1994.
26. Kennish, M. J., Ed., *Practical Handbook of Estuarine and Marine Pollution*, CRC Press, Boca Raton, FL, 1997.
27. NOAA, *A Summary of Data on Tissue Contamination from the First Three Years (1986–1988) of the Mussel Watch Project*, NOAA Tech. Mem. NOS OMA 49, National Oceanic and Atmospheric Administration, Rockville, MD, 1989.
28. DFG, *Selenium Verification Study 1988–1990*, Rept. 91-2- WQ, California Department of Fish and Game, Sacramento, CA, 1991.
29. Wright, D. A. and Phillips, D. J. H., Chesapeake and San Francisco Bays: a study in contrasts and parallels, *Mar. Pollut. Bull.*, 19, 405, 1988.
30. Pereira, W. E., Hostettler, F. D., Cashman, J. R., and Nishioka, R. S., Occurrence and distribution of organochlorine compounds in sediment and livers of striped bass (*Morone saxatilis*) from the San Francisco Bay-Delta estuary, *Mar. Pollut. Bull.*, 28, 434, 1994.
31. Long, E. R., MacDonald, D., Matta, B. B., Van Ness, K. L., Buchman, M., and Harris, H., *Status and Trends in Concentrations of Contaminants and Measures of Biological Stress in San Francisco Bay*, NOAA Tech. Mem. NOS OMA 41, National Oceanic and Atmospheric Administration, Seattle, WA, 1988.
32. Hayes, S. P. and Phillips, P. T., *California State Mussel Watch, Marine Water Quality Monitoring Program 1985–86*, Water Quality Monitoring Report No. 87-2WQ, California Water Research Control Board, Sacramento, CA, 1987.

33. Cain, D. J. and Luoma, S. N., Influence of seasonal growth, age, and environmental exposure on Cu and Ag in a bivalve indicator, *Macoma balthica* in San Francisco Bay, *Mar. Ecol. Prog. Ser.*, 60, 45, 1990.
34. Smith, G. J. and Flegal, A. R., Silver in San Francisco Bay estuarine waters, *Estuaries*, 16, 547, 1993.
35. Mearns, A. J., Matta, M. B., Simecek-Beatty, D., Buchman, M. F., Shigenaka, G., and Wert, W., A., *PCB and Chlorinated Pesticide Contamination in U.S. Fish and Shellfish: A Historical Assessment Report*, NOAA Tech. Mem. NOS OMA 39, National Oceanic and Atmospheric Administration, Seattle, WA, 1988.
36. Hayes, S. P., Phillips, P. T., Martin, M., Stephenson, M., Smith, D., and Linfield, J., *California State Mussel Watch, Marine Water Quality Monitoring Program 1983–84*, Water Quality Monitoring Report No. 85-2WQ, California Water Research Control Board, Sacramento, CA, 1985.
37. Hayes, S. P. and Phillips, P. T., *California State Mussel Watch Marine Water Quality Monitoring Program 1984 to 1985*, Water Quality Monitoring Report, No. 86-3WQ, California Water Research Control Board, Sacramento, CA, 1986.
38. Stephenson, M., Smith, D., Ichikawa, G., Goetzl, J., and Martin, M., *State Mussel Watch Program Preliminary Data Report 1985–86*, Tech. Rept., State Water Resources Control Board, California Department of Fish and Game, Monterey, CA, 1986.
39. Risebrough, R. W., Chapman, J. W., Okazaki, R. K., and Schmidt, T. T., *Toxicants in San Francisco Bay and Estuary*, Tech. Rept., Association of Bay Area Governments, Berkeley, CA, 1978.
40. NOAA, *National Status and Trends Program for Marine Environmental Quality: Progress Report — A Summary of Selected Data on Chemical Contaminants in Tissues Collected During 1984, 1985, and 1986*, Tech. Rept., PAD/OMA/NOS/NOAA, Department of Commerce, Rockville, MD, 1987.
41. Fox, J. P., Mongan, T. R., and Miller, W. J., Trends in freshwater inflow to San Francisco Bay from the Sacramento-San Joaquin Delta, *Water Res. Bull.*, 26, 101, 1990.
42. Cloern, J. E. and Nichols, F. H., Eds., *Temporal Dynamics of an Estuary: San Francisco Bay*, Junk, Dordrecht, The Netherlands, 1985.
43. Caffrey, J. M., Spatial and seasonal patterns in sediment nitrogen remineralization and ammonium concentrations in San Francisco Bay, CA, *Estuaries*, 18, 219, 1995.
44. DWR, *California Water: Looking to the Future*, Bulletin 160-87, California Department of Water Resources, Sacramento, CA, 1987.
45. Walters, R. A. and Gartner, J. W., Subtidal sea level and current variations in the northern reach of San Francisco Bay, *Est. Coastal Shelf Sci.*, 21, 17, 1985.
46. Cloern, J. E., Annual variations in river flow and primary production in the South San Francisco Bay estuary, in *Estuaries and Coasts: Spatial and Temporal Intercomparisons*, Elliott, M. and Ducroty, J. P., Eds., Olsen and Olsen, Fredensborg, Denmark, 1991.
47. Cloern, J. E., Cole, B. E., Wong, R. L. J., and Alpine, A. E., Temporal dynamics of estuarine phytoplankton: a case study of San Francisco Bay, Hydrobiologia, 129, 153, 1985.
48. Lehman, P. W., Environmental factors associated with long-term changes in chlorophyll concentration in the Sacramento-San Jacinto Delta and Suisun Bay, California, *Estuaries*, 15, 335, 1992.
49. Ball, M. D. *Phytoplankton Dynamics and Planktonic Chlorophyll Trends in the San Francisco Bay/Delta Estuary*, Exhibit 103, California State Water Resources Control Board, 1987 Water Quality/Water Rights Proceeding on the San Francisco Bay/Sacramento-San Joaquin Delta, California, U.S. Bureau of Reclamation, Sacramento, CA, 1987.
50. Armor, C. and Herrgesell, P. L., Distribution and abundance of fishes in the San Francisco Bay estuary between 1980 and 1982, *Hydrobiologia*, 129, 211, 1985.
51. Herbold, B., Moyle, P. B., and Jassby, A. D., *Status and Trends Report on Aquatic Resources of the San Francisco Estuary*, Tech. Rept., U.S. Environmental Protection Agency, San Francisco, CA, 1992.
52. Stevens, D. E. and Miller, L. W., Effects of river flow on abundance of young chinook salmon, American shad, longfin smelt, and delta smelt in the Sacramento-San Joaquin River system, *N. Am. J. Fish. Manage.*, 3, 425, 1983.

53. USFWS, *National Estuary Study*, Volume 5, U.S. Fish and Wildlife Service, Washington, D.C., 1970.
54. U.S. Army Corps of Engineers, *Environmental Atlas of the Sacramento/San Joaquin Delta*, U.S. Army Corps of Engineers, San Francisco District, San Francisco, CA, 1979.
55. Cohen, A. N. and Carlton, J. T., Accelerating invasion rate in a highly invaded estuary, *Science*, 279, 555, 1998.
56. Nichols, F. H. and Pamatmat, M. M., *The Ecology of the Soft-Bottom Benthos of San Francisco Bay: A Community Profile*, U.S. Fish and Wildlife Service, Biol. Rept. No. 85(7.19), Washington, D.C., 1988.
57. Herbold, B. and Moyle, P. B., *The Ecology of the Sacramento-San Joaquin Delta: A Community Profile*, U.S. Fish and Wildlife Service, Biol. Rept. No. 85(7.22), 1989.
58. Nichols, F. H., Natural and anthropogenic influences on benthic community structure in San Francisco Bay, in *San Francisco Bay: The Urbanized Estuary*, Conomos, T. J., Ed., Pacific Division, American Association for the Advancement of Science, San Francisco, CA, 1979, 231.
59. U.S. Fish and Wildlife Service, *San Francisco Bay National Wildlife Refuge Predator Management Plan and Final Environmental Assessment*, Tech. Rept., U.S. Fish and Wildlife Service, Newark, CA, 1991.
60. Meiorin, E. C., Josselyn, M. N., Crawford, R., Calloway, J., Miller, K., Richardson, T., and Leidy, R. A., *Status and Trends Report on Wetlands and Related Habitats in the San Francisco Estuary*, Tech. Rept., San Francisco Estuary Project, U.S. Environmental Protection Agency, San Francisco, CA, 1991.
61. Cowardin, L. M., Carter, V., Golet, F. C., and LaRoe, E. T., *Classification of Wetlands and Deepwater Habitats of the United States*, FWS/OBS-79/31, U.S. Fish and Wildlife Service, Office of Biological Services, Washington, D.C., 1979.
62. Leidy, R. A. and Fiedler, P. L., Human disturbance and patterns of fish species diversity in the San Francisco Bay drainage, California, *Biol. Conserv.*, 33, 247, 1985.
63. Cloern, J. E., Alpine, A. E., Cole, B. E., Wong, R. L. J., Arthur, J. F., and Ball, M. D., River discharge controls phytoplankton dynamics in the northern San Francisco Bay estuary, *Est. Coastal Shelf Sci.*, 16, 415, 1983.
64. Cloern, J. E., Temporal dynamics and ecological significance of salinity stratification in an estuary (South San Francisco Bay, U.S.A.), *Oceanol. Acta*, 7, 137, 1984.
65. Cole, B. E., Cloern, J. E., and Alpine, A. E., Biomass and productivity of three phytoplankton size classes in San Francisco Bay, *Estuaries*, 9, 117, 1986.
66. Ambler, J. W., Cloern, J. E., and Hutchinson, A., Seasonal cycles of zooplankton from San Francisco Bay, *Hydrobiologia*, 129, 177, 1985.
67. Orsi, J. J., Bowman, T. E., Marelli, D. C., and Hutchison, A., Recent introduction of the planktonic calanoid copepod *Sinocalanus doerri* (Centropagnidae) from mainland China to the Sacramento-San Joaquin estuary of California, *J. Plank. Res.*, 5, 357, 375.
68. Orsi, J. J. and Mecum, W. L., Zooplankton distribution and abundance in the Sacramento-San Joaquin Delta in relation to certain environmental factors, *Estuaries*, 9, 326, 1986.
69. Nichols, F. H. and Thompson, J. K., Persistence of an introduced mudflat community in South San Francisco Bay, California, *Mar. Ecol. Prog. Ser.*, 24, 83, 1985.
70. Josselyn, M. N. and West, J. A., The distribution and temporal dynamics of the estuarine macroalgal community of San Francisco Bay, *Hydrobiologia*, 129, 139, 1985.
71. Nichols, F. H., Thompson, J. K., and Schemel, L. E., Remarkable invasion of San Francisco Bay (California, USA) by the Asian clam *Potamocorbula amurensis*. II. Displacement of a former community, *Mar. Ecol. Prog. Ser.*, 66, 95, 1990.
72. Carlton, J. T., Introduced invertebrates, of San Francisco Bay, in *San Francisco Bay: The Urbanized Estuary*, Conomos, T. J., Ed., Pacific Division, American Association for the Advancement of Science, San Francisco, CA, 1979, 427.
73. Josselyn, M. N., *The Ecology of San Francisco Bay Tidal Marshes: A Community Perspective*, FWS/OBS-82/23, U.S. Fish and Wildlife Service, Washington, D.C., 1983.
74. SWRCB, *Water Quality Control Plan for Salinity*, Tech. Rept. 91-15WR, San Francisco Bay/Sacramento-San Joaquin Delta Estuary, State Water Resources Control Board, Sacramento, CA, 1991.

75. Jones and Stokes Associates, Inc., *Protection and Restoration of San Francisco Bay Fish and Wildlife Habitat*, Volumes I and II, Tech. Rept., California Department of Fish and Game and U.S. Fish and Wildlife Service, Sacramento, CA, 1979.
76. Harvey, T. E., Miller, K. J., Hothem, R. L., Rauzon, M. J., Page, G. W., and Keck, R. A., *Status and Trends Report on Wildlife of the San Francisco Estuary*, Tech. Rept., U.S. Environmental Protection Agency, San Francisco, CA, 1992.
77. San Francisco Estuary Project, *Comprehensive Conservation and Management Plan*, San Francisco Estuary Project, Oakland, CA, 1993.
78. San Francisco Estuary Project, *CCMP Workbook: Comprehensive Conservation and Management Plan for the Bay-Delta Implementation Progress 1993–1996*, Tech. Rept., San Francisco Estuary Project, Oakland, CA, 1996.

Index

A

C

D

E

F

G

H

I

J

K

L

N

O

R

S

T

U

V

X

Y

Z